CONTROL SYSTEMS

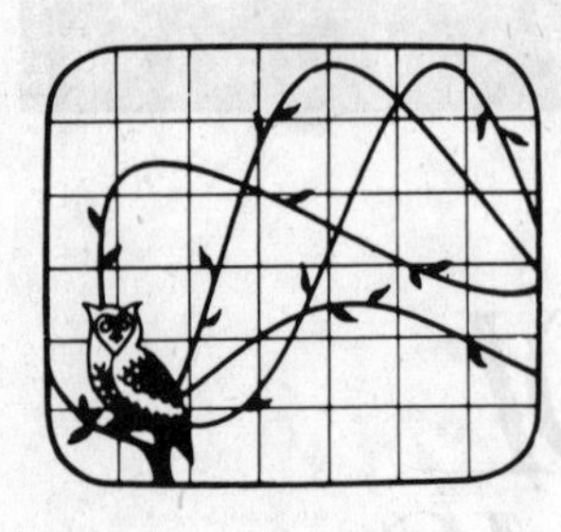

HRW Series in Electrical and Computer Engineering

M. E. Van Valkenburg, Series Editor Electrical Engineering
Michael R. Lightner, Series Editor Computer Engineering

P. R. Bélanger, E. L. Adler, and N. C. Rumin INTRODUCTION TO CIRCUITS WITH ELECTRONICS: AN INTEGRATED APPROACH
L. S. Bobrow ELEMENTARY LINEAR CIRCUIT ANALYSIS
L. S. Bobrow FUNDAMENTALS OF ELECTRICAL ENGINEERING
C. T. Chen LINEAR SYSTEM THEORY AND DESIGN
D. J. Comer DIGITAL LOGIC AND STATE MACHINE DESIGN
D. J. Comer MICROPROCESSOR BASED SYSTEM DESIGN
C. H. Durney, L. D. Harris, C. L. Alley ELECTRIC CIRCUITS: THEORY AND ENGINEERING APPLICATIONS
M. S. Ghausi ELECTRONIC DEVICES AND CIRCUITS: DISCRETE AND INTEGRATED
G. H. Hostetter, C. J. Savant, Jr., R. T. Stefani DESIGN OF FEEDBACK CONTROL SYSTEMS
S. Karni and W. J. Byatt MATHEMATICAL METHODS IN CONTINUOUS AND DISCRETE SYSTEMS
B. C. Kuo DIGITAL CONTROL SYSTEMS
B. P. Lathi MODERN DIGITAL AND ANALOG COMMUNICATION SYSTEMS
C. D. McGillem and G. R. Cooper CONTINUOUS AND DISCRETE SIGNAL AND SYSTEM ANALYSIS Second Edition
D. H. Navon SEMICONDUCTOR MICRODEVICES AND MATERIALS
A. Papoulis CIRCUITS AND SYSTEMS: A MODERN APPROACH
S. E. Schwarz and W. G. Oldham ELECTRICAL ENGINEERING: AN INTRODUCTION
A. S. Sedra and K. C. Smith MICROELECTRONIC CIRCUITS
N. K. Sinha CONTROL SYSTEMS
M. E. Van Valkenburg ANALOG FILTER DESIGN
W. A. Wolovich ROBOTICS: BASIC ANALYSIS AND DESIGN

CONTROL SYSTEMS

Naresh K. Sinha

McMaster University
Hamilton, Ontario, Canada

CBS Publishing Japan Ltd.

New York Chicago San Francisco Philadelphia
Montreal Toronto London Sydney Tokyo
Mexico City Rio de Janeiro Madrid

To the memory of my father.

Acquisitions Editor: Deborah Moore
Production Manager: Paul Nardi
Project Editor: Lila M. Gardner
Designer: Andrea Da Rif
Illustrations: Scientific Illustrators

Address correspondence to:
383 Madison Avenue, New York, NY 10017

Library of Congress Cataloging-in-Publication Data

Sinha, N. K. (Naresh Kumar)
Control systems.

Includes index.
1. Automatic control. 2. Control theory. I. Title.
TJ213.S47444 1986 629.8 85-24796
ISBN 0-03-069357-8 (US College Edition)
ISBN 0-03-910743-4 (CBS International Edition)

This International Edition is not for sale in the United States of America, its dependencies or Canada.

Printed in Japan, 1986

6 7 8 9 038 9 8 7 6 5 4 3 2 1

CBS COLLEGE PUBLISHING
The Dryden Press
Saunders College Publishing

Contents

3. STATE-SPACE METHODS 33

4. CHARACTERISTICS OF CLOSED-LOOP SYSTEMS 59

5. PERFORMANCE OF CONTROL SYSTEMS 74

6. STABILITY OF LINEAR SYSTEMS 100

Preface

Teaching control theory has always been a challenging task. Although real control systems are often quite complicated, they must be presented in a simplified form so that they are mathematically tractable. In the past there has often been a tendency to oversimplify, mainly because real-life problems require a great deal of computation. This is no longer a limitation now that engineers and engineering students have easy access to microcomputers. The main problem in taking full advantage of the situation has been the lack of availability of computer programs suitable for use by students of control theory. This book is an attempt to remedy this situation by providing such programs as well as several challenging problems to which they can be applied in the process of mastering the theory.

This book is intended as an introduction to control systems. The object is to provide the reader with the basic concepts of control theory as developed over the years in both the frequency domain and the time domain. An attempt has been made to retain the classical concepts while introducing modern ideas. The effect of the availability of inexpensive microcomputers has been kept in mind.

The book contains a large number of worked out examples in each chapter in order to help explain the theory and methods developed. Chapters 2 and 3 present a unified treatment of modeling of dynamic systems, including transfer function and state-space models. Chapter 10, on the compensation of control systems, has been presented in a logical and unified manner. In particular, the trial-and-error approach to the design of lead compensators, as found in most books, has been replaced by a direct method developed recently. Moreover, the design of a pole-placement compensator using transfer functions, which is the counterpart of the combined observer and state feedback controller, has been included for the first time in a book appropriate for undergraduates and practicing engineers. Chapter 11, or digital control, is an up-to-date treatment of a rapidly developing and popular area, presented in a manner suitable for a first examination of the subject, with many practical examples to provide motivation. Chapter 12 provides a good treatment of the state-space approach, written in a way that will make it attractive to both undergraduates and professionals. The concepts of controllability and observability are introduced using the theory of discrete-time systems already developed in Chapter 11. This is followed by the theory of state feedback. Both state-space and transfer function approaches are used for determining the state feedback vector. The asymptotic state observer is then introduced and its design described. The combined state feedback and observer design is also discussed and related to the transfer function approach presented in Chapter 10.

It is appreciated that in most engineering disciplines one has to solve many numerical problems in order to learn the various details and subtleties and to be able to handle engineering design problems. To provide motivation, a number of realistic problems have been included at the end of each chapter. Appendix D discusses computational aspects. Program listings have been given for many cases, with thorough documentation. It may be added that in the past most books on control systems were limited to including problems of second-order systems due to the fact that computing facilities were not available. Until about 1978, if a problem required finding the roots of a polynomial of degree higher than 2, one generally gave up. In recognition of the fact that at present all engineers have access to either a programmable calculator or a microcomputer, an attempt has been made to remedy this situation and prepare the reader for real-life problems by including program listings suitable for use on either an HP-41C programmable calculator or an IBM Personal Computer.

The book is intended for all engineering students at the senior level and practicing engineers, without any bias towards a particular branch of engineering. Although it is expected that the reader will have some background in Laplace transforms, z transforms, and matrices, these topics are reviewed in the appendixes. An attempt has been made to retain some mathematical rigor while providing motivation by giving many practical examples.

Earlier versions of this book have been used for a one-semester senior level course on control systems at McMaster University for the past three years. Since our seniors have completed a course on circuit and systems theory at the junior level, they have sufficient background in Laplace transforms, z transforms, state equations for electrical circuits, Bode plots, and theory of matrices. As a result, it has been possible to teach all the material in this book in a 13-week semester. If it is necessary to review some of this background material, there may be insufficient time to cover the entire book. In that case, Chapter 11, Chapter 13, and parts of Chapter 12 may be left out.

ACKNOWLEDGMENTS

It is a pleasure to acknowledge the help that I have received from several individuals in the preparation of this book. In particular, I am indebted to Professor M. E. Van Valkenburg, who is the editor of the Holt, Rinehart and Winston series in Electrical and Computer Engineering, and several other reviewers who helped polish the manuscript in many ways. Among these were Professors

- John Fleming of Texas A & M University
- Susan Reidel of Marquette University
- Edgar Tacker of the University of Tulsa
- Don Kirk of the U.S. Naval Postgraduate School
- Lincoln Jones of San Jose State University
- Madur Sundareshan of the University of Arizona
- Don Pierre of Montana State University
- Violet Haas of Purdue University

Other friends and colleagues who read parts of the manuscript and have made valuable suggestions are Dr. O. Vidal, Mr. S. Puthenpura, and the students in the graduating classes of 1983, 1984, and 1985. I am grateful to Dr. G. J. Lastman of the University of Waterloo for help in improving some of the Pascal programs written for the IBM Personal Computer.

Support and encouragement in preparing this book were received from Ms. Deborah Moore, Senior Editor of Engineering and Computer Science with Holt, Rinehart and Winston. This is gratefully acknowledged. I would also like to thank Mrs. Barbara Petro, Miss L. Honda, and Mrs. J. Arsenault in the Word Processing Centre of the Faculty of Engineering, McMaster University, for their patience in typing and retyping several versions of the text and to Ms. Linda Hunter for typing the captions of the figures.

Just as one does not thank himself, expressing gratitude to one's wife in public is not a Hindu custom. The wife is considered part of the husband and her coauthorship is tacitly assumed in any book that her husband writes. There is little doubt that without Meena's help, patience, and encouragement, this book would not have been possible.

Naresh K. Sinha

CONTROL SYSTEMS

I

Introduction

The subject of control systems is of great inportance to all engineers. The objective is to free human beings from boring repetitive chores that can be done easily and more economically by automatic control devices. The recent developments in the large-scale integration of semiconductor devices and the resulting availability of inexpensive microprocessors has made it practical to use computers as integral parts of control systems, making them cheaper as well as more sophisticated.

Historically, the first automatic control device used in the industry was the Watt fly-ball governor, invented in 1767 by James Watt, who was also the inventor of the steam engine. The object of this device was to keep the speed of the engine nearly constant by regulating the supply of steam to the engine. A schematic diagram is shown in Fig. 1.1. The two fly balls in the governor rotate about a vertical axis at a speed proportional to the speed of the engine. Due to the centrifugal force acting on them, they tend to move out. This movement controls the supply of steam to the engine through a mechanical linkage to the steam flow valve in such a manner that the steam supply is reduced when the speed is high and increased when the speed is low.

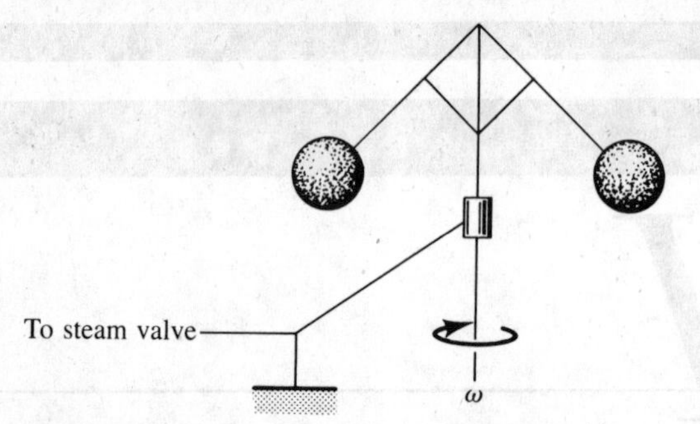

FIGURE 1.1. The Watt fly-ball governor.

It was found that by a proper design of the governor the speed could be kept within narrow limits of a specified value. It was also observed that if one tried to increase the sensitivity of the governor by increasing the gear ratio between the engine shaft and the governor, it tended to "hunt" or oscillate about the desired setting. It was about 100 years later that a complete mathematical analysis was made by James Clerk Maxwell (more well known for his contributions to electromagnetic field theory).

Much later it was realized that all automatic control systems worked on the principle of feedback. By a coincidence, about the same time the theory of feedback amplifiers had been developed by electrical engineers who had been concerned with transmitting telephone signals over long distances. In particular, one may mention the Nyquist theory of stability developed about 1930. A great impetus to the theory of automatic control came during World War II when servomechanisms were utilized for the control of anti-aircraft guns. After World War II many peacetime applications followed. Some of these are the "autopilot" for aircraft, automatic control of machine tools, automatic control of chemical processes, and automatic regulation of voltage at electric power plants. Although originally the theory was based on frequency response and Laplace transform methods, in the 1960s the impact of the digital computer led to the development of time-domain theory using state variables. This was especially useful as more sophisticated multivariable control systems were developed for more complex processes. As computers have become cheaper and more compact, they have been utilized as components of more advanced control systems.

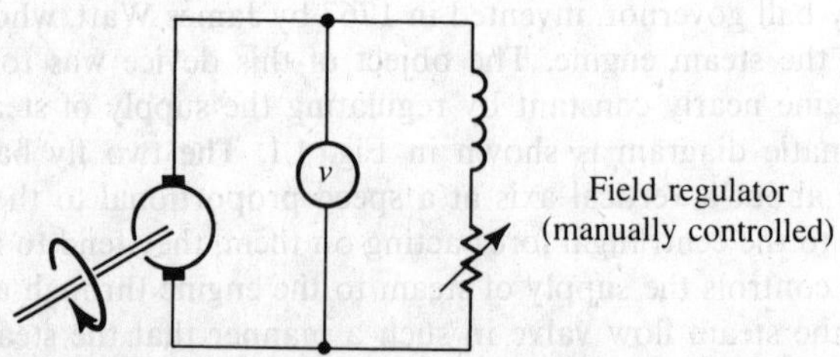

FIGURE 1.2. A voltage control system.

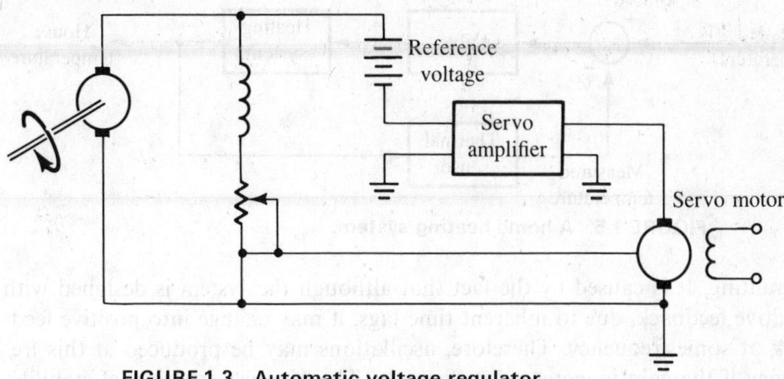

FIGURE 1.3. Automatic voltage regulator.

Let us consider some simple examples of control systems. Figure 1.2 shows the scheme for controlling the voltage at an electric power station in the 1940s. A human operator was required to watch a voltmeter connected to the busbars and adjust the field rheostat in order to keep the voltage close to the specified value. A scheme for automatic voltage regulation is shown in Fig. 1.3 and shows that it works by comparing the actual value of the voltage with the desired value. The difference or "error" is applied to a servomotor, after suitable amplification. This servomotor drives a shaft coupled to the field rheostat to alter the resistance in the field winding in such a manner that the error is reduced. Hence, it may be said that "feedback" is utilized to obtain automatic control. As a matter of fact, all automatic control systems use feedback and can be represented by the block diagram shown in Fig. 1.4. It can be seen that the controlled output is fed back and compared with the reference input. The difference, called the "error," is then utilized to drive the system in such a manner that the output approaches the desired value (i.e., the reference input).

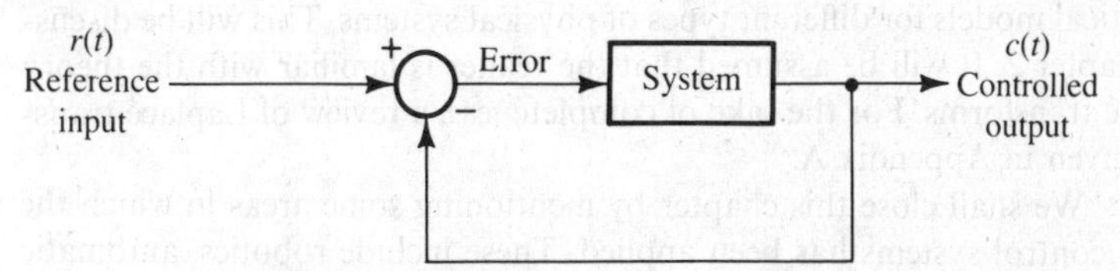

FIGURE 1.4. Block diagram of automatic control system.

Another example is the home heating system. A thermostat senses the temperature and if it is lower than a set value, the furnace is turned on. The furnace is turned off when the temperature exceeds another set value. The block diagram is shown in Fig. 1.5. Although it is similar to Figs. 1.3 and 1.4, it may be noted that this is an on/off-type control system, whereas the voltage regulator is a continuous-type system.

It was noticed at the very outset that if one tried to improve the accuracy of a control system by increasing the loop gain, it led to instability,

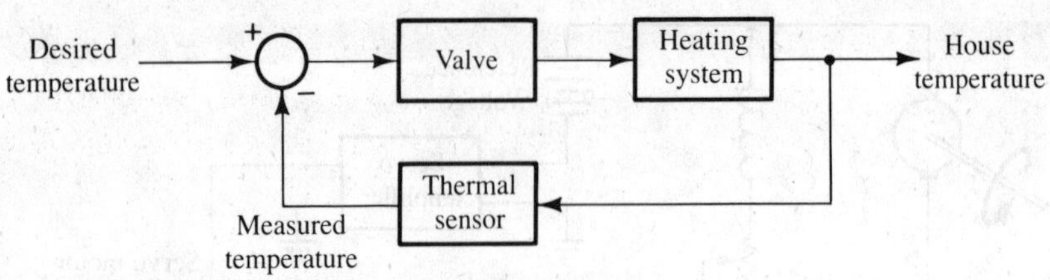

FIGURE 1.5. A home heating system.

or hunting. It is caused by the fact that although the system is designed with negative feedback, due to inherent time lags, it may change into positive feedback at some frequency. Therefore, oscillations may be produced at this frequency if the gain is increased sufficiently. The Nyquist criterion of stability, developed for feedback amplifiers, provides a good understanding of this. We shall later see how one can increase the sensitivity (or accuracy) without causing instability.

The components used for control systems are usually of a wide variety. For example, these may be electromechanical, electronic, thermal, hydraulic, or pneumatic. In order to analyze the response of the various components, we replace them by their mathematical models. Although the input and the output of these devices are generally related through nonlinear differential equations, it is customary to obtain simplified linear models about the operating points because such models are easier to analyze. Transfer function and state-variable models are most commonly employed.

In our development of control theory, we shall generally be carrying out the analysis and design in terms of the mathematical models. Although this approach may sometimes appear abstract, one must appreciate that these models represent real systems. To a certain extent this abstraction is necessary for developing a unified theory of automatic control systems despite the great variety of components. One important aspect is the problem of obtaining mathematical models for different types of physical systems. This will be discussed in Chapter 2. It will be assumed that the reader is familiar with the theory of Laplace transforms. For the sake of completeness, a review of Laplace transforms is given in Appendix A.

We shall close this chapter by mentioning some areas in which the theory of control systems has been applied. These include robotics, automatic control of large-scale power systems, numerical control of machine tools, autopilots for aircraft, prosthetic devices for handicapped persons, and the steering control of ocean liners. An important consequence of the development of control theory has been the increased use of automation in the industry with a view to increasing productivity. The concept of modeling developed by control engineers has been applied to many diverse areas, including biomedical systems, socioeconomical systems, and ecological systems. With the rapid advances in the area of microelectronics and the exciting possibility of using inexpensive computers as parts of control systems, it can truly be said that the applications of control theory are limited only by human imagination.

2 Mathematical Models of Physical Systems

2.1 INTRODUCTION

A crucial problem in engineering design and analysis is the determination of a mathematical model of a given physical system. This model must relate in a quantitative manner the various variables in the system. A model may be defined as "a representation of the essential aspects of a system which presents knowledge of that system in a usable manner." To be useful, a model must not be so complicated that it cannot be understood and thereby be unsuitable for analysis; at the same time it must not be oversimplified and trivial to the extent that predictions of the behavior of the system based on this model are grossly inadequate.

The systems that we shall be concerned with are dynamic in nature, and their behavior will be described in the form of differential equations. Although these will normally be nonlinear, it is customary to linearize them about an operating point to obtain linear differential equations. This is done in order that the analysis can be carried out conveniently. It should, however, be borne in mind that such linear models, though useful for preliminary analysis and design, are valid only over a limited operating range. Nevertheless, they are employed extensively in engineering, utilizing Laplace transform, frequency response, and state-space techniques.

2.2 DIFFERENTIAL EQUATIONS AND TRANSFER FUNCTIONS FOR PHYSICAL SYSTEMS

The components of a control system are diverse in nature and may include electrical, mechanical, thermal, and fluidic devices. The differential equations for these devices are obtained using the basic laws of physics. These include balancing forces, energy, and mass. In practice, some simplifying assumptions are often made to obtain linear differential equations with constant coefficients, although in most cases exact analysis would require the use of nonlinear partial differential equations. For most physical devices one may classify the variables as "through" and "across" variables, in the sense that the former refer to a point and the latter measured between two points. A list of analogous variables for different systems is given in Table 2.1.

TABLE 2.1. Analogous variables for physical systems

System	*Through-Variable*	*Across-Variable*
Electrical	Current, i	Potential difference or voltage, v
Mechanical (translational)	Force, f	Relative velocity, v
Mechanical (rotational)	Torque, T	Relative angular velocity, ω
Thermal	Rate of flow of heat energy, q	Difference in temperature, T
Fluidic	Volumetric rate of flow of fluid flow, Q	Difference in pressure, P

It may be noted that the equations of equilibrium in different systems are based on similar principles. For example, Kirchhoff's current law for an electrical circuit, equating the algebraic sum of currents at a point to zero is analogous to D'Alembert's principle in mechanics, which equates the algebraic sum of forces at a point to zero. In fact, it is often convenient to determine the differential equation for mechanical or thermal systems by first obtaining an equivalent electrical circuit. This is helpful because most people are familiar with the procedures for writing loop or node equations for electrical networks. Laplace transformation of these differential equations yields an algebraic equation, in terms of the complex frequency variables s, relating the various through- and across-variables. These algebraic equations then can easily be manipulated to obtain the transfer function of the system, defined as the ratio of the Laplace transforms of the output and the input under zero initial conditions. This transfer function represents a linear model of the system, and is usually shown in the form of a block diagram as indicated in Fig. 2.1. Note that the block is "unidirectional."

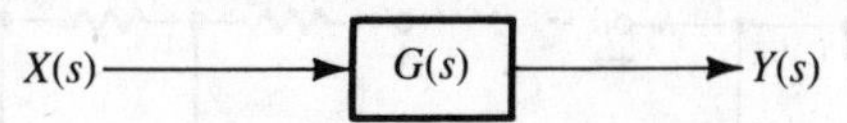

FIGURE 2.1. Block diagram of a linear system.

The input may be regarded as the "cause" and the output as the "effect." The block diagram is undirectional since the "effect" cannot produce the "cause." The transfer function $G(s)$ relates the Laplace transform $Y(s)$ of the output $y(t)$ to the Laplace transform $X(s)$ of the input $x(t)$ through the relationship

$$Y(s) = G(s)X(s) \tag{2.1}$$

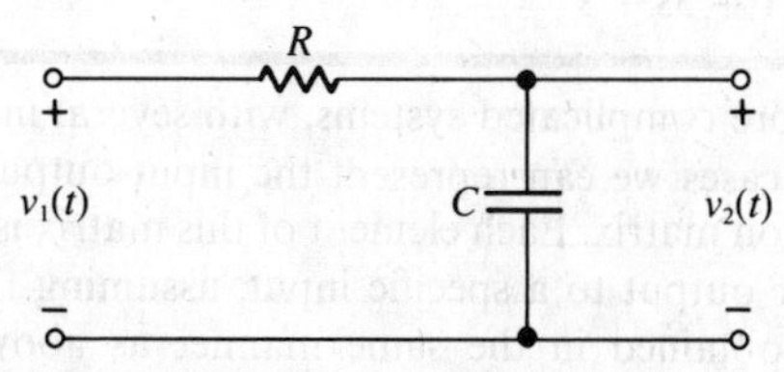

FIGURE 2.2. A simple lag network.

For example, consider the simple electrical circuit shown in Fig. 2.2, in which we apply an input voltage $v_1(t)$ to an RC network. The output voltage $v_2(t)$ is related to the input through the differential equation

$$v_1(t) = RC\frac{dv_2}{dt} + v_2(t) \tag{2.2}$$

which has been obtained by applying Kirchoff's voltage law. Taking the Laplace transform of both sides of Eq. (2.2), assuming zero initial conditions, we have

$$V_1(s) = RCsV_2(s) + V_2(s) \tag{2.3}$$

Solving Eq. (2.3), we get the transfer function

$$G(s) = \frac{V_2(s)}{V_1(s)} = \frac{1}{1 + sCR} \tag{2.4}$$

DRILL PROBLEM 2.1

Determine the transfer functions of the RC networks shown in Fig. 2.3.

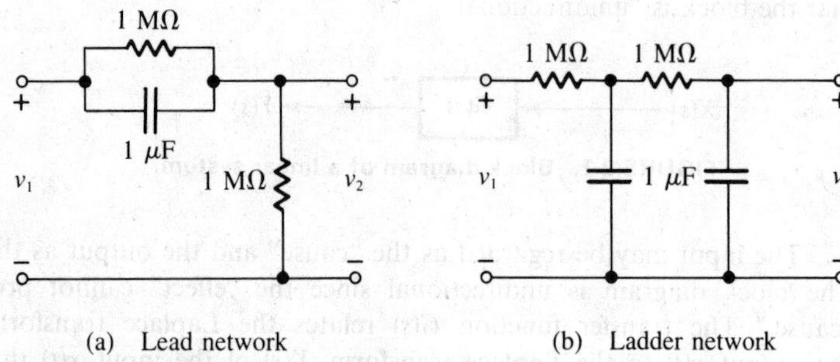

FIGURE 2.3. Networks for Drill Problem 2.1.

Answers:

(a) $\dfrac{1+s}{2+s}$ (b) $\dfrac{1}{1+3s+s^2}$

Often we have more complicated systems, with several independent inputs and outputs. In such cases we can represent the input-output relationship through a transfer function matrix. Each element of this matrix is a transfer function relating a particular output to a specific input, assuming that all the other inputs are zero, and obtained in the same manner as above. This is justified through the principle of superposition for linear systems.

An alternative approach for modeling a system is to write a set of first-order differential equations through a suitable choice of variables. These are called state variables of the system. This is discussed in detail in Chapter 13.

In the next section we shall describe a method for obtaining electrical analogs for mechanical systems and using these for determining the transfer function.

2.3 ELECTRICAL ANALOGS FOR MECHANICAL SYSTEMS

It is often very convenient to obtain an electrical equivalent circuit analogous to a mechanical system. It has the advantage that one can apply Kirchhoff's laws to write the circuit equations, and hence obtain the transfer function. It is also possible to write these equations directly in terms of the Laplace transforms of the currents and voltages. Furthermore, network theorems can often be applied to simplify the circuit. Electrical analogs for mechanical systems have also been used for simulation and analysis.

In particular, if one uses the force-current analogy (or force-torque for a rotational system), the topology of the electrical analog is very similar to that of the mechanical system. Table 2.2 gives the list of the analogous quantities.

TABLE 2.2. Analogous quantities in electrical and mechanical systems

Electrical	*Mechanical (translation)*	*Mechanical (rotation)*
Current, i	Force, f	Torque, T
Voltage, v	Velocity, v	Angular velocity, ω
Flux linkages, $N\phi$	Displacement, x	Angular displacement, h
Capacitance, C	Mass, M	Moment of inertia, J
Conductance, $G = 1/R$	Damping coefficient (of dashpot), D	Rotational damping coefficient (friction), D
Inductance, L	Compliance, $\tau = 1/K$, of spring	Torsional compliance, $\tau = 1/K$, of spring

In each case the through- and across-variables are related in a similar manner. These relationships are shown in Table 2.3.

TABLE 2.3. Relationships between through- and across-variables for analogous system components

Mechanical *Electrical*	*Mechanical (translation)*	*Mechanical (rotation)*
$i = C\dfrac{dv}{dt}$	$f = M\dfrac{dv}{dt}$	$T = J\dfrac{d\omega}{dt}$
$i = Gv$	$f = Dv$	$T = D\omega$
$i = \dfrac{1}{L} \cdot N\phi = \dfrac{1}{L}\int v\,dt$	$f = Kx = K\int v\,dt$	$T = K\theta = K\int \omega\,dt$

Following Cheng [1],* the rule for drawing the equivalent electrical circuit may be stated as follows:

* Numbers in brackets correspond to numbers in the References section at the end of the chapter.

> Each junction in the mathematical system corresponds to a node in the equivalent electrical circuit, joining excitation sources and passive elements. All points on a rigid mass are considered as the same junction, and one terminal of the capacitor analogous to the mass is always connected to the ground in the electrical circuit.

The reason for connecting one terminal of the capacitor to the ground is that the velocity (or displacement) of a mass is always referred to the earth.

The following examples will illustrate the procedure.

EXAMPLE 2.1

The electrical analog of a carriage on wheels, coupled to the wall through a spring, is shown in Fig. 2.4.

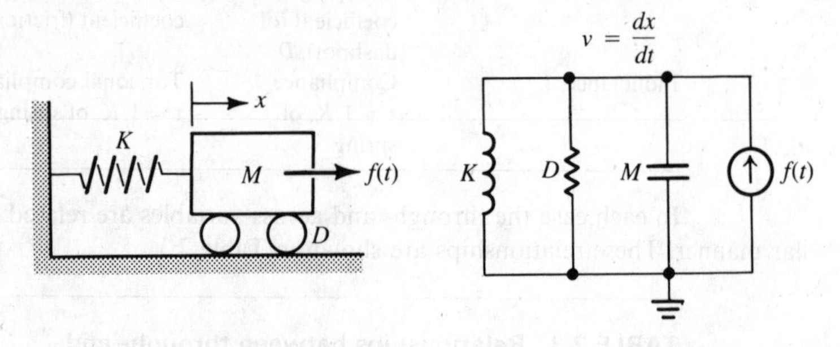

FIGURE 2.4. A mechanical system with movement in one coordinate.

The differential equation for both systems is given as Eq. (2.5). In the case of the electrical network, the equation was obtained by applying Kirchhoff's current law at the node v, and is seen to be identical to the equation that would have been obtained by applying D'Alembert's principle to the mechanical system.

$$M\frac{d^2x}{dt^2} + D\frac{dx}{dt} + Kx = f(t) \tag{2.5}$$

Taking Laplace transforms of both sides of Eq. (2.5), assuming zero initial conditions, we get the transfer function

$$G(s) = \frac{X(s)}{F(s)} = \frac{1}{Ms^2 + Ds + K} \tag{2.6}$$

EXAMPLE 2.2

A mechanical system with two-coordinate movement and its equivalent electrical circuit are shown in Fig. 2.5, where K represents a spring, and D_1 and D_2 represent dashpots.

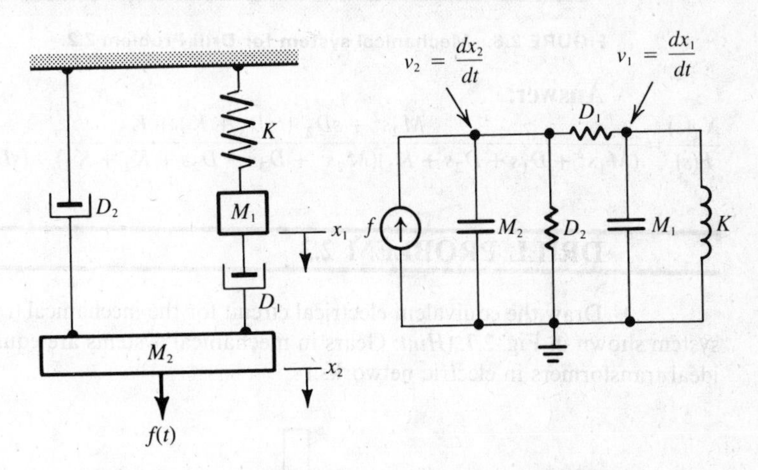

FIGURE 2.5. A two-coordinate mechanical system.

In this case the equations, written directly in terms of the Laplace transform variables, are obtained by applying Kirchhoff's current law at each of the two ungrounded nodes.

$$\begin{aligned} &(s^2M_2 + sD_2 + sD_1)X_2(s) - sD_1X_1(s) = F(s) \\ &-sD_1X_2(s) + (s^2M_1 + sD_1 + K)X_1(s) = 0 \end{aligned} \tag{2.7}$$

The convenience of writing Eqs. (2.7) in terms of node voltages is evident. If we want to obtain the transfer function relating $X_2(s)$ to $F(s)$, we simply solve Eqs. (2.7) for $X_2(s)$. Hence

$$\frac{X_2(s)}{F(s)} = \frac{s^2M_1 + sD_1 + K}{(s^2M_1 + sD_1 + K)(s^2M_2 + sD_2 + sD_1) - s^2D_1^2} \tag{2.8}$$

DRILL PROBLEM 2.2

Determine the transfer function $X_1(s)/F(s)$ for the mechanical system shown in Fig. 2.6 by drawing its equivalent electrical circuit.

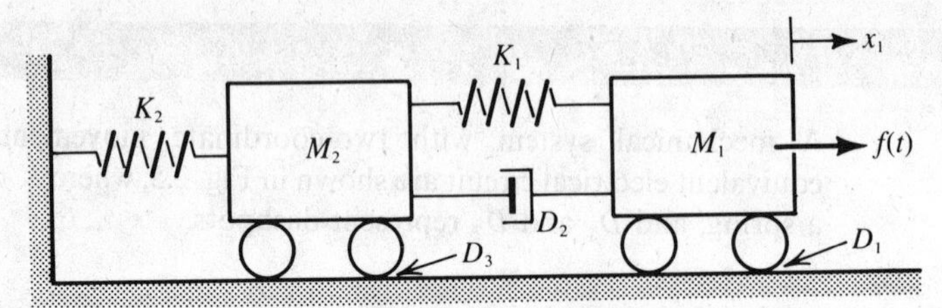

FIGURE 2.6. Mechanical system for Drill Problem 2.2.

Answer:

$$\frac{X_1(s)}{F(s)} = \frac{M_2s^2 + sD_2 + sD_3 + K_1 + K_2}{(M_1s^2 + D_1s + D_2s + K_1)(M_2s^2 + D_3s + D_2s + K_1 + K_2) - (sD_2 + K_1)^2}$$

DRILL PROBLEM 2.3

Draw the equivalent electrical circuit for the mechanical (rotational) system shown in Fig. 2.7. (*Hint*: Gears in mechanical systems are equivalent to ideal transformers in electric networks.)

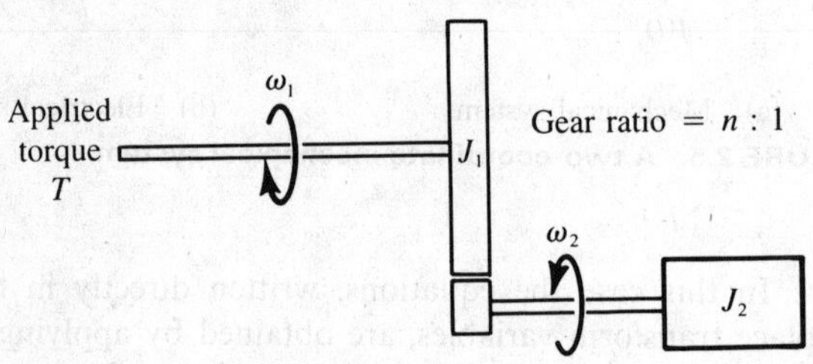

FIGURE 2.7. Rotational system with gear for Drill Problem 2.3.

Answer:

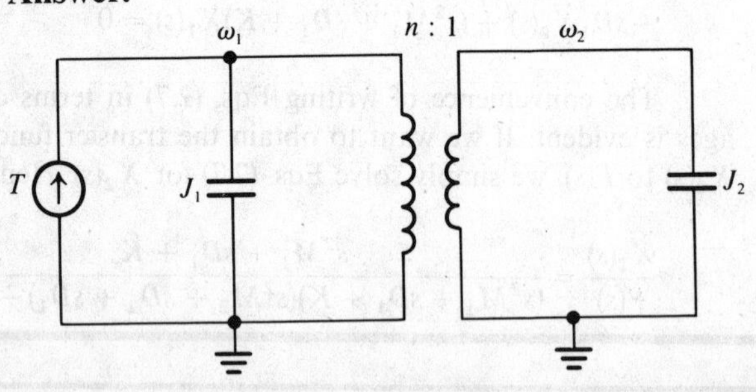

We shall now apply these concepts for obtaining the transfer function of a dc servomotor, which is often used as the "muscle arm" in control systems.

2.4 MODELING AN ARMATURE-CONTROLLED DC SERVOMOTOR

Consider an armature-controlled dc servomotor shown schematically in Fig. 2.8. It will be assumed that the field current is maintained constant, and a voltage $v(t)$ is applied to the armature, which has a resistance R_a and negligible inductance. The effect of the application of the input voltage $v(t)$ will be to cause the armature to rotate. The relationship between $v(t)$ and $\theta(t)$ can be obtained as follows:

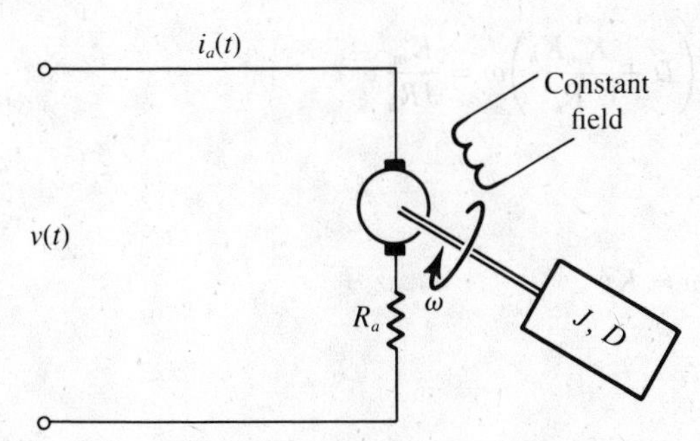

FIGURE 2.8. Armature-controlled dc servomotor.

$$i_a(t) = \frac{1}{R_a}[v(t) - v_b(t)] \tag{2.9}$$

where

$$v_b(t) = \text{back emf} = K_b\omega(t) \tag{2.10}$$

and K_b is the motor back emf constant.

The torque produced by the motor is given by

$$T_m = K_m i_a \tag{2.11}$$

and the load torque

$$T_L = J\frac{d\omega}{dt} + D\omega \tag{2.12}$$

where K_m = motor torque constant

J = moment of inertia of the moving parts about the axis of the armature shaft

D = damping coefficient due to friction

$$\omega = \frac{d\theta}{dt} = \text{angular velocity} \tag{2.13}$$

Note that if we neglect losses in the motor, then $K_b = K_m$, since $v_b i_a = T_m \omega$. The former is the electrical power developed and the latter is mechanical power available. Equating T_m and T_L and substituting for i_a from Eqs. (2.9) and (2.10), we get

$$K_m \left(\frac{v - K_b \omega}{R_a} \right) = J \frac{d\omega}{dt} + D\omega \tag{2.14}$$

which can be rearranged to obtain

$$\frac{d\omega}{dt} + \frac{1}{J} \left(D + \frac{K_m K_b}{R_a} \right) \omega = \frac{K_m}{J R_a} v \tag{2.15}$$

or

$$\frac{d\omega}{dt} + \alpha \omega = K v \tag{2.16}$$

where

$$\alpha = \frac{1}{J} \left(D + \frac{K_m K_b}{R_a} \right) \tag{2.17}$$

and

$$K = \frac{K_m}{J R_a} \tag{2.18}$$

Equation (2.16) may also be written as

$$\frac{d^2\theta}{dt^2} + \alpha \frac{d\theta}{dt} = K v \tag{2.19}$$

and taking Laplace transforms, we get the transfer function

$$\frac{\theta(s)}{V(s)} = \frac{K}{s(s + \alpha)} \tag{2.20}$$

This derivation of the transfer function can be visualized better through the block diagram in Fig. 2.9.

The block diagram has been obtained using the Laplace transforms of Eqs. (2.1) to (2.5). It may be noted that if the armature inductance, L_a, is not negligible, it may be taken into account by simply replacing $1/R_a$ in the armature block by $1/(R_a + sL_a)$.

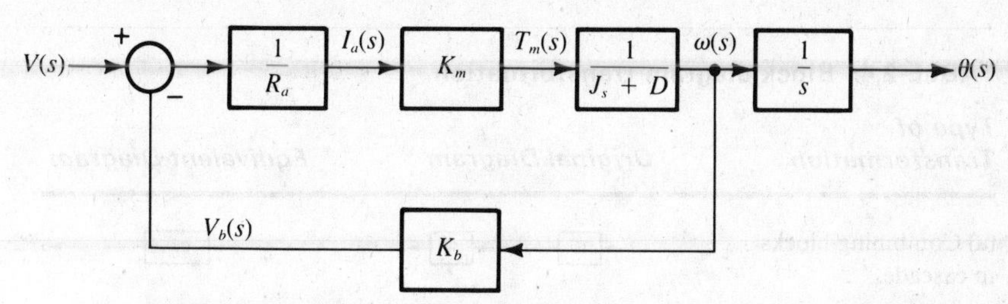

FIGURE 2.9. Block diagram showing the various relationships in the armature-controlled dc motor.

This simple example illustrates the power of the block diagram representation. It can be utilized fully through certain rules of block diagram algebra. These will be discussed in the following section.

It must be emphasized that the transfer function that we have obtained for the dc servomotor is based on several simplifying assumptions. We have neglected the armature reaction in the motor and the voltage drops in the brushes. In addition, we have assumed that the frictional torque is linear and directly proportional to the angular velocity. This is seldom true in practice. A more detailed and accurate model for the dc servomotor is nonlinear and complicated (Sinha, diCenzo, and Szabados, 1974). Nevertheless, the simplified linear model obtained above is useful for a preliminary design of control systems utilizing dc servomotors.

2.5 SIMPLIFICATION OF BLOCK DIAGRAMS

In the analysis of control systems it is very convenient to obtain the block diagrams of different components and their interconnections. If the various components are noninteracting (i.e., there is no "loading" effect of one component on another), it is possible to obtain the overall transfer function of the system through a suitable combination of the transfer functions of the component blocks utilizing some basic rules of block diagram transformations to reduce the original diagram. These are shown in Table 2.4.

These rules can be easily proved. For example, consider the cascade blocks shown in (a). Here, we have the equations

$$X_2 = G_1 X_1 \tag{2.21}$$

and

$$Y = G_2 X_2 \tag{2.22}$$

TABLE 2.4. Block diagram transformation

Type of Transformation	*Original Diagram*	*Equivalent Diagram*
(a) Combining blocks in cascade	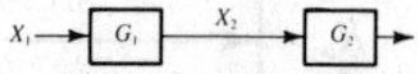	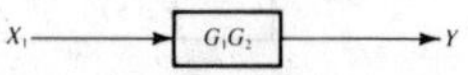
(b) Elimination of feedback loop	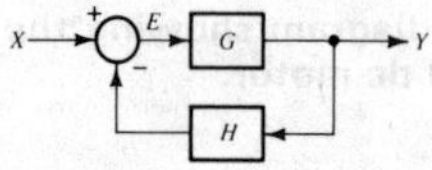	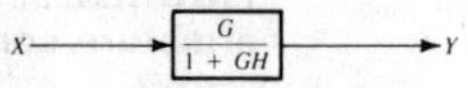
(c) Combining blocks in parallel	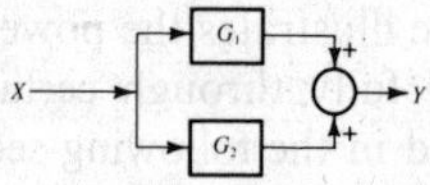	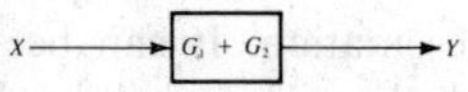
(d) Moving a pick-off point behind a block	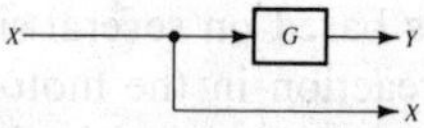	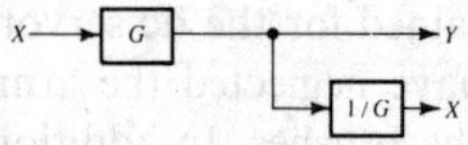
(e) Moving a summing point behind a block	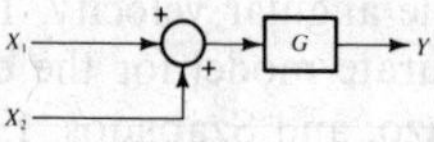	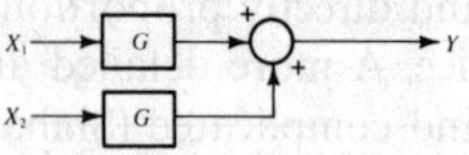
(f) Moving a pick-off point ahead of a summing point	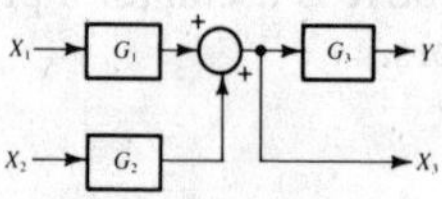	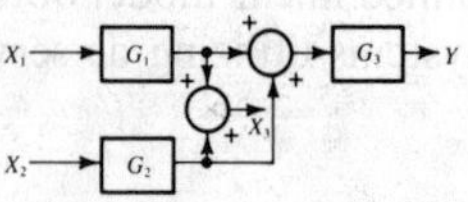
(g) Moving a pick-off point behind a summing point	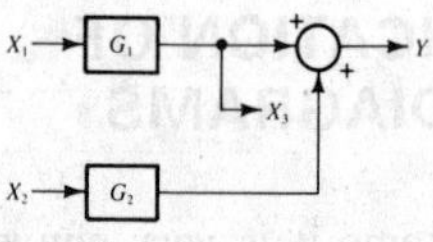	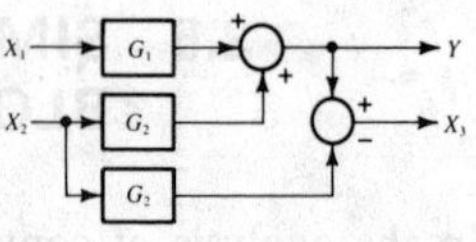

Eliminating X_2 from these two equations, we get

$$Y = (G_1 G_2) X_1 \tag{2.23}$$

which leads to the equivalent diagram.

Similarly, for the feedback loop in (b), we have

$$Y = GE \tag{2.24}$$

and

$$E = X - HY \tag{2.25}$$

Eliminating E from the two equations by substituting Eq. (2.25) into Eq. (2.24), we get, after simplification,

$$Y = \frac{G}{1 + GH} X \tag{2.26}$$

The proofs for the remaining transformations will be left to the reader as an exercise.

Returning to the block diagram in Fig. 2.9, we may apply rules (a) and (b) to obtain

$$\frac{\theta(s)}{V(s)} = \frac{\dfrac{1}{R_a} \cdot K_m \cdot \dfrac{1}{Js + D} \cdot \dfrac{1}{s}}{1 + \dfrac{1}{R_a} \cdot K_m \cdot \dfrac{K_b}{Js + D}} \tag{2.27}$$

which can then be simplified to Eq. (2.20).

We shall discuss some more examples to illustrate the procedure for block diagram simplification.

EXAMPLE 2.3

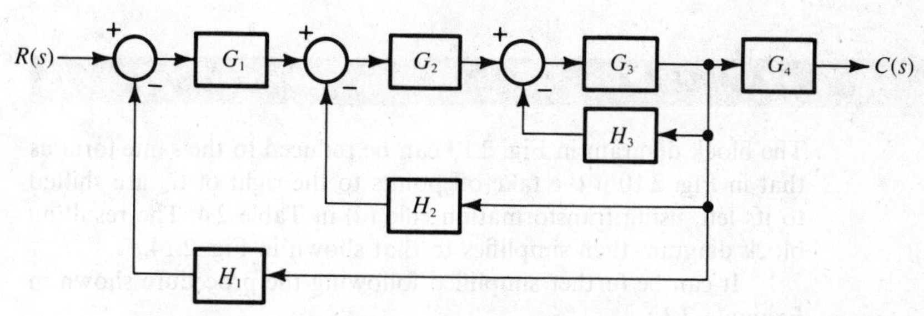

FIGURE 2.10. A mulitple-loop feedback system.

In this case, the first step in simplification is to replace the innermost loop by its equivalent transfer function. The resulting block diagram is shown in Fig. 2.11.

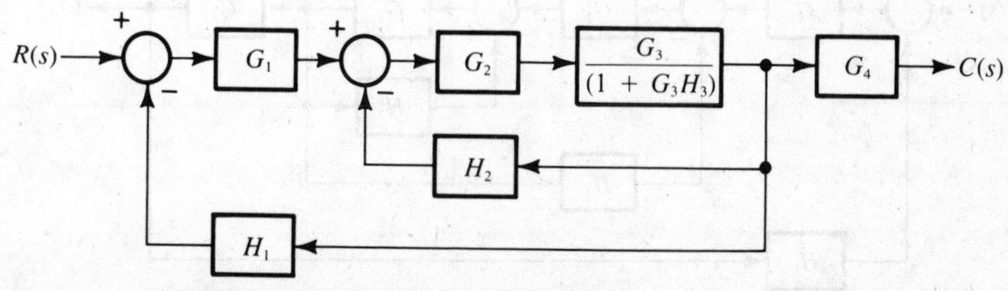

FIGURE 2.11. First step in simplification.

Proceeding in the same manner, we can again remove the inner loop after combining the two blocks in cascade. The resulting block diagram is shown in Fig. 2.12.

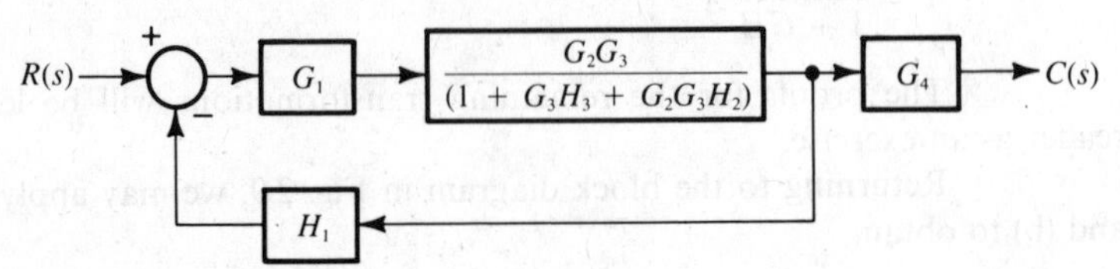

FIGURE 2.12. Second step in simplification.

Finally, removing the last feedback loop, we get the overall transfer function as

$$\frac{C(s)}{R(s)} = \frac{\dfrac{G_1G_2G_3G_4}{1 + G_3H_3 + G_2G_3H_2}}{1 + \dfrac{G_1G_2G_3}{1 + G_3H_3 + G_2G_3H_2} H_1}$$

$$= \frac{G_1G_2G_3G_4}{1 + G_3H_3 + G_2G_3H_2 + G_1G_2G_3H_1} \tag{2.28}$$

EXAMPLE 2.4

The block diagram in Fig. 2.13 can be reduced to the same form as that in Fig. 2.10 if the take-off points to the right of G_4 are shifted to its left, using transformation rule (d) in Table 2.4. The resulting block diagram then simplifies to that shown in Fig. 2.14.

It can be further simplified following the procedure shown in Example 2.3.

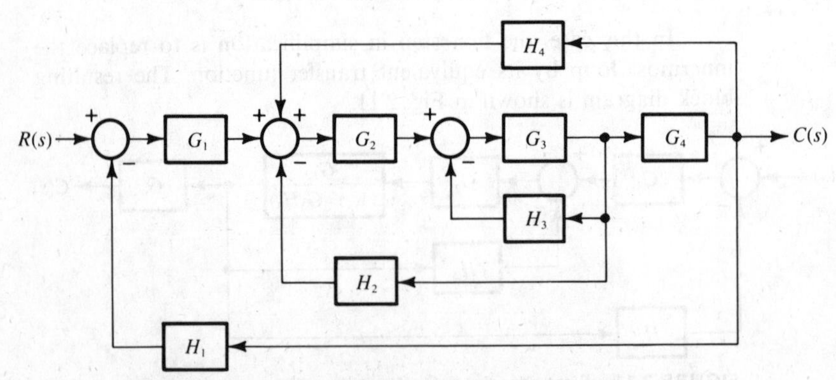

FIGURE 2.13. A multiple-loop feedback system.

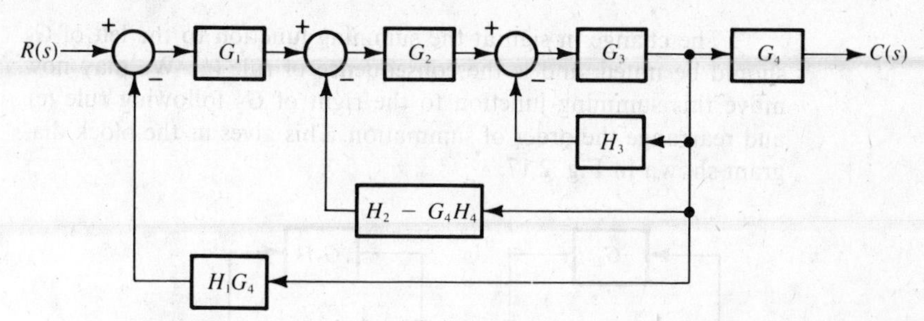

FIGURE 2.14. First step in the reduction of the block diagram shown in Fig. 2.13.

EXAMPLE 2.5

In the case shown in Fig. 2.15 the two forward blocks, G_1 and G_2, cannot be combined due to the summation point to the right of the take-off point. Hence we shall utilize rule (g) from Table 2.4 to obtain the block diagram shown in Fig. 2.16.

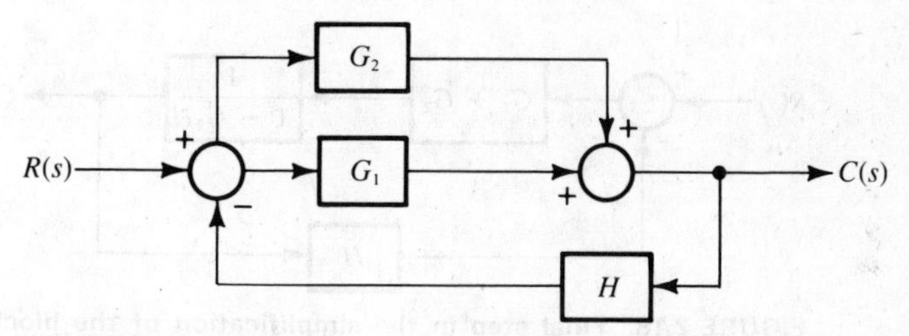

FIGURE 2.15. Block diagram of a system with feedforward and feedback paths.

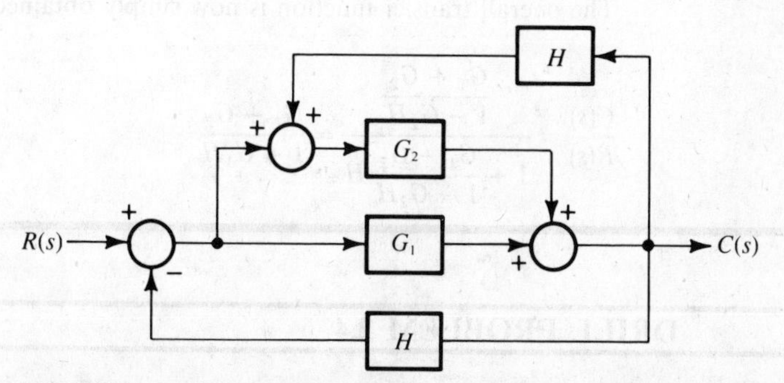

FIGURE 2.16. First step in simplification of the block diagram shown in Fig. 2.15.

The change in sign at the summing junction to the left of G_2 should be noted, and is the consequence of rule (g). We may now move this summing junction to the right of G_2 following rule (e), and rearrange the order of summation. This gives us the block diagram shown in Fig. 2.17.

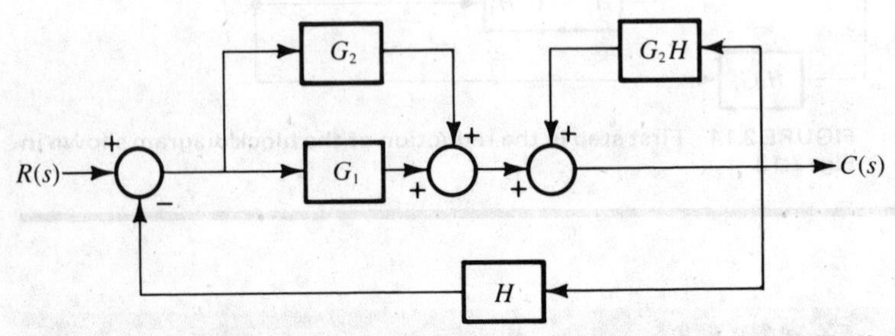

FIGURE 2.17. Second step in simplification of the block diagram shown in Fig. 2.15.

As the final step of simplification, we apply rule (c) to combine the two parallel blocks and rule (b) to eliminate the loop containing G_2H. Hence we get the single-loop block diagram shown in Fig. 2.18.

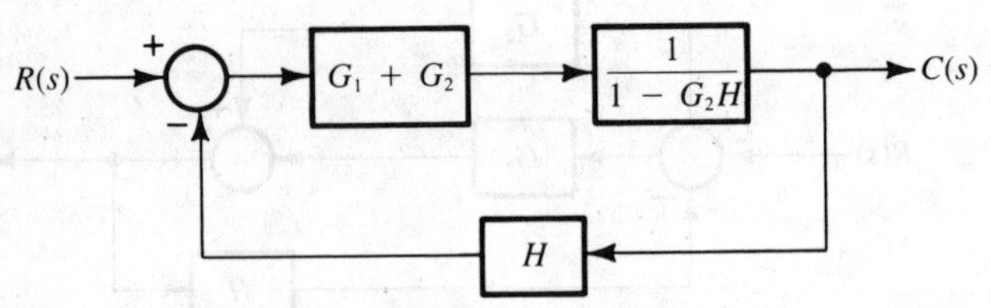

FIGURE 2.18. Final step in the simplification of the block diagram shown in Fig. 2.15.

The overall transfer function is now simply obtained as

$$\frac{C(s)}{R(s)} = \frac{\dfrac{G_1 + G_2}{1 - G_2H}}{1 + \dfrac{G_1 + G_2}{1 - G_2H}H} = \frac{G_1 + G_2}{1 + G_1H} \tag{2.29}$$

DRILL PROBLEM 2.4

Simplify the block diagram shown in Fig. 2.19 and hence determine the overall transfer function.

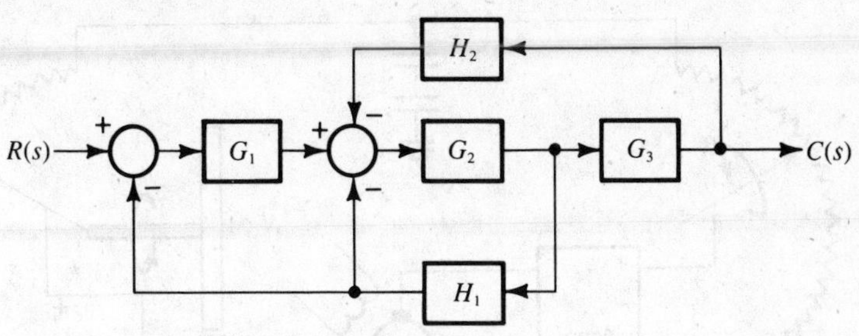

FIGURE 2.19. Block diagram for Drill Problem 2.4.

Answer:

$$\frac{C(s)}{R(s)} = \frac{G_1G_2G_3}{1 + G_2H_1 + G_2G_3H_2 + G_1G_2H_1}$$

DRILL PROBLEM 2.5

Repeat the preceding problem for the block diagram shown in Fig. 2.20.

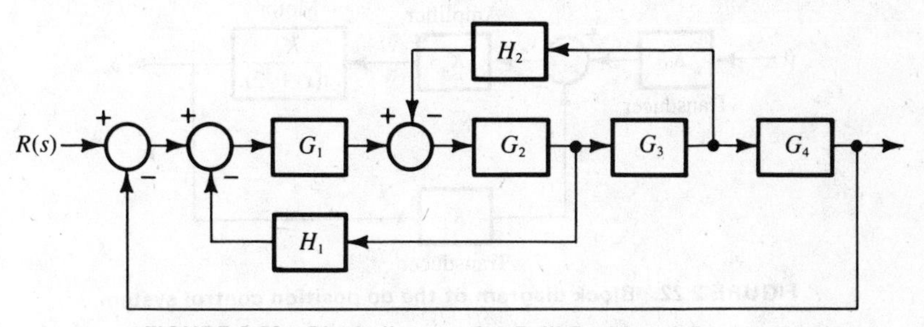

FIGURE 2.20. Block diagram for Drill Problem 2.5.

Answer:

$$\frac{C(s)}{R(s)} = \frac{G_1G_2G_3G_4}{1 + G_1G_2H_1 + G_2G_3H_2 + G_1G_2G_3G_4}$$

2.6 A DC POSITION-CONTROL SYSTEM

A dc position-control system or servomechanism is shown in Fig. 2.21. This device has many practical uses, where the position of the output shaft is required to follow that of the input shaft.

An error voltage, proportional to the difference between the input shaft position θ_i and the output shaft position θ_o, is applied to the amplifier.

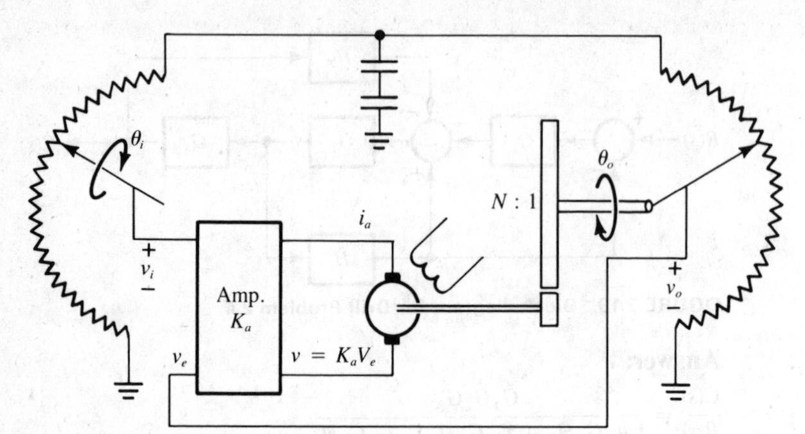

FIGURE 2.21. A dc position-control system.

The output of the amplifier is connected to the armature of a dc servomotor, as shown. The shaft of the motor is connected to the output shaft through a reduction gear of ratio N:1. The block diagram is shown in Fig. 2.22.

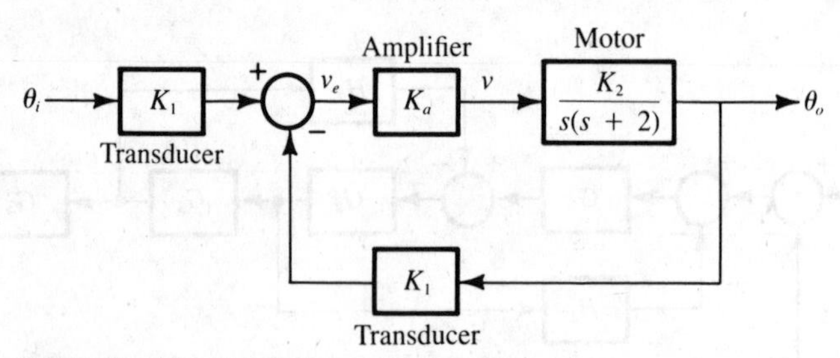

FIGURE 2.22. Block diagram of the dc position control system.

The following relations are easily derived.

$$v_i = K_1\theta_i \tag{2.30}$$

$$v_o = K_1\theta_o \tag{2.31}$$

where K_1 is the constant of the transducer (the potentiometer in this case). Similarly, the load torque on the motor is given by

$$T = J\frac{d^2\theta_o}{dt^2} + D\frac{d\theta_o}{dt} \tag{2.32}$$

where J is the effective moment of inertia and D the effective damping. This torque must equal the electrical torque developed by the motor.

It will be seen that in developing the block diagram we have taken advantage of our knowledge of the transfer function of the motor derived in Sec. 2.4. This is the main point in using the block diagram, since it shows the interconnection of different components for each of which we can determine the transfer function separately. It may be noted that we assume that there is no loading effect of one block on the other. For instance, if the loading effect of the motor on the amplifier could not be neglected, we would have to determine the transfer function of the amplifier-motor combination from the first principles, rather than multiplying the two separate transfer functions.

From the block diagram we get

$$\frac{\theta_o(s)}{\theta_i(s)} = \frac{K_1 K_a \dfrac{K_2}{s(s+\alpha)}}{1 + \dfrac{K_1 K_2 K_a}{s(s+\alpha)}} = \frac{K_1 K_a K_2}{s^2 + \alpha s + K_1 K_a K_2} \tag{2.33}$$

2.7 MASON'S RULE

The overall transfer function can be obtained directly from the block diagram by using Mason's rule, given as follows:

$$G(s) = \frac{C(s)}{R(s)} = \frac{\sum_k T_k(s)\Delta_k(s)}{\Delta(s)} \tag{2.34}$$

where $T_k(s) \triangleq$ transfer function of the kth forward path from the input to the output; a forward path cannot contain any feedback loop

$\Delta(s) \triangleq$ determinant of the block diagram

= 1 − sum of all individual loop transfer functions
+ sum of the products of the transfer functions of all possible sets of two nontouching loops
− sum of the products of the transfer functions of all possible sets of three nontouching loops
+ ⋯

$\Delta_k(s)$ = all terms in $\Delta(s)$ that do not have elements or paths common with an element or path in $T_k(s)$

Note that two loops are said to be nontouching if they do not have a common branch or node. Also, a loop is a closed path in the direction of the arrows that does not retrace itself.

EXAMPLE 2.6

As an example, consider the block diagram shown in Fig. 2.10. Here, we have only one forward path from $R(s)$ to $C(s)$, so that $k = 1$, and three loops with transmittances $-G_3H_3$, $-G_2G_3H_2$, and

$-G_1G_2G_3H_1$. Also, all the loops are touching each other. Thus

$$T_1 = G_1G_2G_3G_4$$
$$\Delta = 1 - (-G_3H_3 - G_2G_3H_2 - G_1G_2G_3H_1)$$
$$= 1 + G_3H_3 + G_2G_3H_2 + G_1G_2G_3H_1$$

Hence

$$G(s) = \frac{T_1}{\Delta} = \frac{G_1G_2G_3G_4}{1 + G_3H_3 + G_2G_3H_2 + G_1G_2G_3H_1}$$

as in Eq. (2.28).

EXAMPLE 2.7

As another example, consider the block diagram shown in Fig. 2.23.

The transfer function $C(s)/R(s)$ can be determined by block diagram simplification. This is done most conveniently by replacing the inner loop by its equivalent and then adding the two parallel branches to obtain a single loop diagram. We shall apply Mason's rule, and the result can be verified by the student by using the procedure suggested above.

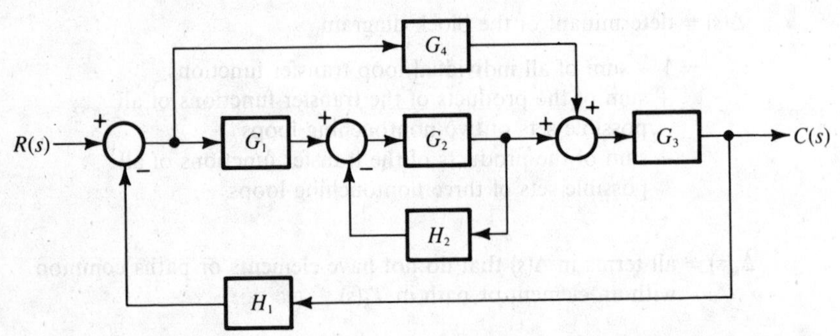

FIGURE 2.23. A block diagram with feedforward and feedback paths.

Here we have two forward paths from $R(s)$ to $C(s)$, and

$$T_1(s) = G_1G_2G_3$$
$$T_2(s) = G_4G_3$$

Also, we have three loops with transmittance $-G_2H_2$, $-G_1G_2G_3H_1$, and $-G_4G_3H_1$. Furthermore, the first and the last loops are nontouching. This gives us

$$\Delta(s) = 1 + (G_2H_2 + G_1G_2G_3H_1 + G_4G_3H_1)$$

$$\Delta_1(s) = 1$$

$$\Delta_2(s) = 1 + G_2H_2$$

Hence

$$\frac{C(s)}{R(s)} = \frac{G_1G_2G_3 + G_4G_3(1 + G_2H_2)}{1 + G_2H_2 + G_1G_2G_3H_1 + G_4G_3H_1 + G_2H_2G_4G_3H_1} \tag{2.35}$$

EXAMPLE 2.8

Consider now the block diagram shown in Fig. 2.15, which we had simplified in Example 2.5. We shall now apply Mason's rule to this diagram.

We see that there are two forward paths from $R(s)$ to $C(s)$, with transmittances $T_1 = G_1$, and $T_2 = G_2$. Furthermore, there is only one loop, with transmittance $-G_1H$, and this is touching both the forward paths. Hence we get

$$\frac{C(s)}{R(s)} = \frac{G_1 + G_2}{1 + G_1H} \tag{2.36}$$

which is seen to be identical to Eq. (2.29).

EXAMPLE 2.9

Consider the block diagram shown in Fig. 2.24.

We have two forward paths from $R(s)$ to $C(s)$, with transmittances

$$T_1 = G_1G_2G_3 \quad \text{and} \quad T_2 = G_4G_3$$

Also, there are five loops, with transmittances $-G_1H_1$, $-G_2H_2$, $-G_1G_2G_3H_3$, $-G_4G_3H_3$, and $G_4H_2H_1$, and they are all touching each other. Hence we have

$$\frac{C(s)}{R(s)} = \frac{G_1G_2G_3 + G_4G_3}{1 + G_1H_1 + G_2H_2 + G_1G_2G_3H_3 + G_4G_3H_3 - G_4H_2H_1} \tag{2.37}$$

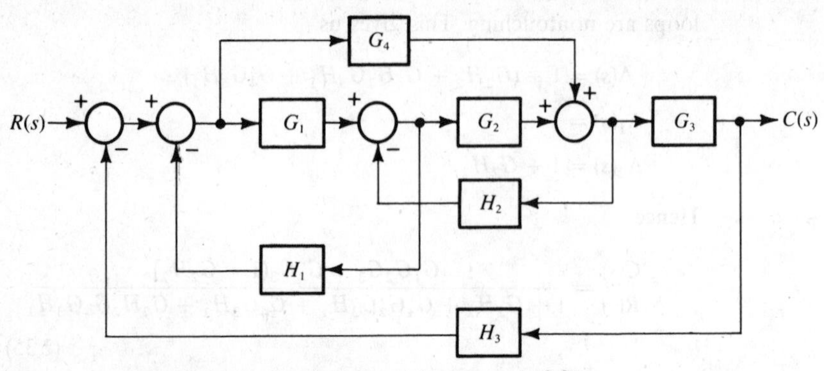

FIGURE 2.24. Block diagram for Example 2.9.

DRILL PROBLEM 2.6

Solve Drill Problem 2.4 by applying Mason's rule.

DRILL PROBLEM 2.7

Verify the answer of Drill Problem 2.5 by applying Mason's rule.

2.8 SUMMARY

In this chapter we have discussed the general procedure for obtaining mathematical models for physical systems. This is often the most important and crucial step in being able to control a system. We may paraphrase Aristotle to say that we can control any system provided that we have a suitable mathematical model for it.

The control engineer often has to obtain the mathematical model for a variety of systems with components of diverse nature, such as electrical, mechanical, fluidic, and thermal, as well as various combinations of these. In all of these cases one must first identify the variables, which may be either through-variables, or across-variables. In most cases it is helpful to draw an analogous electrical circuit for which the differential equations are easily obtained by applying Kirchhoff's laws.

We start by writing the differential equations relating the variables in each component of the system, as well as the interconnection equations. Although in most cases these differential equations are nonlinear, as a first approximation they are linearized about the operating points. In this case it is very convenient to identify one variable as the input, or cause, and another variable as output, or effect. This leads to the block diagram for each component, which relates the Laplace transforms of the input and the output through the transfer function. The overall transfer function of the system is then obtained either by simplifying the block diagram, or directly by using Mason's rule.

2.9 REFERENCES

1. David K. Cheng, *Analysis of Linear Systems*, Addison-Wesley, Reading, Mass., 1959.
2. N. K. Sinha, C. D. diCenzo, and B. Szabados, "Modeling of DC Motors for Control Applications," *IEEE Transactions on Industrial Electronics and Control Instrumentation*, vol. IECI-21, 1974, pp. 84–88.

2.10 PROBLEMS

1. Determine the transfer functions $V_2(s)V_1(s)$ of the electrical networks shown in Figs. 2.25 and 2.26.

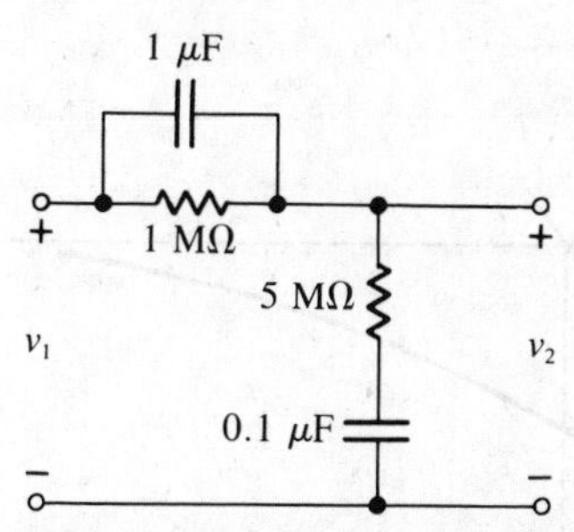

FIGURE 2.25. ***RC*** **network for Problem 1(a).**

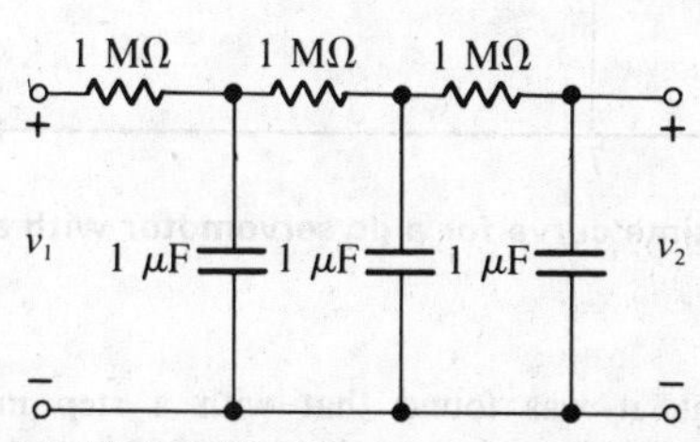

FIGURE 2.26. ***RC*** **network for Problem 1(b).**

2. Compare the transfer function of the network shown in Fig. 2.26 with that obtained by taking the cube of transfer function of the network shown in Fig. 2.27. Why are they different?

3. For each of the networks in Problem 1, determine the steady-state component of $v_2(t)$ if $v_1(t) = 2 + 3\cos(2t + \pi/6)$. (*Hint*: See Appendix A, Sec. A.6.)

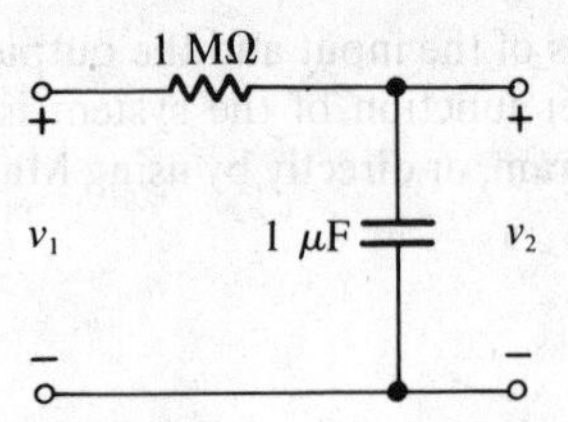

FIGURE 2.27. ***RC* network for Problem 2.**

4. A practical method for determining the transfer function of a dc servomotor is to apply a constant voltage to its armature and, with the field current maintained at a constant value, measure the steady-state speed of the motor. In addition, one must also measure the time required for the motor, starting from rest, to reach a certain fraction of the steady-state speed, say 50 percent or 63.2 percent. A typical speed-time curve is shown in Fig. 2.28. From these two measurements, and assuming that the transfer function of the motor is given by

$$G(s) = \frac{\theta(s)}{V(s)} = \frac{K}{s(s+\alpha)}$$

the values of K and α can be obtained.

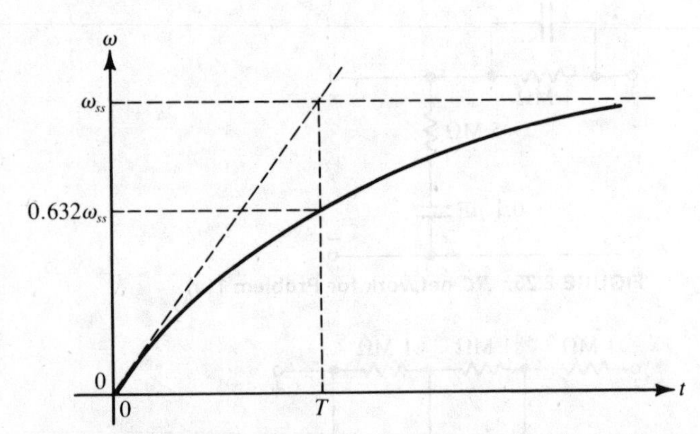

FIGURE 2.28. Speed-time curve for a dc servomotor with a constant applied voltage.

In such an experiment, it was found that with a step input of 10 volts applied to the motor, its steady-state speed was 1200 revolutions per minute, and it required 1.2 seconds to reach 50 percent of this speed. Determine the transfer function of the motor. (*Hint*: Note that

$$\frac{\omega(s)}{V(s)} = \frac{K}{s+\alpha},$$

where $\omega(t) = d\theta/dt$ is the angular velocity in radians per second.)

5. One form of a self-balancing potentiometer is shown in Fig. 2.29. The input is $e_i(t)$ in volts and the output is $x(t)$ in centimeters. The amplifier gain $K = 10$. The lead screw advances 1 mm per revolution. The voltage of the movable arm of the potentiometer increases by 0.4 volt for each centimeter movement of x. With a step input of 10 volts applied to the motor (with the load), its steady-state speed is 1000 revolutions per minute, and it takes 0.5 second to reach 63.2 percent of this value. Draw a block diagram of the system and determine the transfer function $X(s)/E_i(s)$. Also, determine $x(t)$ if $e_i(t)$ is a step of 2 volts.

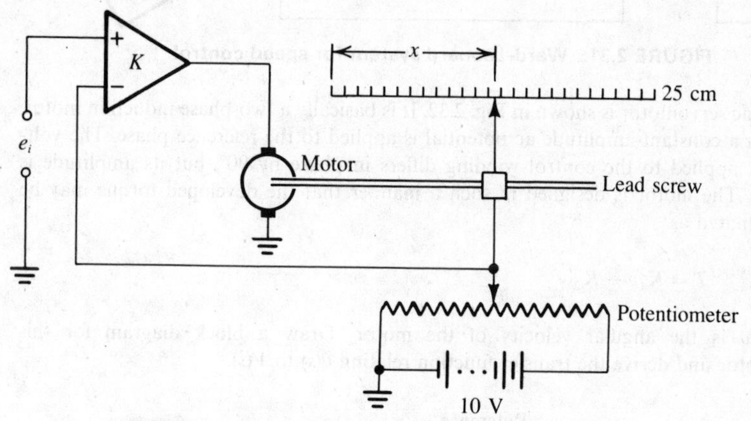

FIGURE 2.29. A self-balancing potentiometer.

6. A dc servomotor is sometimes used in the field-control mode. In this case the armature current is maintained constant and the input voltage is applied to the field winding. A schematic diagram is shown in Fig. 2.30. Determine the transfer function $\theta(s)/V(s)$. (*Hint*: Note that the inductance of the field winding cannot be neglected. Draw a diagram similar to Fig. 2.9.)

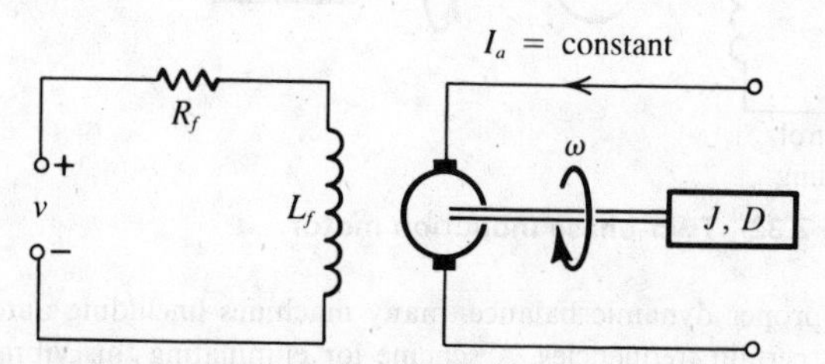

FIGURE 2.30. A field-controlled dc servomotor.

7. The Ward-Leonard scheme for controlling the speed of heavy loads (for example, in a steel rolling mill) is shown in Fig. 2.31. A dc generator, driven at a constant speed by a three-phase synchronous motor, drives an armature-controlled dc motor. Draw a block diagram for this system and determine the transfer function relating ω_0 to v if the constants of the machines are as given below.

Generator: $R_f = 1\ \Omega$, $L_F = 0.2$ H, $R_g = 0.2\ \Omega$, $L_g = 0.01$ H, $K_g = 2$ V/A

Motor: $R_m = 0.5\ \Omega$, $L_m = 0.02$ H, $K_m = 0.1\ \text{V}\cdot\text{rad}^{-1}\cdot\text{s}^{-1}$

Load: $J = 4\ \text{kg}\cdot\text{m}^2$, $D = 0.1\ \text{N}\cdot\text{m}\cdot\text{rad}^{-1}\cdot\text{s}^{-1}$

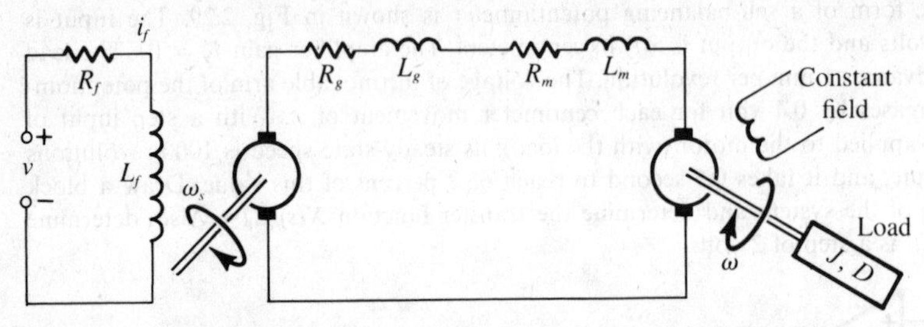

FIGURE 2.31. Ward-Leonard system for speed control.

8. An ac servomotor is shown in Fig. 2.32. It is basically a two-phase induction motor, in which a constant-amplitude ac potential is applied to the reference phase. The voltage, $v(t)$, applied to the control winding differs in phase by 90°, but its amplitude is variable. The motor is designed in such a manner that the developed torque may be approximated as

$$T = K_1 v - K_2 \omega$$

where ω is the angular velocity of the motor. Draw a block diagram for this servomotor and derive the transfer function relating $\theta(s)$ to $V(s)$.

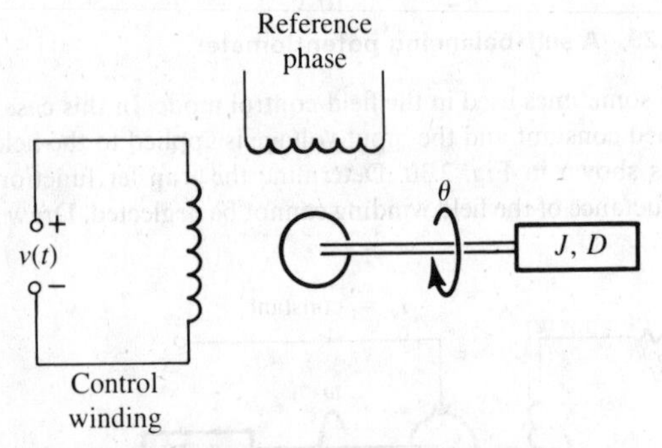

FIGURE 2.32. Two-phase induction motor.

9. Due to lack of proper dynamic balance, many machines (including automobiles) exhibit vibrations at certain frequencies. A scheme for eliminating this vibration at a certain frequency is shown in Fig. 2.33. Here, the object is to prevent the mass M_1 from vibrating at a frequency ω_0, and this is done by including the spring K_2 and the mass M_2 as shown. Obtain the equivalent electrical circuit for the system, and hence determine the relationship between K_2 and M_2 that will meet this objective. Also, determine the transfer function relating $x_1(t)$ and $f(t)$. (*Hint:* Examine the effect of the analogs of M_2 and K_2 on the voltage analogous do dx_1/dt.)

10. An analog computer is often used for simulating dynamic systems and studying the effect of adjusting the values of certain parameters. Its basic components are operational amplifiers, which can be used as adders, integrators, inverters, and potentiometers.

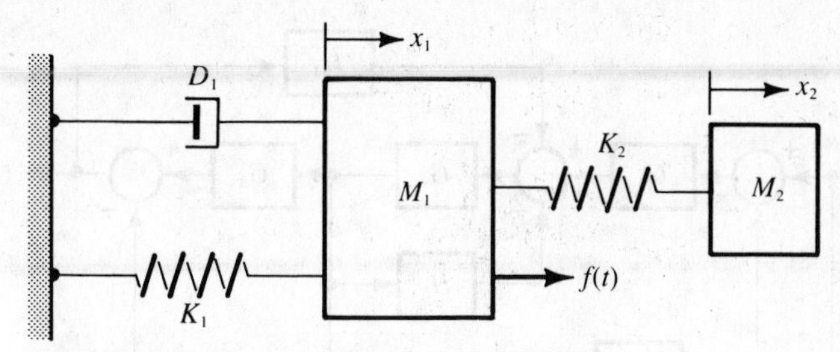

FIGURE 2.33. A dynamic vibration absorber.

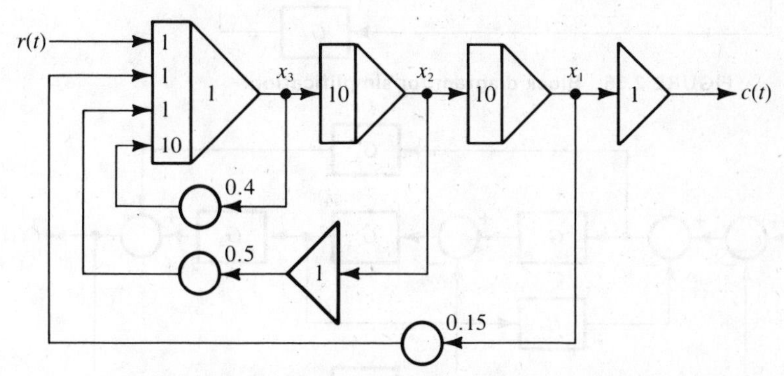

FIGURE 2.34. Analog computer simulation diagram.

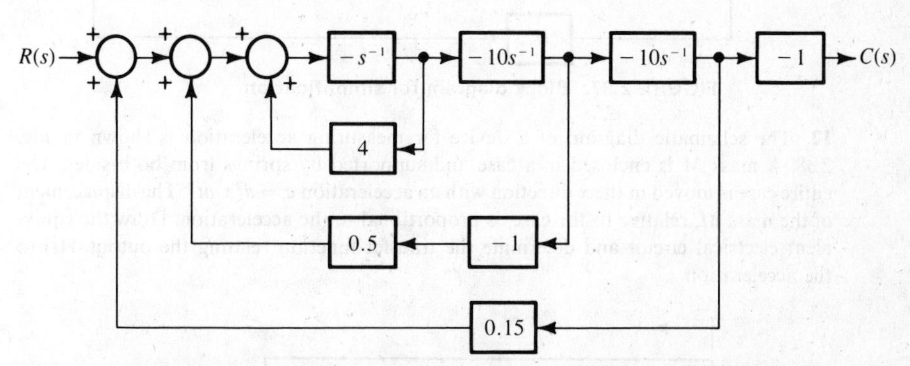

FIGURE 2.35. Block diagram for the analog computer setup shown in Fig. 2.34.

The simulation diagram for a particular system is shown in Fig. 2.34, and the corresponding block diagram is shown in Fig. 2.35. Determine the transfer function $C(s)/R(s)$ from the block diagram.

11. For the block diagrams shown in Figs. 2.36 and 2.37 determine the overall transfer function $C(s)/R(s)$ by block diagram simplification and verify by using Mason's rule.

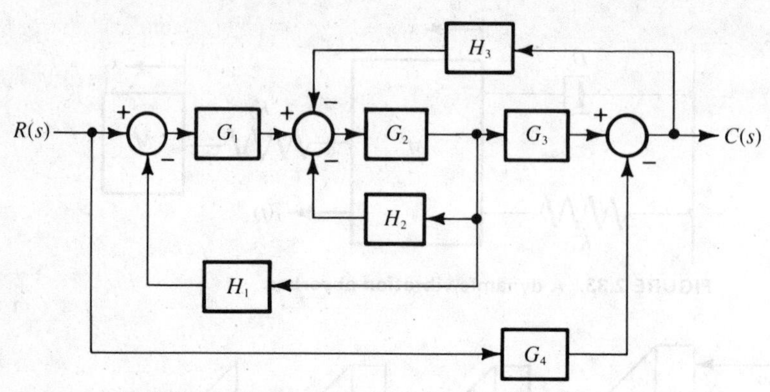

FIGURE 2.36. Block diagram for simplification.

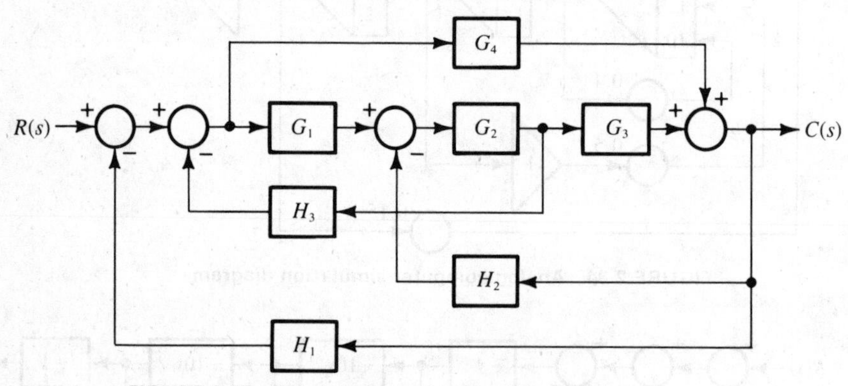

FIGURE 2.37. Block diagram for simplification.

12. The schematic diagram of a device for measuring acceleration is shown in Fig. 2.38. A mass M is enclosed in a case and supported by springs from both sides. The entire case is moved in the x direction with an acceleration $a = d^2x/dt^2$. The displacement of the mass M, relative to the case, is proportional to the acceleration. Draw the equivalent electrical circuit and determine the transfer function relating the output $Y(s)$ to the acceleration.

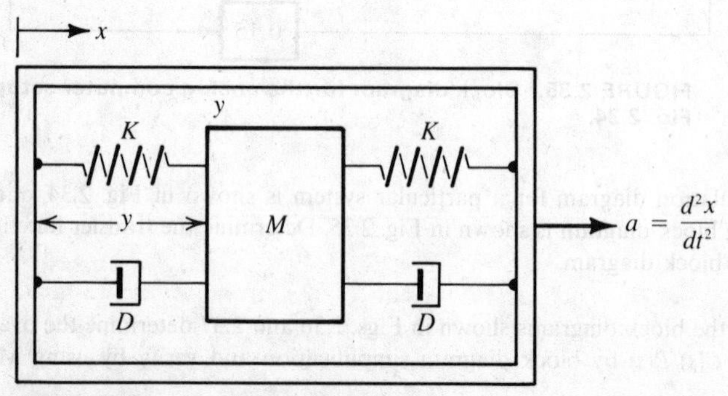

FIGURE 2.38. Accelerometer.

3
State-Space Methods

3.1 INTRODUCTION

Although the transfer function representation of physical systems provides an important tool for the analysis and design of control systems, it has certain basic limitations. This is due to the fact that Laplace transformation converts a differential equation to an algebraic equation in the complex frequency variable, s, only if the differential equation is linear with constant coefficients. Hence its use is limited to systems that are linear and time-invariant.

On the other hand, time-domain methods, based on the state-space representation are more powerful. In addition to being applicable to nonlinear and time-varying systems, they are readily extended to multivariable systems—i.e., systems with several inputs and outputs. Their main attraction is that they are natural and convenient for computer solutions. In particular, it is much easier to study discrete-time systems with the help of digital computers when the state-space representation is employed.

3.2 STATE-SPACE REPRESENTATION

3.2.1 The Concept of State

The state of a system is defined as the smallest set of numbers that must be known in order that its future response to any given input can be calculated from the dynamic equation. Thus the state is a compact representation of the past history of the system, which can be utilized for the prediction of its future behavior in response to any external stimulus. Since the complete solution of a differential equation of order n requires precisely n initial conditions, it follows that the state of such a system will be specified by the values of n quantities, called the state variables.

As an example, in an electrical network we must know the voltage across each capacitor and the current through each inductor at any given instant of time in order to determine its future response for given inputs. It should, however, be emphasized that, in general, the state-space representation of a given system is not unique; there are many different sets of state variables that can be utilized for a given system. This will be illustrated through an example later in this section.

State equations are arranged as a set of first-order differential equations, and have the following form for a linear time-invariant system

$$\dot{x} = Ax + Bu \tag{3.1}$$

where $x(t)$ is the n-dimensional state vector, the overdot represents the derivative with respect to time, and $u(t)$ is the m-dimensional input vector. A and B are $n \times n$ and $n \times m$ matrices, respectively, with constant elements. The output of the system is given by

$$y(t) = Cx(t) + Du(t) \tag{3.2}$$

where $y(t)$ is a vector of dimension p, the number of outputs, and C and D are constant matrices of dimensions $p \times n$ and $p \times m$, respectively.

Although Eqs. (3.1) and (3.2) are meant for the general case of multivariable systems, in this book we shall only study the special and simple case of single-input single-output systems, for which $m = 1$ and $p = 1$. The transfer function of such a system is obtained by taking the Laplace transforms of Eqs. (3.1) and (3.2) for zero initial conditions. Hence

$$G(s) = \frac{Y(s)}{U(s)} = C(sI - A)^{-1}B + D \tag{3.3}$$

For a strictly proper transfer function, the degree of the numerator of $G(s)$ is less than that of the denominator. For such a case, D must be equal to zero.

EXAMPLE 3.1

The differential equation for an armature-controlled dc servomotor was derived in Sec. 2.4. This is given below.

$$\frac{d^2\theta}{dt^2} + \alpha \frac{d\theta}{dt} = Kv \tag{3.4}$$

This is a second-order differential equation. We can change it into two first-order differential equations by defining the variables

$$x_1 = \theta \tag{3.5}$$

and

$$x_2 = \dot{x}_1 = \frac{d\theta}{dt} \tag{3.6}$$

Hence, Eq. (3.4) can be written as

$$\dot{x}_2 = Kv - \alpha x_2 \tag{3.7}$$

Equations (3.6) and (3.7) can now be combined to form the following state equations, which have been expressed in the form of a matrix.

$$\begin{bmatrix} \dot{x}_1 \\ \dot{x}_2 \end{bmatrix} = \begin{bmatrix} 0 & 1 \\ 0 & -\alpha \end{bmatrix} \begin{bmatrix} x_1 \\ x_2 \end{bmatrix} + \begin{bmatrix} 0 \\ K \end{bmatrix} v \tag{3.8}$$

and

$$y = [0 \quad 1] \begin{bmatrix} x_1 \\ x_2 \end{bmatrix} \tag{3.9}$$

Here, the state variable x_1 is the angular position θ and x_2 is the angular velocity ω. Often it is possible to have many choices for the state variables. This will be illustrated in the next example.

EXAMPLE 3.2

For the field-controlled dc servomotor shown in Fig. 3.1 the following differential equations are obtained.

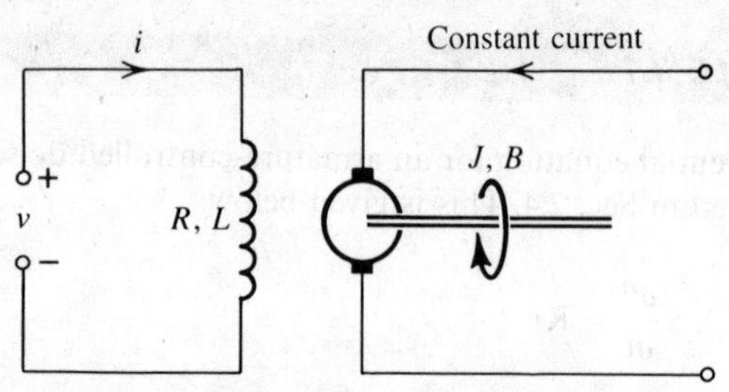

FIGURE 3.1. A field-controlled dc servomotor.

$$v = L\frac{di}{dt} + Ri \tag{3.10}$$

$$Ki = J\frac{d\omega}{dt} + B\omega \tag{3.11}$$

$$\omega = \frac{d\theta}{dt} \tag{3.12}$$

These are already a set of three first-order differential equations. Hence, if we let $x_1 = \theta$, $x_2 = \omega$, and $x_3 = i$, we get

$$\begin{bmatrix} \dot{x}_1 \\ \dot{x}_2 \\ \dot{x}_3 \end{bmatrix} = \begin{bmatrix} 0 & 1 & 0 \\ 0 & -B/J & K/J \\ 0 & 0 & -R/L \end{bmatrix} \begin{bmatrix} x_1 \\ x_2 \\ x_3 \end{bmatrix} + \begin{bmatrix} 0 \\ 0 \\ 1/L \end{bmatrix} v \tag{3.13}$$

and

$$y(t) = \begin{bmatrix} 1 & 0 & 0 \end{bmatrix} \begin{bmatrix} x_1 \\ x_2 \\ x_3 \end{bmatrix} \tag{3.14}$$

In this case it is also possible to select the angular position θ and its first and second derivatives as the state variables. This can be done by differentiating Eq. (3.11) and eliminating i and its derivative by using Eqs. (3.10) and (3.11). The resulting state equations are

$$\begin{bmatrix} \dot{x}_1 \\ \dot{x}_2 \\ \dot{x}_3 \end{bmatrix} = \begin{bmatrix} 0 & 1 & 0 \\ 0 & 0 & 1 \\ 0 & -BR/JL & -(B/J + R/L) \end{bmatrix} \begin{bmatrix} x_1 \\ x_2 \\ x_3 \end{bmatrix} + \begin{bmatrix} 0 \\ 0 \\ K/JL \end{bmatrix} v \tag{3.15}$$

and

$$y = \begin{bmatrix} 1 & 0 & 0 \end{bmatrix} \begin{bmatrix} x_1 \\ x_2 \\ x_3 \end{bmatrix} \tag{3.16}$$

where $x_1 = \theta$, $x_2 = \omega$, and $x_3 = d\omega/dt$.

DRILL PROBLEM 3.1

Write a set of state equations for the mechanical system described in Example 2.1. (*Hint*: Use the state variables $x_1 = x$ and $x_2 = dx/dt$.)

Answer:

$$\begin{bmatrix} \dot{x}_1 \\ \dot{x}_2 \end{bmatrix} = \begin{bmatrix} 0 & 1 \\ -K/M & -D/M \end{bmatrix} \begin{bmatrix} x_1 \\ x_2 \end{bmatrix} + \begin{bmatrix} 0 \\ 1/M \end{bmatrix} f$$

3.2.2 Computation of the Transfer Function from State Equations

The calculation of the transfer function from the state equations requires the inversion of the matrix $(sI - A)$, as shown in Eq. (3.3). This can be done conveniently by using Leverrier's algorithm (also called the Souriau-Frame algorithm).

Let

$$(sI - A)^{-1} = \frac{\operatorname{adj}(sI - A)}{\det(sI - A)} = \frac{P_{n-1}s^{n-1} + P_{n-2}s^{n-2} + \cdots + P_1 s + P_0}{s^n + a_{n-1}s^{n-1} + \cdots + a_1 s + a_0} \tag{3.17}$$

where P_i are $n \times n$ matrices and a_i are scalars.

The algorithm proceeds as follows:

$$\left.\begin{aligned}
&(1)\quad P_{n-1} = I, \text{ the identity matrix} \\
&(2)\quad a_{n-1} = -\operatorname{tr} A, \text{ where tr represents the trace of a matrix} \\
&\qquad \text{(sum of the elements on the main diagonal)} \\
&(3)\quad P_{n-2} = P_{n-1}A + a_{n-1}I \\
&(4)\quad a_{n-2} = -\frac{1}{2}\operatorname{tr}(P_{n-2}A) \\
&\qquad \vdots \\
&\qquad P_k = P_{k+1}A + a_{k+1}I \\
&\qquad a_k = -\frac{1}{n-k}\operatorname{tr}(P_k A)
\end{aligned}\right\} \tag{3.18}$$

Also,

$$P_0 A + a_0 I = 0 \tag{3.19}$$

may be used for checking.

Note that the denominator of the transfer function is the determinant of the matrix $(sI - A)$.

EXAMPLE 3.3

We shall use the above algorithm to determine the transfer function for the following case:

$$A = \begin{bmatrix} -2 & 0 & 1 \\ 1 & -2 & 0 \\ 1 & 1 & -1 \end{bmatrix} \quad B = \begin{bmatrix} 1 \\ 0 \\ 1 \end{bmatrix} \quad C = [2 \quad 1 \quad -1] \tag{3.20}$$

In this case we have $n = 3$. Hence

$$P_2 = I \tag{3.21}$$

$$a_2 = -\operatorname{tr} A = 5 \tag{3.22}$$

$$P_1 = IA + a_2 I = \begin{bmatrix} -2 & 0 & 1 \\ 1 & -2 & 0 \\ 1 & 1 & -1 \end{bmatrix} + \begin{bmatrix} 5 & 0 & 0 \\ 0 & 5 & 0 \\ 0 & 0 & 5 \end{bmatrix} = \begin{bmatrix} 3 & 0 & 1 \\ 1 & 3 & 0 \\ 1 & 1 & 4 \end{bmatrix} \tag{3.23}$$

$$\begin{aligned} a_1 &= -\frac{1}{2} \operatorname{tr}(P_1 A) = -\frac{1}{2} \operatorname{tr} \left\{ \begin{bmatrix} 3 & 0 & 1 \\ 1 & 3 & 0 \\ 1 & 1 & 4 \end{bmatrix} \begin{bmatrix} -2 & 0 & 1 \\ 1 & -2 & 0 \\ 1 & 1 & -1 \end{bmatrix} \right\} \\ &= -\frac{1}{2} \operatorname{tr} \begin{bmatrix} -5 & 1 & 2 \\ 1 & -6 & 1 \\ 3 & 2 & -3 \end{bmatrix} = 7 \end{aligned} \tag{3.24}$$

$$P_0 = P_1 A + a_1 I = \begin{bmatrix} -5 & 1 & 2 \\ 1 & -6 & 1 \\ 3 & 2 & -3 \end{bmatrix} + \begin{bmatrix} 7 & 0 & 0 \\ 0 & 7 & 0 \\ 0 & 0 & 7 \end{bmatrix} = \begin{bmatrix} 2 & 1 & 2 \\ 1 & 1 & 1 \\ 3 & 2 & 4 \end{bmatrix} \tag{3.25}$$

$$\begin{aligned} a_0 &= -\frac{1}{3} \operatorname{tr}(P_0 A) = -\frac{1}{3} \operatorname{tr} \left\{ \begin{bmatrix} 2 & 1 & 2 \\ 1 & 1 & 1 \\ 3 & 2 & 4 \end{bmatrix} \begin{bmatrix} -2 & 0 & 1 \\ 1 & -2 & 0 \\ 1 & 1 & -1 \end{bmatrix} \right\} \\ &= -\frac{1}{3} \operatorname{tr} \begin{bmatrix} -1 & 0 & 0 \\ 0 & -1 & 0 \\ 0 & 0 & -1 \end{bmatrix} = 1 \end{aligned} \tag{3.26}$$

Check:

$$P_0A + a_0I = 0 \tag{3.27}$$

Hence

$$\operatorname{adj}(sI - A) = P_2s^2 + P_1s + P_0$$

$$= \begin{bmatrix} s^2 + 3s + 2 & 1 & s + 2 \\ s + 1 & s^2 + 3s + 1 & 1 \\ s + 3 & s + 2 & s^2 + 4s + 4 \end{bmatrix} \tag{3.28}$$

$$G(s) = C(sI - A)^{-1}B$$

$$= [2 \quad 1 \quad -1] \begin{bmatrix} s^2 + 3s + 2 & 1 & s + 2 \\ s + 1 & s^2 + 3s + 1 & 1 \\ s + 3 & s + 2 & s^2 + 4s + 4 \end{bmatrix}$$

$$\times \begin{bmatrix} 1 \\ 0 \\ 1 \end{bmatrix} \frac{1}{s^3 + 5s^2 + 7s + 1}$$

$$= [2 \quad 1 \quad -1] \begin{bmatrix} s^2 + 4s + 4 \\ s + 2 \\ s^2 + 5s + 7 \end{bmatrix} \frac{1}{s^3 + 5s^2 + 7s + 1}$$

$$= \frac{s^2 + 4s + 3}{s^3 + 5s^2 + 7s + 1} \tag{3.29}$$

EXAMPLE 3.4

Consider the previous problem with A unchanged but

$$B = \begin{bmatrix} 1 & 0 \\ 0 & 1 \\ 1 & 0 \end{bmatrix} \qquad C = \begin{bmatrix} 2 & 1 & -1 \\ 0 & 1 & 0 \end{bmatrix} \tag{3.30}$$

This is a system with two inputs and two outputs. Hence,

$$G(s) = \begin{bmatrix} 2 & 1 & -1 \\ 0 & 1 & 0 \end{bmatrix} \begin{bmatrix} s^2 + 3s + 2 & 1 & s + 2 \\ s + 1 & s^2 + 3s + 1 & 1 \\ s + 3 & s + 2 & s^2 + 4s + 4 \end{bmatrix}$$

$$\times \begin{bmatrix} 1 & 0 \\ 0 & 1 \\ 1 & 0 \end{bmatrix} \frac{1}{s^3 + 5s^2 + 7s + 1}$$

$$= \begin{bmatrix} 2 & 1 & -1 \\ 0 & 1 & 0 \end{bmatrix} \begin{bmatrix} s^2 + 4s + 4 & 1 \\ s + 2 & s^2 + 3s + 1 \\ s^2 + 5s + 7 & s + 2 \end{bmatrix}$$

$$\times \frac{1}{s^3 + 5s^2 + 7s + 1}$$

$$= \frac{1}{s^3 + 5s^2 + 7s + 1} \begin{bmatrix} s^2 + 4s + 3 & s^2 + 2s + 1 \\ s + 2 & s^2 + 3s + 1 \end{bmatrix} \tag{3.31}$$

3.2.3 State Equations from Transfer Functions

Although the transfer function can be uniquely determined from the state equations through Eq. (3.3), the state equations for a given transfer function are not unique and can be obtained in several ways. This will be illustrated by means of several examples.

EXAMPLE 3.5

Consider the transfer function

$$\frac{Y(s)}{U(s)} = \frac{1}{s^3 + 7s^2 + 14s + 8} \tag{3.32}$$

which corresponds to the differential equation

$$\frac{d^3y}{dt^3} + 7\frac{d^2y}{dt^2} + 14\frac{dy}{dt} + 8y = u \tag{3.33}$$

Define

$$\left.\begin{aligned} x_1 &= y \\ x_2 &= \frac{dy}{dt} = \dot{x}_1 \\ \text{and} \quad & \\ x_3 &= \frac{d^2y}{dt^2} = \dot{x}_2 \end{aligned}\right\} \tag{3.34}$$

Then Eq. (3.33) can be written as

$$\dot{x}_3 = -7x_3 - 14x_2 - 8x_1 + u \tag{3.35}$$

and combining this with Eq. (3.34), we get

$$\begin{bmatrix} \dot{x}_1 \\ \dot{x}_2 \\ \dot{x}_3 \end{bmatrix} = \begin{bmatrix} 0 & 1 & 0 \\ 0 & 0 & 1 \\ -8 & -14 & -7 \end{bmatrix} \begin{bmatrix} x_1 \\ x_2 \\ x_3 \end{bmatrix} + \begin{bmatrix} 0 \\ 0 \\ 1 \end{bmatrix} u \tag{3.36}$$

$$y = [1 \quad 0 \quad 0] \begin{bmatrix} x_1 \\ x_2 \\ x_3 \end{bmatrix} \tag{3.37}$$

The state equations (3.36) and (3.37) lead to a simple analog computer simulation, as shown in Fig. 3.2.

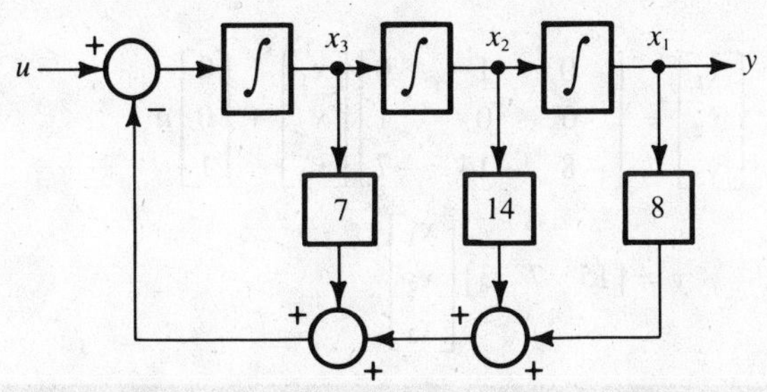

FIGURE 3.2. Block diagram for analog computer simulation of Eqs. (3.36) and (3.37).

EXAMPLE 3.6

Consider the transfer function

$$\frac{Y(s)}{U(s)} = \frac{3s^2 + 7s + 15}{s^3 + 7s^2 + 14s + 8} \tag{3.38}$$

It will be seen that this transfer function has the same denominator as Eq. (3.32). The numerator of Eq. (3.32), however, was one, whereas in Eq. (3.38) it is a polynomial of degree 2 (less than the degree of the denominator, to make it a strictly proper transfer function). Hence, if we define a variable $v(t)$ such that

$$\frac{V(s)}{U(s)} = \frac{1}{s^3 + 7s^2 + 14s + 8} \tag{3.39}$$

we get exactly the transfer function in Eq. (3.32). Now, we note that

$$Y(s) = (3s^2 + 7s + 15)V(s) \tag{3.40}$$

or

$$y(t) = 3\frac{d^2v}{dt^2} + 7\frac{dv}{dt} + 15v(t) \tag{3.41}$$

If we use Eqs. (3.34) and (3.35) to obtain the state equations for (3.39) with y replaced by v, it is easily seen that differentiation is avoided by writing Eq. (3.41) as

$$y(t) = 3x_3 + 7x_2 + 15x_1 \tag{3.42}$$

Hence the state equations for the transfer function of Eq. (3.38) are obtained as

$$\begin{bmatrix} \dot{x}_1 \\ \dot{x}_2 \\ \dot{x}_3 \end{bmatrix} = \begin{bmatrix} 0 & 1 & 0 \\ 0 & 0 & 1 \\ -8 & -14 & -7 \end{bmatrix} \begin{bmatrix} x_1 \\ x_2 \\ x_3 \end{bmatrix} + \begin{bmatrix} 0 \\ 0 \\ 1 \end{bmatrix} u \tag{3.43}$$

$$y = [15 \quad 7 \quad 3] \begin{bmatrix} x_1 \\ x_2 \\ x_3 \end{bmatrix} \tag{3.44}$$

It should be noted that one may generalize this procedure to obtain the state equations for an nth-order transfer function by inspection. For example, consider

$$G(s) = \frac{b_{n-1}s^{n-1} + b_{n-2}s^{n-2} + \cdots + b_1 s + b_0}{s^n + a_{n-1}s^{n-1} + a_{n-2}s^{n-2} + \cdots + a_1 s + a_0} \tag{3.45}$$

The state equations for this transfer function may be written as

$$\dot{x} = \bar{A}x + \bar{B}u \tag{3.46}$$

$$y = \bar{C}x \tag{3.47}$$

where

$$\bar{A} = \begin{bmatrix} 0 & 1 & 0 & \cdots & 0 & 0 \\ 0 & 0 & 1 & \cdots & 0 & 0 \\ \vdots & \vdots & \vdots & \cdots & \vdots & \vdots \\ 0 & 0 & 0 & \cdots & 0 & 1 \\ -a_0 & -a_1 & -a_2 & \cdots & -a_{n-2} & -a_{n-1} \end{bmatrix} \tag{3.48}$$

$$\bar{B} = \begin{bmatrix} 0 \\ 0 \\ \vdots \\ 0 \\ 1 \end{bmatrix} \tag{3.49}$$

and

$$\bar{C} = [b_0 \quad b_1 \quad \cdots \quad b_{n-2} \quad b_{n-1}] \tag{3.50}$$

This form of the state equations has been called the controllable canonical form in the literature. The reason for this name will be discussed in Chapter 12.

The block diagram for analog computer simulation for this general case is shown in Fig. 3.3.

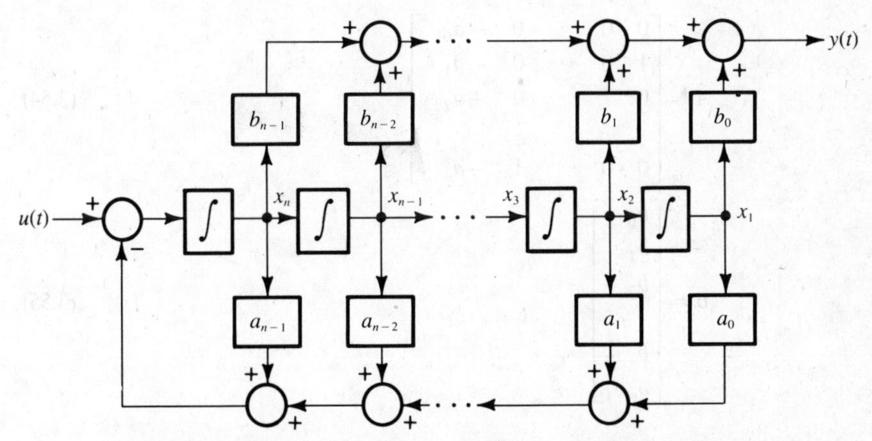

FIGURE 3.3. Block diagram for analog computer simulation of state equations in the controllable canonical form.

EXAMPLE 3.7

The transfer function in Eq. (3.38) can also be represented by the following state equations.

$$\begin{bmatrix} \dot{x}_1 \\ \dot{x}_2 \\ \dot{x}_3 \end{bmatrix} = \begin{bmatrix} 0 & 0 & -8 \\ 1 & 0 & -14 \\ 0 & 1 & -7 \end{bmatrix} \begin{bmatrix} x_1 \\ x_2 \\ x_3 \end{bmatrix} + \begin{bmatrix} 15 \\ 7 \\ 3 \end{bmatrix} u \tag{3.51}$$

$$y = [0 \quad 0 \quad 1] \begin{bmatrix} x_1 \\ x_2 \\ x_3 \end{bmatrix} \tag{3.52}$$

The validity of this representation is evident if we note that the A matrix for this case is the transpose of the A matrix in Eq. (3.43). Similarly, the B matrix and the C matrix for this representation are the transpose of the C matrix and the B matrix, respectively, for the representation in Eqs. (3.43) and (3.44). It follows that for the scalar case

$$G(s) = C(sI - A)^{-1}B = [C(sI - A)^{-1}B]^T$$
$$= B^T(sI - A^T)^{-1}C^T \tag{3.53}$$

These results are readily extended to the general case of the nth-order transfer function given in Eq. (3.45) to obtain the alternative formulation of state equations by inspection.

$$\bar{A} = \begin{bmatrix} 0 & 0 & \cdots & 0 & -a_0 \\ 1 & 0 & \cdots & 0 & -a_1 \\ 0 & 1 & \cdots & 0 & -a_2 \\ \vdots & \vdots & & \vdots & \vdots \\ 0 & 0 & \cdots & 1 & -a_{n-1} \end{bmatrix} \tag{3.54}$$

$$\bar{B} = \begin{bmatrix} b_0 \\ b_1 \\ b_2 \\ \vdots \\ b_{n-2} \\ b_{n-1} \end{bmatrix} \tag{3.55}$$

and

$$\bar{C} = [0 \;\; 0 \;\; \cdots \;\; 0 \;\; 1] \tag{3.56}$$

This form of the state equations is called the observable canonical form. The block diagram for the analog computer simulation for this case is shown in Fig. 3.4.

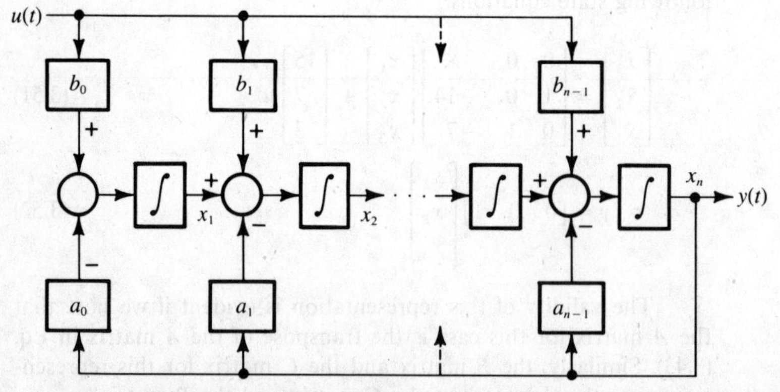

FIGURE 3.4. Block diagram for analog computer simulation of state equations in the observable canonical form.

EXAMPLE 3.8

The transfer function of Eq. (3.38) can be expanded into partial fractions to obtain

$$\frac{Y(s)}{U(s)} = \frac{11/3}{s+1} - \frac{13/2}{s+2} + \frac{35/6}{s+4} \tag{3.57}$$

Defining

$$X_1(s) = \frac{1}{s+1} U(s)$$

$$X_2(s) = \frac{1}{s+2} U(s) \tag{3.58}$$

and

$$X_3(s) = \frac{1}{s+4} U(s)$$

we get the following state equations

$$\begin{bmatrix} \dot{x}_1 \\ \dot{x}_2 \\ \dot{x}_3 \end{bmatrix} = \begin{bmatrix} -1 & 0 & 0 \\ 0 & -2 & 0 \\ 0 & 0 & -4 \end{bmatrix} \begin{bmatrix} x_1 \\ x_2 \\ x_3 \end{bmatrix} + \begin{bmatrix} 1 \\ 1 \\ 1 \end{bmatrix} u(t) \tag{3.59}$$

$$y(t) = \begin{bmatrix} \frac{11}{3} & \frac{-13}{2} & \frac{35}{6} \end{bmatrix} \begin{bmatrix} x_1 \\ x_2 \\ x_3 \end{bmatrix} \tag{3.60}$$

It should be noted that the A matrix is diagonal in this case. This is called the diagonal canonical form.*

DRILL PROBLEM 3.2

The transfer function of a linear system is given by

$$\frac{Y(s)}{U(s)} = \frac{10(s+5)}{(s+1)(s+2)(s+4)}$$

* The diagonal canonical form is possible only if the poles are distinct. For a transfer function with multiple poles (repeated roots of the denominator) we can get Jordan forms. These are described in Appendix B (Sec. B.4).

Write the state equations in (a) the controllable canonical form, (b) the observable canonical form, and (c) the diagonal canonical form.

Answers:

$$\text{(a)}\ A = \begin{bmatrix} 0 & 1 & 0 \\ 0 & 0 & 1 \\ -8 & -14 & -7 \end{bmatrix} \qquad B = \begin{bmatrix} 0 \\ 0 \\ 1 \end{bmatrix} \qquad C = [50 \quad 10 \quad 0]$$

$$\text{(b)}\ A = \begin{bmatrix} 0 & 0 & -8 \\ 1 & 0 & -14 \\ 0 & 1 & -7 \end{bmatrix} \qquad B = \begin{bmatrix} 50 \\ 10 \\ 0 \end{bmatrix} \qquad C = [0 \quad 0 \quad 1]$$

$$\text{(c)}\ A = \begin{bmatrix} -1 & 0 & 0 \\ 0 & -2 & 0 \\ 0 & 0 & -4 \end{bmatrix} \qquad B = \begin{bmatrix} 1 \\ 1 \\ 1 \end{bmatrix} \qquad C = \begin{bmatrix} \dfrac{40}{3} & -15 & \dfrac{5}{3} \end{bmatrix}$$

3.2.4 Linear Transformations and Canonical Forms

From the examples in the previous section, it is evident that the state-space description of a transfer function is not unique. The equations are all related, however, through linear transformations. Consider a linear transformation of the state vector, given by

$$\mathbf{x} = Px \tag{3.61}$$

where P is an $n \times n$ nonsingular constant matrix. Hence,

$$\begin{aligned} \dot{\mathbf{x}} &= \bar{A}\mathbf{x} + \bar{B}u \\ \mathbf{y} &= \bar{C}\mathbf{x} + Du \end{aligned} \tag{3.62}$$

where

$$\begin{aligned} \bar{A} &= PAP^{-1} \\ \bar{B} &= PB \\ \bar{C} &= CP^{-1} \end{aligned} \tag{3.63}$$

Thus, Eq. (3.62) represents another set of state equations for the same system. It is evident that with different choices of P, numerous sets of state equations may be obtained.

Linear transformations of state equations are often employed for obtaining some special forms of the matrices A, B, and C. These are called canonical forms and are very useful for analysis and design. Equations (3.48), (3.54), and (3.59) are three of these canonical forms, often called the "controllable," "observable," and "diagonal" forms, respectively. The matrices for transforming the state equations to these canonical forms can be obtained if certain conditions are satisfied. These will be discussed in Chapter 12 and Appendix B.

It may also be noted that a linear transformation does not alter the transfer function of a system, since

$$C(sI - A)^{-1}B = \bar{C}(sI - \bar{A})^{-1}\bar{B} \tag{3.64}$$

as can be proved by direct substitution from Eq. (3.63).

3.2.5 Solution of State Equations

The solution of the state equation (3.1) for a given initial condition $x(t_0)$ is obtained as

$$x(t) = e^{A(t-t_0)}x(t_0) + \int_{t_0}^{t} e^{A(t-\tau)}Bu(\tau)\, d\tau \tag{3.65}$$

where

$$e^{At} \triangleq I + At + \frac{1}{2!}(At)^2 + \frac{1}{3!}(At)^3 + \cdots \tag{3.66}$$

Alternatively, by taking Laplace transforms of both sides of Eq. (3.1), and solving for $X(s)$, we get (for $t_0 = 0$)

$$X(s) = [sI - A]^{-1}[x(0) + BU(s)] \tag{3.67}$$

Hence, we recognize that

$$e^{At} = \mathscr{L}^{-1}[(sI - A)^{-1}] \tag{3.68}$$

and is called the state transition matrix of the system.

A numerical solution of Eq. (3.65) is easily obtained provided that the state transition matrix is known. Some methods for calculating e^{At} are described in Appendix B.

3.2.6 State Transition Equation for a Discrete-Time System

Consider a linear system described by Eqs. (3.7) and (3.2), where the input $u(t)$ is allowed to vary only at the sampling instants $t = kT$, $k = 1, 2, \ldots$. In this case, making $t_0 = kT$ and $t = (k+1)T$, Eq. (3.65) takes the form

$$x[(k+1)T] = e^{AT}x(kT) + \int_{kT}^{(k+1)T} e^{A(kT+T-\tau)}Bu(\tau)\, d\tau \tag{3.69}$$

Since $u(\tau) = u(kT)$ for $kT < \tau \leq (k+1)T$, Eq. (3.69) may be rewritten as

$$x(kT + T) = F(T)x(kT) + G(T)u(kT) \tag{3.70}$$

where $F(T) = e^{AT}$ and

$$G(T) = \int_0^T e^{At}\, dt\, B \tag{3.71}$$

The last result is obtained as follows. From Eqs. (3.69) and (3.70)

$$\begin{aligned} G(T) &= \int_{kT}^{(k+1)T} e^{A(kT+T-\tau)} B\, d\tau \\ &= -\int_T^0 e^{A\xi} B\, d\xi \qquad \text{where } \xi = kT + T - \tau \\ &= \int_0^T e^{A\xi}\, d\xi \cdot B \end{aligned}$$

The calculation of $F(T)$ and $G(T)$ can be carried out on a computer using the power series expansions

$$F(T) = I + AT + \frac{1}{2!}(AT)^2 + \frac{1}{3!}(AT)^3 + \cdots \tag{3.72}$$

and

$$G(T) = \left(IT + \frac{1}{2!} AT^2 + \frac{1}{3!} A^2 T^3 + \cdots \right) B \tag{3.73}$$

where the series are truncated after a certain number of terms to obtain a specified accuracy. A program for calculating F and G is given in Appendix D.

Equation (3.70) is called the "state transition equation" of the discrete-time system, and shows how conveniently the response of the system can be obtained if the matrices $G(T)$ and $F(T)$ are known. This equation is suitable for use on a digital computer, since it only requires matrix multiplication.

EXAMPLE 3.9

Consider the discrete-time system shown in Fig. 3.5. A state-space representation for the continuous-time portion of the system is obtained as

$$\begin{bmatrix} \dot{x}_1 \\ \dot{x}_2 \end{bmatrix} = \begin{bmatrix} 0 & 1 \\ 0 & -2 \end{bmatrix} \begin{bmatrix} x_1 \\ x_2 \end{bmatrix} + \begin{bmatrix} 0 \\ 1 \end{bmatrix} u(t) \tag{3.74}$$

$$y = [4 \quad 0]\begin{bmatrix} x_1 \\ x_2 \end{bmatrix} \tag{3.75}$$

$$e^{At} = \mathscr{L}^{-1}[(sI - A)^{-1}] = \begin{bmatrix} 1 & \frac{1}{2}(1 - e^{-2t}) \\ 0 & e^{-2t} \end{bmatrix} \tag{3.76}$$

Hence,

$$F(T) = e^{AT} = \begin{bmatrix} 1 & \frac{1}{2}(1 - e^{-0.4}) \\ 0 & e^{-0.4} \end{bmatrix} = \begin{bmatrix} 1 & 0.1648 \\ 0 & 0.6703 \end{bmatrix} \tag{3.77}$$

and

$$G(T) = \int_0^T e^{AT} B \, dt = \int_0^{0.2} \begin{bmatrix} \frac{1}{2}(1 - e^{-2t}) \\ e^{-2t} \end{bmatrix} dt$$

$$= \begin{bmatrix} 0.01758 \\ 0.16484 \end{bmatrix} \tag{3.78}$$

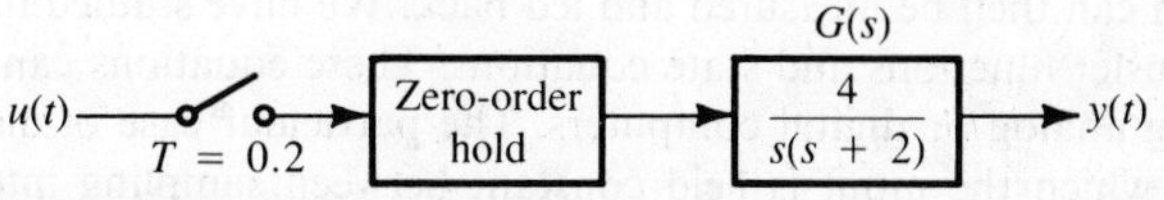

$u(t)$ — $T = 0.2$ — Zero-order hold — $G(s)$: $\dfrac{4}{s(s+2)}$ — $y(t)$

FIGURE 3.5. A discrete-time system.

From Eqs. (3.77) and (3.78), the state transition equation is obtained as

$$x[(k+1)T] = \begin{bmatrix} 1 & 0.16484 \\ 0 & 0.67032 \end{bmatrix} x(kT) + \begin{bmatrix} 0.01758 \\ 0.16484 \end{bmatrix} u(kT) \tag{3.79}$$

DRILL PROBLEM 3.3

The state equations for a helicopter near hover are as follows:

$$\begin{bmatrix} \dot{x}_1 \\ \dot{x}_2 \\ \dot{x}_3 \end{bmatrix} = \begin{bmatrix} -0.02 & -1.4 & 9.8 \\ -0.01 & -0.4 & 0 \\ 0 & 1.0 & 0 \end{bmatrix} \begin{bmatrix} x_1 \\ x_2 \\ x_3 \end{bmatrix} + \begin{bmatrix} 9.8 \\ 6.3 \\ 0 \end{bmatrix} u$$

where x_1 = horizontal velocity, x_2 = pitch rate, x_3 = pitch angle, and u = rotor tilt angle.

Determine the state transition equation if the sampling period is 0.5 second and the input is held constant between sampling instants.

Answer:

$$F(T) = \begin{bmatrix} 0.989744 & 0.512262 & 4.87589 \\ -0.004509 & 0.818414 & -0.011435 \\ -0.001167 & 0.453201 & 0.998061 \end{bmatrix}$$

$$G(t) = \begin{bmatrix} 5.06819 \\ 2.84373 \\ 0.73565 \end{bmatrix}$$

3.3 SUMMARY

In this chapter we have discussed the state-space representation of linear systems as an alternative to the transfer function representation. The differential equation for the system is rewritten into a set of first-order differential equations, called state equations. For any given system the state-variable description is not unique. In practice, it is desirable to use physical variables as state variables, which can then be measured and fed back. We have studied methods for relating transfer functions and state equations. These equations can be readily solved using analog or digital computers. The particular case of discrete-time systems, in which the input is held constant between sampling intervals, can be solved very conveniently on a digital computer using the state transition equation.

3.4 PROBLEMS

1. Write a set of state equations for the dynamic vibration absorber shown in Fig. 2.33 (see Chapter 2, Problem 9).

2. Write a set of state equations for the analog computer simulation diagram shown in Fig. 2.34 (see Chapter 2, Problem 10), using x_1, x_2, and x_3 as state variables. Hence, determine the transfer function relating $C(s)$ to $R(s)$.

3. The state equations of a linear system are as follows:

$$\dot{x} = \begin{bmatrix} -2 & 0 & 1 \\ 1 & -3 & 0 \\ 1 & 1 & -1 \end{bmatrix} x + \begin{bmatrix} 1 \\ 0 \\ 1 \end{bmatrix} u$$

$$y = [2 \quad 1 \quad -1] x$$

a. Determine the transfer function $Y(s)/U(s)$.
b. Determine the state transition equation if the input is held constant between the sampling instants 0.5 second apart.

4. Draw an analog computer simulation diagram for the diagonal canonical form of state equations obtained in Example 3.8.

5. The diagonal canonical form discussed in Example 3.8 would have complex elements in the matrices A and C if the transfer function has complex poles. A simple transformation leads to a block diagonal structure for A, with each pair of complex conjugate eigenvalues replaced by 2×2 blocks of real numbers. For example, consider the transfer function

$$G(s) = \frac{Y(s)}{U(s)} = \frac{s+2}{s^2+2s+5} = \frac{0.5 - j0.25}{s+1-j2} + \frac{0.5 + j0.25}{s+1+j2}$$

Let

$$X_1(s) = \frac{U(s)}{s+1-j2} \quad \text{and} \quad X_2(s) = \frac{U(s)}{s+1+j2}$$

Then we get the following state equations

$$\begin{bmatrix} \dot{x}_1 \\ \dot{x}_2 \end{bmatrix} = \begin{bmatrix} -1+j2 & 0 \\ 0 & -1-j2 \end{bmatrix} \begin{bmatrix} x_1 \\ x_2 \end{bmatrix} + \begin{bmatrix} 1 \\ 1 \end{bmatrix} u(t)$$

$$y = [0.5 - j0.25 \quad 0.5 + j0.25] \begin{bmatrix} x_1 \\ x_2 \end{bmatrix}$$

If we now apply the transformation $\mathbf{x} = Px$, where

$$P = \begin{bmatrix} 1 & j \\ 1 & -j \end{bmatrix} \quad \text{and} \quad P^{-1} = \begin{bmatrix} 0.5 & 0.5 \\ -j0.5 & j0.5 \end{bmatrix}$$

then we get

$$\bar{A} = P^{-1}AP = \begin{bmatrix} -1 & -2 \\ 2 & -1 \end{bmatrix}$$

$$\bar{B} = P^{-1}B = \begin{bmatrix} 1 \\ 0 \end{bmatrix} \quad \text{and} \quad \bar{C} = CP = [1 \quad 0.5]$$

Use the above procedure to show that for the transfer function

$$G(s) = \frac{Y(s)}{U(s)} = \frac{s+2}{s^2+2s+5} + \frac{1}{s} + \frac{2}{s+2}$$

the following state equations are obtained

$$\dot{x} = \begin{bmatrix} -1 & -2 & 0 & 0 \\ 2 & -1 & 0 & 0 \\ 0 & 0 & 0 & 0 \\ 0 & 0 & 0 & -2 \end{bmatrix} x \begin{bmatrix} 1 \\ 0 \\ 1 \\ 1 \end{bmatrix} u$$

$$y = [1 \quad 0.5 \quad 1 \quad 2]\, x$$

and the transformation matrix is given by

$$P = \begin{bmatrix} 1 & j & 0 & 0 \\ 1 & -j & 0 & 0 \\ 0 & 0 & 1 & 0 \\ 0 & 0 & 0 & 1 \end{bmatrix}$$

6. The diagonal form can be obtained only if the transfer function does not have multiple poles. For the more general case, one can obtain the Jordan form. For example, consider

$$G(s) = \frac{Y(s)}{U(s)} = \frac{2s + 8}{(s + 2)^2} = \frac{2}{(s + 2)^2} + \frac{4}{s + 2}$$

Let

$$X_1(s) = \frac{1}{s + 2} X_2(2) \quad \text{or} \quad \dot{x}_1 = -2x_1 + x_2$$

and

$$X_2(s) = \frac{1}{s + 2} U(s) \quad \text{or} \quad \dot{x}_2 = -2x_2 + u(t)$$

Since

$$Y(s) = 2X_1(s) + 4X_2(s)$$

we get the following state equations.

$$\begin{bmatrix} \dot{x}_1 \\ \dot{x}_2 \end{bmatrix} = \begin{bmatrix} -2 & 1 \\ 0 & -2 \end{bmatrix} \begin{bmatrix} x_1 \\ x_2 \end{bmatrix} + \begin{bmatrix} 0 \\ 1 \end{bmatrix} u(t)$$

$$y = [2 \quad 4] \begin{bmatrix} x_1 \\ x_2 \end{bmatrix}$$

Use the above procedure to show that the Jordan form for the state equations for the transfer function

$$G(s) = \frac{Y(s)}{U(s)}$$

$$= \frac{2}{(s+2)^3} + \frac{4}{(s+2)^2} + \frac{5}{s+2} + \frac{3}{s+1}$$

is

$$\dot{x} = \begin{bmatrix} -2 & 1 & 0 & 0 \\ 0 & -2 & 1 & 0 \\ 0 & 0 & -2 & 0 \\ 0 & 0 & 0 & -1 \end{bmatrix} x + \begin{bmatrix} 0 \\ 0 \\ 1 \\ 1 \end{bmatrix} u$$

$$y = [2 \quad 4 \quad 5 \quad 3]\, x$$

7. **a.** For the fourth-order transfer function in the previous problem, determine the state equations in the controllable canonical form and the observable canonical form.

   b. Draw the analog computer simulation diagram for each of the three canonical state-space representations (i.e., Jordan, controllable, and observable).

8. Write the state equations for the electrical networks shown in Fig. 3.6.

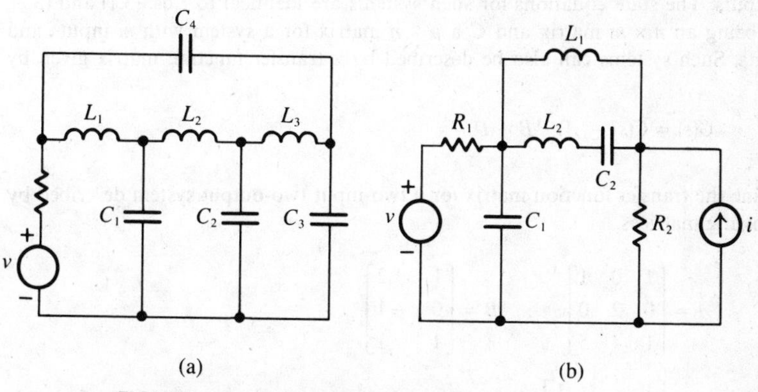

FIGURE 3.6. Electrical networks.

9. Obtain a set of state equations for the Ward-Leonard speed control system shown in Fig. 2.31 with the parameters given in Chapter 2, Problem 7. Physical variables should be selected as state variables.

10. Draw an analog computer simulation diagram for the state equations of the fourth-order system given in Problem 5. Use Mason's rule to determine the transfer function.

11. Obtain a set of state equations for the ac servomotor shown in Fig. 2.32 and described in Problem 8 in Chapter 2.

12. The linearized equations for a satellite in a circular equatorial orbit are given by $\dot{x} = Ax + Bu$, where

$$A = \begin{bmatrix} 0 & 1 & 0 & 0 \\ 3 & 0 & 0 & 2 \\ 0 & 0 & 0 & 1 \\ 0 & -2 & -3 & 0 \end{bmatrix} \quad \text{and} \quad B = \begin{bmatrix} 0 \\ 1 \\ 0 \\ 0 \end{bmatrix}$$

with the following state variables.

x_1 = the distance from the center of the earth
x_2 = the rate of change of x_1
x_3 = angular displacement in the equatorial plane
x_4 = the rate of change of x_3

The input $u(t)$ is the thrust produced by a rocket engine. Given this information, determine the transfer function relating $X_1(s)$ to $U(s)$.

13. Determine the state transition equation for the system described in Problem 12 assuming a sampling interval of 0.2 second and that the input is held constant between sampling instants.

14. Many control systems are of the multivariable type; i.e., they have several inputs and outputs. The state equations for such systems are identical to Eqs. (3.1) and (3.2), with B being an $n \times m$ matrix and C a $p \times n$ matrix for a system with m inputs and p outputs. Such systems can also be described by a transfer function matrix given by

$$G(s) = C(sI - A)^{-1}B + D$$

Determine the transfer function matrix for a two-input two-output system described by the following matrices.

$$A = \begin{bmatrix} 1 & 0 & 1 \\ 0 & 2 & 0 \\ 1 & 4 & 5 \end{bmatrix} \qquad B = \begin{bmatrix} 1 & 2 \\ 0 & -1 \\ 1 & 1 \end{bmatrix}$$

$$C = \begin{bmatrix} 1 & -2 & 1 \\ 0 & 1 & 0 \end{bmatrix} \qquad D = 0$$

(*Hint*: Use Leverrier's algorithm to evaluate the inverse of the matrix $sI - A$.)

15. An important consideration in the design of the suspension system for an automobile is to increase the comfort of the passengers by absorbing the vibrations due to the terrain of the road. A model for the vertical suspension is shown in Fig. 3.7.

a. Obtain an equivalent electrical network.
b. Write a set of state equations for the system.

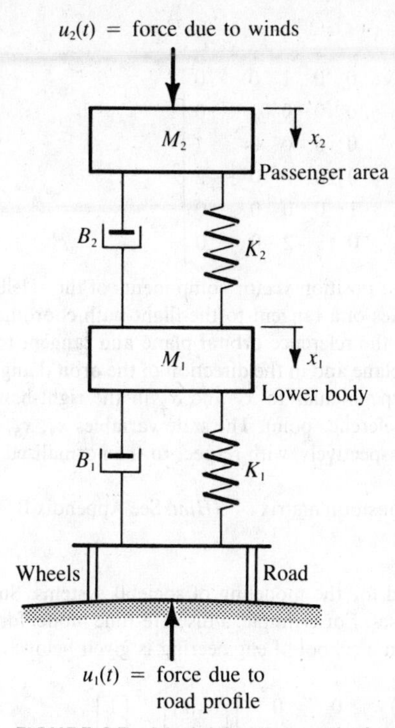

FIGURE 3.7. Model for the vertical suspension of an automobile.

c. Determine the transfer function matrix relating x_1 and x_2 to the inputs u_1 and u_2.

16. The approximate state equations of a spherical satellite are given below (after linearization)

$$I\ddot{\theta}_1 + \omega_0 I\dot{\theta}_3 = L_1$$

$$I\ddot{\theta}_2 = L_2$$

$$I\ddot{\theta}_3 - \omega_0 I\dot{\theta}_1 = L_2$$

where θ_1, θ_2, and θ_3 represent the angular deviations from a set of axes with fixed orientation; ω_0 is the angular frequency of the oriented axis; I represents the moment of inertia; and L_1, L_2, and L_3 represent applied torques.

a. Determine a set of state equations for the satellite.

b. Determine the transfer function matrix relating the angular deviations to the applied torques.

17. The linearized state equations for a satellite in orbit, with no external force acting, on it, are given by

$$\dot{x} = Ax$$

where

$$x = \begin{bmatrix} x_1 \\ x_2 \\ x_3 \\ x_4 \\ x_5 \\ x_6 \end{bmatrix} \qquad A = \begin{bmatrix} 0 & 0 & 0 & 1 & 0 & 0 \\ 0 & 0 & 0 & 0 & 1 & 0 \\ 0 & 0 & 0 & 0 & 0 & 1 \\ 0 & 0 & 0 & 0 & 0 & -2 \\ 0 & -1 & 0 & 0 & 0 & 0 \\ 0 & 0 & 3 & 2 & 0 & 0 \end{bmatrix}$$

In the above, x_1, x_2, and x_3 refer to position vector components of the satellite relative to the reference point along the axes of a tangent-to-the-flight-path coordinate system, in which the x_1 axis is forward in the reference orbital plane and tangent to the flight path, the x_2 axis is normal to this plane and in the direction of the orbital angular velocity ω, the x_3 axis is mutually perpendicular to x_1 and x_2 in the right-handed sense, and the origin coincides with the reference point. The state variables x_4, x_5, and x_6 are the derivatives of x_1, x_2, and x_3, respectively, with respect to the normalized time given by $\tau = \omega t$.

Determine the state transition matrix $e^{A\tau}$. (*Hint*: See Appendix B for methods for computing e^{At}.)

18. State equations are often used for the modeling of societal systems. Such models are very useful for planning purposes. For example, a discrete-time model for the population of undergraduate students in a school of engineering is given below:

$$\begin{bmatrix} x_1(k+1) \\ x_2(k+1) \\ x_3(k+1) \\ x_4(k+1) \end{bmatrix} = \begin{bmatrix} 0.1 & 0 & 0 & 0 \\ 0.7 & 0.1 & 0 & 0 \\ 0 & 0.82 & 0.08 & 0 \\ 0 & 0 & 0.85 & 0.02 \end{bmatrix} \begin{bmatrix} x_1(k) \\ x_2(k) \\ x_3(k) \\ x_4(k) \end{bmatrix} + \begin{bmatrix} 1 \\ 0 \\ 0 \\ 0 \end{bmatrix} u(k+1)$$

$$y(k) = [0 \;\; 0 \;\; 0 \;\; 0.97]\, x[k]$$

In the above, $x_i(k)$ is the number of students in the ith year of the four-year program in academic year k, $u(k)$ represents the input (that is, the enrollment into the freshman class in year k) and $y[k]$ is the output (that is, the number of students graduating in year k).

In September 1978 the school had 188 students in the freshman class, 126 in the sophomore class, 103 in the junior class, and 72 in the senior class. Assuming the following history of admissions to the freshman class in the subsequent years, use this model to determine the number of students graduating in May of each year from 1980 to 1988.

Year	1979	1980	1981	1982	1983	1984	1985
Admissions	275	320	350	400	450	450	430

19. The transfer function of a discrete-time system, described by the state equations

$$x(k+1) = Ax(k) + Bu(k)$$

$$y(k) = Cx(k)$$

where u and y are scalars, can be defined using z transforms as below (see Appendix C for a review of z transforms).

$$\frac{Y(z)}{U(z)} = C(zI - A)^{-1}B$$

Determine the transfer function for Problem 18.

20. The schematic diagram for a beam-block transducer system, which involves both electrical and mechanical components, is shown in Fig. 3.8. The state equations are of the form

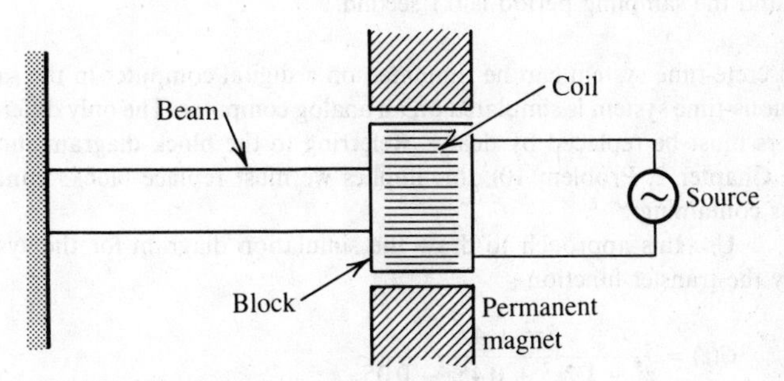

FIGURE 3.8. Beam-block transducer system.

$$\dot{x} = Ax + Bu$$

$$y = Cx$$

where

$$A = \begin{bmatrix} 0 & 0 & -1.5 & 2 & 0 \\ 0 & 0 & 1 & 0 & 0 \\ 6.5 & -3 & 0 & 0 & 0 \\ -7 & -4 & 0 & 0 & 15 \\ 0 & 0 & 0 & -10 & -18 \end{bmatrix} \qquad B = \begin{bmatrix} 0 \\ 0 \\ 0 \\ 0 \\ 1 \end{bmatrix}$$

$$C = [1 \quad 1.5 \quad 0 \quad 0 \quad 0]$$

The state variables are as follows:

x_1 = vertical displacement of the beam
x_2 = rotation of the beam
x_3 = angular momentum of the block
x_4 = vertical momentum of the block
x_5 = flux linkage in the coil

The input $u(t)$ is the voltage applied to the coil and the output $y(t)$ is the displacement of the block.

Determine the transfer function relating $Y(s)$ to $U(s)$.

21. The linearized state equations for the longitudinal flight control system of the F-4 aircraft are of the form

$$\dot{x} = Ax + Bu$$

where

$$A = \begin{bmatrix} 0 & 0 & 0 & 0 & 1 & 0 \\ 1 & 0 & 0 & -1 & 0 & 0 \\ -0.0475 & 0 & -0.0096 & -0.0399 & 0 & 0 \\ 0 & 0.0014 & -0.1024 & -0.3143 & 0.9974 & -0.0338 \\ 0 & 0.000186 & -0.02668 & -2.005 & -0.4125 & -5.39 \\ 0 & 0 & 0 & 0 & 0 & 20 \end{bmatrix}, \quad B = \begin{bmatrix} 0 \\ 0 \\ 0 \\ 0 \\ 0 \\ 20 \end{bmatrix}$$

Determine the state transition equation if the input is held constant between the sampling instants and the sampling period is 0.1 second.

22. A discrete-time system can be simulated on a digital computer in the same way as a continuous-time system is simulated on an analog computer. The only difference is that integrators must be replaced by delays. Referring to the block diagram shown in Fig. 2.35 (see Chapter 2, Problem 10), this implies we must replace blocks containing s^{-1} by blocks containing z^{-1}.

Use this approach to draw the simulation diagram for the system represented by the transfer function

$$G(z) = \frac{3z^2 + 4z}{z^3 - 1.2z^2 + 0.45z - 0.05}$$

23. From the transfer function, $G(z)$, of a discrete-time system, one can obtain state equations of the form

$$x(k+1) = Ax(k) + Bu(k)$$

$$y(k) = Cx(k)$$

where $x(k)$ represents the state, $u(k)$ represents the input, and $y(k)$ represents the output at the kth sampling instant. This is done following a procedure similar to that discussed in Sec. 3.2.3 for the continuous-time case.

Derive a set of state equations for the system described by the transfer function

$$G(z) = \frac{0.2z^2 + 0.65z}{z^3 - 1.2z^2 + 0.5z - 0.06}$$

24. State-space formulation of compartmental analysis has been very useful for the study of physiology with the primary objective of optimization of drug administration. The matrices A, B, and C for such a model for the hepatobiliary system are as follows:

$$A = \begin{bmatrix} -0.4 & 0.5 & 0 \\ 0 & -0.8 & 0 \\ 0 & 0.3 & 0.5 \end{bmatrix} \qquad B = \begin{bmatrix} 1 \\ 0 \\ 0 \end{bmatrix}$$

$$C = [1 \quad 0 \quad 1]$$

a. Determine the transfer function of the system.
b. Determine the state transition matrix for the system.

4 Characteristics of Closed-Loop Systems

4.1 INTRODUCTION

As mentioned in Chapter 1, all automatic control systems are of the closed-loop type. This is necessitated by the introduction of feedback for comparing the reference input, $r(t)$, with the controlled output, $c(t)$. A system without feedback is called an open-loop system. Both of these are shown in Fig. 4.1. In this chapter we shall study some characteristics of closed-loop systems.

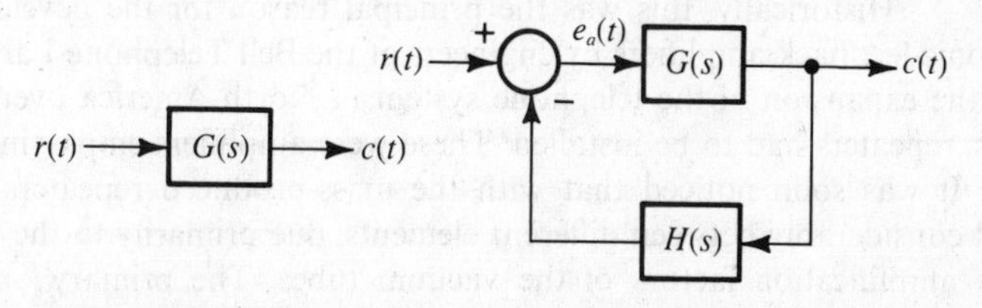

(a) Open-loop system (b) Closed-loop system

FIGURE 4.1. Block diagrams of open- and closed-loop systems.

For the open-loop system,

$$C(s) = G(s)R(s) \tag{4.1}$$

For the closed-loop system,

$$\begin{aligned} C(s) &= G(s)E_a(s) \\ &= G(s)[R(s) - H(s)C(s)] \end{aligned}$$

Hence

$$C(s) = \frac{G(s)}{1 + G(s)H(s)} \cdot R(s) \tag{4.2}$$

Also, the actuating error is given by

$$E_a(s) = \frac{1}{1 + G(s)H(s)} R(s) \tag{4.3}$$

To reduce the error, we must make the loop gain $G(s)H(s)$ large over the range of frequencies of interest; that is,

$$|G(s)H(s)| \gg 1 \tag{4.4}$$

Note that due to the low-pass nature of physical systems, this inequality will not be satisfied for high frequencies. Also, if $G(s)H(s)$ turns out to be a negative number close to one at some value of the frequency, then we may have a large error.

4.2 SENSITIVITY TO PARAMETER VARIATIONS

One important property of negative feedback is the reduction in the sensitivity to variations in the parameters of the forward path.

Historically, this was the principal reason for the development of electronic feedback amplifiers by engineers at the Bell Telephone Laboratories. With the expansion of the telephone system in North America over long distances, repeaters had to be installed. These were amplifiers employing vacuum tubes. It was soon noticed that with the mass-produced repeaters, the gain varied considerably between different elements, due primarily to the variations in the amplification factors of the vacuum tubes. The primary objective of using negative feedback amplifiers was to make their gain nearly insensitive to the variation in the parameters of the vacuum tube. For reasons of economy, one must always allow some tolerance in the manufacture of components;

otherwise the products will be too expensive and unable to compete in the market. Hence, in the design of control systems, it is important that the closed-loop system transfer function be relatively insensitive to small changes in the values of the parameters.

Let α be a parameter of $G(s)$. Then the sensitivity of $G(s)$ with respect to the parameter α is defined by Bode [1] as

$$S_\alpha^G \triangleq \frac{d \ln G}{d \ln \alpha} = \frac{dG/G}{d\alpha/\alpha} = \frac{\alpha}{G} \cdot \frac{dG}{d\alpha} \tag{4.5}$$

Note that Eq. (4.5) can also be written as

$$S_\alpha^G = \lim_{\alpha \to 0} \frac{\dfrac{\Delta G}{G}}{\dfrac{\Delta \alpha}{\alpha}} \tag{4.6}$$

which can be considered as the percentage change in G due to a very small percentage change in α. This will be the open-loop sensitivity to variations in α. Since, the closed-loop transfer function is given by

$$T(s) \triangleq \frac{C(s)}{R(s)} = \frac{G(s)}{1 + G(s)H(s)} \tag{4.7}$$

we have

$$S_\alpha^T = \frac{d \ln T}{d \ln \alpha} = \frac{\alpha}{T} \cdot \frac{dT}{d\alpha} = \frac{\alpha(1 + GH)}{G} \cdot \frac{dT}{dG} \cdot \frac{dG}{d\alpha}$$

$$= \frac{\alpha(1 + GH)}{G} \cdot \frac{dG}{d\alpha} \cdot \frac{1}{(1 + GH)^2} \overset{*}{=} \frac{\alpha}{G} \cdot \frac{dG}{d\alpha} \cdot \frac{1}{(1 + GH)}$$

Hence

$$S_\alpha^T = \frac{1}{1 + G(s)H(s)} \cdot S_\alpha^G \tag{4.8}$$

Thus, feedback has reduced the sensitivity to variations in α by the factor $1/[1 + G(s)H(s)]$, which is small over the range of frequencies of interest. It may be emphasized that this important advantage of closed-loop systems was the main reason for the development of feedback amplifiers by the telephone industry.

---

* Since

$$\frac{dT}{dG} = \frac{1}{(1 + GH)^2}$$

It may be verified that there is no reduction in the sensitivity to variations in the parameters of the feedback path.

Let α be a parameter of the feedback transfer function $H(s)$. Then

$$S_\alpha^T = \frac{\alpha}{T}\cdot\frac{dT}{d\alpha} = \frac{\alpha}{T}\cdot\frac{dT}{dH}\cdot\frac{dH}{d\alpha} = \frac{\alpha}{H}\cdot\frac{dH}{d\alpha}\cdot\frac{H}{T}\frac{dT}{dH}$$

$$= S_\alpha^H\cdot\frac{H}{T}\cdot\frac{(-G)\cdot G}{(1+GH)^2} = -S_\alpha^H\cdot\frac{H(1+GH)}{G}\cdot\frac{G^2}{(1+GH)^2}$$

$$= -S_\alpha^H\cdot\frac{GH}{1+GH} \tag{4.9}$$

It follows that the magnitudes of S_α^T and S_α^H are nearly equal when the loop gain $G(s)H(s)$ is much greater that one.

EXAMPLE 4.1

Consider the simple closed-loop system shown in Fig. 4.2, which represents a position-control system using a dc servomotor in the armature-control mode. We shall assume that the nominal value of the gain constant K is 10, that of α is 2, and the feedback parameter β is equal to 1. Hence we have

$$G(s) = \frac{K}{s(s+\alpha)} \tag{4.10}$$

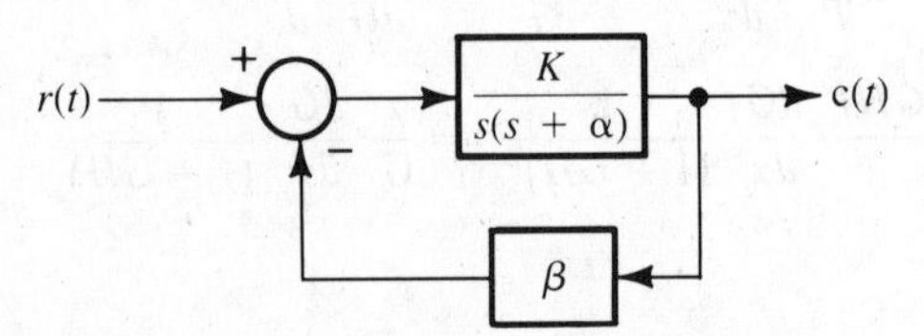

FIGURE 4.2. Block diagram of a position-control system.

$$H(s) = \beta \tag{4.11}$$

The sensitivity of G with respect to K is given by

$$S_K^G = \frac{K}{G}\cdot\frac{dG}{dK} = s(s+\alpha)\cdot\frac{1}{s(s+\alpha)} = 1 \tag{4.12}$$

Also, the sensitivity of G with respect to α is given by

$$S_\alpha^G = \frac{\alpha}{G}\cdot\frac{dG}{d\alpha} = \frac{\alpha s(s+\alpha)}{K}\cdot\frac{-K}{s(s+\alpha)^2} = -\frac{\alpha}{s+\alpha} \tag{4.13}$$

The sensitivity of the feedback transfer function $H(s)$ with respect to β is given by

$$S_\beta^H = \frac{\beta}{H} \cdot \frac{dH}{d\beta} = 1 \tag{4.14}$$

We shall now evaluate the closed-loop sensitivities.

$$S_K^T = \frac{S_K^G}{1 + G(s)H(s)} = \frac{1}{1 + G(s)H(s)} = \frac{s(s+\alpha)}{s(s+\alpha)+K}$$

$$= \frac{s^2 + 2s}{s^2 + 2s + 10} \tag{4.15}$$

$$S_\alpha^T = \frac{S_\alpha^G}{1 + G(s)H(s)} = -\frac{\alpha}{(s+\alpha)} \cdot \frac{s(s+\alpha)}{s(s+\alpha)+K}$$

$$= -\frac{2s}{s^2 + 2s + 10} \tag{4.16}$$

$$S_\beta^T = -S_\beta^H \cdot \frac{GH}{1 + GH} = -\frac{K}{s^2 + \alpha s + K}$$

$$= -\frac{10}{s^2 + 2s + 10} \tag{4.17}$$

The sensitivities of $T(s)$ with respect to K and α are seen to be small at low frequencies and are zero at zero frequency (dc). On the other hand, the sensitivity of $T(s)$ with respect to β has the magnitude of 1 at zero frequency that is equal to the sensitivity of $H(s)$ with respect to β. Thus, feedback has not introduced any reduction in the sensitivity to variation of a parameter in the feedback path.

Now consider the effect of a 5 percent change in the parameter K. The resulting change in the closed-loop transfer function is obtained as

$$\frac{\Delta T(s)}{T(s)} \approx S_K^{T(s)} \cdot \frac{\Delta K}{K} \approx \frac{s^2 + 2s}{(s^2 + 2s + 10)^2} \cdot 0.05 \tag{4.18}$$

Hence

$$\Delta T(s) \approx \frac{0.5s(s+2)}{s^2 + 2s + 10} \tag{4.19}$$

If the reference input to the system is given by

$$r(t) = 2 \cos 0.5t \tag{4.20}$$

then, for the nominal case,

$$T(0.5j) = \left. \frac{10}{s^2 + 2s + 10} \right|_{s=0.5j} = 1.02\underline{/-0.102} \tag{4.21}$$

That is, the steady-state response is given by

$$c_{ss}(t) = 2.04 \cos (0.5t - 0.102) \tag{4.22}$$

Also, for a 5 percent change in K, the change in the closed-loop transfer function is given by

$$\Delta T(0.5j) = \left. \frac{0.5s(s + 2)}{(s^2 + 2s + 10)^2} \right|_{s=0.5j} = 0.005\underline{/-4.672} \tag{4.23}$$

and the resulting change in the steady-state response is

$$\Delta c_{ss}(t) = 0.01 \cos (0.5t - 4.672) \tag{4.24}$$

which is less than 0.5 percent of the steady-state response for the nominal case.

DRILL PROBLEM 4.1

For the system described in Example 4.1 determine the change in $T(s)$ for a 5 percent change in the parameter α. Calculate this value for $\omega = 0.5$, and hence determine the change in the steady-state response to the input $r(t) = 2 \cos 0.5t$.

Answer:

$$\Delta T(s) = \frac{-s}{(s^2 + 2s + 10)^2} \qquad \Delta T(0.5j) = 0.005\underline{/-1.775}$$

$$\Delta c_{ss}(t) = 0.01 \cos (0.5t - 1.775)$$

DRILL PROBLEM 4.2

The block diagram of the speed-control system of an internal combustion engine is shown in Fig. 4.3. The gain constant K of the system has a nominal value of 10, with a tolerance of ± 5 percent. Determine the tolerance in the speed of the engine for step changes in the reference speed setting.

Answer: 0.413 percent

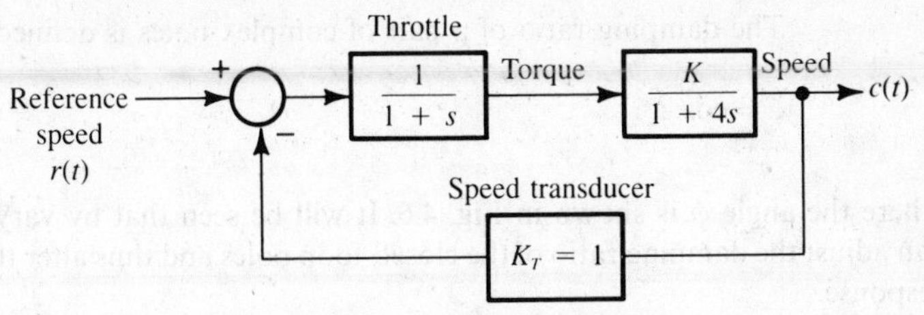

FIGURE 4.3. Speed-control system for a gasoline engine.

4.3 TRANSIENT RESPONSE

The transient response of a closed-loop system is easily altered by changing the loop gain. For example, consider the closed-loop system shown in Fig. 4.4, which represents a position-control system.

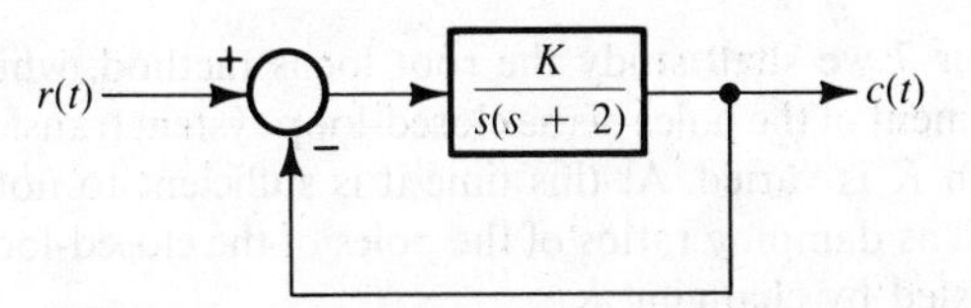

FIGURE 4.4. A simple closed-loop system.

For the system shown,

$$\frac{C(s)}{R(s)} = \frac{K}{s^2 + 2s + K} \tag{4.25}$$

where the value of K can be adjusted by changing the gain of an amplifier.

The response $c(t)$ to a unit step input, $r(t)$, is shown in Fig. 4.5 for $K = 1$, $K = 2$, and $K = 10$. For these three cases, the closed-loop poles are located at -1, -1, at $-1 \pm j1$, and at $-1 \pm j3$, respectively.

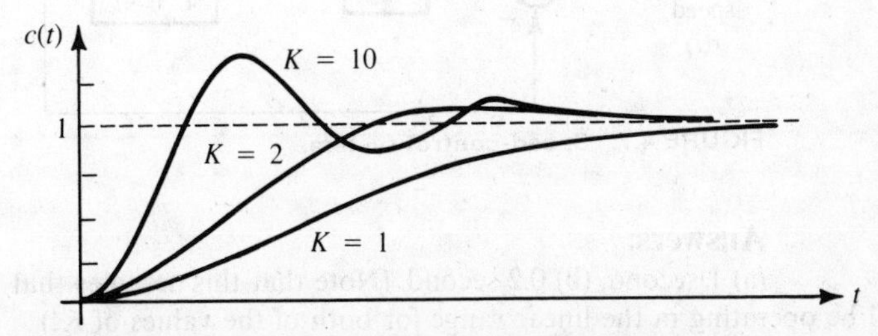

FIGURE 4.5. Step responses for different values of K.

The damping ratio of a pair of complex poles is defined as

$$\zeta = \cos \phi \tag{4.26}$$

where the angle ϕ is shown in Fig. 4.6. It will be seen that by varying K, one can adjust the damping ratio of the closed-loop poles and thus alter the transient response.

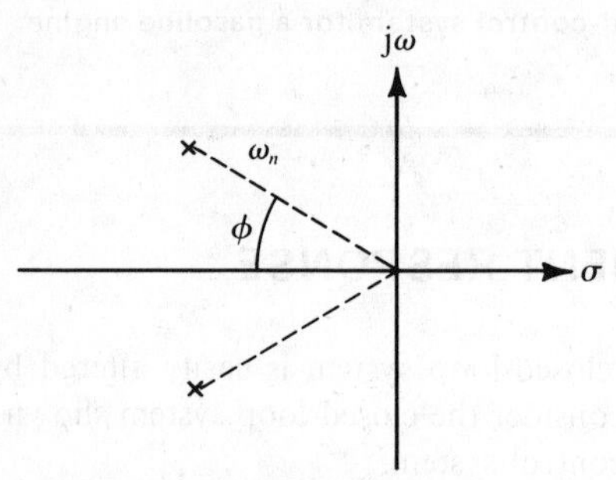

FIGURE 4.6. Complex poles.

In Chapter 7 we shall study the root locus method, which is concerned with the movement of the poles of the closed-loop system transfer function as the open-loop gain K is varied. At this time it is sufficient to note that the time constants as well as damping ratios of the poles of the closed-loop transfer function can be adjusted by changing K.

DRILL PROBLEM 4.3

The block diagram of a speed-control system using a dc motor with a time constant of 4 seconds is shown in Fig. 4.7. Determine the time constant of the closed-loop system for (a) $K = 3$ and (b) $K = 19$. (*Hint:* The time constant of a first-order system is the reciprocal of the magnitude of the location of the pole.)

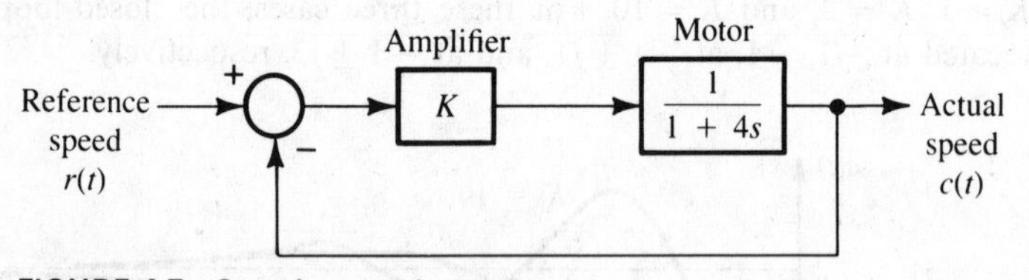

FIGURE 4.7. Speed-control system.

Answers:

(a) 1 second, (b) 0.2 second. (Note that this assumes that the motor will be operating in the linear range for both of the values of K!)

4.4 EFFECT OF DISTURBANCE SIGNALS

Most control systems are subject to unwanted disturbance signals. For example, we may have noise generated in electronic amplifiers, gusts of wind affecting radar antennas, load fluctuations, and so forth. The block diagram of such a system is shown in Fig. 4.8, where $N(s)$ is the disturbance (or noise) signal.

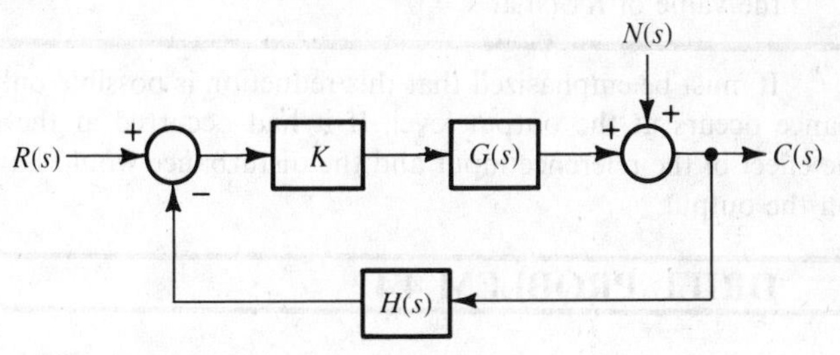

FIGURE 4.8. System with disturbance.

Applying superposition, we get

$$C(s) = \frac{KG(s)}{1 + KG(s)H(s)} R(s) + \frac{1}{1 + KG(s)H(s)} N(s) \tag{4.27}$$

Hence, the effect of the noise is reduced by the ratio $1/[1 + KG(s)H(s)]$ over the frequency range of interest. Thus by increasing the loop gain, we can alleviate the effect of disturbances occurring at the output level. Load disturbances fall under this category.

EXAMPLE 4.2

As an example let us reconsider the speed-control system for a gasoline engine with load disturbance, as shown in Fig. 4.9, where $K = 10$.

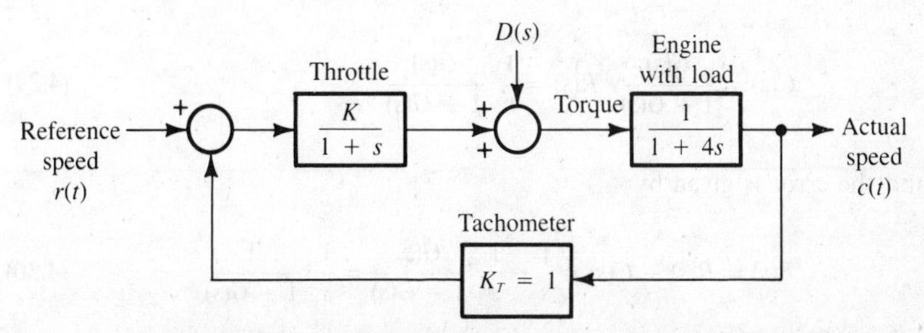

FIGURE 4.9. Speed-control system of a gasoline engine with load disturbance.

In this case, we have

$$C(s) = \frac{2.5}{s^2 + 1.25s + 2.75} R(s) + \frac{0.25(s+1)}{s^2 + 1.25s + 2.75} D(s) \quad (4.28)$$

The reduction in the disturbance is evident by noting that the dc gain of $C(s)/R(s)$ is 10 times that of $C(s)/D(s)$, which is precisely the value of $KG(s)$ at $s = 0$.

It must be emphasized that this reduction is possible only when the disturbance occurs at the output level. If it had occurred at the input level then the effect of the reference input and the disturbance would have been the same on the output.

DRILL PROBLEM 4.4

For the system described in Example 4.2, compare the frequency response of the output with respect to the reference input and the disturbance at 0.5 rad/s.

Answers:

$$\frac{C(0.5j)}{R(0.5j)} = 0.97\underline{/-0.245}, \qquad \frac{C(0.5j)}{D(0.5j)} = 0.1085\underline{/0.2187}$$

4.5 STEADY-STATE ERROR

Another effect of feedback is that it provides us with some control on the steady-state error to standard inputs by adjustment of the open-loop gain. As an example, we shall discuss the steady-state error to a unit step input.

Consider again the closed-loop system shown in Fig. 4.1(b). We shall now assume that $H(s) = 1$; that is, it is a unity-feedback system. For a unit step input, the output is given by

$$C(s) = \frac{G(s)}{1 + G(s)} \cdot R(s) = \frac{1}{s} \cdot \frac{G(s)}{1 + G(s)} \quad (4.29)$$

and the error is given by

$$E(s) = R(s) - C(s) = \frac{1}{s} - \frac{1}{s} \cdot \frac{G(s)}{1 + G(s)} = \frac{1}{s} \cdot \frac{1}{1 + G(s)} \quad (4.30)$$

Hence, applying the final-value theorem, the steady-state error is given by

$$e_{ss} = \lim_{s \to 0} [sE(s)] = \frac{1}{1 + G(0)} \tag{4.31}$$

It is evident that the steady-state error may be made as small as desired by increasing the dc loop gain, $G(0)$, of the system.

It is important to note that the above derivation assumes that the closed-loop system transfer function $T(s)$ has all its poles in the left half of the s-plane, with the exception of a simple pole at the origin; otherwise the final-value theorem cannot be applied. Furthermore, in many cases, a large increase in the loop gain may lead to instability, and violate the condition mentioned above. Hence, the possible decrease in the steady-state error is often limited.

DRILL PROBLEM 4.5

For the speed-control system described in Drill Problem 4.4 determine the steady-state error to a step input as well as the damping ratio of the poles of the closed-loop transfer function for (a) $K = 4$ and (b) $K = 20$.

Answers:

(a) $e_{ss} = 0.2$, damping ratio = 0.559.
(b) $e_{ss} = 0.048$, damping ratio = 0.273.

4.6 DISADVANTAGES OF FEEDBACK

Like most good things in life, the introduction of feedback in a system is not free of cost. The main disadvantages are given below.

1. Since feedback decreases the overall gain, this must be made up by an increase in the loop gain of the system, requiring additional hardware and increased complexity.
2. The components in the feedback path must be made more accurate because feedback does not reduce the sensitivity to the variations in these parameters. The result is a further increase in the overall cost.
3. The sensors required for the feedback path may introduce a small amount of noise in the system, thereby reducing the overall accuracy.
4. The introduction of feedback may lead to instability of the closed-loop system, even though the open-loop system may be stable. This is caused by inherent time lags within the system, with the result that what was intended as negative feedback may turn out to be positive feedback at some higher frequency. This was one of the first features of feedback observed when

automatic control was introduced. A thorough understanding of the Nyquist criterion of stability leads to the solution of this problem. This will be discussed in detail in Chapter 9.

4.7 SUMMARY

In this chapter we have studied the various advantages and disadvantages of the introduction of feedback necessary for automatic control systems.

The main advantages are (1) reduction in sensitivity to variations of parameters in the forward path, (2) control over the transient response as well as steady-state accuracy by adjusting the loop gain, and (3) reduction of noise or disturbance at the output level. The main disadvantages are the need for additional hardware with consequent increase in price as well as the possibility of instability due to phase lags in the feedback loop. The latter disadvantage is not serious and can be offset by proper design. The cost of additional hardware is often worthwhile due to the overall improvement in performance. Hence, it is fair to say that the advantages of feedback are much more significant than the disadvantages.

4.8 REFERENCES

1. H. D. Bode, *Network Analysis and Feedback Amplifier Design*, Van Nostrand, Princeton, N.J., 1945.
2. Paul M. Frank, *Introduction to System Sensitivity Theory*, Academic Press, New York, 1978.

4.9 PROBLEMS

1. The block diagram of a scheme for automatic control of the speed of an automobile is shown in Fig. 4.10. Determine (a) the sensitivity of the transfer function $C(s)/R(s)$ to changes in the engine gain K_e, (b) the effect of the load disturbance on the speed, and (c) the steady-state speed when the input setting $r(t)$ is 60 km/h if $K_e = 100$ and $d(t) = 0$.

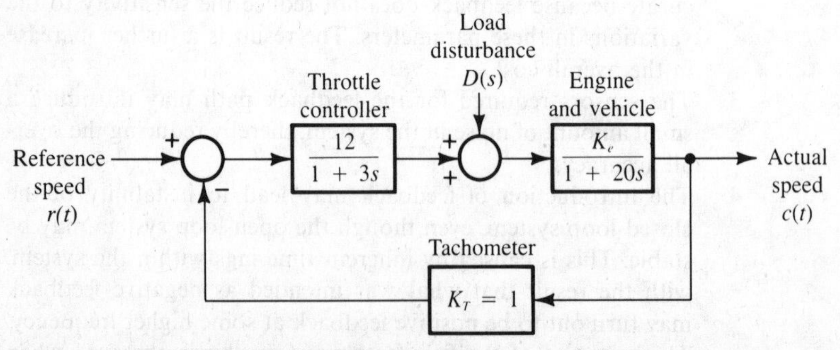

FIGURE 4.10. An automatic speed-control system.

2. Repeat part (c) of the previous problem if there is a 5 percent variation in K_e and if there is a 5 percent variation in K_T.

3. The block diagram of a speed-control system is shown in Fig. 4.11.

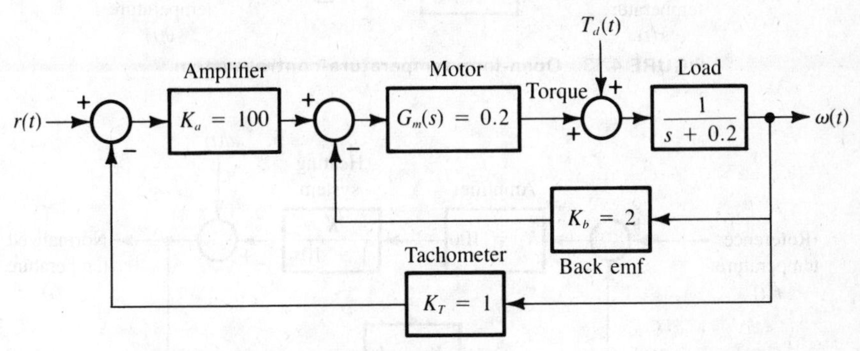

FIGURE 4.11. A speed-control system.

a. Compare the open- and closed-loop sensitivities of ω to small changes in the amplifier gain K_a.
b. Determine the steady-state error in speed control if $r(t)$ is a unit step and if $T_d(t) = 0$.
c. Determine the error in the system when it is subjected to a disturbance $T_d(s) = 1/s$ and $r(t) = 0$.

4. The block diagram of an electronic pacemaker system for controlling the rate of heartbeats is shown in Fig. 4.12.

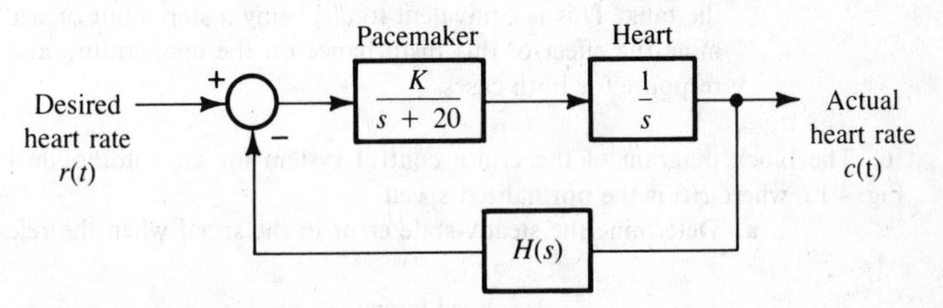

FIGURE 4.12. A heart rate control system.

a. Determine the sensitivity of the closed-loop transfer function to small changes in K at the normal heart rate of 72 beats per minute if $H(s) = 1$ and the nominal value of $K = 400$.
b. Repeat if $H(s) = 1/(1 + 0.1s)$.

5. In many chemical processes the temperature in a tank has to be maintained within a good degree of accuracy. The block diagram of an open-loop temperature-control system is shown in Fig. 4.13 and that of a closed-loop system in Fig. 4.14. Assume that the nominal value of K is 1 for both cases.

a. Determine and sketch the variation of θ with time if $r(t)$ is a unit step

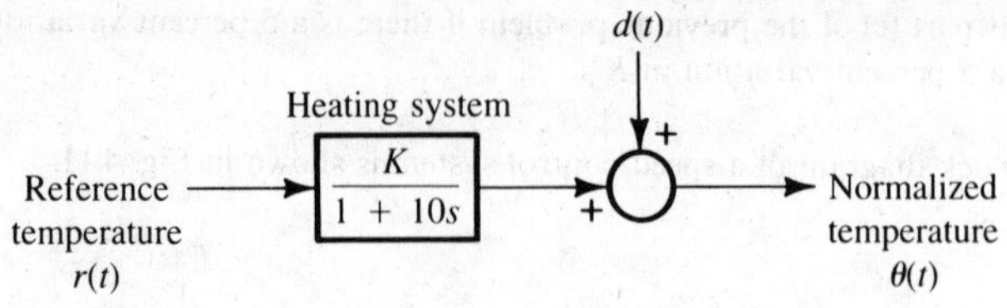

FIGURE 4.13. Open-loop temperature-control system.

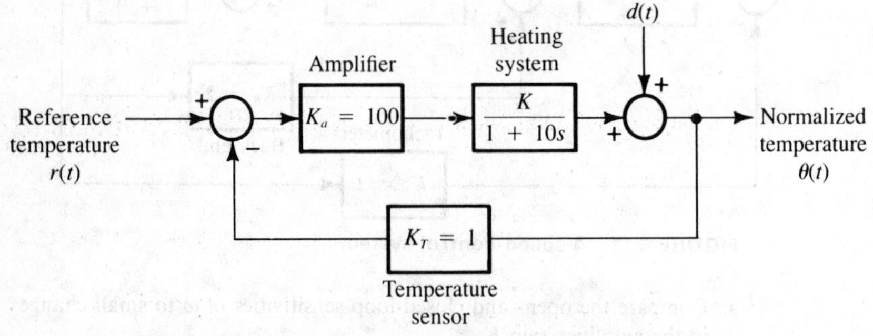

FIGURE 4.14. Closed-loop temperature-control system.

and the disturbance is zero. What time must elapse before the temperature reaches 63.2 percent of its final value in each case?

b. Determine the percentage change in the output temperature for both cases if there is a 5 percent change in K.

c. After the temperature in the tank has reached the steady-state value, a disturbance occurs due to a sudden increase in the volume of fluid in the tank. This is equivalent to $d(t)$ being a step input of size 0.1. Determine the effect of this disturbance on the temperature and sketch the response for both cases.

6. The block diagram of the cruise control system for an automobile is shown in Fig. 4.15, where $c(t)$ is the normalized speed.

a. Determine the steady-state error in the speed when the reference input

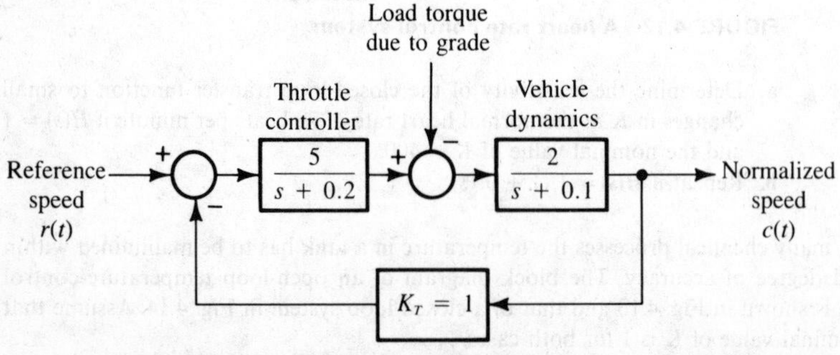

FIGURE 4.15. Cruise control system for an automobile.

is a unit step corresponding to the normalized value of the desired speed. Express this error as percentage relative to the desired speed.

b. With the automobile moving at this steady speed along a level road, suddenly the grade changes. This corresponds to introducing a load torque disturbance of 0.1 times the unit step. What will be the new steady-state error in the speed as a percentage of the desired speed? (*Hint:* Use superposition.)

7. The Ward-Leonard scheme for controlling the speed of heavy loads is described in Chapter 2, Problem 7. Determine the percentage change in the speed if the inertia of the load is increased by 5 percent.

8. The forward transfer function of a unity feedback system is given by

$$G(s) = \frac{K(s+2)}{s(s+1)(1+sT_0)}$$

Determine the relative change in the output $\Delta C/C$ for the relative change $\Delta T_0/T_0$ of the time constant.

9. Repeat for the relative change $\Delta K/K$ in the forward gain of the system described in Problem 8.

10. The forward transfer function of a unity-feedback system is

$$G(s) = \frac{K(s+2)}{s^2(s+4)}$$

a. For $K = 8$, determine the steady-state error when the input to the system is

$$r(t) = 4 - t - 2t^2$$

b. What will be the change in the steady-state error if there is a 5 percent change in the value of K?

c. What will be the change in the steady-state error if there is a 5 percent change in the location of the zero at $s = -2$?

11. The forward transfer function of a position control system with velocity feedback is given by

$$G(s) = \frac{K}{s(s+p)}$$

The transfer function of the feedback path is

$$H(s) = 1 + \alpha s$$

Determine the sensitivity of the transfer function of the closed-loop system to changes in K, p, and α if the nominal values are $K = 10$, $p = 2$, and $\alpha = 0.14$.

12. In the previous problem determine the percentage change in the steady-state error of the system to a unit ramp input for a 5 percent change in the value of (a) K, (b) p, and (c) α.

5 Performance of Control Systems

5.1 INTRODUCTION

As discussed in the previous chapter, it is possible to alter the transient as well as the steady-state behavior of a closed-loop system by adjusting the open-loop gain. In practice, however, the two requirements are often contradictory, and in trying to improve the transient performance one may introduce a deterioration in the steady-state behavior, and vice versa. This fact was observed even in the earliest days of automatic control when it was found that if one tried to improve the "sensitivity" (or steady-state accuracy) of the Watt fly-ball governor, it led to "hunting" or degradation in the transient performance. It was then felt that the engineer must make a compromise between steady-state accuracy and the transient performance. Although it is true that such a compromise is necessary if the open-loop gain is the only adjustable parameter, in Chapter 10 we shall study compensation methods that will enable us to obtain both high steady-state accuracy and desirable transient response without sacrificing one for the other.

In general, we do not know the inputs to which a particular control system will be subjected while under operation. Hence, for the purposes of analysis and design, it is customary to consider the effect of the application of certain standard inputs, which subject the system to sudden changes. The most commonly used inputs for this purpose are (1) the unit step input, (2) the unit ramp input, and (3) the unit parabolic input. In terms of a position control system, these correspond to (1) a step position input, (2) a step velocity input, and (3) a step acceleration input. Another standard test input is the sinusoid, which is of considerable importance since it provides a great deal of information about the system. Early work in the development of control theory was based on the steady-state frequency response of systems. This will be discussed in detail in Chapters 8 and 9.

The complete response of the system to any of these test inputs will have a transient component, determined by the natural frequencies (or poles of the transfer function) and the initial conditions, as well as a steady-state component determined by the input signal. These discussions assume that the system is stable; otherwise it will not reach a steady state. The stability of a system will be studied in Chapter 6.

5.2 STANDARD TEST INPUTS

Some of the standard test inputs will be described in this section.

5.2.1 The step input

This is one of the most common and convenient test inputs because it can be implemented by a sudden change in the value of the input signal. This corresponds to switching on a constant voltage in an electrical circuit, applying a sudden change in the position of an input shaft, or suddenly opening or closing a valve.

Mathematically, the unit step function is described by the equation

$$\begin{aligned} r(t) &= 1 \qquad \text{for } t > 0 \\ &= 0 \qquad \text{for } t < 0 \end{aligned} \tag{5.1}$$

and is shown in Fig. 5.1.

Its Laplace transform is given by

$$R(s) = \frac{1}{s} \tag{5.2}$$

It should be noted that the function is undefined at $t = 0$, where its value suddenly changes from 0 to 1. In most practical cases this can only be approximately correct, since the time for transition of the level can at best

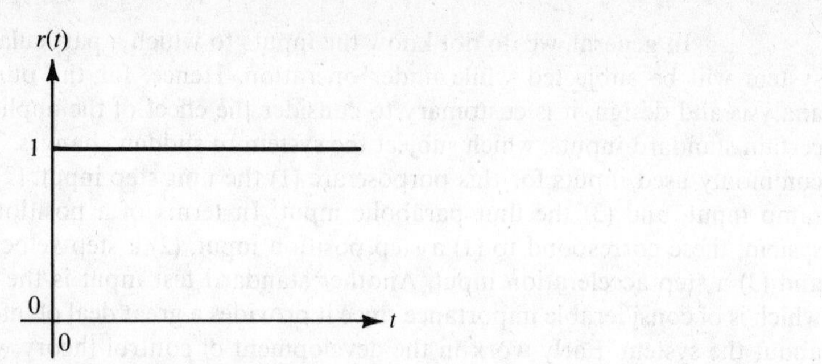

FIGURE 5.1. The unit step input.

be made very small, but never zero. For control system applications this approximation will not cause any problem as long as the time for transition is considerably smaller than the smallest time constant of the system.

5.2.2 The ramp input

The ramp input starts with zero at $t = 0$ and increases linearly with time. The unit ramp function is shown in Fig. 5.2, and is described by the equation

$$\begin{aligned} r(t) &= t \quad \text{for } t > 0 \\ &= 0 \quad \text{for } t \leq 0 \end{aligned} \tag{5.3}$$

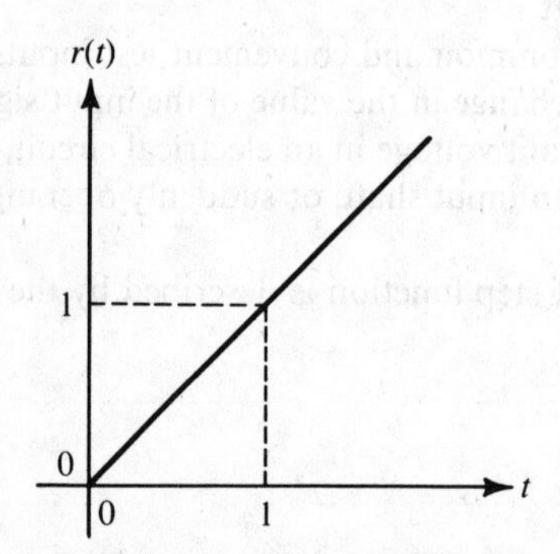

FIGURE 5.2. The unit ramp.

Its Laplace transform is given by

$$R(s) = \frac{1}{s^2} \tag{5.4}$$

and it is seen to be the integral of the unit step.

5.2.3 The parabolic input

The parabolic input may be regarded as the integral of the ramp input. The unit parabola is shown in Fig. 5.3 and is described by the equation

$$\begin{aligned} r(t) &= \frac{1}{2}t^2 \qquad \text{for } t > 0 \\ &= 0 \qquad \text{for } t \leq 0 \end{aligned} \tag{5.5}$$

Its Laplace transform is obtained as

$$R(s) = \frac{1}{s^3} \tag{5.6}$$

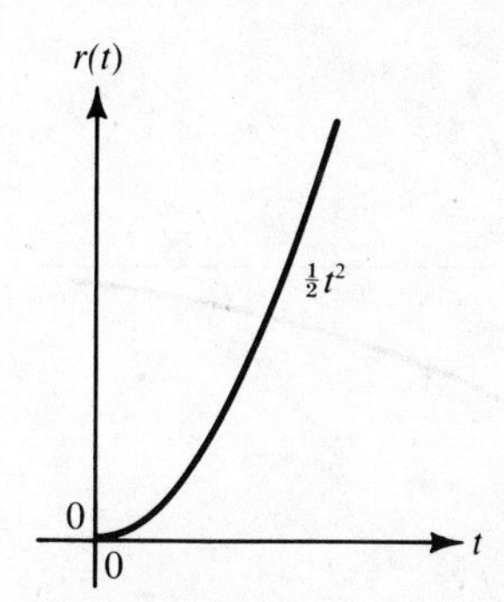

FIGURE 5.3. The unit parabola.

5.3 RESPONSE OF FIRST-ORDER SYSTEMS

Consider the first-order system shown in Fig. 5.4, which represents a speed-control system, where T is the time constant of the motor.

The transfer function of the system is given by

$$\frac{C(s)}{R(s)} = \frac{K}{1 + sT + K} = \frac{K/T}{s + (K+1)/T} \tag{5.7}$$

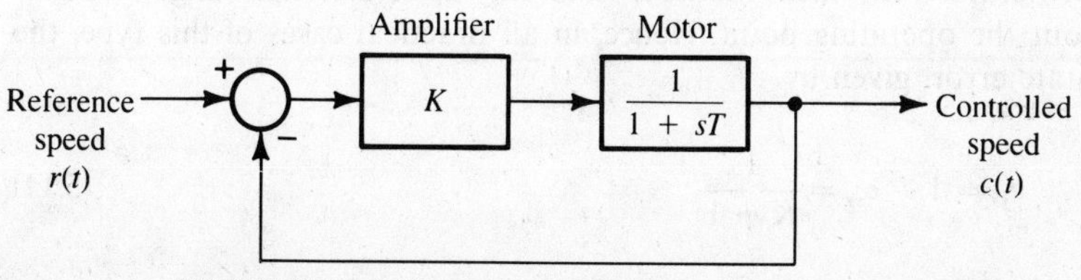

FIGURE 5.4. A first-order system.

We shall first obtain the response to a unit step input. This is given by

$$C(s) = \frac{K/T}{s[s + (K + 1)/T]} = \frac{K/(K + 1)}{s} - \frac{K/(K + 1)}{s + (K + 1)/T} \tag{5.8}$$

and

$$c(t) = \frac{K}{K + 1} - \frac{K}{K + 1} e^{-(K+1)t/T} \tag{5.9}$$

The response of the system is shown in Fig. 5.5. It will be seen that the steady-state value of the response is given by

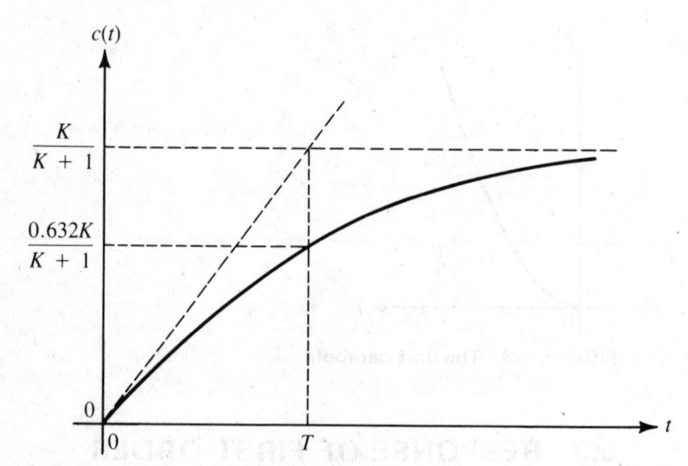

FIGURE 5.5. Response of a first-order system to a unit step input.

$$c_{ss} = \frac{K}{K + 1} \tag{5.10}$$

which is less than one, the value of the input, but can be made close to one by increasing the amplifier gain, K. In practice, K cannot be made infinitely large since practical amplifiers also generate internal noise, which increases with the gain. Furthermore, the linear model is valid only up to a certain range of the inputs about the operating point. Hence, in all practical cases of this type, the steady-state error, given by

$$e_{ss} = 1 - c_{ss} = \frac{1}{K + 1} \tag{5.11}$$

cannot be made zero.

The time constant of the system is given by

$$\tau = \frac{T}{K+1} \tag{5.12}$$

which can be defined as the time required by the response to reach 63.2 percent of its steady-state value. Alternatively, the time constant may be defined as the time that the response would have taken to reach the steady-state if it had continued to increase at the initial rate. We have

$$\frac{dc}{dt} = \frac{K}{T} e^{-(K+1)t/T} \tag{5.13}$$

and

$$\frac{dc}{dt}(0) = \frac{K}{T} \tag{5.14}$$

so

$$\tau = \frac{K}{K+1} \div \frac{K}{T} = \frac{T}{K+1} \tag{5.15}$$

Thus we see that this system will track the unit step input with a steady-state error of $1/(K+1)$ and will have a time constant $T/(K+1)$, where T is the time constant of the open-loop system.

If we apply a unit ramp input to the system, we get

$$\begin{aligned} C(s) &= \frac{K/T}{s^2[s+(K+1)/T]} \\ &= \frac{K/(K+1)}{s^2} - \frac{KT/(K+1)^2}{s} + \frac{KT/(K+1)^2}{s+(K+1)/T} \end{aligned} \tag{5.16}$$

and

$$c(t) = \frac{K}{K+1} t - \frac{KT}{(K+1)^2} + \frac{KT}{(K+1)^2} e^{-(K+1)t/T} \tag{5.17}$$

For large values of t we can neglect the last term. Hence the error between the reference input and the controlled output for such large values of t is given by

$$e(t) = r(t) - c(t) \simeq \frac{1}{K+1} t + \frac{KT}{(K+1)^2} \tag{5.18}$$

Hence the error increases with time, and the steady-state error to a ramp input will approach infinity as t approaches infinity. This is an important property of such a system. We shall be discussing this in more detail in Sec. 5.7.

DRILL PROBLEM 5.1

Determine the response to a unit step input of the speed-control system shown in Fig. 5.6, and hence the values of the steady-state error and the time constant.

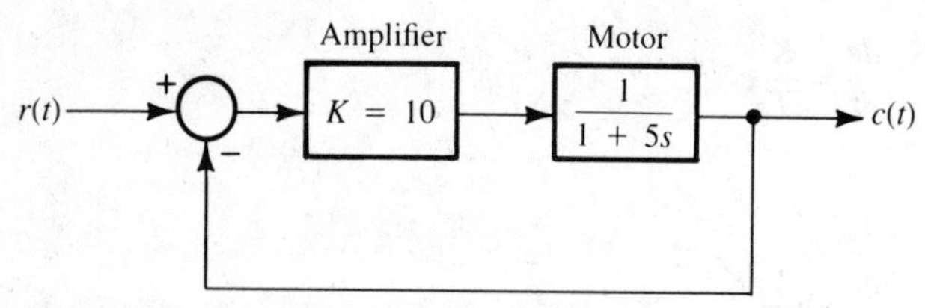

FIGURE 5.6. Speed-control system.

Answer: $e_{ss} = \frac{1}{11}, \tau = \frac{5}{11}$ s

5.4 RESPONSE OF A SECOND-ORDER SYSTEM

Consider the block diagram of the second-order system shown in Fig. 5.7. This represents a position-control system as discussed in Chapter 2 (Sec. 2.6).

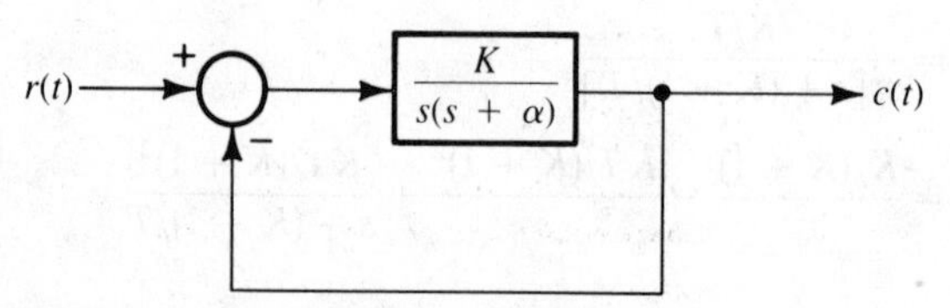

FIGURE 5.7. A second-order system.

The transfer function is obtained as

$$\frac{C(s)}{R(s)} = \frac{K}{s^2 + \alpha s + K} \tag{5.19}$$

The denominator of the transfer function is a quadratic in s. Hence the poles may be either real or complex. We shall first discuss the case of complex poles. Equation (5.19) may be rewritten as

$$\frac{C(s)}{R(s)} = \frac{\omega_n^2}{s^2 + 2\zeta\omega_n s + \omega_n^2} \tag{5.20}$$

where

$$\omega_n = \sqrt{K} \triangleq \text{undamped natural frequency} \tag{5.21}$$

and

$$\zeta = \frac{\alpha}{2\omega_n} \triangleq \text{damping ratio} \tag{5.22}$$

It should be noted that the poles will be complex only if $\zeta < 1$. Figure 5.8 shows the plot of the poles of the transfer function.

Define

$$\phi = \cos^{-1} \zeta \tag{5.23}$$

and

$$\beta = \sin \phi = \sqrt{(1 - \zeta^2)} \tag{5.24}$$

as shown in Fig. 5.8.

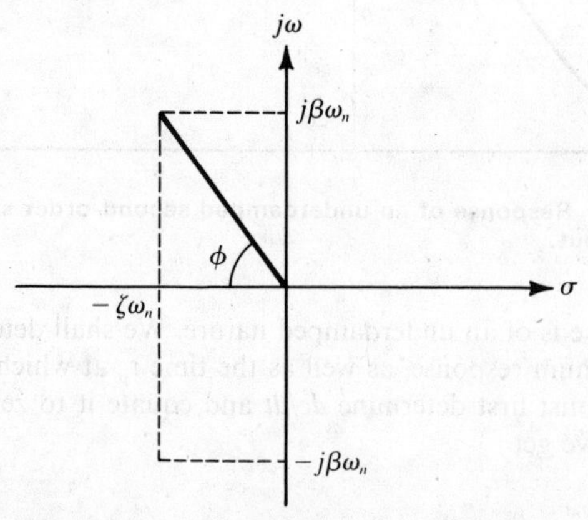

FIGURE 5.8. Poles of the transfer function.

The response to a unit step is given by

$$\begin{aligned} C(s) &= \frac{\omega_n^2}{s(s^2 + 2\zeta\omega_n s + \omega_n^2)} \\ &= \frac{1}{s} - \frac{s + 2\zeta\omega_n}{s^2 + 2\zeta\omega_n s + \omega_n^2} \\ &= \frac{1}{s} - \frac{(s + \zeta\omega_n) + \zeta\omega_n}{(s + \zeta\omega_n)^2 + (\beta\omega_n)^2} \end{aligned} \tag{5.25}$$

Taking inverse Laplace transforms we get

$$c(t) = 1 - e^{-\zeta\omega_n t}\left(\cos \beta\omega_n t + \frac{\zeta}{\beta} \sin \beta\omega_n t\right)$$

$$= 1 - \frac{1}{\beta} e^{-\zeta\omega_n t} \sin (\beta\omega_n t + \phi) \tag{5.26}$$

The response is shown in Fig. 5.9.

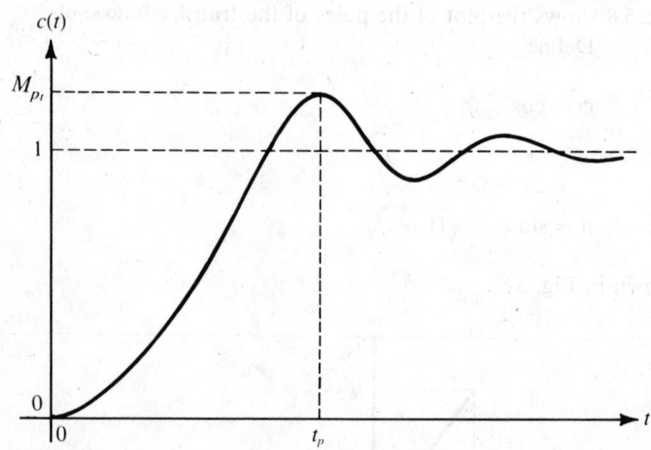

FIGURE 5.9. Response of an underdamped second-order system to a unit step input.

The response is of an underdamped nature. We shall determine the value M_{p_t} of the maximum response, as well as the time t_p at which it occurs. For this purpose, we must first determine dc/dt and equate it to zero. By differentiating Eq. (5.26) we get

$$\frac{dc}{dt} = \frac{\omega_n e^{-\zeta\omega_n t}}{\beta} [\zeta \sin (\beta\omega_n t + \phi) - \beta \cos (\beta\omega_n t + \phi)]$$

$$= \frac{\omega_n e^{-\zeta\omega_n t}}{\beta} \sin \beta\omega_n t \tag{5.27}$$

Alternatively, we may use the Laplace transform relationship

$$\frac{dc}{dt} = \mathscr{L}^{-1}[sC(s) - c(0)] = \mathscr{L}^{-1}\left(\frac{\omega_n^2}{s^2 + 2\zeta\omega_n s + \omega_n^2}\right)$$

$$= \frac{\omega_n}{\beta} e^{-\zeta\omega_n t} \sin \beta\omega_n t \tag{5.28}$$

Equating dc/dt to zero, the smallest nonzero value of t is obtained as

$$t_p = \frac{\pi}{\beta\omega_n} = \frac{\pi}{\omega_n\sqrt{1-\zeta^2}} \tag{5.29}$$

Substituting in Eq. (5.29) we now obtain

$$M_{p_t} = 1 + e^{-\zeta\pi/\beta} = 1 + e^{-\pi/\tan\phi} = 1 + e^{-\pi\zeta/\sqrt{1-\zeta^2}} \tag{5.30}$$

Expressing the difference $M_{p_t} - 1$ as a percentage, we get

$$\text{Maximum overshoot} = 100e^{-\pi\zeta/\sqrt{1-\zeta^2}} \text{ percent} \tag{5.31}$$

Thus, we see that the maximum overshoot depends only on the value of the damping ratio.

We also note that the steady-state value of the response is given by

$$c_{ss} = \lim_{t\to\infty} c(t) = 1 \tag{5.32}$$

and hence the steady-state error is zero.

We could also have determined the steady-state error without inverse Laplace transformation by first obtaining

$$E(s) = R(s) - C(s) = R(s)\cdot\left[1 - \frac{C(s)}{R(s)}\right]$$

$$= \frac{1}{s}\left[1 - \frac{\omega_n^2}{s^2 + 2\zeta\omega_n s + \omega_n^2}\right]$$

$$= \frac{s + 2\zeta\omega_n}{s^2 + 2\zeta\omega_n s + \omega_n^2} \tag{5.33}$$

Since $E(s)$ does not have any pole on the $j\omega$-axis or the right half of the s-plane, we can apply the final-value theorem to obtain

$$e_{ss} = \lim_{t\to\infty} e(t) = \lim_{s\to 0} [sE(s)] = 0 \tag{5.34}$$

We shall now consider the effect of applying the unit ramp input to this system. Since $R(s) = 1/s^2$, we get

$$C(s) = \frac{\omega_n^2}{s^2(s^2 + 2\zeta\omega_n s + \omega_n^2)}$$

$$= \frac{1}{s^2} - \frac{2\zeta/\omega_n}{s} + \frac{(2\zeta s/\omega_n) - (1 - 4\zeta^2)}{s^2 + 2\zeta\omega_n s + \omega_n^2} \tag{5.35}$$

and

$$c(t) = t - \frac{2\zeta}{\omega_n} + e^{-\zeta\omega_n t}\left(\frac{2\zeta}{\omega_n}\cos\beta\omega_n t - \frac{1-2\zeta^2}{\beta\omega_n}\sin\beta\omega_n t\right) \tag{5.36}$$

A plot of the response $c(t)$ is shown in Fig. 5.10.

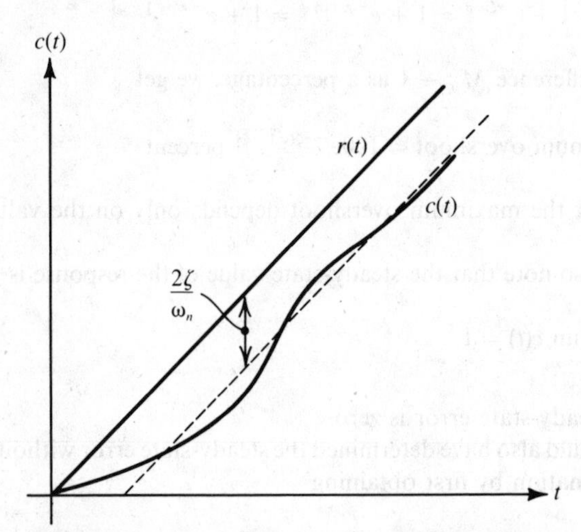

FIGURE 5.10. Response of an underdamped second-order system to a unit ramp input.

It is seen that the output $c(t)$ follows the input $r(t)$ with the steady-state error

$$e_{ss} = \frac{2\zeta}{\omega_n} \tag{5.37}$$

Again, the steady-state error could have been obtained directly from the transfer function using the final value theorem. This will be left as an exercise to the student.

The transfer function of the second-order system has real poles when ζ is greater than or equal to one. The system is said to be "critically damped" when ζ is equal to one, and "overdamped" when ζ is greater than one. For the critically damped case, the two real roots of the denominator are equal. The step responses for the two cases are shown in Fig. 5.11.

The responses are seen to be much slower than for the underdamped case. The calculation of the step responses will be left as an exercise for the student.

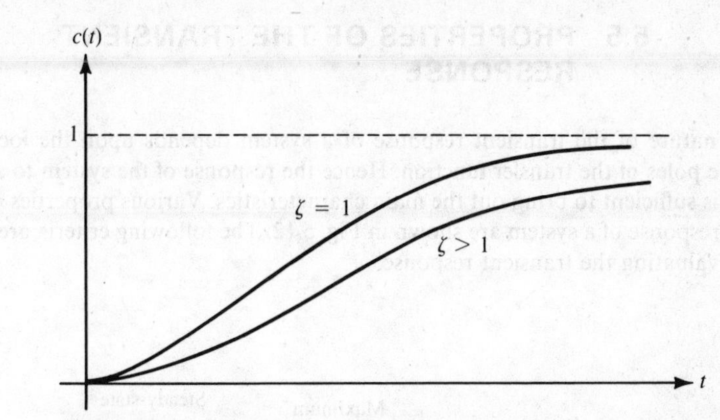

FIGURE 5.11. Response of an overdamped second-order system to a unit step.

DRILL PROBLEM 5.2

Using the final-value theorem, show that the steady-state error to a unit ramp input of the second-order system described by the transfer function in Eq. (5.20) is given by $e_{ss} = 2\zeta/\omega_n$.

DRILL PROBLEM 5.3

The transfer function of a second-order system is given by Eq. (5.20) with ζ equal to one. Determine its response to a unit step input.

Answer:
$c(t) = 1 - e^{-\omega_n t} - t\omega_n e^{-\omega_n t}$

DRILL PROBLEM 5.4

Repeat the previous problem for $\zeta = 1.5$.

Answer:
$c(t) = 1 - 1.1708e^{-0.382\omega_n t} + 0.1708e^{-2.618\omega_n t}$

DRILL PROBLEM 5.5

The step response of a second-order system described by Eq. (5.20) is required to have maximum overshoot of 10 percent. What must be the value of the damping ratio in order to achieve this? [*Hint*: Use Eq. (5.31).]

Answer: $\zeta = 0.5912$

5.5 PROPERTIES OF THE TRANSIENT RESPONSE

The nature of the transient response of a system depends upon the location of the poles of the transfer function. Hence the response of the system to a unit step is sufficient to bring out the main characteristics. Various properties of the step response of a system are shown in Fig. 5.12. The following criteria are used for evaluating the transient response.

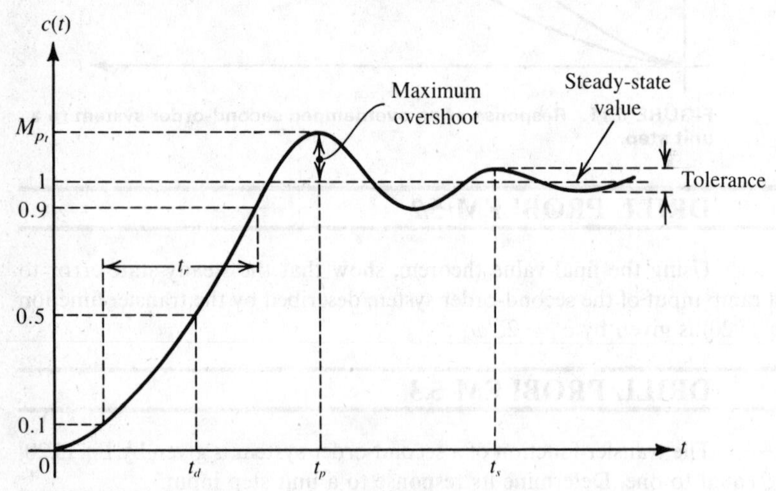

FIGURE 5.12. Properties of the transient response.

1. *Swiftness.* The speed with which the system responds to sudden changes are important. The common measures of the swiftness of the response are (a) the 10 to 90 percent rise time and (b) the time required to reach the first peak (for the case of underdamped systems).

   For an underdamped second-order system with no zeros, the time required to reach the first peak was derived in Eq. (5.29) and is given by

$$t_p = \frac{\pi}{\omega_n\sqrt{1-\zeta^2}} \tag{5.38}$$

2. *Settling Time.* This is defined as the time required for the response to stay within a specified tolerance of its final value. The tolerance is usually taken as 2 or 5 percent. Considering only the exponentially decaying envelope of the response of

the underdamped second-order system, we get the following approximation for 2 percent tolerance.

$$t_s \approx \frac{4}{\zeta\omega_n} \tag{5.39}$$

which is based on the relationship $e^{-4} \approx 0.02$.

3. *Maximum Overshoot.* This is a measure of the oscillatory nature of the response of an underdamped system. For the second-order case (with no zeros), we had shown earlier that the maximum overshoot depends only on the value of the damping ratio ζ. [Refer to Eq. (5.31).]

In view of the above, it is seen that for a second-order transfer function with no zeros, the various characteristics depend entirely upon the values of the parameters ω_n and ζ. Hence we can expect only two of these to be independent.

For a system with more poles and zeros, the transient response is more difficult to calculate. As a result, the various characteristics of the transient response cannot be determined directly. The only exception is the case when the system transfer function has a pair of complex poles much closer to the $j\omega$-axis in the s-plane than all the other poles and zeros. This is called the case of "dominant" poles, since the response is dominated by this pair of complex poles and the contributions of other poles are negligible except for the initial part of the transient response. Such systems can be approximated by the second-order system consisting of these two poles only, and hence one can obtain the various characteristics of the transient response. As we shall see later, in control system design it is often convenient to force the system to have a pair of dominant poles in order to have desired transient response characteristics.

DRILL PROBLEM 5.6

It is specified that the system described by Eq. (5.20) should have a maximum overshoot of 20 percent, and the settling time must not exceed 2 seconds. Determine the values of ζ and ω_n that will just satisfy these specifications.

Answer: $\zeta = 0.4559$ $\quad \omega_n = 4.3864$

DRILL PROBLEM 5.7

A first-order system was described in Sec. 5.3. Show that the 10 to 90 percent rise time t_r and the settling time t_s of this system are given by $t_r = 2.2\tau$ and $t_s = 4\tau$, where $\tau = T/(K+1)$ is the time constant of the system [as defined in Eq. (5.12)].

5.6 STEADY-STATE PERFORMANCE

Consider the closed-loop system shown in Fig. 5.13. The error is defined as

$$e(t) = r(t) - c(t) \tag{5.40}$$

Taking Laplace transforms of both sides,

$$\begin{aligned} E(s) &= R(s) - C(s) \\ &= R(s)\left[1 - \frac{C(s)}{R(s)}\right] \\ &= R(s)\left[1 - \frac{G(s)}{1 + G(s)H(s)}\right] \\ &= \frac{1 + G(s)H(s) - G(s)}{1 + G(s)H(s)} R(s) \end{aligned} \tag{5.41}$$

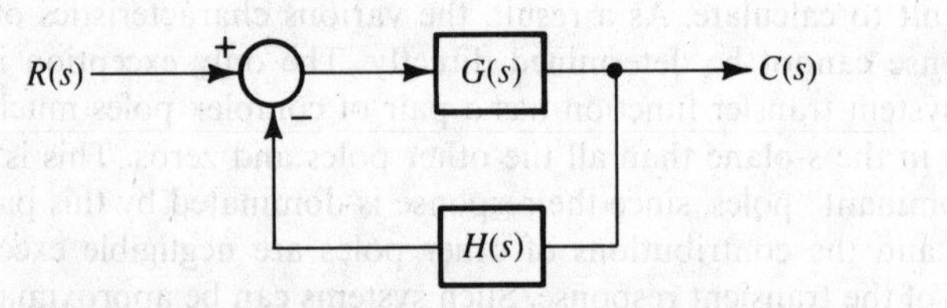

FIGURE 5.13. A general closed-loop system.

Applying the final-value theorem (assuming that we have a stable system), we get the steady-state error

$$\begin{aligned} e_{ss} &= \lim_{t \to \infty} e(t) = \lim_{s \to 0} [sE(s)] \\ &= \lim_{s \to 0} \left[\frac{1 + G(s)H(s) - G(s)}{1 + G(s)H(s)} \cdot sR(s)\right] \end{aligned} \tag{5.42}$$

Equation (5.42) can be utilized for determining the steady-state response for the general case.

We shall now consider the special case of unity feedback—that is, $H(s) = 1$—which is of considerable importance. This reduces Eq. (5.42) to

$$e_{ss} = \lim_{s \to 0} \left[\frac{sR(s)}{1 + G(s)}\right] \tag{5.43}$$

Let us now evaluate the steady-state errors for some of the standard inputs for the unity-feedback case.

1. For unit step input, $R(s) = 1/s$ and

$$e_{ss} = \lim_{s \to 0} \left[\frac{1}{1 + G(s)} \right] = \frac{1}{1 + G(0)} = \frac{1}{1 + K_p} \tag{5.44}$$

where $K_p = G(0)$ is called the position constant.

Thus, $e_{ss} = 0$ if and only if $G(0) = \infty$; that is, $G(s)$ has one or more poles at the origin.

A type 0 system has no pole at the origin, and

$$e_{ss} = \frac{1}{1 + G(0)} \neq 0 \tag{5.45}$$

A type 1 system has one pole at the origin, and a type 2 system has two poles at the origin. For such systems, the steady-state error to a step input is zero since $K_p = G(0) = \infty$.

2. For a unit ramp input, $r(t) = t$ and

$$R(s) = \frac{1}{s^2} \tag{5.46}$$

The steady-state error is as follows:

$$e_{ss} = \lim_{s \to 0} \left[\frac{1}{s + sG(s)} \right] = \lim_{s \to 0} \left[\frac{1}{sG(s)} \right] \tag{5.47}$$

For a type 0 system, from Eq. (5.47),

$$e_{ss} = \infty \tag{5.48}$$

For a type 1 system, we have

$$\lim_{s \to 0} [sG(s)] \triangleq K_v \quad \text{and} \quad e_{ss} = \frac{1}{K_v} \tag{5.49}$$

It is customary to call K_v the velocity constant of the system since it is a measure of the steady-state accuracy of a position-control system subjected to a unit step change in velocity. The reason for this name is mainly historical, going back to the 1940s when the theory of control systems was applied in the development of gun control systems (or servomechanisms, as they were called in those days).

It follows that for a type 2 system, $K_v = \infty$.

3. For a parabolic input, $r(t) = \frac{1}{2}t^2$ and

$$R(s) = \frac{1}{s^3} \tag{5.50}$$

In this case, substitution in Eq. (5.43) leads to

$$e_{ss} = \lim_{s \to 0} \left[\frac{1}{s^2 G(s)} \right] \tag{5.51}$$

Hence e_{ss} will be infinite unless $G(s)$ has two or more poles at the origin of the s-plane. In that case we have

$$e_{ss} = \frac{1}{K_a} \tag{5.52}$$

where

$$K_a = \lim_{s \to 0} [s^2 G(s)] \tag{5.53}$$

is called the acceleration constant for the system.

The steady-state error to a parabolic input will be zero for systems with $G(s)$ having more than two poles at the origin of the s-plane.

A simple interpretation of these results is obtained if we appreciate that every pole of $G(s)$ at the origin of the s-plane corresponds to an integration. Hence, if we want a constant output with zero actuating error, we must have at least one integration in the forward path. Similarly, if we want a ramp as the output with zero actuating error, we must have at least two integrations in the forward path.

5.7 STEADY-STATE ERROR IN TERMS OF THE CLOSED-LOOP TRANSFER FUNCTION

The steady-state error is often more conveniently obtained from the closed-loop transfer function.

Define

$$T(s) \triangleq \frac{C(s)}{R(s)} \tag{5.54}$$

Hence the Laplace transform of the error is obtained as

$$E(s) \triangleq R(s) - C(s) = R(s)[1 - T(s)] \tag{5.55}$$

Then the steady-state error,

$$e_{ss} = \lim_{s \to 0} [sE(s)] = \lim_{s \to 0} \{sR(s)[1 - T(s)]\} \tag{5.56}$$

We shall now consider the special inputs.

STEP INPUT. For this case we have $R(s) = 1/s$ and $sR(s) = 1$. Hence the steady-state error is given by

$$e_{ss} = \lim_{s \to 0} [1 - T(s)] = 1 - T(0) \tag{5.57}$$

Thus $e_{ss} = 0$ if and only if $T(0) = 1$; that is, the dc gain of the closed-loop system is 1. This also follows from the phyiscal interpretation of a constant as a dc or zero-frequency input.

RAMP INPUT. First we note that the steady-state error of a system to a ramp input is finite only if its steady-state error to a step input is zero—that is, only if $T(0) = 1$. Here, $r(t) = t$, $R(s) = 1/s^2$, and $sR(s) = 1/s$.

Hence the steady-state error to the unit ramp input is given by

$$e_{ss} = \lim_{s \to 0} \left[\frac{1 - T(s)}{s}\right]$$

which will be finite only if we have $T(0) = 1$, so that it takes the form of 0/0.

Using L'Hospital's rule,

$$e_{ss} = \lim_{s \to 0} \left[-\frac{dT(s)}{ds}\right] = \lim_{s \to 0} \left[-\frac{1}{T(s)} \frac{dT(s)}{ds}\right] = \lim_{s \to 0} \left[-\frac{d \ln T(s)}{ds}\right] \tag{5.58}$$

since $T(0) = 1$.

Let

$$T(s) = \frac{K(s - z_1)(s - z_2) \cdots (s - z_m)}{(s - p_1)(s - p_2) \cdots (s - p_n)} \tag{5.59}$$

Then

$$\ln T(s) = \ln K + \sum_{i=1}^{m} \ln (s - z_i) - \sum_{i=1}^{n} \ln (s - p_i) \tag{5.60}$$

and

$$e_{ss} = \lim_{s\to 0}\left[-\frac{d \ln T(s)}{ds}\right] = \lim_{s\to 0}\left[-\sum_{i=1}^{m}\frac{1}{s-z_i} + \sum_{i=1}^{n}\frac{1}{s-p_i}\right]$$

$$= \sum_{i=1}^{m}\frac{1}{z_i} - \sum_{i=1}^{n}\frac{1}{p_i} = \frac{1}{K_v} \tag{5.61}$$

Example: For a second-order system,

$$T(s) = \frac{\omega_n{}^2}{s^2 + 2\zeta\omega_n s + \omega_n{}^2} = \frac{\omega_n{}^2}{(s+\zeta\omega_n \pm j\beta\omega_n)} \qquad \text{where } \beta = \sqrt{1-\zeta^2}$$

$$T(0) = 1$$

Hence, the steady-state error to a step input = 0. The steady-state error to a ramp input is given by

$$\frac{1}{K_v} = -\left[\frac{1}{-\zeta\omega_n + j\beta\omega_n} + \frac{1}{-\zeta\omega_n - j\beta\omega_n}\right]$$

$$= -\left[\frac{-\zeta\omega_n - j\beta\omega_n - \zeta\omega_n + j\beta\omega_n}{(\zeta^2+\beta^2)\omega_n{}^2}\right] = \frac{2\zeta}{\omega_n} \tag{5.62}$$

or

$$K_v = \frac{\omega_n}{2\zeta} \tag{5.63}$$

PARABOLIC INPUT. We shall now consider the case of the parabolic input, for which $R(s) = 1/s^3$. In this case, the steady-state error is given by

$$e_{ss} = \lim_{s\to 0}\left[\frac{1-T(s)}{s^2}\right] \tag{5.64}$$

Hence, it follows, as for the case of the ramp input, that the steady-state error will be infinite unless

$$T(0) = 1 \tag{5.65}$$

and

$$\lim_{s\to 0}\frac{dT}{ds} = \lim_{s\to 0}\left[\frac{d \ln T(s)}{ds}\right] = 0 \tag{5.66}$$

The latter is satisfied if

$$\sum_{i=1}^{m}\frac{1}{z_i} = \sum_{i=1}^{n}\frac{1}{P_i} \tag{5.67}$$

When these two conditions are met we get, by applying L'Hospital's rule twice,

$$e_{ss} = \lim_{s \to 0} \left[-\frac{1}{2} \frac{d^2 \ln T(s)}{ds^2} \right] \tag{5.68}$$

To evaluate this limit we shall return to Eq. (5.60) and differentiate it twice. Hence we get

$$\frac{d^2 \ln T(s)}{ds^2} = -\sum_{i=1}^{m} \frac{1}{(s - z_i)^2} + \sum_{i=1}^{n} \frac{1}{(s - p_i)^2} \tag{5.69}$$

and

$$e_{ss} = \frac{1}{K_a} = \frac{1}{2} \left[\sum_{i=1}^{m} \frac{1}{z_i^2} - \sum_{i=1}^{n} \frac{1}{p_i^2} \right] \tag{5.70}$$

DRILL PROBLEM 5.8

The closed-loop transfer function of a speed-control system with proportional-plus-derivative control is given by

$$T(s) = \frac{K(s + z)}{s^2 + 8s + 65} \tag{5.71}$$

Determine the values of the adjustable parameters K and z in order that the steady-state error to a ramp input may be zero. What will be the steady-state error of this system to the unit parabolic input $r(t) = \frac{1}{2}t^2$?

Answers:

$$K = 8 \qquad z = \frac{65}{8} \qquad e_{ss} = \frac{1}{K_a} = \frac{1}{65}$$

5.8 INTEGRAL PERFORMANCE CRITERIA

Often it is convenient to assess the quality of a control system by evaluating a performance index that can either be calculated or measured. An important application is in the area of adaptive control where one can adjust certain parameters that will minimize the value of this index, also known as the "cost function." A number of such cost functions are used by control engineers.

The most popular cost function is the integral of the square of the error (ISE), defined as

$$J_1 = \int_0^T e^2(t)\,dt \tag{5.72}$$

where $e(t) = r(t) - c(t)$ is the error between the desired output and the actual output. The upper limit T is often taken somewhat larger than the settling time of the system in order that it can be measured experimentally. In most cases the computation of J_1 is convenient if the upper limit T is taken as infinity, since it is then possible to apply Parseval's theorem and calculate J_1 from the poles of the Laplace transform of $E(s)$, since

$$\int_0^\infty e^2(t)\,dt = \frac{1}{2\pi j}\int_{-j\infty}^{j\infty} E(s)E(-s)\,ds$$
$$= \text{sum of the residues of } E(s)E(-s) \text{ at all of its left-half-plane poles} \tag{5.73}$$

Another criterion that is quite popular is the integral of the absolute error (IAE), defined as

$$J_2 = \int_0^T |e(t)|\,dt \tag{5.74}$$

The main advantage of this criterion is the ease with which it can be measured using an analog computer.

In many other cases there is some advantage in providing a lower weighting to errors occurring later in the response than during the initial part. One criterion of this type is the integral of the time multiplied by the square of the error (ITSE), defined as

$$J_3 = \int_0^T te^2(t)\,dt \tag{5.75}$$

Another is the integral of the time multiplied by the absolute error (ITAE),

$$J_4 = \int_0^T t|e(t)|\,dt \tag{5.76}$$

As expected, the optimum response will be different depending upon the criterion used. For example, it was seen earlier that in the case of second-order transfer functions with no zeros, the nature of the step response depends entirely on the value of the damping ratio ζ. It can be shown theoretically that the ISE criterion gives the optimum value of the damping ratio as 0.5, whereas the ITSE cost function is minimized for $\zeta = 0.6$. On the other hand the ITAE cost function is minimized for $\zeta = 0.7$. The values of the cost functions calculated for different values of ζ for step function inputs are shown in Fig. 5.14. It is seen that the ITAE criterion is the most selective of the four criteria considered here.

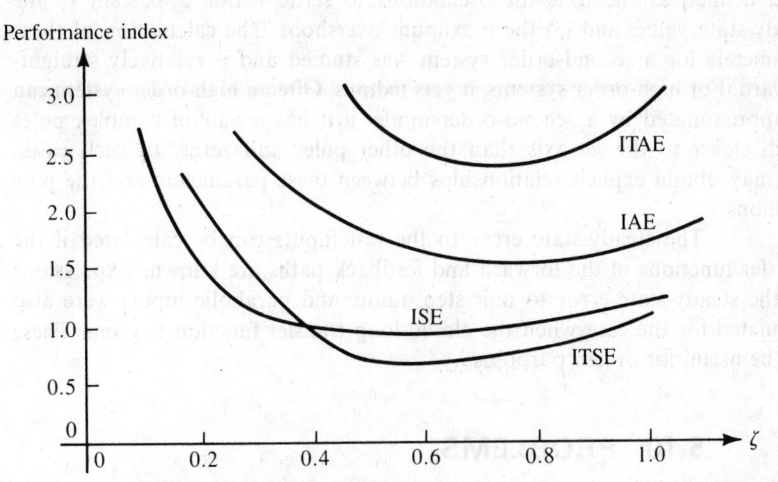

FIGURE 5.14. Performance indices for the second-order system $T(s) = \omega_n^2/(s^2 + 2\zeta\omega_n s + \omega_n^2)$.

DRILL PROBLEM 5.9

Show that the ISE criterion applied to the step response of the second-order system with transfer function $\omega_n^2/(s^2 + 2\zeta\omega_n s + \omega_n^2)$ is obtained as

$$J_1 = \int_0^\infty e^2(t)\,dt = \frac{1}{2\omega_n}\left(\frac{1}{2\zeta} + 2\zeta\right) \tag{5.77}$$

Hence show that, for a fixed value of ω_n, J_1 is minimized by making $\zeta = 0.5$. [*Hint*: Use the expression for $c(t)$ in Eq. (5.26) to obtain $e(t) = 1 - c(t)$].

5.9 SUMMARY

Some ideas for analyzing the performance of a control system have been discussed. These relate to both the transient and steady-state responses to certain standard inputs. The most commonly used test inputs are the unit step, the unit ramp, and the unit parabola. For a position-control system, these amount to a step change in (1) position, (2) velocity, and (3) acceleration.

The transient performance of a system can be quantified in terms of the following properties of its response to a unit step: (1) speed of response (either 10 to 90 percent rise time, or time to reach the first peak); (2) settling time, defined as the time for oscillations to settle within 2 percent of the steady-state value; and (3) the maximum overshoot. The calculation of these parameters for a second-order system was studied and is relatively straightforward. For high-order systems, it gets tedious. Often a high-order system can be approximated by a second-order model if it has a pair of complex poles much closer to the $j\omega$ axis than the other poles and zeros. In such cases, one may obtain explicit relationships between these parameters and the pole locations.

The steady-state error to the test inputs can be calculated if the transfer functions of the forward and feedback paths are known. Expressions for the steady-state error to unit step, ramp, and parabolic inputs were also calculated for the case when the closed-loop transfer function is given. These will be useful for design purposes.

5.10 PROBLEMS

1. The open-loop transfer function of a unity feedback system is

$$\frac{K}{s(s+2)}$$

It is specified that the response of the system to a unit step input should have a maximum overshoot of 10 percent, and the settling time should be less than 1 second.

a. Is it possible to satisfy both the specifications simultaneously?

b. If not, determine the value of K that will satisfy the first specification. What will be the settling time and the time to reach the first peak for this case?

2. Find the steady-state error to (i) a unit step input, (ii) a unit ramp input, and (iii) a unit parabolic input ($r = \frac{1}{2}t^2$) for unity feedback systems that have the following forward transfer functions.

a. $G(s) = \dfrac{20}{s(s+2)(s^2+2s+2)}$

b. $G(s) = \dfrac{108}{s^2(s+4)(s^2+3s+12)}$

3. For the systems described in Problem 2, determine the steady-state error if (i) $r(t) = 8 + 3t$, and (ii) $r(t) = 2 + 5t + 2t^2$.

4. The block diagram of a position control system with velocity feedback is shown in Fig. 5.15. Determine the value of α so that the step response has maximum overshoot of 10 percent. What is the steady-state error to a unit ramp input for this value of α?

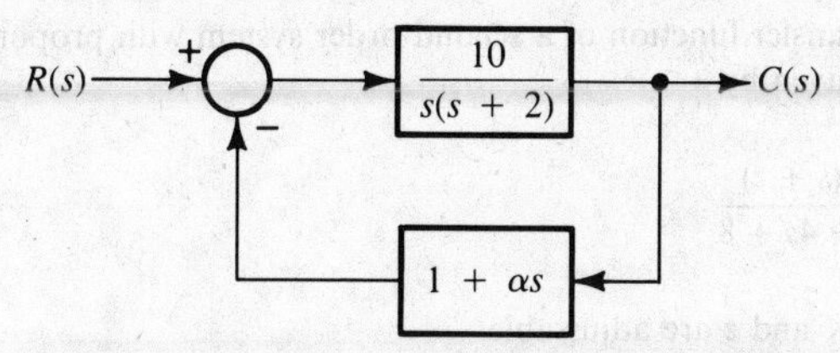

FIGURE 5.15. Position-control system with velocity feedback.

5. A position-control system with velocity feedback damping is shown in Fig. 5.16.

a. Determine the setttling time and maximum overshoot in the response to a step input as well as the steady-state error to a unit ramp input in the absence of velocity feedback, i.e., for $\alpha = 0$.

b. Repeat if $\alpha = 2$.

c. Determine the value of α so that the damping ratio of the poles of the closed-loop transfer function may be increased to 0.6. What are the values of the settling time, maximum overshoot, and the steady-state error to a unit ramp input in this case?

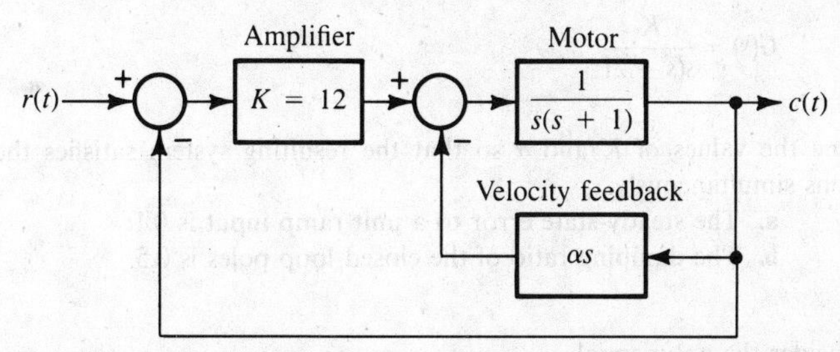

FIGURE 5.16. Position-control system with velocity feedback damping.

6. The attitude control system for a missile is shown in Fig. 5.17. Determine the values of K_1 and K_2 in order that the closed-loop system may have undamped natural frequency $\omega_n = 10$ rad/s and damping ratio $\zeta = 0.5$. What will be the settling time and maximum overshoot for a unit step input?

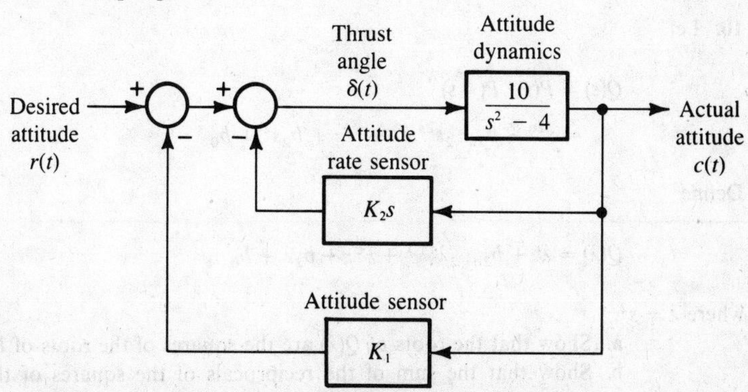

FIGURE 5.17. An attitude control system for a missile.

7. The closed-loop transfer function of a second-order system with proportional plus error-rate feedback is given by

$$\frac{C(s)}{R(s)} = \frac{K(s+z)}{s^2 + 4s + 8}$$

where the parameters K and z are adjustable.

- **a.** If $r(t) = t$, determine the values of K and z such that the steady-state error is zero.
- **b.** For these values of K and z, determine the steady-state error to a unit parabolic input; that is, $r(t) = \frac{1}{2}t^2$.
- **c.** Determine the integral of the square of the error if a unit step input is applied to the system [with the values of K and z determined in (a)].

8. Reconsider the position-control system with velocity feedback described in Problem 4 and shown in Fig. 5.15. If the forward-path gain is adjustable so that the forward-path transfer function is given by

$$G(s) = \frac{K}{s(s+2)}$$

determine the values of K and α so that the resulting system satisfies the following conditions simultaneously.

- **a.** The steady-state error to a unit ramp input is 0.1.
- **b.** The damping ratio of the closed-loop poles is 0.5.

9. Consider the polynomial

$$P(s) = s^n + a_{n-1}s^{n-1} + \cdots + a_1 s + a_0$$

- **a.** Show that the sum of the roots of $P(s) = -a_{n-1}$.
- **b.** Show that the sum of the reciprocal of the roots of $P(s) = -a_1/a_0$.

10. Let

$$Q(s) = P(s) \cdot P(-s)$$
$$= s^{2n} + b_{2n-2}s^{2n-2} + \cdots + b_2 s^2 + b_0$$

Define

$$Q(\lambda) = \lambda^n + b_{2n-2}\lambda^{n-1} + \cdots + b_2\lambda + b_0$$

where $\lambda = s^2$.

- **a.** Show that the roots of $Q(\lambda)$ are the squares of the roots of $P(s)$.
- **b.** Show that the sum of the reciprocals of the squares of the roots of $P(s) = -b_2/b_0$.

11. The closed-loop transfer function of a position-control system is given by

$$G(s) = \frac{K(s + z)(s + 4)}{(s + p)(s^2 + 6s + 25)}$$

where the parameters K, z, and p are adjustable. Is it possible to select them in such a manner that the system has zero steady-state error to a step input, a ramp input, and a parabolic input? If it is possible, determine the steady-state error when the input is t^3.

12. The linearized dynamic model relating changes in the angle of attack of a ship to changes in the angular position of the rudder is given by the transfer function

$$G(s) = \frac{K(1 + 2.5s)}{s(1 + 4s)(1 + 10s)}$$

Assume that this represents the forward path of a unity-feedback system and $K = 2$.

a. Calculate the response to a unit step input.

b. Calculate the integral of the square of the error for a unit step input.

c. Determine the steady-state error to a unit ramp input.

13. Repeat Problem 12 with velocity feedback added so that the transfer function of the feedback path is changed to $(1 + 0.1s)$. In this case the error is defined as the difference between the reference input and the desired output.

14. The block diagram of an electronic pacemaker for controlling the rate of heartbeats is shown in Fig. 4.12 (see Chapter 4, Problem 4). Assume that $K = 400$ and $H(s)$ is unity.

a. Calculate the output $c(t)$ for a unit step input.

b. Calculate the integral of the square of the error between $r(t)$ and $c(t)$ for this case.

c. Determine the steady-state error to a unit ramp input.

d. Determine the value of K for which the steady-state error to a ramp input will be 0.02.

15. Repeat Problem 14 with the transfer function of the feedback path changed to $H(s) = 1 + 0.12s$.

16. It is often possible to approximate a high-order system by a model of lower order. For example, consider a system described by the third-order transfer function

$$G(s) = \frac{s + 4}{s^3 + 6s^2 + 11s + 6}$$

Compare the step response of this system to that of the second-order model

$$\hat{G}(s) = \frac{4/3}{s^2 + 3s + 2}$$

6 Stability of Linear Systems

6.1 INTRODUCTION

To be useful, a control system must be stable. A stable system is defined as one that will have a bounded response if the input is bounded in magnitude. This definition is not suitable for control systems because it does not rule out oscillatory systems.

A linear system will be defined as stable if and only if all the poles of the system transfer function have negative real parts. Note that this definition is stronger, inasmuch as it does not allow simple poles on the imaginary axis.

This follows from the fact that all the components of the natural response will then decrease with time.

Some alternative definitions of stability, which are equivalent to the above, are given in Eqs. (6.1) and (6.2) in terms of the impulse response, $w(t)$, of the system.

$$\lim_{t \to \infty} w(t) = 0 \tag{6.1}$$

$$\int_0^\infty w^2(t)\,dt < \infty \tag{6.2}$$

The most direct approach for investigating the stability of a linear system is to determine the locations of the poles of its transfer function. Often this is not very convenient as it requires the evaluation of the roots of a polynomial, the degree of which may be high. A direct method for calculating the roots of polynomials of degree greater than five is not known. In the general case, therefore, one must use search techniques to find the roots. These are generally slow and do not always give accurate results.

It should be recognized that for investigating the stability of a linear system, it is not necessary to determine the actual location of all the poles of the transfer function. In fact, we only need to know whether the number of poles in the right half of the s-plane is zero.

A procedure for determining the number of roots of a polynomial $P(s)$ in the right half of the s-plane without actually evaluating the roots will be discussed in the next section.*

6.2 THE ROUTH-HURWITZ CRITERION

The Routh-Hurwitz criterion was developed independently by A. Hurwitz (1895) in Germany and E. J. Routh (1892) in the United States. Let the characteristic polynomial (the denominator of the transfer function after the cancellation of factors common with the numerator) be given by

$$\Delta(s) = a_n s^n + a_{n-1} s^{n-1} + \cdots + a_1 s + a_0 \tag{6.3}$$

Then the Routh table is obtained as follows:

$$\begin{array}{l|llll}
s^n & a_n & a_{n-2} & a_{n-4} & \cdots \\
s^{n-1} & a_{n-1} & a_{n-3} & a_{n-5} & \cdots \\
s^{n-2} & b_{n-1} & b_{n-3} & b_{n-5} & \cdots \\
s^{n-3} & c_{n-1} & c_{n-3} & c_{n-5} & \cdots \\
\vdots & & & & \\
s^0 & h_{n-1} & & &
\end{array}$$

where the first two rows are obtained from the coefficients of $\Delta(s)$. The elements of the following rows are obtained as follows:

$$b_{n-1} = \frac{a_{n-1}a_{n-2} - a_n a_{n-3}}{a_{n-1}} \tag{6.4}$$

* It may be noted that if any coefficient of the polynomial is zero or negative, it is implied that there are roots with zero or positive real parts.

$$b_{n-3} = \frac{a_{n-1}a_{n-4} - a_n a_{n-5}}{a_{n-1}} \tag{6.5}$$

$$\vdots$$

$$c_{n-1} = \frac{b_{n-1}a_{n-3} - a_{n-1}b_{n-3}}{b_{n-1}} \tag{6.6}$$

etc.

In preparing the Routh table for a given polynomial as suggested above, some of the elements may not exist. In calculating the entries in the line that follows, these elements are considered to be zero. The procedure will be clearer from the examples that follow.

The Routh-Hurwitz criterion states that the number of roots of $\Delta(s)$ with positive real parts is equal to the number of changes in the sign of the first column of the Routh table.

Hence the system is stable if and only if there are no sign changes in the first column of the table.

EXAMPLE 6.1

$$\Delta(s) = s^4 + 5s^3 + 20s^2 + 40s + 50 \tag{6.7}$$

The Routh table is as follows:

s^4	1	20	50
s^3	5	40	
s^2	12	50	
s^1	230/12		
s^0	50		

No sign change in the first column indicates no root in the right half of the s-plane, and hence a stable system.

EXAMPLE 6.2

$$\Delta(s) = s^3 + s^2 + 2s + 24 \tag{6.8}$$

The Routh table is as follows:

s^3	1	2
s^2	1	24
s^1	-22	
s^0	24	

Two sign changes in the first column indicate two roots in the right half of the s-plane. Hence this is unstable.

DRILL PROBLEM 6.1

Utilize the Routh table to determine the number of roots of each of the following polynomials in the right half of the s-plane.

(a) $s^3 + 2s^2 + 4s + 9$
(b) $s^4 + 6s^3 + 23s^2 + 40s + 50$
(c) $5s^6 + 8s^5 + 12s^4 + 20s^3 + 100s^2 + 150s + 200$

Answers:
(a) Two, (b) none, (c) two.

6.3 SPECIAL CASES

Two special difficulties may arise while obtaining the Routh table for a given characteristic polynomial.

CASE I. If an element in the first column turns out to be zero, it should be replaced by a small positive number, ε, in order to complete the array. The sign of the elements of the first column is then examined as ε approaches zero.

EXAMPLE 6.3

$$\Delta(s) = s^5 + 2s^4 + 3s^3 + 6s^2 + 10s + 15 \tag{6.9}$$

The Routh table is shown below, where zero in the third row of the first column is replaced by ε.

s^5	1	3	10
s^4	2	6	15
s^3	ε	$5/2$	
s^2	$6 - 5/\varepsilon$	15	
s^1	$\dfrac{30\varepsilon - 25 - 30\varepsilon^2}{12\varepsilon - 10}$		
s^0	15		

As ε approaches zero, the first column of the table may be simplified to obtain

s^5	1
s^4	2
s^3	ε
s^2	$-5/\varepsilon$
s^1	2.5
s^0	15

and two sign changes indicate two roots in the right half of the s-plane.

CASE II. Sometimes we may find that an entire row of the Routh table is zero. This indicates the presence of some roots that are negative of each other. In such cases we should form an auxiliary polynomial from the row preceding the zero row. This auxiliary polynomial has only alternate powers of s, starting with the highest power indicated by the power shown in the leftmost column of the row. This auxiliary polynomial is a factor of the characteristic polynomial. The number of roots of the characteristic polynomial in the right half of the s-plane will be the sum of the number of right-half-plane roots of the auxiliary polynomial and the number determined from the Routh table of the lower-order polynomial obtained by dividing the characteristic polynomial by the auxiliary polynomial. The following examples will clarify the procedure.

EXAMPLE 6.4

$$\Delta(s) = s^5 + 6s^4 + 15s^3 + 30s^2 + 44s + 24 \tag{6.10}$$

The Routh table is as follows:

s^5	1	15	44	
s^4	6	30	24	
s^3	10	40		
s^2	6	24		
s^1	0			← zero row

The auxiliary row gives the equation

$$6s^2 + 24 = 0 \tag{6.11}$$

Hence $s^2 + 4$ is a factor of the characteristic equation, and we get

$$q(s) = \frac{\Delta(s)}{s^2 + 4} = s^3 + 6s^2 + 11s + 6 \qquad (6.12)$$

The Routh table for $q(s)$ is as follows:

s^3	1	11
s^2	6	6
s^1	10	
s^0	6	

Since this shows no change of sign in the first column, it follows that all the roots of $\Delta(s)$ are in the left half of the s-plane, with one pair of roots $s = \pm j2$ on the imaginary axis. Hence $\Delta(s)$ is the characteristic polynomial of an oscillatory system. According to our definition, this system will be considered unstable.

EXAMPLE 6.5

$$\Delta(s) = s^6 + 4s^5 + 12s^4 + 16s^3 + 41s^2 + 36s + 72 \qquad (6.13)$$

The Routh table is as follows:

s^6	1	12	41	72
s^5	4	16	36	
s^4	8	32	72	
s^3	0	0		

The auxiliary polynomial

$$\begin{aligned} p(s) &= 8s^4 + 32s^2 + 72 \\ &= 8(s^2 + \sqrt{2}s + 3)(s^2 - \sqrt{2}s + 3) \end{aligned} \qquad (6.14)$$

and has two roots in the right half of the s-plane. Also,

$$q(s) = \frac{\Delta(s)}{p(s)} = s^2 + 4s + 8 \qquad (6.15)$$

The Routh table for $q(s)$, shown below, indicates that it has no root in the right half of the s-plane.

s^2	1	8
s^1	4	
s^0	8	

Hence, $\Delta(s)$ has two roots in the right half of the s-plane and is the characteristic polynomial of an unstable system.

DRILL PROBLEM 6.2

How many roots does each of the following polynomials have in the right half of the s-plane?

(a) $s^4 + 2s^3 + 4s^2 + 8s + 15$
(b) $s^5 + 2s^4 + 6s^3 + 10s^2 + 8s + 12$
(c) $s^5 + 2s^4 + 4s^3 + 8s^2 + 16s + 32$
(d) $s^6 + 4s^5 + 11s^4 + 12s^3 + 26s^2 + 84s + 16$

Answers:

(a) Two, (b) none (two on the $j\omega$-axis), (c) two, (d) none (two pairs of roots on the $j\omega$-axis).

6.4 RELATIVE STABILITY

Application of the Routh criterion, as discussed above, only tells us whether a system is stable or not. In many cases we need more information. For example, if the Routh test shows that a system is stable, we often like to know how close it is to instability—i.e., how far from the $j\omega$-axis is the pole closest to it. The Routh criterion can be utilized for obtaining this information by shifting the vertical axis in the s-plane to obtain the p-plane, as shown in Fig. 6.1.

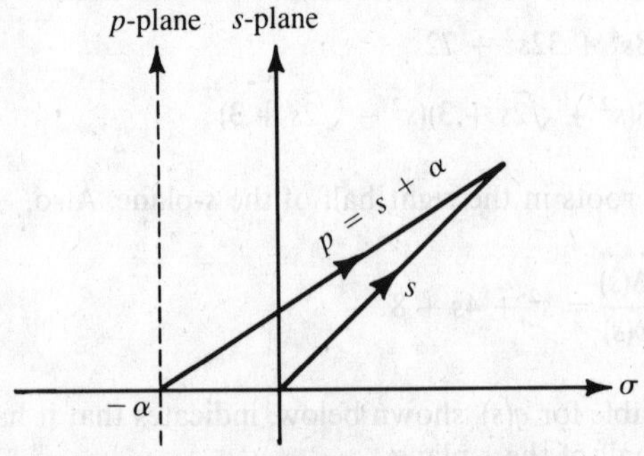

FIGURE 6.1. Shift of the axis to the left by α.

Hence in the polynomial $\Delta(s)$, if we replace s by $p - \alpha$, we get a new polynomial $\Delta(p)$. Applying the Routh test to this polynomial will tell us how many roots $\Delta(p)$ has in the right half of the p-plane. This is also the number of roots of $\Delta(s)$ to the right of the line $s = -\alpha$ in the s-plane.

The procedure is illustrated in the following example.

EXAMPLE 6.6

Consider the polynomial

$$\Delta(s) = s^3 + 10.1s^2 + 21s + 2 \tag{6.16}$$

The Routh table shown below tells us that the system is stable.

s^3	1	21
s^2	10.1	2
s^1	20.802	
s^0	2	

Now shift the axis to the left by 0.2; that is,

$$p = s + 0.2 \tag{6.17}$$

or

$$s = p - 0.2 \tag{6.18}$$

Hence

$$\begin{aligned}\Delta(p) &= (p - 0.2)^3 + 10.1(p - 0.2)^2 + 21(p - 0.2) + 2 \\ &= p^3 + 9.5p^2 + 17.08p - 1.804\end{aligned} \tag{6.19}$$

The Routh table given below shows one root in the right half of the p-plane. This implies one root in the s-plane between 0 and -0.2. It follows that shifting by a smaller amount would enable us to get a better idea of the location of this root.

p^3	1	17.08
p^2	9.5	-1.804
p^1	17.27	
p^0	-1.804	

Hence, although $\Delta(s)$ is the characteristic polynomial of a stable system, the presence of a root between 0 and -0.2 in the s-plane indicates a large time constant, of the order of more than 5 seconds.

DRILL PROBLEM 6.3

The characteristic polynomial of a third-order system is given by

$$\Delta(s) = s^3 + 4s^2 + 8s + 30$$

(a) Does $\Delta(s)$ have any root in the right half of the s-plane?

(b) By suitable shifts of the axis, determine the approximate location of the dominant poles of the system.

Answers:
(a) No, (b) two poles between $\sigma = 0$ and $\sigma = -0.05$.

DRILL PROBLEM 6.4

The closed-loop transfer function of a position control system utilizing an ac servomotor is given by

$$T(s) = \frac{20}{s^3 + 2s^2 + 11s + 20}$$

It is desired that all the poles of the transfer function lie to the left of the line $\sigma = -0.1$. Is this condition satisfied?

Answer:
No. There are two poles between $\sigma = 0$ and $\sigma = -0.1$.

6.5 APPLICATION TO DESIGN

The Routh criterion may also be utilized to determine the range of the open-loop gain for which a feedback system may be stable. This is done by finding the limiting values of K that will cause changes in the signs of the elements of the first column of the Routh table. It will be illustrated by means of an example.

EXAMPLE 6.7

Consider the feedback system shown in Fig. 6.2. Its characteristic equation is obtained as

$$\Delta(s) = s(s + 3)(s + 10) + K = s^3 + 13s^2 + 30s + K \qquad (6.20)$$

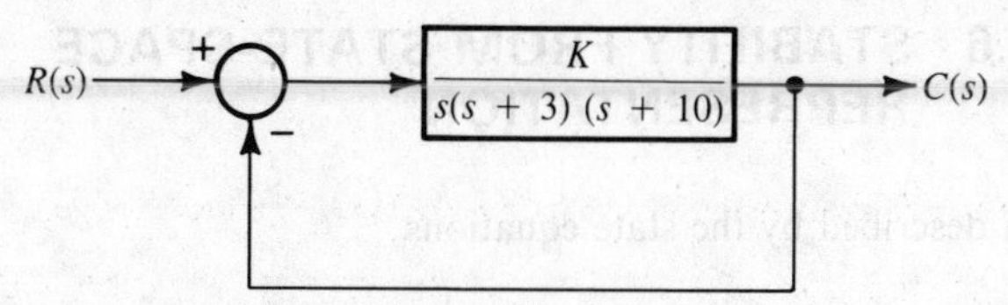

FIGURE 6.2. A feedback control system.

The Routh table is shown below:

s^3	1	30
s^2	13	K
s^1	$\frac{390 - K}{13}$	
s^0	K	

From the Routh table it is obvious that the sytem will be stable if and only if

$$0 < K < 390 \tag{6.21}$$

For $K = 390$ we get a zero row, and the auxiliary equation is

$$13s^2 + 390 = 0 \tag{6.22}$$

Hence, for this case,

$$\Delta(s) = (s^2 + 30)(s + 13) \tag{6.23}$$

DRILL PROBLEM 6.5

The closed-loop transfer function of an antenna control system is given by

$$T(s) = \frac{K}{s^4 + 6s^3 + 30s^2 + 60s + K}$$

(a) Determine the range in which K must lie for the system to be stable.

(b) What should be the upper limit on K if all the closed-loop poles are required to be to the left of the line $\sigma = -1$?
[*Hint*: Shift the axis suitably before applying the method discussed in this section.)

Answers:
(a) $0 < K < 200$, (b) $K < 112$.

6.6 STABILITY FROM STATE-SPACE REPRESENTATION

For a system described by the state equations

$$\left.\begin{aligned} \dot{x} &= Ax + Bu \\ y &= Cx \end{aligned}\right\} \tag{6.24}$$

it was seen in Sec. 3.2 that the denominator of the resulting transfer function is given by

$$\Delta(s) = \det(sI - A) \tag{6.25}$$

This is commonly called the characteristic polynomial of matrix A.

The roots of this polynomial are the poles of the system as well as the eigenvalues of A. This also follows from the denominator of Eq. (3.3).

The system is stable if and only if all the roots of the characteristic polynomial, $\Delta(s)$, are in the left half of the s-plane.

Alternatively, the system described by Eq. (6.24) is stable if and only if all the eigenvalues of A have negative real parts.

Instead of actually finding all the roots of $\Delta(s)$, one may use the Routh-Hurwitz criterion to determine the number of roots in the right half of the s-plane. Relative stability may be investigated by shifting the axis as discussed in Sec. 6.4.

EXAMPLE 6.8

The state equations for a third-order system are given below. Use the Routh criterion to determine whether it is stable.

$$\dot{x} = \begin{bmatrix} 0 & 1 & -8 \\ 1 & 0 & -5 \\ 0 & 2 & -6 \end{bmatrix} x + \begin{bmatrix} 2 \\ 1 \\ 0 \end{bmatrix} u \tag{6.26}$$

$$y = [1 \quad 2 \quad 1]\, x$$

The characteristic polynomial is obtained as

$$\Delta(s) = \det(sI - A) = \begin{vmatrix} s & -1 & 8 \\ -1 & s & 5 \\ 0 & -2 & s+6 \end{vmatrix}$$

$$= s(s^2 + 6s + 10) + 1(-s - 6 + 16)$$

$$= s^3 + 6s^2 + 9s + 10 \tag{6.27}$$

The Routh table is as follows:

$$\begin{array}{c|cc} s^3 & 1 & 9 \\ s^2 & 6 & 10 \\ s^1 & \dfrac{54-10}{6} & \\ s^0 & 10 & \end{array}$$

Since every term in the first column is positive, the system is stable.

DRILL PROBLEM 6.6

The A matrix in the state equation for a helicopter near hover is given below. Determine whether any eigenvalue of A has a positive real part.

$$A = \begin{bmatrix} -0.02 & -1.4 & 9.8 \\ -0.01 & -0.4 & 0 \\ 0 & 1 & 0 \end{bmatrix}$$

Answer:
A has two eigenvalues with positive real parts.

6.7 SUMMARY

The stability of a linear system can be obtained conveniently by using the Routh criterion, which determines the number of roots of the characteristic polynomial in the right half of the s-plane without actually requiring the evaluation of the roots. By shifting the axis, it is also possible to determine whether the system is on the verge of instability. The criterion may be utilized to determine the range of the open-loop gain over which a feedback system may be stable. When a system is described by state equations, the Routh criterion may be applied to determine the number of eigenvalues of the A matrix with positive real parts.

6.8 PROBLEMS

1. Using the Routh criterion of stability, determine whether the following characteristic polynomials represent stable systems.

a. $s^3 + 5s^2 + 10s + 20$
b. $s^4 + 3s^3 + 10s^2 + 20s + 100$
c. $s^4 + 2s^3 + s^2 + 2s + 3$
d. $s^5 + 2s^4 + 3s^3 + 4s^2 + 6s + 12$

2. Determine the range of K over which the following characteristic polynomials belong to stable systems.

a. $s^3 + 6s^2 + 11s + K$
b. $s^4 + 5s^3 + 9s^2 + 20s + K$
c. $s^5 + 8s^4 + 15s^3 + 32s^2 + 60s + K$

3. For Problem 2 determine the value of K in each case so that all the roots of the polynomials are to the left of the line $\sigma = -0.5$.

4. The block diagram of a type 2 system stabilized by a lead compensator is shown in Fig. 6.3.

a. Determine the range of values of K for which the system will be stable if $\alpha = 1$.
b. Determine the maximum value of K for which all the closed-loop poles will be to the left of the line $\sigma = -0.2$, assuming $\alpha = 1$.

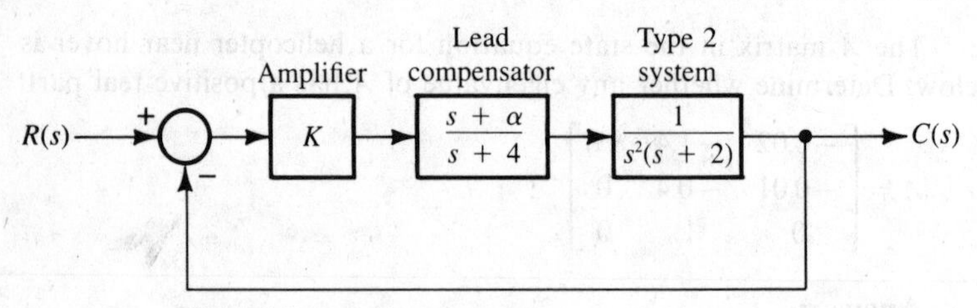

FIGURE 6.3. A type 2 system.

5. Suppose that in the previous problem the gain K is fixed at 12, but the location of the zero at α is adjustable.

a. Determine the range of values of α so that the system may be stable.
b. Determine the range of values of α so that all the closed-loop poles are to the left of the line $\sigma = -0.2$.

6. The forward transfer function of a unity feedback system is given by

$$G(s) = \frac{K(s+1)}{s(s+2)(s+3)}$$

a. Determine the range of values of K so that the system may be stable.
b. Determine the maximum value of K for which all the poles of the closed-loop system are to the left of the line $\sigma = -0.5$.

7. The state equations of a linear system are

$$\dot{x} = \begin{bmatrix} 0 & 1 & -6 \\ 1 & 0 & 5 \\ 0 & -1 & 2 \end{bmatrix} x + \begin{bmatrix} 1 \\ 1 \\ 0 \end{bmatrix} u$$

$$y = [0 \quad 1 \quad 2]\, x$$

Is this a stable system?

8. For each of the matrices given below, determine the number of eigenvalues with positive real parts.

a. $\begin{bmatrix} -1.2 & -5 & -2.0 \\ 10 & -8 & 0 \\ 0 & 1 & 0 \end{bmatrix}$

b. $\begin{bmatrix} -1 & 2 & 0 \\ 0 & -5 & -3 \\ 2 & 1 & -6 \end{bmatrix}$

(*Hint*: Leverrier's algorithm, described in Sec. 3.2, can be utilized for calculating the characteristic polynomial of a matrix, if desired.)

9. The forward-path transfer function of a unity-feedback system is given by

$$G(s) = \frac{Ks(s+12)}{(s+5)(s+3)(s+6)}$$

Determine the range of values of K for which the closed-loop system will be stable.

10. For Problem 9 determine the range of values of K for which all the poles of the closed-loop system will be to the left of the line $s = -1$.

11. The forward transfer function of a position control system with velocity feedback is given by

$$G(s) = \frac{K}{s(s+4)}$$

The transfer function of the feedback path is

$$H(s) = 1 + 0.14s$$

Determine the range of values of K for which all the poles of the transfer function of the closed-loop system will be to the left of the line $\sigma = -1$.

12. The characteristic polynomial of a closed-loop system is given by

$$P(s) = s^4 + 5s^3 + (K+12)s^2 + (2K+8)s + 2K$$

a. Determine the range of values of K for which all the roots of $P(s)$ will be in the left half of the s-plane.

b. Determine the range of values of K for which all the roots of $P(s)$ will be to the left of the line $\sigma = -1$.

13. The steady-state response of a stable system to a step input can be obtained directly from the state equations. This follows from the fact that, for such an input, all the states must be constant in the steady state; that is, $\dot{x}$ must approach zero as t approaches infinity. Hence the steady-state value of x is obtained as $A^{-1}Bu$, where u is the constant input.

Use this approach to determine the steady-state output to a unit step of the system described by the state equations

$$\dot{x} = \begin{bmatrix} -2 & 0 & 1 \\ 1 & -2 & 0 \\ 1 & 1 & -1 \end{bmatrix} x + \begin{bmatrix} 1 \\ 0 \\ 1 \end{bmatrix} u$$

$$y = [2 \quad 1 \quad -1]\, x$$

Verify your answer by calculating the transfer function of the system. (Note that this requires that all the eigenvalues of A must have negative real parts. This also implies nonsingularity of A.)

14. The result of Problem 13 can be easily extended to the case when the input is a ramp function. For such an input the state x in the steady state must have the form

$$x = \alpha t + \beta$$

where α and β are constant vectors. Substituting this into the state equation, one can solve for α and β by equating the corresponding coefficients of t.

Use this approach to determine the steady-state output of the system described by the state equations in Problem 13 when the input is a unit ramp.

7 The Root Locus Method

7.1 INTRODUCTION

In the previous chapter we saw that it is possible to adjust the location of the poles of the transfer function of a closed-loop system by varying the loop gain. Since the nature of the transient response is closely related to the location of the poles, it is very important to know how they move in the s-plane as the gain, or some other parameter, is varied. Furthermore, since for economy in manufacture we must allow some tolerance in the values of the parameters, it is also desirable to determine the movement of the poles with small changes in these parameters.

The root locus method, introduced by W. R. Evans in 1948, is a graphical procedure for sketching the movement in the s-plane of the poles of the closed-loop transfer function as a parameter is varied from zero to infinity. The method was developed as an alternative to repeatedly determining the roots of a polynomial using numerical methods. Although originally the method was intended for the case when the open-loop gain was the variable parameter, it can be generalized as follows. Starting with the characteristics polynomial of the system, rewrite it as

$$KP(s) + Q(s) = 0 \tag{7.1}$$

where K is the variable parameter. This can be done by separating all terms that contain K from those that do not.

Our objective is to plot the locus of the roots of Eq. (7.1) as K varies from 0 to ∞. We shall first consider the example of a second-order system to understand the implications and the power of the method.

7.2 ROOT LOCI FOR A SECOND-ORDER SYSTEM

Consider the second-order system shown in Fig. 7.1, which represents a typical position control system with velocity feedback. By applying Mason's rule, we obtain the transfer function of the closed-loop system as

$$\frac{C(s)}{R(s)} = \frac{K}{s^2 + (K\alpha + 2)s + K} \tag{7.2}$$

The characteristic polynomial is

$$\Delta(s) = s^2 + (K\alpha + 2)s + K \tag{7.3}$$

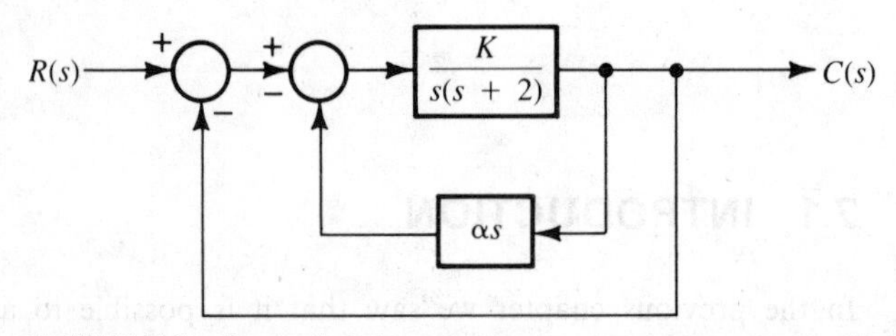

FIGURE 7.1. Second-order system.

We shall consider two cases. First we shall assume that α is zero—i.e., there is no velocity feedback—and investigate the effect of varying the loop gain K from zero to infinity. Next we shall assume that K is equal to 25 and examine the effect of varying α from zero to infinity. In both cases our main interest is to determine how the closed-loop poles will move in the s-plane.

EXAMPLE 7.1

We shall consider the case of variable K, with $\alpha = 0$. In this case, we have to determine the roots of the characteristic polynomial

$$\Delta(s) = s^2 + 2s + K = 0 \tag{7.4}$$

First consider the case when $K = 0$. From Eq. (7.4), the roots of the characteristic polynomial are obtained as $s = 0$ and $s = -2$, which are also the poles of the open-loop transfer function with $\alpha = 0$.

As K is increased from 0 to 1, we have real roots, but for K greater than one, we get a pair of complex conjugate roots with real part equal to -1. This movement of the roots of Eq. (7.4) is shown in Fig. 7.2.

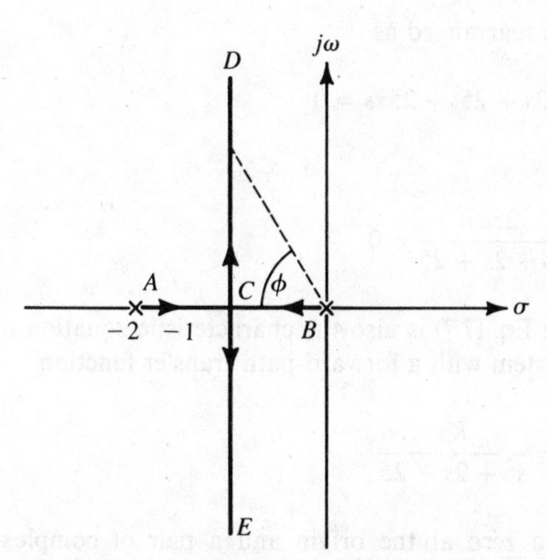

FIGURE 7.2. Root locus for Eq. (7.4).

It will be seen that the root locus consists of the two straight lines, *AB* and *DE*, intersecting at *C*. One may visualize the loci starting from the two open-loop poles (*A* and *B*). The locus starting from *A* moves to the right, while the locus starting from *B* moves to the left on the real axis until they meet at *C* for $K = 1$. For larger values of K, the loci move along the lines *CD* and *CE*, as shown in Fig. 7.2.

From the plot of the root locus, we can conclude that this system will be stable for all values of K, and the poles of the closed-loop transfer function will be complex only if K is greater than one. Furthermore, for large values of K, although the system will be stable, it will be very lightly damped, the damping ratio being given by

$$\zeta = \frac{1}{\sqrt{K}} \tag{7.5}$$

EXAMPLE 7.2 CASE OF VARIABLE VELOCITY FEEDBACK

Let us now consider the case when the open-loop gain K is fixed at 25 and the variable parameter is the velocity feedback coefficient α. From Eq. (7.3), the characteristic polynomial can now be written as

$$\Delta(s) = s^2 + (25\alpha + 2)s + 25 = 0 \tag{7.6}$$

This can be rearranged as

$$(s^2 + 2s + 25) + 25\alpha s = 0 \tag{7.7}$$

or

$$1 + \frac{25\alpha s}{s^2 + 2s + 25} = 0 \tag{7.8}$$

Hence Eq. (7.7) is also the characteristic equation of a unity-feedback system with a forward-path transfer function

$$G'(s) = \frac{Ks}{s^2 + 2s + 25} \tag{7.9}$$

which has a zero at the origin and a pair of complex poles at $s = -1 \pm j\sqrt{24}$ with $K = 25\alpha$. The pole-zero plot of the open-loop transfer function is shown in Fig. 7.3. We note that for $\alpha = 0$ the

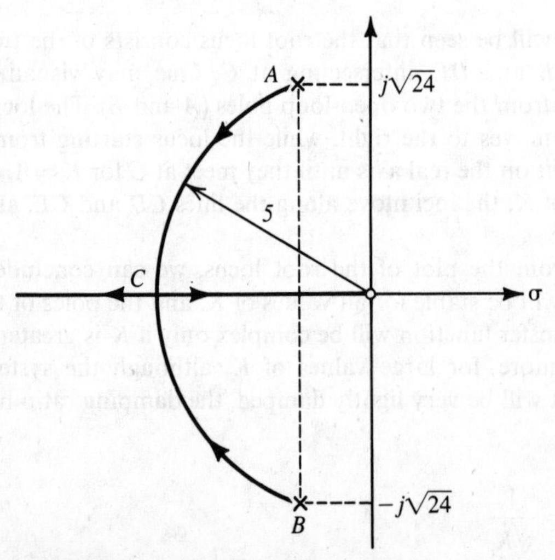

FIGURE 7.3. Root locus for Eq. (7.7).

roots of $\Delta(s)$ are located at $s = -1 \pm j\sqrt{24}$, labeled A and B in Fig. 7.3. As α is increased from zero to 0.32, the roots move along a circular arc of radius 5 with the center at the origin. This follows from the fact that Eq. (7.6) has complex roots for these values of α. As α is further increased, we get real roots, with one moving to the left and the other to the right.

From the root locus plot we can conclude that the effect of velocity feedback is to increase the damping ratio of the system without altering the undamped natural frequency.

DRILL PROBLEM 7.1

For the problem considered in Example 7.2 determine the value of α so that the poles of the closed-loop system transfer function have a damping ratio of 0.6.

Answer: 0.16

7.3 BASIC PRINCIPLES OF THE ROOT LOCUS

In the previous section we considered two examples of a second-order system where the root locus can be obtained from analytical considerations. In most other cases it is not so straightforward. For example, if we had assumed in Example 7.1 that α was fixed at 0.1 and K was to be varied from zero to infinity, we would have obtained the characteristic polynomial

$$\Delta(s) = s^2 + (0.1K + 2)s + K \tag{7.10}$$

In this case the root locus is not obtained as simply as before, even though we have a polynomial of the second degree. For higher-order systems the problem gets more involved.

Hence we shall look at Eq. (7.1) again and try to develop some basic conditions that apply to the root locus. These then will be utilized to derive certain properties of the root locus that will be useful in the construction of the root locus plot.

Equation (7.1) may be written as

$$1 + \frac{KP(s)}{Q(s)} = 0 \tag{7.11}$$

which may be rearranged as

$$K\frac{P(s)}{Q(s)} = -1 = 1\underline{/180^\circ} \tag{7.12}$$

Hence we derive the following basic conditions that must be satisfied for all points on the root locus.

CONDITION 1 (ANGLE CONDITION). For some value of K, any point in the s-plane will be a root of Eq. (7.12) if and only if the angle condition is satisfied; i.e., at this point the algebraic sum of the angles of vectors drawn to it from the open-loop poles and zeros is an odd multiple of 180°. [Note that the roots of $P(s)$ are the open-loop zeros and the roots of $Q(s)$ are the open-loop poles of the transfer function $P(s)/Q(s)$.]

CONDITION 2 (MAGNITUDE CONDITION). At any point on the root locus (with the angle condition satisfied) the value of K is obtained as

$$K = \frac{\text{product of lengths of vectors from poles}}{\text{product of lengths of vectors from zeros}} \tag{7.13}$$

Thus we first apply the angle condition to determine whether any point in the s-plane lies on the root locus. If this condition is satisfied, then we can use the magnitude condition to obtain the value of K for which this point will be a root of the characteristic equation.

EXAMPLE 7.3

We shall now use these principles to obtain the root locus for Eq. (7.10). Since we can rearrange it as

$$\Delta(s) = (s + 1 + 0.05K)^2 + K - (1 + 0.05K)^2 = 0 \tag{7.14}$$

it follows that the roots will be complex if and only if

$$K - (1 + 0.05K)^2 > 0 \tag{7.15}$$

To determine these values of K, let us solve Eq. (7.15) as an equality. This gives two values of K, as follows:

$$K = 1.1146 \text{ or } 358.8854 \tag{7.16}$$

It can be verified that the inequality (7.15) is satisfied if

$$1.1146 < K < 358.8854 \tag{7.17}$$

For the locus of the complex roots, we shall rearrange Eq. (7.10) in the form of Eq. (7.11) so that we have

$$1 + \frac{0.1K(s+10)}{s(s+2)} = 0 \tag{7.18}$$

The pole-zero plot of the corresponding forward-path transfer function $G'(s)$ is shown in Fig. 7.4, where

$$G'(s) = \frac{K'(s+10)}{s(s+12)} \tag{7.19}$$

and

$$K' = 0.1K \tag{7.20}$$

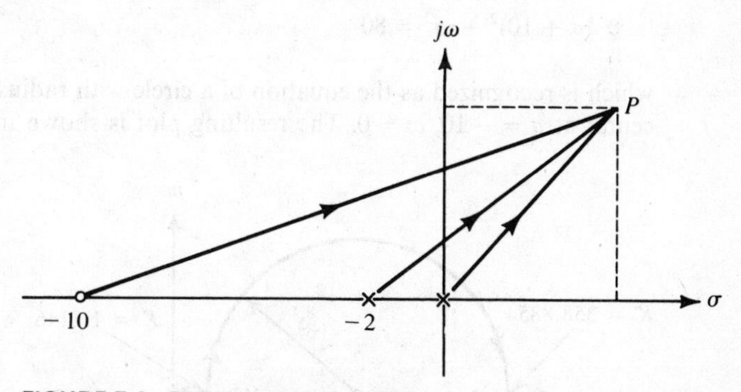

FIGURE 7.4. Pole-zero plot for $G'(s)$.

Consider any point $P(s)$ in the s-plane, where $s = \sigma + j\omega$. If P is on the root locus, then the angle condition must be satisfied; that is,

$$\tan^{-1}\frac{\omega}{\sigma} + \tan^{-1}\frac{\omega}{\sigma+2} - \tan^{-1}\frac{\omega}{\sigma+10} = \pm\pi \tag{7.21}$$

It can be rearranged as

$$\tan^{-1}\frac{\omega}{\sigma+2} - \tan^{-1}\frac{\omega}{\sigma+10} = \pm\pi - \tan^{-1}\frac{\omega}{\sigma} \tag{7.22}$$

or

$$\tan^{-1}\frac{\dfrac{\omega}{\sigma+2} - \dfrac{\omega}{\sigma+10}}{1 + \dfrac{\omega^2}{(\sigma+2)(\sigma+10)}} = \pm\pi - \tan^{-1}\frac{\omega}{\sigma} \tag{7.23}$$

Taking the tangent of both sides of Eq. (7.23), we get

$$\frac{\dfrac{\omega}{\sigma+2}-\dfrac{\omega}{\sigma+10}}{1+\dfrac{\omega^2}{(\sigma+2)(\sigma+10)}}=\tan\left(\pm\pi-\tan^{-1}\frac{\omega}{\sigma}\right)=-\frac{\omega}{\sigma} \tag{7.24}$$

or

$$\frac{8\omega}{\sigma^2+12\sigma+20+\omega^2}=-\frac{\omega}{\sigma} \tag{7.25}$$

Equation (7.25) can be simplified further to obtain (for $\omega \neq 0$)

$$(\sigma+10)^2+\omega^2=80 \tag{7.26}$$

which is recognized as the equation of a circle with radius $\sqrt{80}$ and center at $\sigma = -10$, $\omega = 0$. The resulting plot is shown in Fig. 7.5.

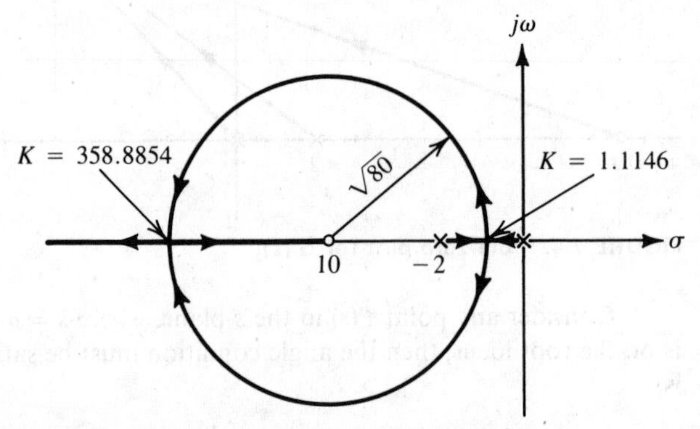

FIGURE 7.5. Root locus plot for Eq. (7.18).

7.4 SOME PROPERTIES OF THE ROOT LOCUS

In the previous section we were able to obtain the root locus using the basic principles for determining whether a point in the s-plane lies on the root locus. This will not be suitable for the general case, since we shall be obtaining polynomials of higher order. Furthermore, it is not practical to use a trial-and-error procedure to determine the set of all points satisfying the angle condition.

Hence we shall now discuss some properties of the root locus that will enable us to obtain an approximate plot with much less effort. These are given in the following paragraphs.

1. The root locus is symmetrical about the real axis of the s-plane.

This follows since the coefficients of the polynomials P and Q are real with the result that the polynomial $KP(s) + Q(s)$ can have complex roots only in conjugate pairs.

2. A branch of the root locus starts from each open-loop pole and terminates at each open-loop zero at infinity. The number of branches terminating at infinity is equal to the number of open-loop poles minus the number of open-loop zeros.

This property is evident if we examine Eq. (7.1), which was also written as Eq. (7.11), by dividing both sides by $Q(s)$. For $K = 0$, the equation takes the form

$$Q(s) = 0 \tag{7.27}$$

That is, for this case the roots of the characteristic polynomial are the roots of $Q(s)$, which are also the poles of the open-loop transfer function $G(s) = KP(s)/Q(s)$.

We may divide both sides of Eq. (7.1) by K to obtain

$$P(s) + \frac{Q(s)}{K} = 0 \tag{7.28}$$

Hence it follows that for K equal to infinity the roots of the characteristic polynomial will coincide with the roots of

$$P(s) = 0 \tag{7.29}$$

which are the zeros of $G(s)$.

3. Any point on the real axis is a part of the root locus if and only if the number of poles and zeros to its right is odd.

This property follows by applying the angle condition at any point on the real axis of the s-plane. Since the open-loop poles and zeros are symmetrical about the real axis (i.e., all poles and zeros occur in conjugate pairs), the angles of the vectors drawn to any point on the real axis from each pair of complex poles and zeros will add up to zero. Hence we need consider only the poles and zeros on the real axis. Furthermore, the angle subtended at any point on the real axis by a real pole or zero to its left is always zero. Hence, only the poles and zeros on the real axis that are to the right of the point should be considered.

Since each of these subtends an angle of 180°, we must have an odd number to satisfy the angle condition.

4. If the number of finite zeros m, is less than the number of finite poles, n, then $n - m$ branches of the root locus must end at zeros at infinity. The asymptotes to these branches intersect at a common point on the real axis, given by

$$\sigma_A = \frac{\text{sum of poles} - \text{sum of zeros}}{n - m} \tag{7.30}$$

and are inclined to the real axis at angles ϕ_A, given by

$$\phi_A = \frac{2q + 1}{n - m} \cdot 180^\circ \qquad q = 0, 1, 2, \ldots, n - m - 1 \tag{7.31}$$

Equations (7.30) and (7.31) can be proved by rewriting Eq. (7.1) as

$$-K = \frac{Q(s)}{P(s)} = \frac{(s - p_1)(s - p_2) \cdots (s - p_n)}{(s - z_1)(s - z_2) \cdots (s - z_m)}$$

$$\approx \frac{s^n - \sum_{i=1}^{n} p_i \cdot s^{n-1}}{s^m - \sum_{i=1}^{m} z_i \cdot s^{m-1}} \qquad \text{(for large } s\text{)} \tag{7.32}$$

or

$$K\underline{/\pi} \approx s^{n-m} - \left(\sum_{i=1}^{n} p_i - \sum_{i=1}^{m} z_i\right) s^{n-m-1} \qquad \text{(by long division)} \tag{7.33}$$

Taking the $(n - m)$th root of both sides, we get

$$K^{1/(n-m)} \underline{\Big/ \frac{(2q + 1)\pi}{n - m}} = s - \frac{\sum_{i=1}^{n} p_i - \sum_{i=1}^{m} z_i}{n - m} \tag{7.34}$$

From Eq. (7.34) it is seen that all the asymptotes intersect at σ_A, as given by Eq. (7.30). Furthermore, the angles ϕ_A that the asymptotes make with the positive real axis are given by Eq. (7.31).

5. If the root locus crosses the $j\omega$-axis for some value of K, this is readily determined through the Routh-Hurwitz criterion.

From the Routh table we can determine the value of K that causes the system to be just unstable, as well as the locations of the resulting roots on the $j\omega$-axis. This was discussed in Sec. 6.5 and will be illustrated by means of an example.

EXAMPLE 7.4

Consider a unity-feedback system with an open-loop transfer function given by

$$GH(s) = \frac{K}{s(s+4)(s+16)} \tag{7.35}$$

The characteristic polynomial for this system is given by

$$\Delta(s) = s^3 + 20s^2 + 64s + K \tag{7.36}$$

and the Routh table is shown below.

$$\begin{array}{c|cc} s^3 & 1 & 64 \\ s^2 & 20 & K \\ s^1 & \dfrac{1280 - K}{20} & \\ s^0 & K & \end{array}$$

Hence, we get a zero row if $K = 1280$. Substituting this value in the auxiliary equation obtained from the row above the zero row, we get

$$20s^2 + 1280 = 0 \tag{7.37}$$

giving roots at

$$s = \pm j8 \tag{7.38}$$

DRILL PROBLEM 7.2

Determine the $j\omega$-axis crossover points of the root locus for a unity-feedback system with an open-loop transfer function

$$G(s) = \frac{K}{s(s^2 + 6s + 16)} \tag{7.39}$$

where K is the variable parameter. What is the value of K for this crossover?

Answers: $s = \pm j4$, $K = 96$

6. The break-away and re-entry points on the root locus are determined from the roots of $dK/ds = 0$. If r branches meet at a point, they break away at an angle of $\pm 180°/r$.

In the examples of root locus that we have discussed so far, we saw instances where two branches of the root locus met at a point on the real axis, and then moved away from the real axis at right angles to it. These are called "break-away" points and are characterized by the fact that the characteristic polynomial has multiple roots at these points. The dual situation occurs when the branches of the root loci enter the real axis. These may be called "break-in" or "re-entry" points. Normally, a break-away point occurs on a segment of the root locus on the real axis connecting two poles and a re-entry point occurs between two zeros, as shown in Fig. 7.6. Later we shall see some cases where loci can enter and leave the real axis where a segment of the root locus connects a pole and a zero.

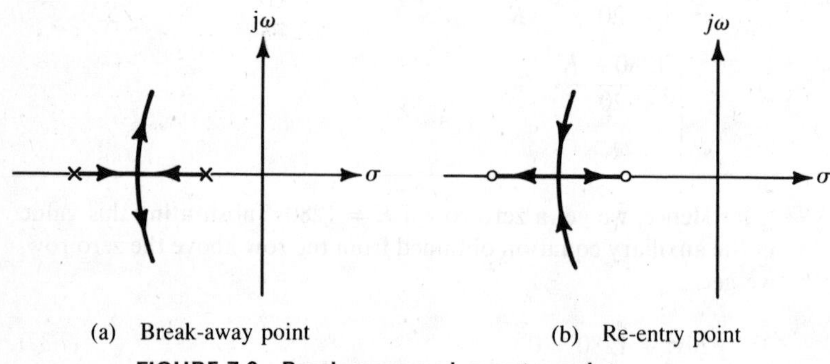

FIGURE 7.6. Break-away and re-entry points.

There may be points on the real axis where more than two branches of the root locus meet. The angles at which the branches will leave the real axis are then given by

$$\phi = \pm \frac{180°}{r} \tag{7.40}$$

where r is the multiplicity of the roots. Equation (7.40) follows from the fact that the angle condition must be satisfied at this point in all the branches of the root locus. Two such cases are shown in Fig. 7.7.

A special case arises when two or more poles occur at a point on the real axis. If the number of poles and zeros to the right of this point is even, then the angles at which the branches leave the real axis are given by Eq. (7.40); otherwise the angles will be $360°/r$, where r is the multiplicity of poles at that point.

Break-away (and re-entry) points are not confined to the real axis. If the characteristic equation has multiple complex roots, then we get complex

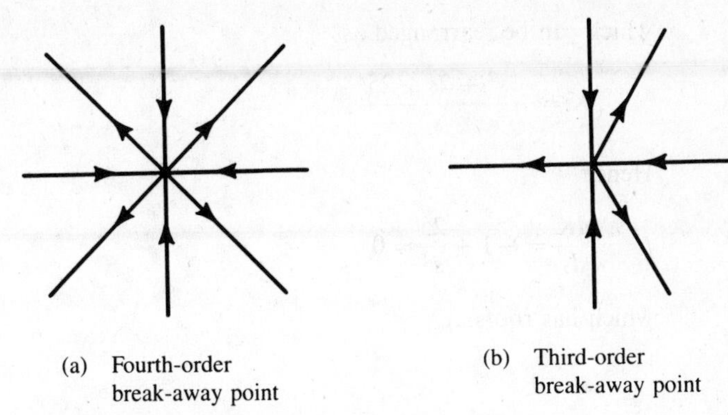

(a) Fourth-order break-away point

(b) Third-order break-away point

FIGURE 7.7. Higher-order break-away points.

break-away points. However, such points must occur in conjugate pairs in order to satisfy the property of symmetry about the real axis.

Several methods can be utilized for determining the location of break-away points. The simplest of these is based on the fact that a break-away point represents the maximum value of K for which the roots can be real. Similarly, a re-entry point represents the minimum value of K for which the roots can be real. Hence the roots of

$$\frac{dK}{ds} = 0 \tag{7.41}$$

where

$$K = -\frac{Q(s)}{P(s)} \tag{7.42}$$

obtained by rearranging Eq. (7.1), are the break-away or re-entry points, provided that they also satisfy the angle condition (i.e., that the number of poles and zeros to the right of the point under consideration is odd). The following example will illustrate this method.

EXAMPLE 7.5

Consider the problem discussed in Example 7.2. As seen in Fig. 7.3, we have a re-entry point at C. We shall now determine the exact location of C by using the method described above. From Eq. (7.9), all points on the root locus must satisfy the relationship

$$G'(s) = \frac{Ks}{s^2 + 2s + 25} = -1 \tag{7.43}$$

which can be rearranged as

$$K = -\frac{s^2 + 2s + 25}{s} \tag{7.44}$$

Hence

$$\frac{dK}{ds} = -1 + \frac{25}{s^2} = 0 \tag{7.45}$$

which has roots at

$$s = \pm 5 \tag{7.46}$$

Since the point $s = +5$ does not satisfy the angle condition, the only possible solution is

$$s = -5 \tag{7.47}$$

which is a re-entry point, as shown in Fig. 7.3.

The main difficulty with this approach is that the determination of the break-away (or re-entry) points requires obtaining the roots of dK/ds, which in some cases, may turn out to be a polynomial of high degree. Since the purpose of the root locus method was to avoid this very problem, it does not appear to be very helpful.

In such cases, a more practical way to locate the break-away (or re-entry) points is to make a table of the values of K for different real values of s using Eq. (7.42), and hence determine the values of s for which K is either a minimum or maximum. This search need only be made for the admissible regions of the real axis, and can be carried out easily with the help of a small programmable pocket calculator.

DRILL PROBLEM 7.3

Determine the break-away and re-entry point in the root locus shown in Fig. 7.5 by solving for the roots of dK/ds obtained from Eq. (7.18).

Answer: $s = -10 \pm \sqrt{80}$

---

7. The angle of departure of the root locus from a complex pole is given by

$$\phi_d = 180° - \text{sum of angles of vectors drawn to this pole from other poles} + \text{sum of angles of vectors drawn to this pole from the zeros} \tag{7.48}$$

The angle of arrival at a complex zero may be obtained in a similar manner; that is,

$$\phi_a = 180° - \text{sum of angles of vectors drawn to this zero from other zeros}$$
$$+ \text{sum of angles of vectors drawn to this zero from the poles} \quad (7.49)$$

Both of these properties follow directly from the application of the angle condition at a point on the root locus very close to the pole or zero under consideration.

The properties of the root locus are summarized in Table 7.1.

We shall now discuss several examples of sketching the root locus using these properties. For the sake of clarity, the procedure will be described in a step-by-step fashion.

EXAMPLE 7.6

Consider a unity-feedback system with forward-path transfer function

$$G(s) = \frac{K}{s(s+4)(s+5)} \quad (7.50)$$

It is desired to sketch the locus of the poles of the transfer function of the closed-loop system as K varies from 0 to infinity.

Solution

(i) The first step is to show the plot of the open-loop poles and zeros. In this case, we have no zeros, but three poles, at $s = 0$, -4, and -5. Also, we note that $n = 3$ and $m = 0$.

(ii) The second step is to mark the real axis segments of the root locus. These are between 0 and -4 and to the left of -5.

(iii) There are $n - m = 3$ branches going to infinity. The asymptotes intersect at

$$\sigma_A = \frac{0 - 4 - 5}{3} = -3$$

and the angles ϕ_A are 60°, 180°, and 300°. These are then drawn in the s-plane.

(iv) For $G(s)$ given in Eq. (7.50) to be equal to -1, we have

$$K = -s(s+4)(s+5) = -s^3 - 9s^2 - 20s$$

TABLE 7.1 Properties of the root locus

Property Number	Property
1	The root locus is symmetrical about the real axis.
2	Each branch of the root locus starts from an open-loop pole (for $K = 0$) and terminates at an open-loop zero (for $K = \infty$) or at infinity. The number of branches terminating at infinity is equal to $n - m$, for a system with n poles and m zeros.
3	Segments of the real axis are part of the root locus if and only if the total number of real poles and zeros to their right is odd,
4	The $n - m$ branches terminating at infinity are asymptotic to straight lines intersecting at the real axis at the point $$\sigma_A = \frac{\text{sum of poles} - \text{sum of zeros}}{n - m}$$ and inclined to the real axis at angles $$\phi_A = \left(\frac{2q+1}{n-m}\right)180°, \qquad q = 0, 1, 2, \ldots, n - m - 1$$
5	The points at which the root loci intersect the imaginary axis can be determined by using the Routh criterion.
6	The break-away and re-entry points on the root locus are determined from the roots of $dK/ds = 0$. If r branches meet at a point, they break away at an angle of $\pm 180°/r$.
7	The angle of departure from an open-loop (complex) pole is given by $$\phi_d = 180° - \text{sum of angles of vectors drawn from other poles to this pole} + \text{sum of angles of vectors drawn from the zeros to this pole}$$ The angle of arrival at an open-loop (complex) zero is given by $$\phi_a = 180° - \text{sum of angles of vectors drawn from other zeros to this zero} + \text{sum of angles of vectors drawn from the poles to this zero.}$$

Setting the derivative dK/ds to zero, we get

$$-3s^2 - 18s - 20 = 0$$

which has roots at $s = -1.4725$ and -4.5275. Of these two roots, only the one at $s = -1.4725$ is admissible as a break-away point since the segment between $s = -4$ and $s = -5$ is not a part of the root locus.

(v) We now determine the $j\omega$-axis crossover point using the Routh criterion. The characteristic polynomial is

$$\Delta(s) = s^3 + 9s^2 + 20s + K$$

and the Routh table is as follows:

s^3	1	20
s^2	9	K
s^1	$\dfrac{180 - K}{9}$	
s^0	K	

Hence, for $K = 180$, we get roots at $s = \pm j\sqrt{20}$.

(vi) A sketch of the complete root locus is shown in Fig. 7.8.

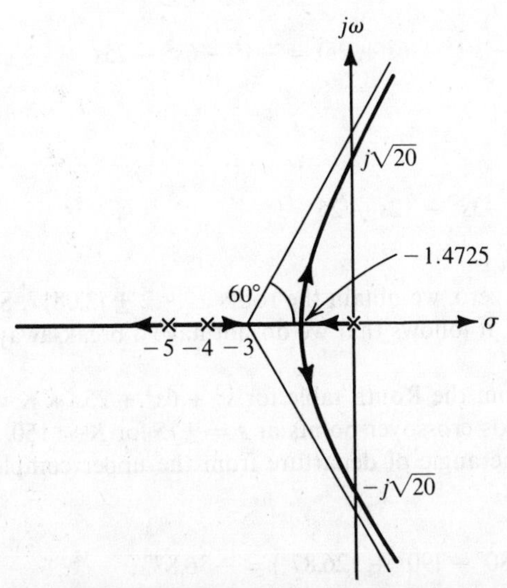

FIGURE 7.8. Root locus for Eq. (7.50).

EXAMPLE 7.7

Obtain the root locus for a unity-feedback system with the open-loop transfer function

$$G(s) = \frac{K}{s(s^2 + 6s + 25)} \tag{7.51}$$

Solution

We shall obtain the root locus for this case in the same way as in the previous example.

(i) We have no zero and three poles. Hence $m = 0$ and $n = 3$. The open-loop poles are located at $s = 0$ and $-3 \pm j4$. These are then plotted in the s-plane.

(ii) Since there is only one real pole, the entire left half of the real axis is part of the root locus.

(iii) Since $n - m = 3$, we have three branches going to infinity. The asymptotes intersect at

$$\sigma_A = \frac{0 - 3 - 3}{3} = -2$$

and the angles ϕ_A are 60°, 180°, and 300°.

(iv) From the open-loop transfer function we get

$$K = -s(s^2 + 6s + 25) = -s^3 - 6s^2 - 25s$$

and

$$\frac{dK}{ds} = -3s^2 - 12s - 25$$

Setting it to zero, we obtain the roots as $-2 \pm j2.0817$. Since these are complex, it follows that we do not have a break-away point on the real axis.

(v) From the Routh table for $s^3 + 6s^2 + 25s + K$ we get the imaginary-axis crossover points at $s = \pm j5$ for $K = 150$.

(vi) The angle of departure from the upper complex pole is calculated as

$$\phi_d = 180° - (90° + 126.87°) = -36.87°$$

(vii) A sketch of the complete root locus is shown in Fig. 7.9.

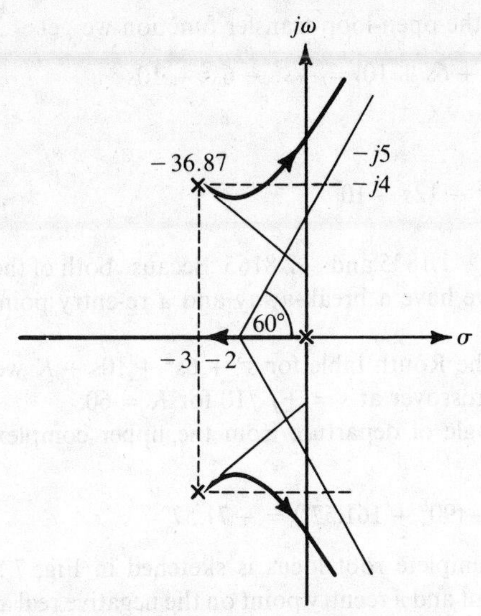

FIGURE 7.9. Root locus for Eq. (7.51).

EXAMPLE 7.8

Obtain the root locus for a unity feedback system with the open-loop transfer function

$$G(s) = \frac{K}{s(s^2 + 6s + 10)} \tag{7.52}$$

Solution

We shall see that although it is similar to Eq. (7.51), the root locus plot for $0 \leq K \leq \infty$ is quite different.

(i) As before, we have $m = 0$ and $n = 3$. Also, the open-loop poles are located at $s = 0$ and $-3 \pm j1$.

(ii) Since there is only one real pole at $s = 0$, the entire left half of the real axis is part of the root locus.

(iii) Since $n - m = 3$, we have three branches going to infinity. The asymptotes intersect at

$$\sigma_A = \frac{0 - 3 - 3}{3} = -2$$

and the angles ϕ_A are 60°, 180°, and 300°.

(iv) From the open-loop transfer function we get

$$K = -s(s^2 + 6s + 10) = -s^3 - 6s^2 - 10s$$

and

$$\frac{dK}{ds} = -3s^2 - 12s - 10$$

with roots at $s = -1.1835$ and -2.8165. Because both of these roots are admissible, we have a break-away and a re-entry point on the negative real axis.

(v) From the Routh table for $s^3 + 6s^2 + 10s + K$ we get the imaginary axis crossover at $s = \pm j\sqrt{10}$ for $K = 60$.

(vi) The angle of departure from the upper complex pole is calculated as

$$\phi_d = 180° - (90° + 161.57°) = -71.57°$$

(vii) The complete root locus is sketched in Fig. 7.10. Note a break-away point and a reentry point on the negative real axis. This follows from the fact that d^2K/ds^2 is negative for $s = -1.1835$ (making it a break-away point) and positive for $s = -2.8165$ (making it a break-in point).

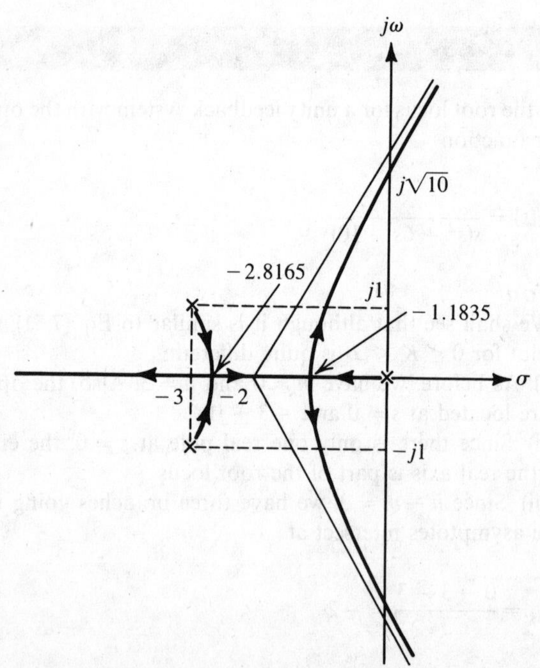

FIGURE 7.10. Root locus for Eq. (7.52).

EXAMPLE 7.9

The open-loop transfer function of a control system is given by

$$G(s) = \frac{K}{s(s+2)(s^2+6s+25)} \tag{7.53}$$

Sketch the root locus for $0 < K < \infty$.

Solution

(i) We have $m = 0$ and $n = 4$. The open-loop poles are located at $s = 0$, -2, and $-3 \pm j4$.

(ii) The segment of the real axis between $s = 0$ and $s = -2$ is part of the root locus.

(iii) Since $n - m = 4$, we have four branches of the root locus going to infinity. The asymptotes intersect at

$$\sigma_A = \frac{0-2-6}{4} = -2$$

and the angles ϕ_A are 45°, 135°, 225°, and 315°.

(iv) From the open-loop transfer function we get

$$K = -s(s+2)(s^2+6s+25) = -(s^4+8s^3+37s^2+50s)$$

and

$$\frac{dK}{ds} = -(4s^3+24s^2+74s+50)$$

To determine the break-away points we would have to solve for the roots of the cubic. To avoid doing this, we shall calculate the values of K for different values of s in the range 0 to -2, since we know that if break-away points exist, they must lie between 0 and -2. Table 7.2 is calculated by using Horner's rule, which is described in detail in Appendix D. The table shows that the maximum occurs between $s = -0.8$ and $s = -1.0$. A finer search between these two points revealed that the maximum value of K occurs at $s = -0.8981$, with $K = 20.206$.

TABLE 7.2. Determination of break-away point

s	−0.2	−0.4	−0.6	−0.8	−1.0	−1.2	−1.4	−1.6	−1.7	−1.8
K	8.58	14.57	18.28	20.01	20.0	18.47	15.59	11.49	9.02	6.28

(v) The Routh table for the characteristic polynomial

$$\Delta(s) = s^4 + 8s^3 + 37s^2 + 50s + K$$

is shown below.

s^4	1	37	K
s^3	8	50	
s^2	30.75	K	
s^1	$\dfrac{1537.5 - 8K}{30.75}$		
s^0	K		

Hence, for $K = 192.1875$, we get roots at $s = \pm j2.5$.

(vi) The angle of departure from the upper complex pole is given by

$$\phi_d = 180° - (90° + 126.87° + 104.04°)$$
$$= -140.91°$$

The root locus is shown in Fig. 7.11.

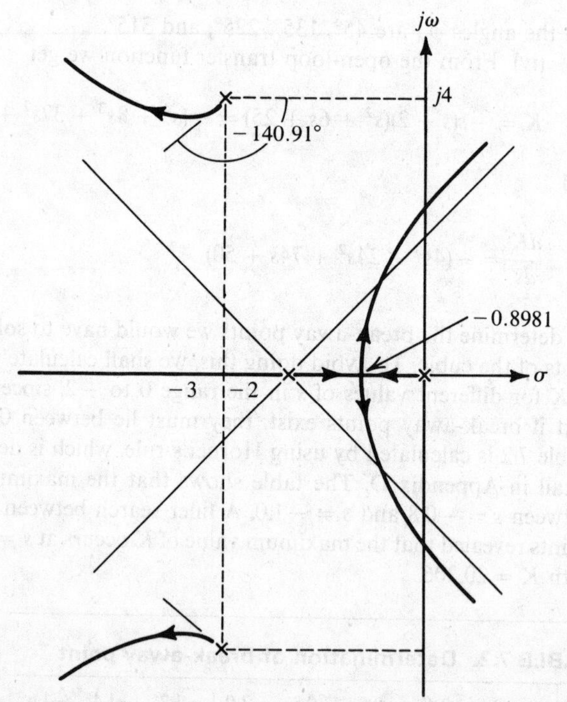

FIGURE 7.11. Root locus for Eq. (7.53).

DRILL PROBLEM 7.4

Sketch the root locus for each of the following open-loop transfer functions as K varies from zero to infinity. Be sure to label all the important points.

(a) $\dfrac{K}{s(s^2 + 6s + 12)}$ (b) $\dfrac{K}{s(s + 2)(s^2 + 8s + 20)}$ (c) $\dfrac{K(s^2 + 6s + 25)}{s(s + 1)(s + 2)}$

Answers:

(a)

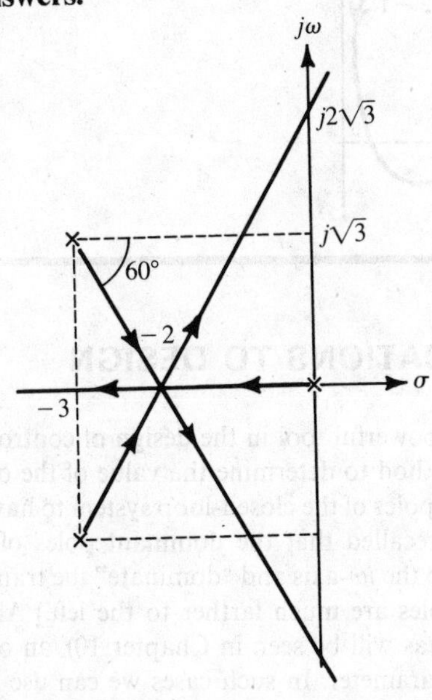

(b)

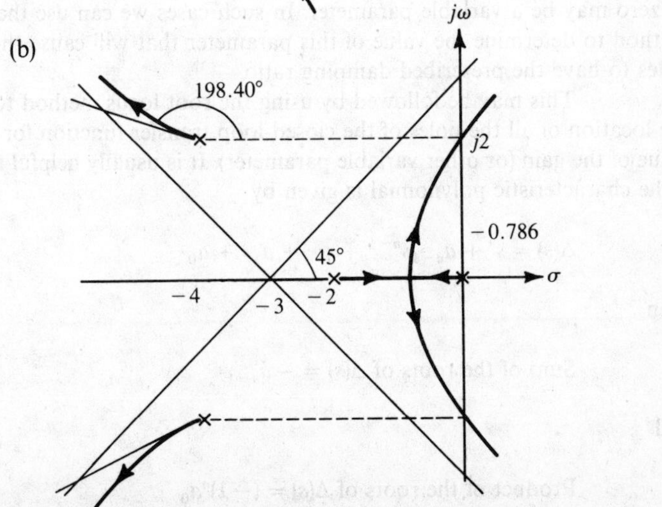

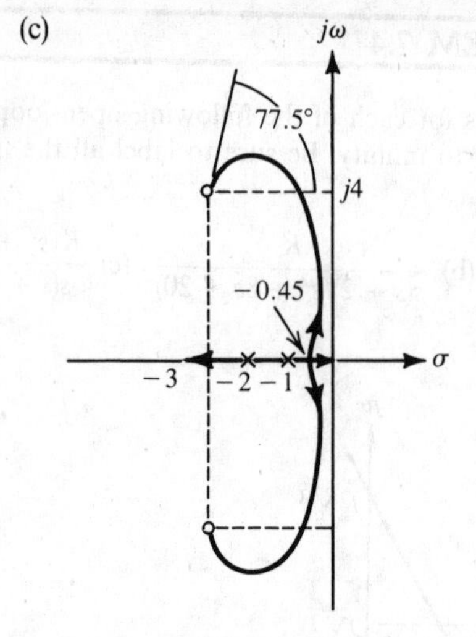

7.5 APPLICATIONS TO DESIGN

The root locus method is a powerful tool in the design of control systems. For example, we can use this method to determine the value of the open-loop gain that will cause the dominant poles of the closed-loop system to have a prescribed damping ratio. (It may be recalled that the dominant poles of a system are defined as the poles closest to the $j\omega$-axis and "dominate" the transient response of the system if all other poles are much farther to the left.) Alternatively, in the design of compensators (as will be seen in Chapter 10), an open-loop pole or zero may be a variable parameter. In such cases we can use the root locus method to determine the value of this parameter that will cause the dominant poles to have the prescribed damping ratio.

This may be followed by using the root locus method to determine the location of all the poles of the closed-loop transfer function for the desired value of the gain (or other variable parameter). It is usually helpful to note that if the characteristic polynomial is given by

$$\Delta(s) = s^n + a_{n-1}s^{n-1} + \cdots + a_1 s + a_0 \tag{7.54}$$

then

$$\text{Sum of the roots of } \Delta(s) = -a_{n-1} \tag{7.55}$$

and

$$\text{Product of the roots of } \Delta(s) = (-1)^n a_0 \tag{7.56}$$

The following examples will illustrate some applications of the root locus method to design.

EXAMPLE 7.10

Consider the third-order position-control system with velocity feedback shown in Fig. 7.12. Determine the value of the open-loop gain K such that the dominant poles of the closed-loop transfer function have a damping ratio of 0.5. What will be the response of the system to a unit step for this value of K?

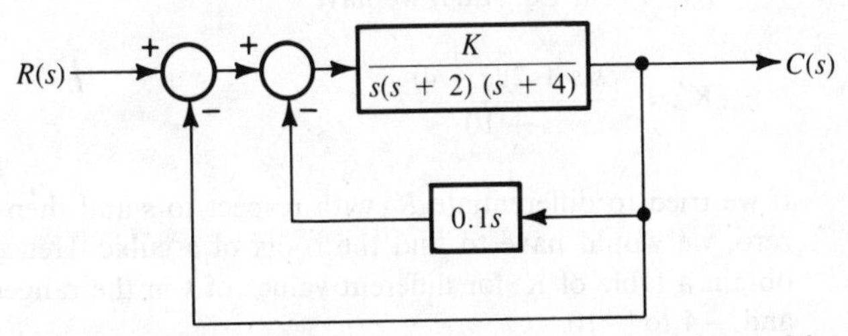

FIGURE 7.12. A third-order position-control system with velocity feedback.

Solution

Applying Mason's rule, the transfer function of the closed-loop system is obtained as

$$\frac{C(s)}{R(s)} = \frac{K}{s(s+2)(s+4) + 0.1Ks + K} \tag{7.57}$$

and the characteristic polynomial is given by

$$\Delta(s) = s^3 + 6s^2 + 8s + 0.1K(s+10) = 0 \tag{7.58}$$

which can be rearranged as

$$1 + \frac{K'(s+10)}{s(s+2)(s+4)} = 0 \tag{7.59}$$

where $K' = 0.1K$.

Hence we shall sketch the root locus for the auxiliary system

$$G'(s) = \frac{K'(s+10)}{s(s+2)(s+4)} \tag{7.60}$$

which has the same characteristic polynomial as our original system. We shall proceed as in the previous section.

(i) The open-loop poles are at $s = 0, -2$, and -4, and there is an open-loop zero at $s = -10$.

(ii) The real-axis segments of the root locus are obtained between $s = 0$ and -2 and also between $s = -4$ and -10.

(iii) Here $n = 3$ and $m = 1$. Hence, two branches of the root locus go to infinity. The asymptotes intersect at

$$\sigma_A = \frac{0 - 2 - 4 - (-10)}{2} = 2$$

and the angles ϕ_A are $\pm 90°$.

(iv) From Eq. (7.60) we have

$$K' = -\frac{s(s+2)(s+4)}{s+10}$$

If we tried to differentiate K' with respect to s and then equate to zero, we would have to find the roots of a cubic. Hence we shall obtain a table of K' for different values of s in the ranges 0 to -2 and -4 to -10.

TABLE 7.3 Determination of break-away point

s	−0.2	−0.4	−0.6	−0.8	−1.0	−1.2	−1.4	−1.6	−1.8
K'	0.1396	0.24	0.3038	0.3339	0.3333	0.3055	0.254	0.1829	0.0966

s	−4.5	−5.0	−5.5	−6.0	−6.5	−7.0	−7.5	−8.0	−8.5	−9.0	−9.5
K'	1.0227	3.0	6.4167	12.00	20.8929	35.00	57.75	96.0	167.75	315.00	783.7

From the first part of the table we locate a maximum between -0.8 and -1.0. A finer search can be used to determine the break-away point at $s = -0.8951$. The second part of the table shows K' increasing monotonically from 1.0227 to 783.7, and hence there is no break-away or break-in point in this segment.

(v) The Routh table for the characteristic polynomial is shown below.

$$\begin{array}{c|cc}
s^3 & 1 & 8 + K' \\
s^2 & 6 & 10K' \\
s^1 & \dfrac{48 - 4K'}{6} & \\
s^0 & 10K' &
\end{array}$$

Hence, the imaginary-axis crossover occurs at $s = \pm j\sqrt{20}$ for $K' = 12$. The sketch of the complete root locus is shown in Fig. 7.13.

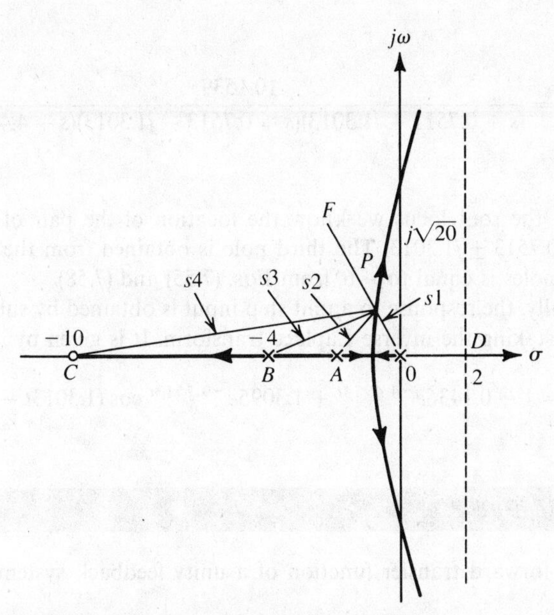

FIGURE 7.13. Root locus for Eq. (7.59).

We now draw the line OP making an angle of $60° = \cos^{-1} \zeta$ with the negative real axis, intersecting the root locus at P. It is important to note that since we have drawn the root locus only approximately, with the knowledge of the break-away point, the $j\omega$-axis crossover and the asymptotes, we might not have obtained the root precisely. This may be verified by measuring the angles of the various vectors and checking that

$$\angle POD + \angle PAD + \angle PBD - \angle PCD = 180°$$

If necessary, one should move along the line OP by a small amount to locate P such that the angle condition is satisfied. This may be done either graphically, using a ruler and a protractor, or one may use a calculator to obtain the angles and add them. Using the latter procedure, the point P was located at $s = -0.7513 + j1.3013$. Then the value of K' was obtained as

$$K' = \frac{s_1 s_2 s_3}{s_4} = 1.01539$$

where, again for convenience, a calculator was used.

Hence

$$K = 10K' = 10.1539$$

and

$$\frac{C(s)}{R(s)} = \frac{10.1539}{(s + 0.7513 + j1.3013)(s + 0.7513 - j1.3013)(s + 4.4974)} \tag{7.61}$$

Note that from the root locus we know the location of the pair of complex poles at $s = -0.7513 \pm j1.3013$. The third pole is obtained from the fact that the sum of the poles is equal to -6, from Eqs. (7.55) and (7.58).

Finally, the response to a unit step input is obtained by substituting $1/s$ for $R(s)$ and taking the inverse Laplace transform. It is given by

$$c(t) = 1 - 0.1436e^{-4.4974t} + 1.3095e^{-0.7513t} \cos(1.3013t - 3.9995) \tag{7.62}$$

EXAMPLE 7.11

The forward transfer function of a unity feedback system is given by

$$G(s) = \frac{50(s + \alpha)}{s(s + 10)(s + 30)} \tag{7.63}$$

Determine the value of α so that the closed-loop poles will have a damping ratio of 0.5. For this value of α, obtain the response to a unit step.

Solution

The closed-loop transfer function is obtained as

$$\frac{C(s)}{R(s)} = \frac{50(s + \alpha)}{s(s + 10)(s + 30) + 50(s + \alpha)} \tag{7.64}$$

Hence the characteristic polynomial is

$$\Delta(s) = s^3 + 40s^2 + 350s + 50\alpha = 0$$

which can be rearranged as

$$1 + \frac{50\alpha}{s^3 + 40s^2 + 350s} = 0 \tag{7.65}$$

Let

$$G'(s) = \frac{50\alpha}{s(s^2 + 40s + 350)} = \frac{50\alpha}{s(s + 12.93)(s + 27.07)} \tag{7.66}$$

We have $n = 3$ and $m = 0$. Hence, there will be three asymptotes with

$$\sigma_A = -\frac{40}{3} = -13.33, \qquad \phi_A = 60°, 180°, 300°$$

To obtain the break-away point, we note $K = -(s^3 + 40s^2 + 350s)$. Solving for $dK/ds = 0$ gives $s = -5.516$ or -21.15. Hence the break-away point is at -5.516. The $j\omega$-axis crossover is at $s = \pm j18.7$ (from Routh table). The root locus is shown in Fig. 7.14.

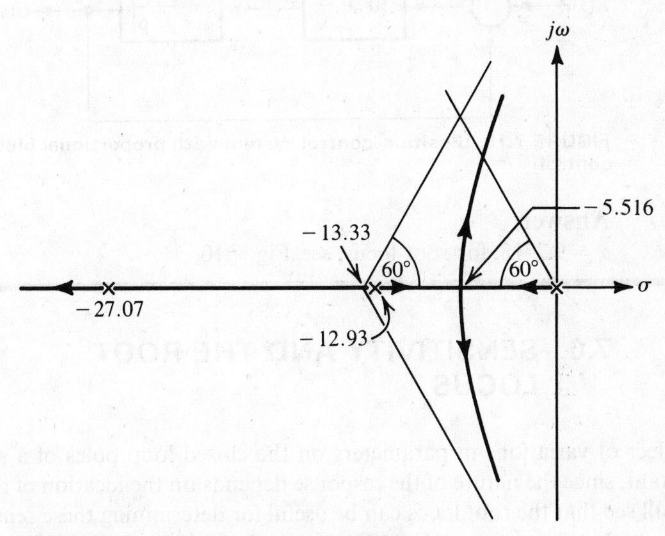

FIGURE 7.14. Root locus for Eq. (7.63).

For $\zeta = 0.5$, the root is obtained at $s = -4.375 \pm j7.578$ with $K = 2{,}392.6$. Hence, $\alpha = K/50 = 47.8521$. For the original system, using this value of α, we get

$$\frac{C(s)}{R(s)} = \frac{50(s + 47.8521)}{(s + 31.25)(s + 4.375 + j7.578)(s + 4.375 - j7.578)} \tag{7.67}$$

(Note that the third pole has been obtained by utilizing the fact that the sum of the roots is -40.)

For $R(s) = 1/s$,

$$c(t) = 1 - 0.0341e^{-31.25t} + 1.1918e^{-4.375t} \cos(7.578t - 3.767) \tag{7.68}$$

DRILL PROBLEM 7.5

The block diagram of a position-control system with integral control is shown in Fig. 7.15. Draw the root locus for the system as the parameter α varies from zero to infinity and hence determine the value of α so that the dominant poles have a damping ratio of 0.6.

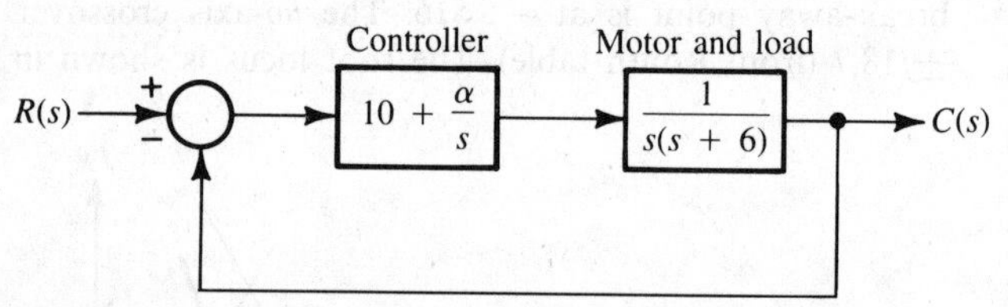

FIGURE 7.15. Position-control system with proportional plus integral control.

Answer:
$\alpha = 9.7713$; for root locus, see Fig. 7.10.

7.6 SENSITIVITY AND THE ROOT LOCUS

The effect of variations in parameters on the closed-loop poles of a system is important, since the nature of the response depends on the location of the poles. We shall see that the root locus can be useful for determining these sensitivities.

Let us rearrange Eq. (6.1) as

$$K(s - z_1)(s - z_2) \cdots (s - z_m) + (s - p_1)(s - p_2) \cdots (s - p_n) = (s - q_1)(s - q_2) \cdots (s - q_n) \tag{7.69}$$

where K is the variable parameter, z_i are the open-loop zeros, p_i are the open-loop poles, and q_i are the roots (i.e., the poles of the closed-loop system transfer function).

Define

$$T(s) = \frac{K(s - z_1)(s - z_2) \cdots (s - z_m)}{(s - q_1)(s - q_2) \cdots (s - q_n)} \tag{7.70}$$

as the closed-loop transfer function.

The sensitivities ("root sensitivities") are defined below.

$$S_K^i = \frac{\partial q_i}{\partial K/K} \triangleq \text{the sensitivity of the } i\text{th root to } K \tag{7.71}$$

$$S_{p_j}^i = \frac{\partial q_i}{\partial p_j} \triangleq \text{the sensitivity of the } i\text{th root to the open-loop pole } p_j \tag{7.72}$$

and

$$S_{z_j}^i = \frac{\partial q_i}{\partial z_j} \triangleq \text{the sensitivity of the } i\text{th root to the open-loop zero } z_j \tag{7.73}$$

The total change Δq_i in the root q_i due to small variations in the open-loop gain and the locations of the open-loop zeros and poles can be obtained as

$$\begin{aligned}\Delta q_i &= \frac{\partial q_i}{\partial K}\Delta K + \sum_{j=1}^{m}\frac{\partial q_i}{\partial z_j}\Delta z_j + \sum_{j=1}^{n}\frac{\partial q_i}{\partial p_j}\Delta p_j \\ &\approx S_k^i\frac{\Delta K}{K} + \sum_{j=1}^{m} S_{z_j}^i\,\Delta z_j + \sum_{j=1}^{n} S_{p_j}^i\,\Delta p_j\end{aligned} \tag{7.74}$$

which is valid for small changes.

It can be shown [by differentiation of Eq. (7.69) and simplification] that

$$S_k^i = [-(s-q_i)T(s)]_{s=q_i} \tag{7.75}$$

$$S_{z_j}^i = \left[\frac{s-q_i}{s-z_j}T(s)\right]_{s=q_i} = \frac{S_K^i}{z_j-q_j} \tag{7.76}$$

$$S_{p_j}^i = \left[-\frac{s-q_i}{s-p_j}T(s)\right]_{s=q_i} = \frac{S_K^i}{q_i-p_j} \tag{7.77}$$

Hence we see that the sensitivities of q_i with respect to z_j and p_j are easily determined after S_K^i is calculated. The latter is obtained from the root locus plot and can be interpreted graphically as

$$S_K^i = -\frac{K \times \text{product of distances from the zeros } z_j \text{ to } q_i}{\text{product of distances from the other roots } q_j \text{ to } q_i} \tag{7.78}$$

Since these distances are complex numbers, S_K^i will also be a complex number, denoting the direction in which the root will move if K varies by a small amount.

The following example will illustrate the procedure.

EXAMPLE 7.12

Determine the sensitivity of the complex poles to α in Example 7.11.

Solution

$$T'(s) = \frac{50\alpha}{(s-q_1)(s-q_2)(s-q_3)}$$

$$= \frac{2{,}392.6}{(s+31.25)(s+4.375+j7.578)(s+4.375-j7.578)}$$

$$= \frac{K}{(s-q_1)(s-q_2)(s-q_3)} \qquad (7.79)$$

$$S_K^{q_3} = -[(s-q_3)T'(s)]_{s=q_3} = \frac{2{,}392.6}{(-4.375+j7.578+31.25)\times j7.578\times 2}$$

$$= 5.654\underline{/74.25^\circ} \qquad (7.80)$$

$$S_\alpha^{q_3} = \alpha\frac{\partial q_3}{\partial \alpha} = \frac{K}{50}\cdot\frac{\partial q_3}{\partial K}\cdot\frac{\partial K}{\partial \alpha} = K\frac{\partial q_3}{\partial K} = S_K^{q_3} \qquad (7.81)$$

since $K = 50\alpha$.

DRILL PROBLEM 7.6

Determine the sensitivity of the dominant poles in Drill Problem 7.5 to small changes in α for the nominal value $\alpha = 9.7713$.

Answer: $S_\alpha^i = 1.15\underline{/11.21^\circ}$

7.7 SUMMARY

In this chapter we have studied the root locus method for determining the movement of the poles of the closed-loop transfer function as a parameter of the open-loop transfer function is varied. This is very important information since the transient response of the system depends on the location of these poles. The method is, therefore, very useful for the design of control systems, where one may adjust either the open-loop gain or the location of a pole or zero in a compensator in order to obtain desirable response. These design procedures will be discussed in Chapter 10.

Another important application of the root locus method is in determining the sensitivity of the poles of the closed-loop transfer function to variations in certain parameters of the open-loop transfer function. This is of great practical value, since in actual design we must allow for some tolerance in component values in order to make the product economical.

The procedure discussed in this chapter enables us to obtain a rough sketch of the root locus. In many practical situations we require a more accurate plot of the root locus. This can be obtained by searching for points in the vicinity of the rough sketch at which the angle condition is satisfied precisely. In the earlier days this was done by means of a ruler or protractor or a spirule. With the availability of modern electronic pocket calculators, it is now possible to locate these points more easily. A short program for doing this on a programmable calculator is given in Appendix D. Furthermore, computer programs are available for obtaining accurate plots of the root locus. Such a program, written in Pascal for the IBM Personal Computer is also included in Appendix D.

7.8 REFERENCES

1. R. H. Ash and G. R. Ash, "Numerical Computation of Root Loci using the Newton-Raphson Technique," *IEEE Transactions on Automatic Control*, vol. AC-13, 1968, pp. 576–582.
2. W. R. Evans, "Graphical Analysis of Control Systems," *Transactions of the AIEE*, vol. 67, 1948, pp. 547–551.
3. W. R. Evans, "Control System Synthesis by Root Locus Method," *Transactions of the AIEE*, vol. 69, 1950, pp. 66–69.
4. W. R. Evans, *Control System Dynamics*, McGraw-Hill, New York, 1954.
5. N. K. Sinha, "The Impact of the Electronic Pocket Calculator on Electrical Engineering Education," *IEEE Transactions on Education*, vol. E-20, 1977, pp. 6–9.
6. G. J. Thaler, *Design of Feedback Systems*, Dowden, Hutchinson & Ross, Inc., Stroudsburg, Pa., 1973.

7.9 PROBLEMS

1. Sketch the root locus diagram for each of the following open-loop transfer functions as K is varied from 0 to infinity. Be sure to label all the critical points.

a. $\dfrac{K}{s(s+5)(s+7)}$

b. $\dfrac{K}{s(s^2+10s+34)}$

c. $\dfrac{K}{s(s+1)(s^2+4s+20)}$

d. $\dfrac{K}{s(s+4)(s^2+4s+20)}$

e. $\dfrac{K}{s^2(s^2+6s+25)(s+10)}$

2. The block diagram of a position-control system with velocity feedback is shown in Fig. 7.16. Use the root locus method to determine the value of α so that the damping ratio of the closed-loop poles may be 0.6. What is the sensitivity of these poles to α for this case?

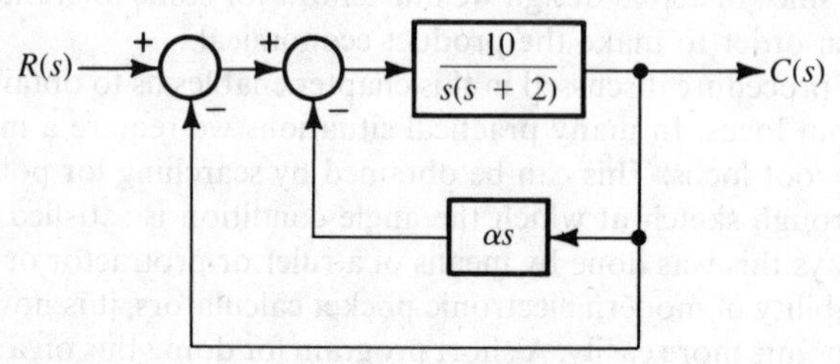

FIGURE 7.16. Block diagram for Problem 2.

3. Consider the block diagram shown in Fig. 7.17.

a. Draw the locus of the poles of the overall system as z is varied from 0 to infinity.

b. Determine the value of z so that the damping ratio of the dominant poles may be 0.4. (*Hint*: There may be two answers!)

c. Determine the response of the system to a unit step input for this value of z. What is the maximum overshoot?

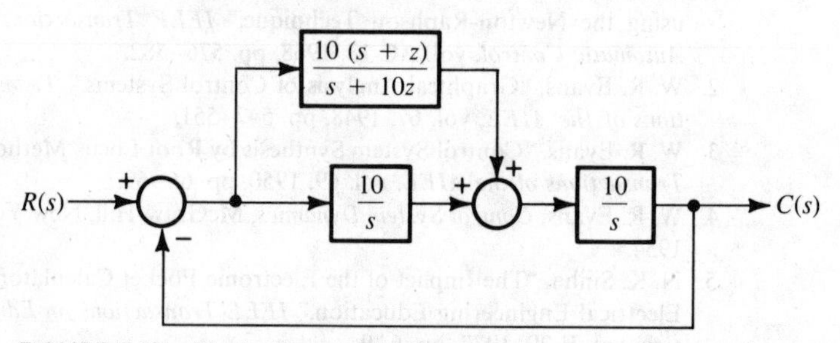

FIGURE 7.17. Block diagram for Problem 3.

4. A third-order position-control system with velocity feedback is shown in Fig. 7.18.

a. What value of α will give the maximum damping ratio for the complex poles?

b. What value of α will give a damping ratio of 0.5 for the complex poles?

c. Determine the response of the system to a unit step for α obtained in (b).

d. Determine the sensitivity of the complex poles in (b) to changes in α.

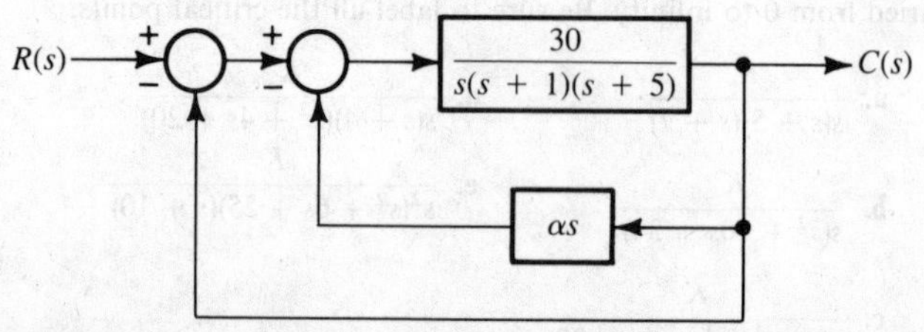

FIGURE 7.18. A position-control system with velocity feedback.

5. For the system shown in Fig. 7.19 determine the value of α so that the complex poles may have damping ratio of 0.5. What will be the steady-state error of this system to a unit ramp input?

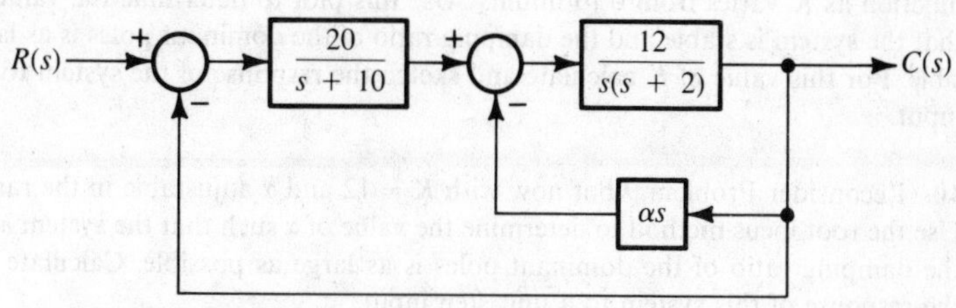

FIGURE 7.19. Block diagram for Problem 5.

6. The open-loop transfer function of a control system with unity feedback is given by

$$G(s) = \frac{25(s+\alpha)}{s^2(s+6)}$$

Sketch the root locus as α varies from zero to infinity and determine the value of α so that the damping ratio of complex poles may be 0.5. Determine the sensitivity of these poles to small changes in α.

7. The simplified block diagram of the pitch control system of a supersonic aircraft, flying at an altitude of 60,000 feet, is shown in Fig. 7.20. Sketch the root locus as K varies from zero to infinity and determine the value of K so that the damping ratio of the complex poles may be 0.707. What will be the response of this system to a unit step input?

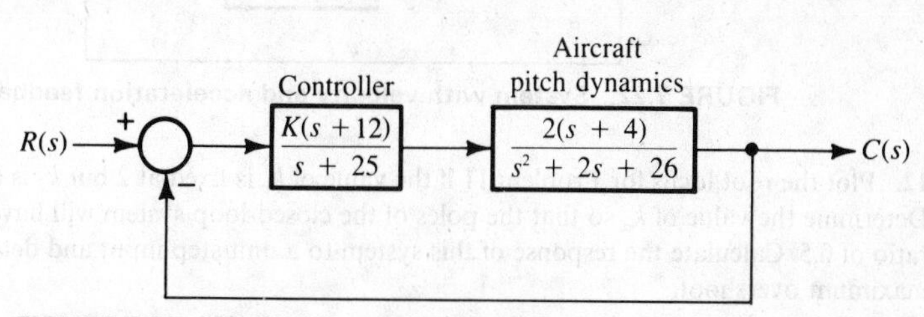

FIGURE 7.20. Block diagram for Problem 7.

8. The block diagram of a position-control system with a lead compensator is shown in Fig. 7.21. Sketch the root locus as p varies from zero to infinity, and determine the value of p so that the complex poles may have damping ratio of 0.5. What will be the response of the system to a unit step input for this value of p?

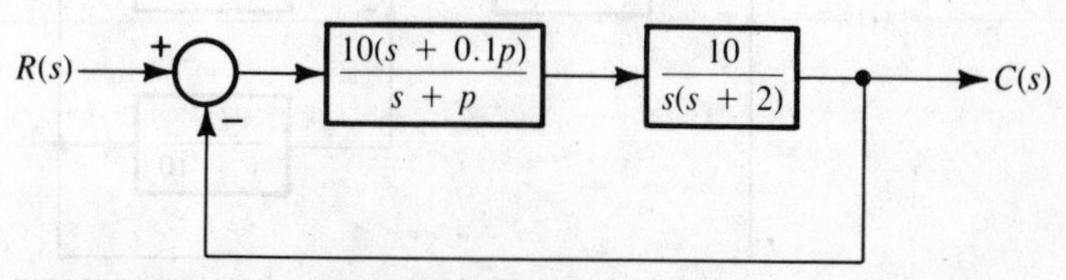

FIGURE 7.21. Block diagram for Problem 8.

9. Consider the block diagram of a type 2 system shown in Fig. 6.3 (see Chapter 6, Problem 4). Assuming that $\alpha = 1$, plot the locus of the poles of the closed-loop transfer function as K varies from 0 to infinity. Use this plot to determine the value of K such that the system is stable and the damping ratio of the dominant poles is as large as possible. For this value of K calculate and sketch the response of the system to a unit step input.

10. Reconsider Problem 9 but now with $K = 12$ and α adjustable in the range 0 to 10. Use the root locus method to determine the value of α such that the system is stable and the damping ratio of the dominant poles is as large as possible. Calculate and sketch the response of this system to a unit step input.

11. The block diagram shown in Fig. 7.22 represents a system for controlling the position of a radar-tracking antenna. It is desired to feed back both the velocity and acceleration in order to obtain a better transient response.

Assuming that the value of k_a is fixed at 1, use the root locus method to determine the value of k_v such that the damping ratio of the closed-loop poles will be 0.45. What will be the steady-state error of the system to a unit ramp input for this value of k_v? Also calculate the maximum overshoot in the response to a unit step input.

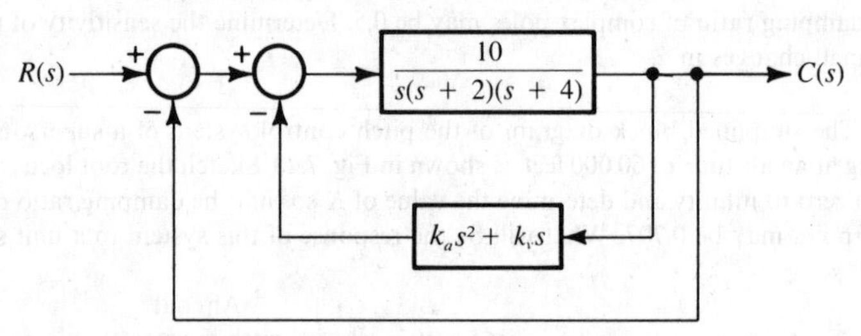

FIGURE 7.22. System with velocity and acceleration feedback.

12. Plot the root locus for Problem 11 if the value of k_v is fixed at 2 but k_a is adjustable. Determine the value of k_a so that the poles of the closed-loop system will have damping ratio of 0.5. Calculate the response of this system to a unit step input and determine the maximum overshoot.

13. The block diagram shown in Fig. 7.23 represents a more realistic position-control system with velocity feedback, including the time constant of the tachometer and also a

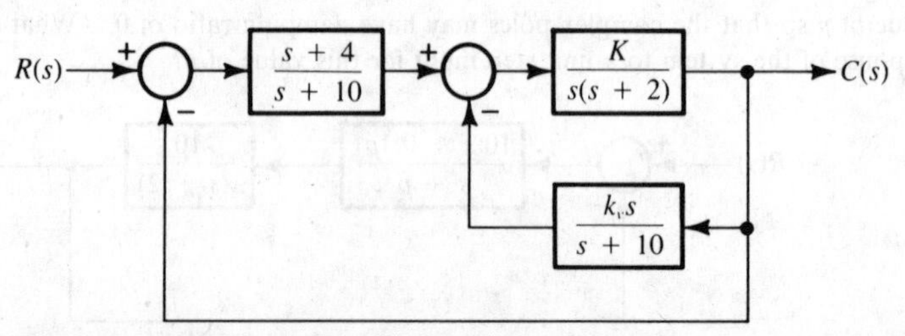

FIGURE 7.23. System with tachometer feedback.

compensator. Assuming that the forward gain K has been fixed at 10, use the root locus plot to determine the value of k_v such that the damping ratio of the dominant poles of the closed-loop system transfer function is 0.5. Determine the steady-state error of this system to a unit ramp input.

14. Plot the root locus for Problem 13 if the value of k_v is fixed at 2 but the gain K is adjustable. Determine the value of K for which the damping ratio of the dominant poles of the closed-loop system transfer function will be 0.707.

15. Determine the sensitivity of the dominant poles of the closed-loop system transfer function in Problem 14 to changes in K from the nominal value calculated.

8 Frequency Response Plots

8.1 INTRODUCTION

In the previous chapters we discussed several methods for studying the stability of a closed-loop system. The Routh-Hurwitz criterion enables us to determine the stability from the characteristic polynomial without actually determining the roots, whereas the root locus method shows how the roots of the characteristic polynomial move in the s-plane as one parameter is varied from zero to infinity. Both of these methods are very helpful in providing an insight into the transient performance of the closed-loop system. However, they are based on the assumption that the transfer function of the system is known precisely. In many practical situations this is not the case.

An alternative procedure is to utilize the frequency response of the open-loop system to determine the stability of the closed-loop system. As we shall see later, this may also provide us with information about the transient performance of the closed-loop system. The main advantage of this approach is that the frequency response of a stable open-loop system can be determined experimentally and it is not necessary to know the transfer function. We simply apply a sine-wave input to the system, and after waiting until steady-state conditions are reached, we measure the ratio of the magnitudes of the output and the input as well as the phase difference between them. These measurements are easily carried out using digital gain-phase meters currently available in the market, although it is possible to utilize more sophisticated (and expensive) transfer-function analyzers. The measurements are repeated at different frequencies until the entire frequency range of interest is covered.

Another advantage of frequency response analysis is that it can also be applied to systems that do not have rational transfer functions. Systems with pure delay belong to this category because for these the transfer function is the ratio of two polynomials multiplied by the factor $e^{-\beta s}$, where β is the transport lag (in seconds) in the system. For such cases it is not possible to utilize the Routh-Hurwitz criterion. The root locus method can be applied only after the irrational factor $e^{-\beta s}$ is replaced by an approximate rational function.

In this chapter we shall first consider the relationship between the transfer function and the frequency response of a linear system. This will be followed by the study of the various types of frequency response plots that are used by control engineers. Finally we shall discuss the estimation of the transfer function of a linear system from its frequency response.

8.2 TRANSFER FUNCTION AND FREQUENCY RESPONSE

If we apply a sine-wave input to a stable linear system the steady-state output will also be a sine wave of the same frequency. This output will, in general, differ from the input in both magnitude and phase. As shown in Appendix A, the frequency response can be calculated by replacing s in the transfer function $G(s)$ by $j\omega$; that is

$$G(j\omega) = G(s)\Big|_{s=j\omega} = M(\omega)e^{j\phi(\omega)} = M\underline{/\phi} \tag{8.1}$$

In Eq. (8.1), M is the ratio of the amplitudes of the output and input sinusoids and is called the magnitude ratio or gain. Also, ϕ is the phase shift between the output and the input, and is the angle by which the output leads the input. Both M and ϕ are functions of the angular frequency ω.

The function $G(j\omega)$ is called the frequency response of the system and can be given the same graphical interpretation as for the calculation of residues. Thus if

$$G(s) = \frac{K(s - z_1)(s - z_2)\cdots(s - z_m)}{(s - p_1)(s - p_2)\cdots(s - p_n)} \tag{8.2}$$

then for $s = j\omega_1$ we get

$$G(j\omega_1) = \frac{K(j\omega_1 - z_1)(j\omega_1 - z_2)\cdots(j\omega_1 - z_m)}{(j\omega_1 - p_1)(j\omega_1 - p_2)\cdots(j\omega_1 - p_n)}$$

$$= K \cdot \frac{\text{product of directed distances from each zero to } s = j\omega_1}{\text{product of directed distances from each pole to } s = j\omega_1} \tag{8.3}$$

If each of the directed distances in Eq. (8.3) is expressed in the polar form, then we can separate the moduli and the arguments to obtain

$$M = K \cdot \frac{\text{product of lengths of vectors from the zeros}}{\text{product of lengths of vectors from the poles}} \tag{8.4}$$

and

$$\phi = \text{sum of arguments of vectors from the zeros}$$
$$- \text{sum of arguments of vectors from the poles} \tag{8.5}$$

where it is understood that all of these vectors (or directed distances) are drawn to the point $s = j\omega_1$ on the imaginary axis.

EXAMPLE 8.1

Consider a linear system described by the transfer function

$$G(s) = \frac{10(s+1)}{(s^2 + 4s + 13)} = \frac{10(s+1)}{(s+2+j3)(s+2-j3)} \tag{8.6}$$

The pole-zero plot of the transfer function is shown in Fig. 8.1, which also illustrates the calculation of $G(j2)$.

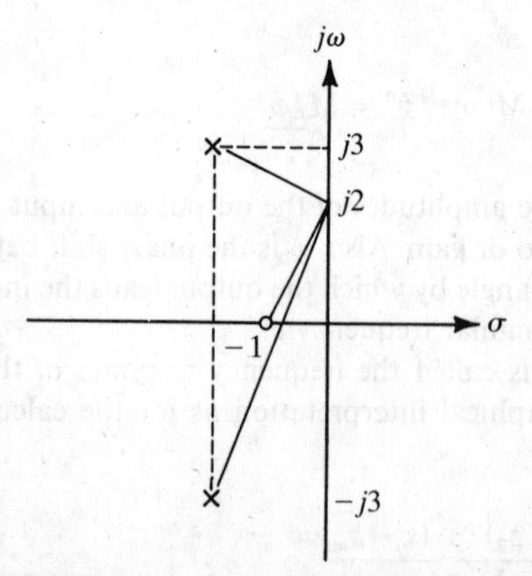

FIGURE 8.1. Frequency response calculation for the transfer function given by Eq. (8.6).

Hence

$$G(j2) = \frac{10(j2+1)}{(j2+2+j3)(j2+2-j3)}$$

$$= \frac{10 \times \sqrt{5}\underline{/63.4^\circ}}{\sqrt{29}\underline{/68.2^\circ} \times \sqrt{5}\underline{/-26.6^\circ}}$$

$$= 1.857\underline{/21.8^\circ} \tag{8.7}$$

and thus $M(2) = 1.857$ and $\phi(2) = 21.8°$.

This calculation can be repeated for different values of ω. Table 8.1 shows the values of M and ϕ for different values of ω for this example.

TABLE 8.1. Frequency response for the transfer function given by Eq. (8.6)

ω	M	ϕ	ω	M	ϕ
0	0.769	0°	5	2.186	−42.3°
1	1.118	26.6°	6	1.830	−53.2°
2	1.857	21.8°	7	1.550	−60.3°
3	2.5	0°	8	1.339	−65.0°
3.5	2.596	−12.9°	10	1.050	−71.0°
4	2.533	−24.7°			

Since we have two dependent variables, M and ϕ, we need two separate plots to indicate their variation with ω. These are shown in Fig. 8.2.

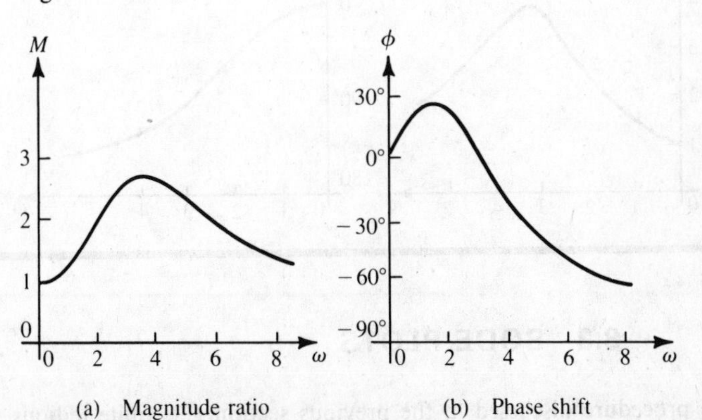

(a) Magnitude ratio (b) Phase shift

FIGURE 8.2. Frequency response curves for the transfer function given by Eq. (8.6).

Equation (8.4) can be given a nice pictorial interpretation. Imagine a flexible rubber sheet suspended over the complex frequency plane. At each pole location, the sheet is poked up by a thin rod of infinite height, and at the location of each zero the sheet is tacked down to the plane. The height of the rubber sheet above any point in the s-plane will then represent the magnitude of $G(s)$ for that value of s. The magnitude ratio curve is then represented by the cross section of the sheet along the imaginary axis of the s-plane.

DRILL PROBLEM 8.1

Sketch the frequency response curves for the following transfer functions.

(a) $G(s) = \dfrac{4(s+1)}{s+4}$ (b) $G(s) = \dfrac{10}{s^2 + 2s + 10}$

Answers:

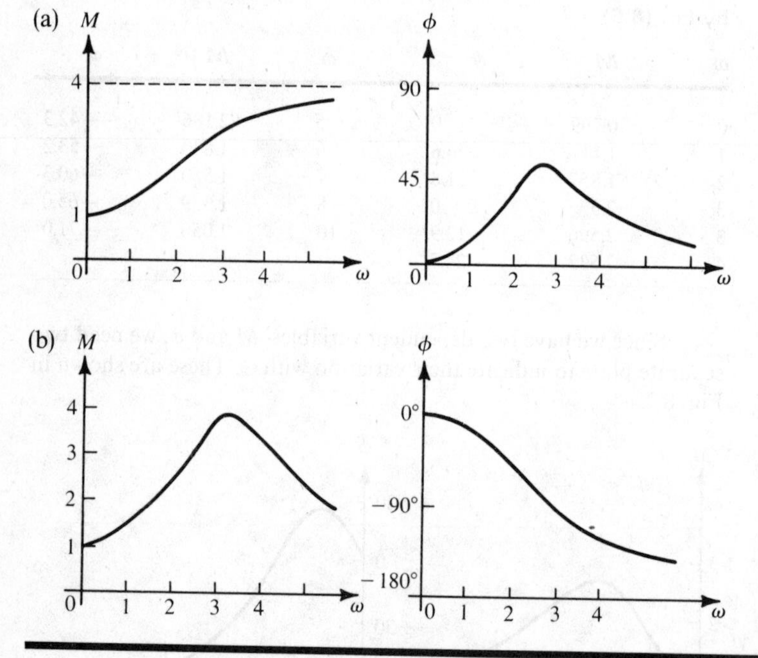

8.3 BODE PLOTS

The procedure discussed in the previous section gets quite tedious when the transfer function has several poles and zeros. The drudgery in computation is considerably reduced by using logarithmic coordinates. The resulting plots are called Bode plots, honoring H. W. Bode, who used them in the study of feedback amplifiers. These plots require semilog graph paper, where the logarithmic scale is used for the ω axis. Two graphs are required, one for the gain in decibels, defined as 20 log M plotted against frequency on the log scale, and the other for the phase shift in degrees plotted against frequency on the log scale. The simplification in Bode plots results partly due to the basic advantage of logarithmic representation that multiplication and division are replaced by addition and subtraction, respectively. There are several other advantages, as we shall see later.

8.3.1 Logarithmic Scales

The logarithmic scale, used for frequency in the Bode plots, has some interesting properties. First we observe that the logarithmic scale is nonlinear; that is, the distance between 1 and 2 is greater than the distance between 2 and 3, and so on. As a result, use of this scale enables us to cover a greater range of frequencies. Semilog graph paper comes in one, two, three, or four cycles, indicating the range of coverage. For example, a two-cycle graph paper has the range from 1 to 10 and 10 to 100. It is interesting to note that on the log scale, the distance between 1 and 10 is equal to the distance between 10 to 100. This distance is called a decade. In fact, the distance between k and $10k$, where k is any positive (nonzero) number is equal to the decade. This follows because

$$\log 10k - \log k = \log 10 = 1 \tag{8.8}$$

Similarly, the distance between k and $2k$ is equal to the constant log 2, and is called an octave. It may be noted that we cannot locate the point $\omega = 0$ on the log scale since $\log 0 = -\infty$.

8.3.2 Magnitude Plot: Straight-Line Approximation

We shall now discuss how the magnitude of the frequency response function can be sketched with very little effort by using logarithmic coordinates. In general, the transfer function can be factored into four types of terms:

1. Constants
2. Poles or zeros at the origin of the s-plane
3. Poles or zeros on the real axis of the s-plane
4. Complex conjugate pairs of poles or zeros

Consider first the transfer function, $G(s)$, given in the factored form shown below.

$$G(s) = \frac{K(s+a)}{s(s+b)(s^2 + 2\zeta\omega_n s + \omega_n^2)} \tag{8.9}$$

where $\zeta < 1$.

This transfer function has one real zero, two real poles (including the origin), and a pair of complex conjugate poles. The first step will be to rearrange it by making the constant term of each factor equal to one. Hence

$$G(s) = \frac{Ka}{b\omega_n^2} \frac{1 + s/a}{s\left(1 + \frac{s}{b}\right)\left(1 + 2\zeta\frac{s}{\omega_n} + \frac{s^2}{\omega_n^2}\right)} \tag{8.10}$$

If we now replace s by $j\omega$ and take the magnitude of both sides, we have

$$M = |G(j\omega)| = \frac{Ka}{b\omega_n^2} \frac{|1 + j\omega/a|}{|j\omega| \cdot \left|1 + j\frac{\omega}{b}\right| \cdot \left|1 + 2\zeta j\frac{\omega}{\omega_n} + \left(\frac{j\omega}{\omega_n}\right)^2\right|} \tag{8.11}$$

If we now take the logarithm of both sides and multiply by 20, we get

$$\begin{aligned} 20 \log M &= 20 \log \left(\frac{Ka}{b\omega_n^2}\right) + 20 \log \left|1 + \frac{j\omega}{a}\right| - 20 \log |j\omega| \\ &\quad - 20 \log \left|1 + \frac{j\omega}{b}\right| - 20 \log \left|1 + 2\zeta\frac{j\omega}{\omega_n} + \left(\frac{j\omega}{\omega_n}\right)^2\right| \\ &= 20 \log M_1 + 20 \log M_2 - 20 \log M_3 \\ &\quad - 20 \log M_4 - 20 \log M_5 \end{aligned} \tag{8.12}$$

The left-hand side of Eq. (8.12), the gain in decibels, is seen to be the algebraic sum of the gain resulting from each factor of the transfer function shown in Eq. (8.10). The first term on the right-hand side of Eq. (8.12) is a constant and is easily evaluated. Let us now examine the second term on the right-hand side.

In particular, we shall study the asymptotic behavior of the second term for both small and large ω. First consider the case when

$$\frac{\omega}{a} \ll 1 \tag{8.13}$$

Hence

$$M_2 = \left|1 + j\frac{\omega}{a}\right| = \sqrt{1 + \left(\frac{\omega}{a}\right)^2} \approx 1 \tag{8.14}$$

and

$$20 \log M_2 = 20 \log \left|1 + j\frac{\omega}{a}\right| \approx 20 \log 1 \approx 0 \tag{8.15}$$

Similarly, for

$$\frac{\omega}{a} \gg 1 \tag{8.16}$$

we have

$$M_2 = \left|1 + j\frac{\omega}{a}\right| \approx \frac{\omega}{a} \tag{8.17}$$

and

$$20 \log M_2 \approx 20 \log \frac{\omega}{a} \tag{8.18}$$

For the intermediate case, when $\omega/a = 1$, we have

$$\left|1 + j\frac{\omega}{a}\right| = |1 + j1| = \sqrt{2}$$

and

$$20 \log M_2 = 20 \log \sqrt{2} \approx 3 \text{ dB} \tag{8.19}$$

A plot of the asymptotes is shown in Fig. 8.3, where the exact curve is shown dotted.

It is seen that the asymptotes consist of a pair of straight lines, intersecting at $\omega = a$, which is called the break or corner frequency. For $\omega < a$, the asymptote is a straight line of zero slope and zero height, represented by Eq. (8.15). For $\omega > a$, the asymptote is a straight line of slope 20 dB/decade

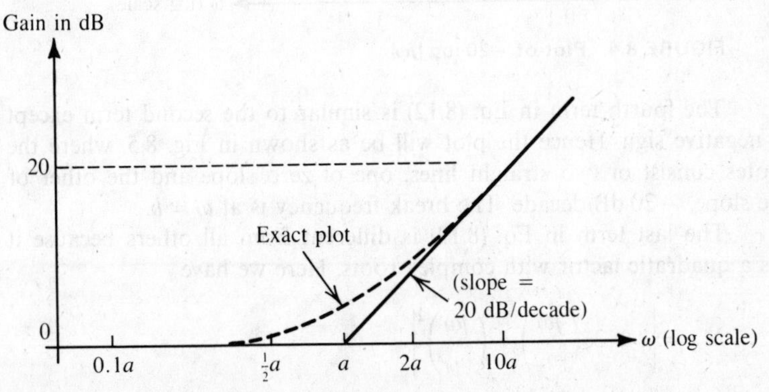

FIGURE 8.3. Plot of $20 \log \left|1 + \frac{j\omega}{a}\right|$.

(or 6 dB/octave), as given by Eq. (8.18). The maximum difference between the exact curve and the asymptotic approximation occurs at the break frequency, and is equal to 3 dB. If we move one octave away from the break frequency the difference between the exact curve and the asymptotes is approximately 1 dB, and one decade away from the break frequency the error is less than 0.05 dB. Hence, for most applications, the straight-line approximations are often of sufficient accuracy. However, if a better approximation is desired, one may introduce a correction of 3 dB at the break frequency, and corrections of 1 dB at points one octave above and below the break frequency.

Let us now consider the third term of Eq. (8.12). Here,

$$M_3 = |j\omega| = \omega \tag{8.20}$$

and

$$-20 \log M_3 = -20 \log \omega \tag{8.21}$$

The resulting plot is shown in Fig. 8.4. This is simply a straight line of slope -20 dB/decade, crossing the zero-decibel line at $\omega = 1$. Note that this is exact.

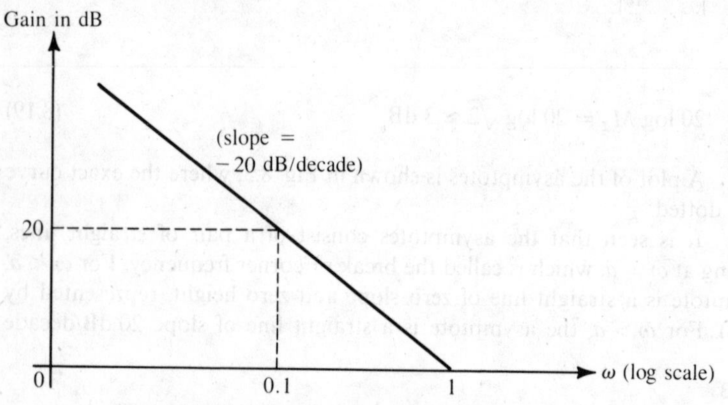

FIGURE 8.4. Plot of $-20 \log |j\omega|$.

The fourth term in Eq. (8.12) is similar to the second term except for the negative sign. Hence the plot will be as shown in Fig. 8.5, where the asymptotes consist of two straight lines, one of zero slope and the other of negative slope, -20 dB/decade. The break frequency is at $\omega = b$.

The last term in Eq. (8.12) is different from all others because it involves a quadratic factor with complex roots. Here we have

$$\begin{aligned} M_5 &= \left|1 + 2\zeta\left(\frac{j\omega}{\omega_n}\right) + \left(\frac{j\omega}{\omega_n}\right)^2\right| \\ &= \left|\left(1 - \frac{\omega^2}{\omega_n{}^2}\right) + j2\zeta\frac{\omega}{\omega_n}\right| \end{aligned} \tag{8.22}$$

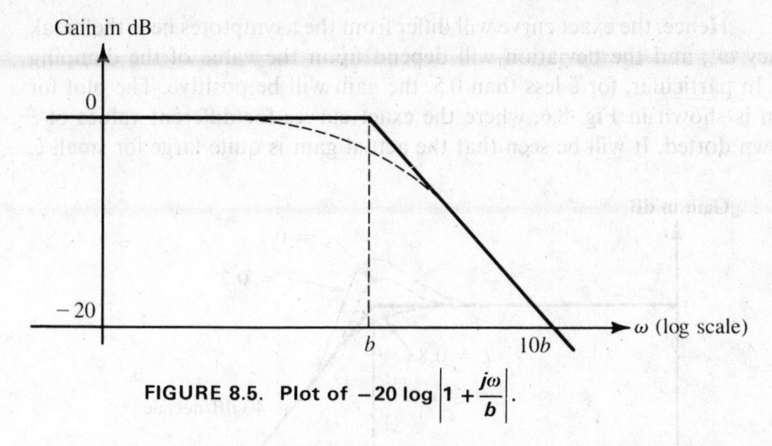

FIGURE 8.5. Plot of $-20 \log \left|1 + \frac{j\omega}{b}\right|$.

We shall again study the asymptotic behavior for the two cases $\omega/\omega_n \ll 1$ and $\omega/\omega_n \gg 1$. For the former we get

$$M_5 \approx 1 \tag{8.23}$$

and

$$-20 \log M_5 \approx 0 \tag{8.24}$$

which is a straight line of zero slope, coinciding with the zero-decibel line. For $\omega/\omega_n \gg 1$, we have

$$M_5 \approx \frac{\omega^2}{\omega_n^{\ 2}} \tag{8.25}$$

and

$$-20 \log M_5 = -20 \log \left(\frac{\omega}{\omega_n}\right)^2 \tag{8.26}$$

which is a straight line of slope -40 dB/decade, crossing the zero-decibel line at $\omega = \omega_n$.

Hence, it is seen that the asymptotes consist of a pair of straight lines, intersecting at $\omega = \omega_n$, the undamped natural frequency. For $\omega < \omega_n$, the asymptote is the zero-decibel line of slope zero, whereas for $\omega > \omega_n$, the asymptote is a line of slope -40 dB/decade. At the intermediate point, $\omega = \omega_n$, we get

$$M_5 = 2\zeta \tag{8.27}$$

and

$$-20 \log M_5 = -20 \log 2\zeta \tag{8.28}$$

Hence, the exact curve will differ from the asymptotes near the break frequency ω_n, and the deviation will depend upon the value of the damping ratio ζ. In particular, for ζ less than 0.5, the gain will be positive. The plot for the gain is shown in Fig. 8.6, where the exact curves for different values of ζ are shown dotted. It will be seen that the actual gain is quite large for small ζ.

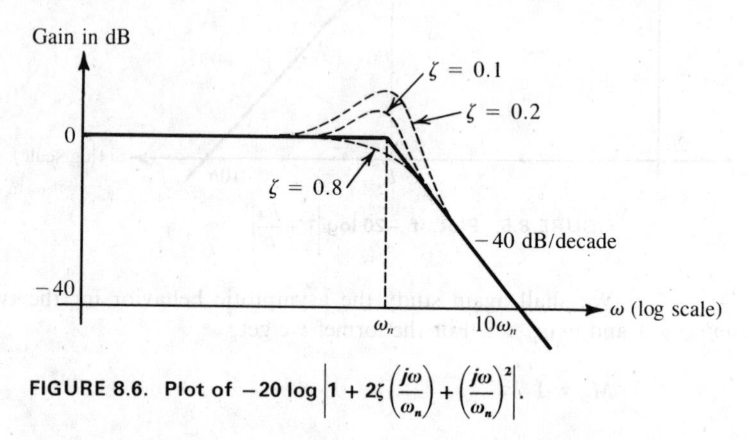

FIGURE 8.6. Plot of $-20 \log \left| 1 + 2\zeta \left(\frac{j\omega}{\omega_n} \right) + \left(\frac{j\omega}{\omega_n} \right)^2 \right|$

By equating $dM_5/d\omega$ to zero, it can be shown that the exact curve has a peak value for

$$\omega_{\max} = \omega_n \sqrt{1 - 2\zeta^2} \tag{8.29}$$

This indicates that a maximum occurs only if $\zeta < 1/\sqrt{2}$. If this condition is satisfied, then the maximum value of M_5 is obtained by substituting for in Eq. (8.22),

$$M_5(\omega_{\max}) = 2\zeta\sqrt{1 - \zeta^2} \qquad \text{for } \zeta \leq \sqrt{\frac{1}{2}} \tag{8.30}$$

Hence

$$-20 \log M_5 = 20 \log \frac{1}{2\zeta\sqrt{1 - \zeta^2}} \tag{8.31}$$

is the peak gain in decibels.

The overall gain is now obtained as the algebraic sum of the gains for the different factors. For the asymptotic plot, this summation can be carried out very easily if we realize that any contribution to the gain is made only for the frequencies above the break frequency for the factor. Hence, if we start with the plot for ω less than the lowest break frequency, then all we have to do is to account for the change in slope at the various break frequencies. For the transfer function in Eq. (8.10), the overall gain is shown in Fig. 8.7, where it is assumed that $a < b < \omega_n$.

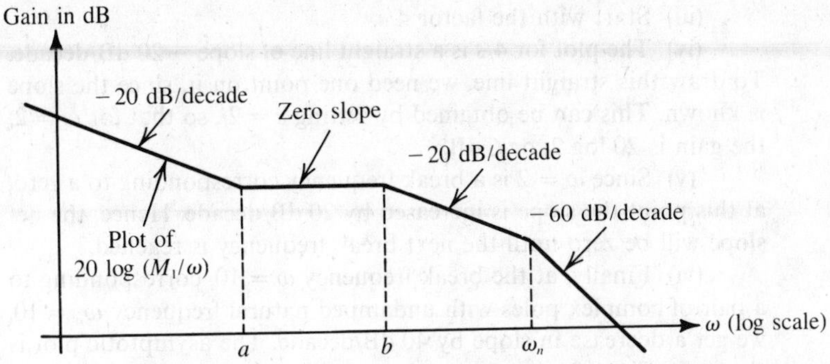

FIGURE 8.7. Asymptotic plot of gain for the transfer function given in Eq. (8.10).

For $\omega < a$, only the first and the third terms on the right-hand side of Eq. (8.12) are significant, and they give us a straight line of slope −20 dB/decade, corresponding to 20 log M_1/ω, where $M_1 = Ka/b\omega_n^2$.

As the frequency is increased, we first get the effect of the zero at $s = -a$ when ω reaches the value a. The linear factor causes the slope to be increased by 20 dB/decade, resulting in a straight line of zero slope. At the next break frequency, $\omega = b$, the effect of the pole is to reduce the slope by −20 dB/decade. Finally at the break frequency ω_n, the effect of the complex poles is to further reduce the slope by −40 dB/decade.

EXAMPLE 8.2

To get a clear idea of the procedure we shall go over each step for the transfer function

$$G(s) = \frac{200(s+2)}{s(s^2+10s+100)} \tag{8.32}$$

Solution

(i) First, rewrite it as

$$G(s) = \frac{4(1+s/2)}{s[1+s/10+(s/10)^2]} \tag{8.33}$$

(ii) Order the poles and zeros in sequence as frequency increases. These are:

(a) a pole at the origin of multiplicity one,
(b) a zero at $s = -2$, corresponding to a break frequency at $\omega = 2$,
(c) a pair of complex poles with undamped natural frequency $\omega_n = 10$, and damping ratio 0.5.

(iii) Start with the factor $4/s$.

(iv) The plot for $4/s$ is a straight line of slope -20 dB/decade. To draw this straight line, we need one point on it, since the slope is known. This can be obtained by setting $s = 2j$, so that for $\omega = 2$, the gain is 20 log 2, or 6 dB.

(v) Since $\omega = 2$ is a break frequency corresponding to a zero, at this point the slope is increased by 20 dB/decade. Hence, the net slope will be zero until the next break frequency is reached.

(vi) Finally, at the break frequency $\omega = 10$, corresponding to a pair of complex poles with undamped natural frequency $\omega_n = 10$, we get a decrease in slope by 40 dB/decade. The asymptotic plot is shown in Fig. 8.8.

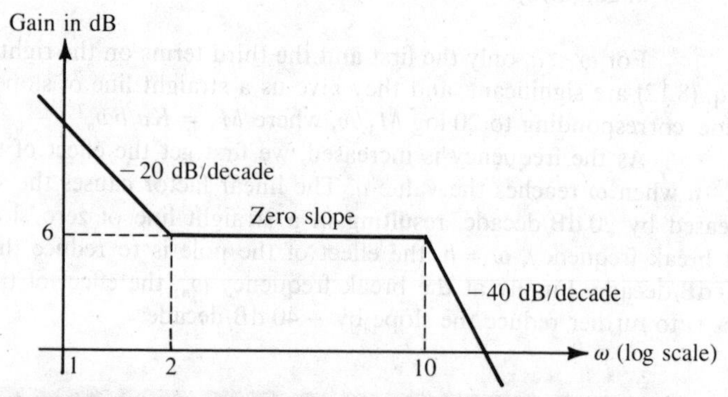

FIGURE 8.8. Asymptotic plot for the gain for Eq. (8.32).

DRILL PROBLEM 8.2

Sketch the asymptotic gain plots for the following transfer functions.

(a) $\dfrac{20s}{(s+1)(s+10)}$ (b) $\dfrac{300(s^2+2s+4)}{s(s+10)(s+20)}$

Answers:

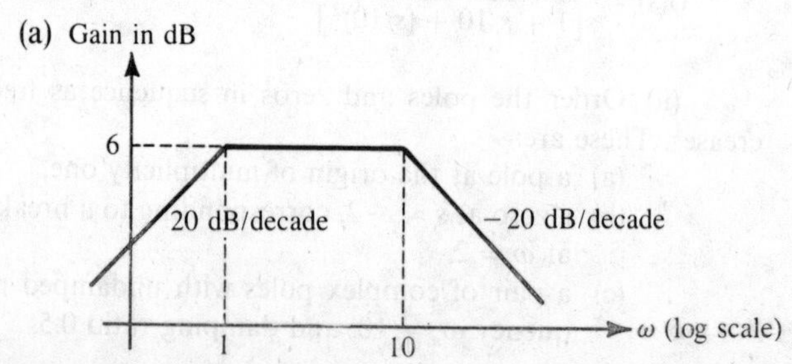

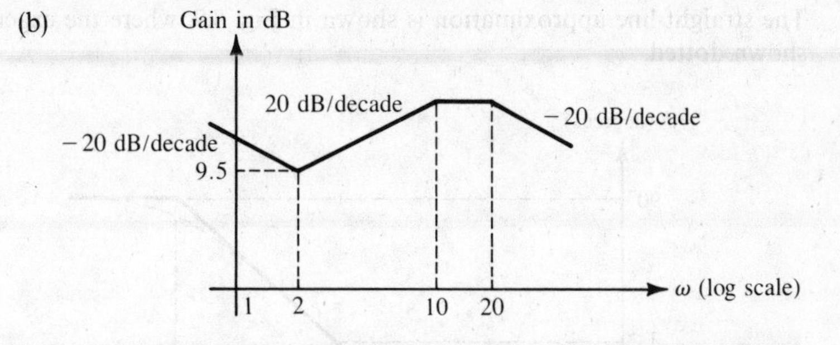

8.3.3 Phase Plot

The phase shift at any frequency can be obtained as the algebraic sum of the phase shift due to different factors in the transfer function, in a manner similar to that used for obtaining magnitude plots. To illustrate the procedure in detail, let us again consider the transfer function given in Eq. (8.9) and rearranged in Eq. (8.10). The argument of $G(s)$ for $s = j\omega$ is given by

$$\begin{aligned}\phi &= \arg G(j\omega)\\ &= \arg\left(\frac{Ka}{b{\omega_n}^2}\right) + \arg\left(1 + j\frac{\omega}{a}\right) - \arg\,(j\omega) - \arg\left(1 + j\frac{\omega}{b}\right)\\ &\quad - \arg\left(1 - \frac{\omega^2}{{\omega_n}^2} + j2\zeta\frac{\omega}{\omega_n}\right)\\ &= \phi_1 + \phi_2 - \phi_3 - \phi_4 - \phi_5\end{aligned} \tag{8.34}$$

The first term on the right-hand side, the phase-shift due to a real constant, will be identically zero. Hence we shall now examine the second term, given by

$$\phi_2 = \arg\left(1 + j\frac{\omega}{a}\right) = \tan^{-1}\frac{\omega}{a} \tag{8.35}$$

Asymptotic approximations can again be obtained by considering two cases: $\omega/a \ll 1$ and $\omega/a \gg 1$. For the former, we have

$$\phi_2 \approx 0 \tag{8.36}$$

Similarly, for $\omega/a \gg 1$, we may approximate

$$\phi_2 \approx 90° \tag{8.37}$$

For the intermediate case, $\omega/a = 1$, we get

$$\phi_2 = \tan^{-1} 1 = 45° \tag{8.38}$$

The straight-line approximation is shown in Fig. 8.9, where the exact curve is shown dotted.

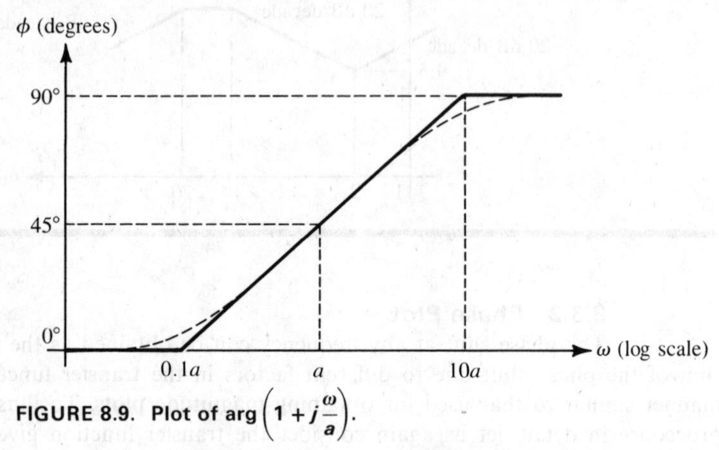

FIGURE 8.9. Plot of $\arg\left(1 + j\dfrac{\omega}{a}\right)$.

The asymptotic plot consists of three straight-line segments. For ω up to one-tenth of the break frequency, the phase shift is approximated as 0°, and for ω above ten times the break frequency, the phase shift is approximated as 90°. From one-tenth of the break frequency to ten times the break frequency, there is a slope of 45° per decade. The maximum error in the approximation is equal to $\tan^{-1} 0.1$, or about 5.7°, and occurs at the two corners, $0.1a$ and $10a$. Note that there is no error at the break frequency, $\omega = a$.

The third term on the right-hand side is the phase shift due to a purely imaginary expression, and is identically equal to 90°.

The fourth term in Eq. (8.34) is, again, similar to the second term, except for the negative sign. Hence, it can be approximated by phase shift of 0°, up to $\omega = 0.1b$, a straight line of slope $-45°$/decade from $0.1b$ to $10b$, and phase shift equal to $-90°$ for ω greater than $10b$.

We shall now look at the fifth term of Eq. (8.34), given by

$$\phi_5 = \tan^{-1} \frac{2\zeta \dfrac{\omega}{\omega_n}}{1 - \left(\dfrac{\omega}{\omega_n}\right)^2} \tag{8.39}$$

For $\omega/\omega_n \ll 1$, we have

$$\phi_5 \approx 0 \tag{8.40}$$

and for $\omega/\omega_n \gg 10$, we get

$$\phi_5 \approx 180° \tag{8.41}$$

At the intermediate point, $\omega/\omega_n = 1$, we have

$$\phi_5 = 90° \tag{8.42}$$

The plot of the asymptotic approximation is shown in Fig. 8.10, where the exact curves for different values of the damping ratio ζ are also shown. The straight-line approximation has a slope of 90° per decade between $0.1\omega_n$ and $10\omega_n$. It will be seen that for small values of ζ, the actual phase shift changes very rapidly with ω near the undamped natural frequency ω_n.

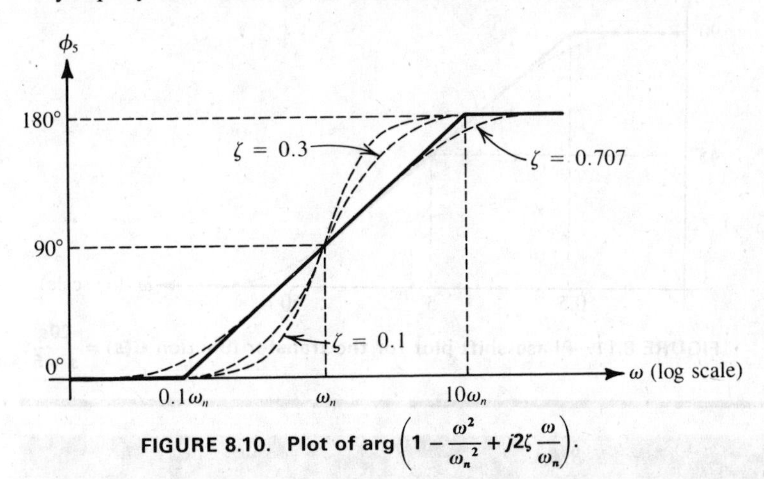

FIGURE 8.10. Plot of $\arg\left(1 - \frac{\omega^2}{\omega_n^2} + j2\zeta\frac{\omega}{\omega_n}\right)$.

The overall phase shift is now obtained as the algebraic sum of the phase shift due to each factor. The addition, however, is not as simple as in the case of the gain plot due to the fact that the straight-line approximation consists of three parts over two decades centered around the break frequency. The only exception arises when the various poles and zeros are separated by more than two decades. In this case, the phase-shift plots for the various factors can be superimposed on each other to obtain the overall phase shift.

For the more general case, it is usually advisable to make a table of phase shift against frequency for each factor and then obtain the total phase shift as the algebraic sum. Better accuracy can be obtained by using the exact values in the table instead of those obtained through the asymptotic approximation.

EXAMPLE 8.3

Consider the transfer function

$$G(s) = \frac{20s}{s+5} = \frac{4s}{1+s/5} \tag{8.43}$$

In this case, we have one zero at the origin, and one real pole at $s = -5$. The two break frequencies are separated by more than two decades. Hence, we can obtain the overall plot by adding the straight-line approximation of the phase shift due to the real pole to the 90° phase lead caused by the zero at the origin. The resulting phase-shift plot is shown in Fig. 8.11, where the exact curve is shown by the dashed line.

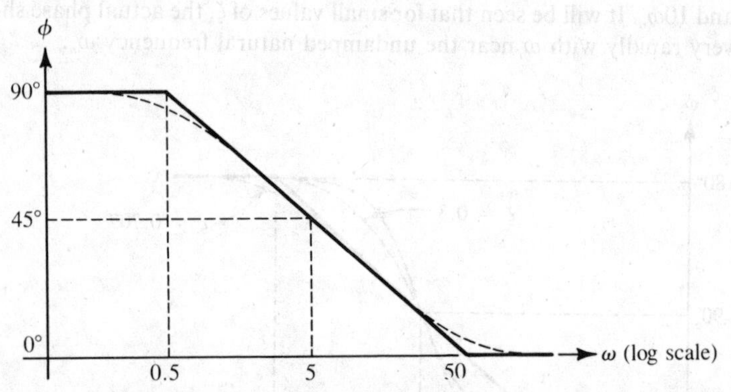

FIGURE 8.11. Phase-shift plot for the transfer function $G(s) = \dfrac{20s}{s+5}$.

EXAMPLE 8.4

Consider the transfer function given by Eq. (8.32), for which the gain curve was plotted in Fig. 8.8. We shall now plot the phase-shift curve.

TABLE 8.2. Phase shift due to different factors in the transfer function given by Eq. (8.32)

ω	*Pole at the 0 Origin*	*Zero at* $s = -2$	*Complex Conjugate Poles, with* $\omega_n = 10, \zeta = 0.5$	*Total Phase Shift*
0.1	−90°	2.86°	−0.57°	−87.71°
0.2	−90°	5.71°	−1.15°	−85.44°
0.5	−90°	14.04°	−2.87°	−78.83°
1	−90°	26.57°	−5.77°	−69.20°
2	−90°	45°	−11.77°	−56.77°
5	−90°	68.20°	−33.69°	−55.49°
10	−90°	78.69°	−90°	−101.31°
20	−90°	84.29°	−146.31°	−152.02°
50	−90°	87.71°	−168.23°	−170.52°
100	−90°	88.85°	−174.23°	−175.68°

This transfer function has a pole at the origin, a zero at $s = -2$, and a pair of complex conjugate poles with undamped natural frequency $\omega_n = 10$ and damping ratio $\zeta = 0.5$. The phase shift due to each factor for different values of ω is given in Table 8.2 and plotted in Fig. 8.12.

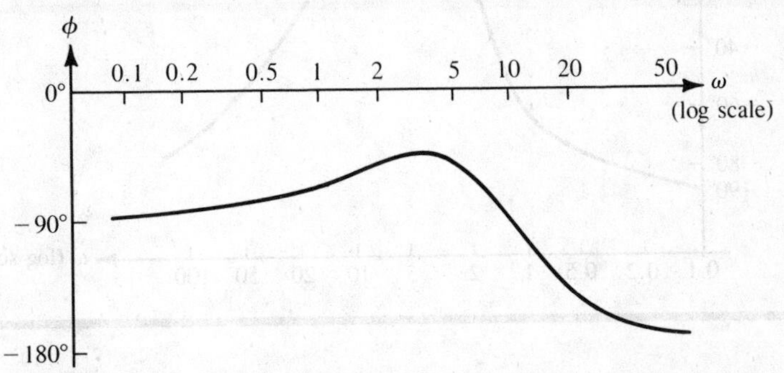

FIGURE 8.12. Phase-shift plot for the transfer function given by Eq. (8.32).

DRILL PROBLEM 8.3

Sketch the phase-shift curves for the following transfer functions.

(a) $\dfrac{20s}{(s+1)(s+10)}$

(b) $\dfrac{300(s^2+2s+4)}{s(s+10)(s+20)}$

Answers:

(a)

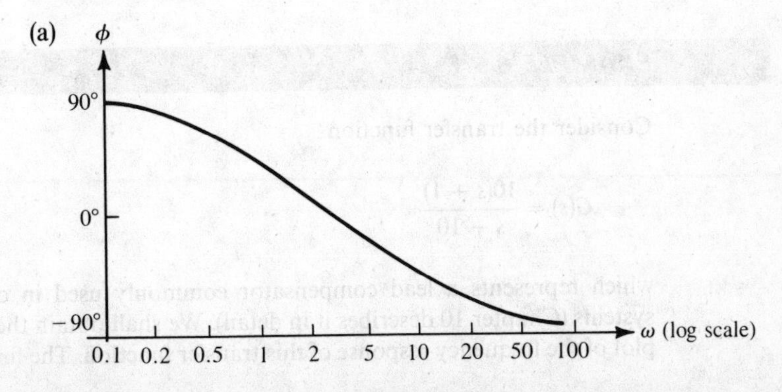

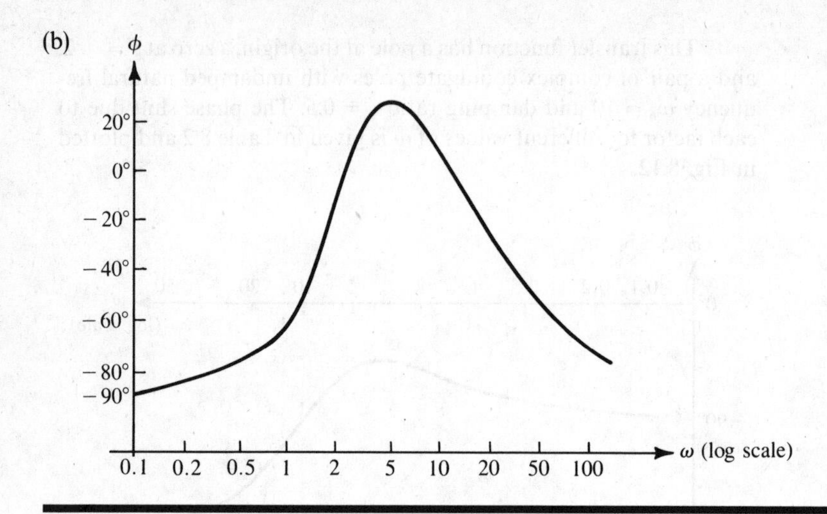

8.4 POLAR PLOTS

One disadvantage of Bode plots is that we have two separate curves showing the variation of the gain and the phase shift with frequency. The information contained in these two plots can be combined in one curve, called the polar plot. This can be regarded as representing the frequency response $Me^{j\phi}$ for a given ω as the tip of a vector of length M and making an angle ϕ with the positive real axis. As ω is varied, the tip of this vector moves in the G-plane. The value of ω at selected points can be labeled on the plot to show the variation with frequency. Mathematically, the polar plot can be considered as the mapping of the positive ω-axis of the s-plane into a curve in the G-plane.

This type of plot is very useful in determining the stability of a closed-loop system from its open-loop frequency response, as will be seen in Chapter 9 when we study the Nyquist criterion of stability.

EXAMPLE 8.5

Consider the transfer function

$$G(s) = \frac{10(s+1)}{s+10} \tag{8.44}$$

which represents a lead compensator commonly used in control systems (Chapter 10 describes it in detail). We shall obtain the polar plot of the frequency response of this transfer function. The first step is to calculate the magnitude ratio M and the phase shift ϕ for different values of ω. These are given in Table 8.3.

TABLE 8.3. Frequency response of the transfer function given by Eq. (8.44)

ω	M	ϕ	ω	M	ϕ
0	1	0°	5	4.56	52.13°
0.1	1.00	5.14°	10	7.1	39.3°
0.2	1.02	10.16°	20	8.96	23.7°
0.5	1.12	23.7°	50	9.81	10.16°
1	1.41	39.29°	100	9.95	5.14°
2	2.19	52.13°	200	9.99	2.58°
3.16	3.16	54.9°	1000	10.00	0.52°

The corresponding polar plot is shown in Fig. 8.13, where ω is shown as a parameter along the curve. This curve is a semicircle due to the bilinear nature of the mapping represented by Eq. (8.44).

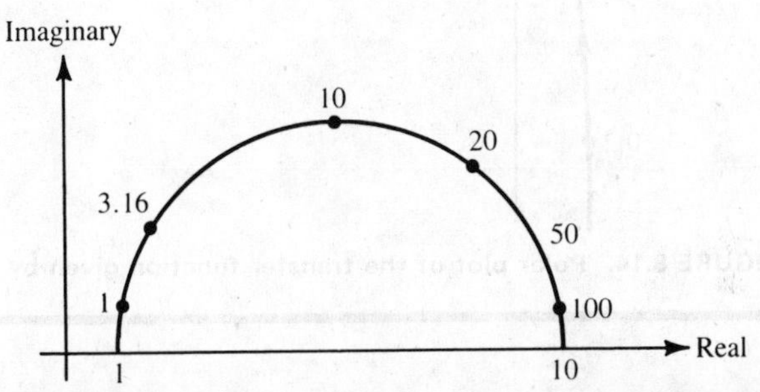

FIGURE 8.13. Polar plot of the transfer function given by Eq. (8.44).

EXAMPLE 8.6

Consider the transfer function

$$G(s) = \frac{10}{s(s + 1)} \tag{8.45}$$

which represents the forward path of a position control system. The frequency response of this transfer function is given in Table 8.4 and its polar plot is shown in Fig. 8.14.

TABLE 8.4. Frequency response of the transfer function given by Eq. (8.45)

ω	M	ϕ	ω	M	ϕ
0	∞	$-90°$	5.0	0.39	$-168.69°$
0.1	99.5	$-95.71°$	7	0.202	$-171.87°$
0.2	49.03	$-101.3°$	10	0.100	$-174.29°$
0.5	17.89	$-116.57°$	20	0.025	$-177.14°$
1.0	7.07	$-135°$	50	0.004	$-178.85°$
2.0	2.24	$-153.43°$	70	0.002	$-179.18°$

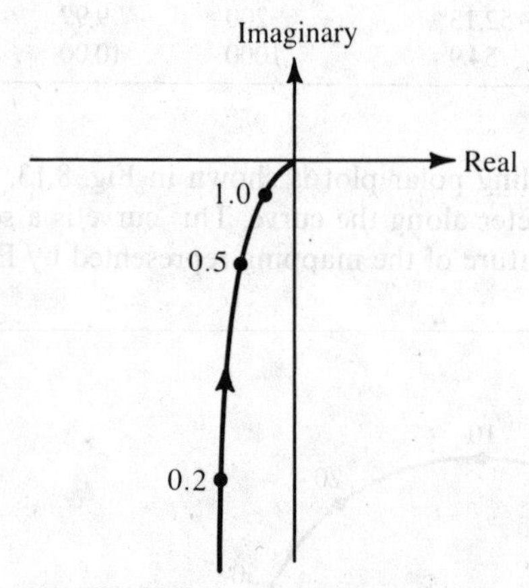

FIGURE 8.14. Polar plot of the transfer function given by Eq. (8.45).

DRILL PROBLEM 8.4

Sketch the polar plots of the frequency response of each of the following transfer functions.

(a) $\dfrac{20s}{(s+1)(s+10)}$

(b) $\dfrac{10}{s(s+1)(s+4)}$

(c) $\dfrac{300(s^2+2s+4)}{s(s+10)(s+20)}$

Answers:

(a)

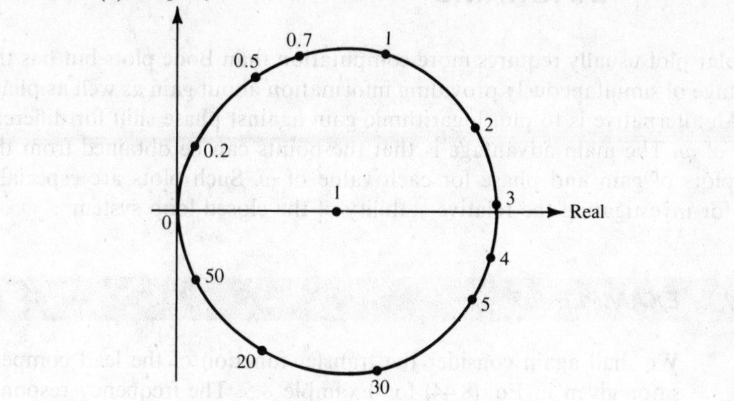
Imaginary
0.7
1
0.5
2
0.2
1
3
Real
0
4
50
5
20
30

(b)

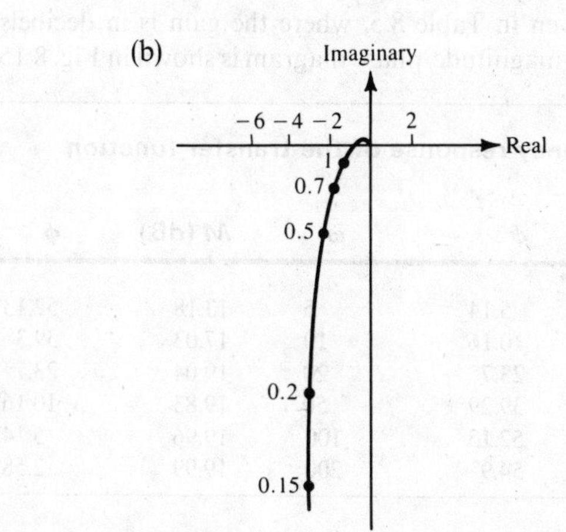
Imaginary
−6 −4 −2
2
Real
1
0.7
0.5
0.2
0.15

(c)

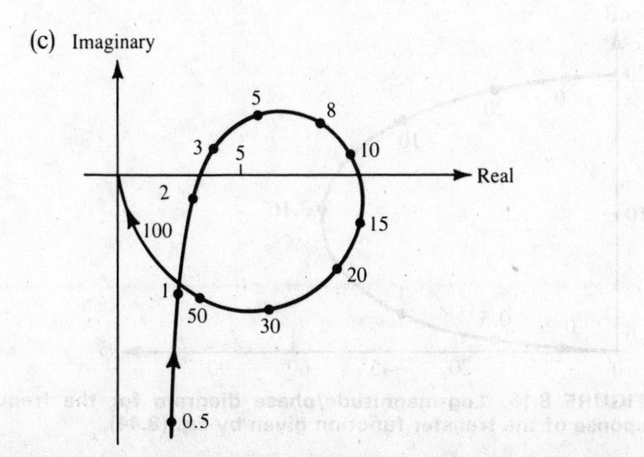
Imaginary
5
8
3
5
10
Real
2
100
15
20
1
50
30
0.5

8.5 LOG-MAGNITUDE AND PHASE DIAGRAMS

The polar plot usually requires more computation than Bode plots but has the advantage of simultaneously providing information about gain as well as phase shift. An alternative is to plot logarithmic gain against phase shift for different values of ω. The main advantage is that the points can be obtained from the Bode plots of gain and phase for each value of ω. Such plots are especially useful for investigating the relative stability of the closed-loop system.

EXAMPLE 8.7

We shall again consider the transfer function of the lead compensator given in Eq. (8.44) for Example 8.5. The frequency response for this system is given in Table 8.5, where the gain is in decibels, and the resulting log-magnitude/phase diagram is shown in Fig. 8.15.

TABLE 8.5. Frequency response of the transfer function given by Eq. (8.44)

ω	M (dB)	ϕ	ω	M (dB)	ϕ
0.1	0.0	5.14°	5	13.18	52.13°
0.2	0.17	10.16°	10	17.03	39.3°
0.5	0.96	23.7°	20	19.04	23.7°
1	2.79	39.29°	50	19.83	10.16°
2	6.82	52.13°	100	19.96	5.14°
3.16	9.99	54.9°	200	19.99	2.58°

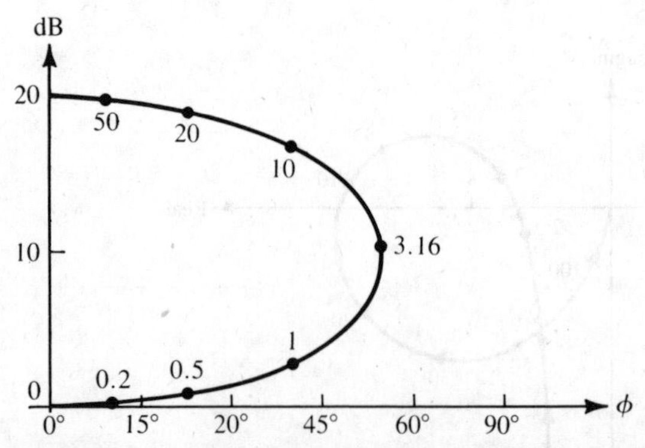

FIGURE 8.15. Log-magnitude/phase diagram for the frequency response of the transfer function given by Eq. (8.44).

EXAMPLE 8.8

We shall now consider the transfer function of the position control system given in Eq. (8.45) and discussed in Example 8.6. The frequency response for this system is given in Table 8.6, with the gain in decibels. The resulting log-magnitude/phase diagram is shown in Fig. 8.16.

TABLE 8.6. Frequency response of the transfer function given by Eq. (8.45)

ω	M (dB)	ϕ	ω	M (dB)	ϕ
0.1	39.97	−95.71°	5.0	−8.13	−168.69°
0.2	33.8	−101.3°	7.0	−13.89	−171.87°
0.5	25.05	−116.57°	10.0	−20.04	−174.29°
1.0	16.99	−135°	20.0	−32.05	177.14°
2.0	6.99	−153.43°	50.0	−47.96	178.85°

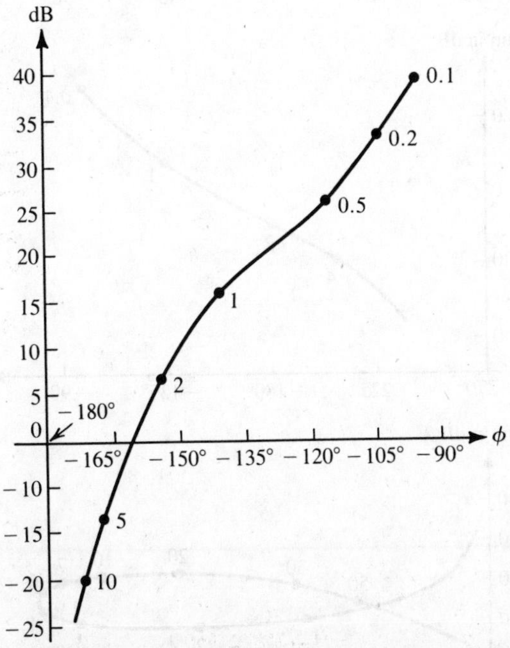

FIGURE 8.16. Log-magnitude/phase diagram for the frequency response of the transfer function given by Eq. (8.45).

DRILL PROBLEM 8.5

Plot the log-magnitude/phase diagrams for the following transfer functions.

(a) $\dfrac{20s}{(s+1)(s+10)}$ (b) $\dfrac{10}{s(s+1)(s+4)}$ (c) $\dfrac{300(s^2+2s+4)}{s(s+10)(s+20)}$

Answers:

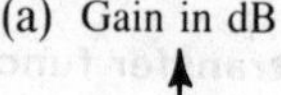

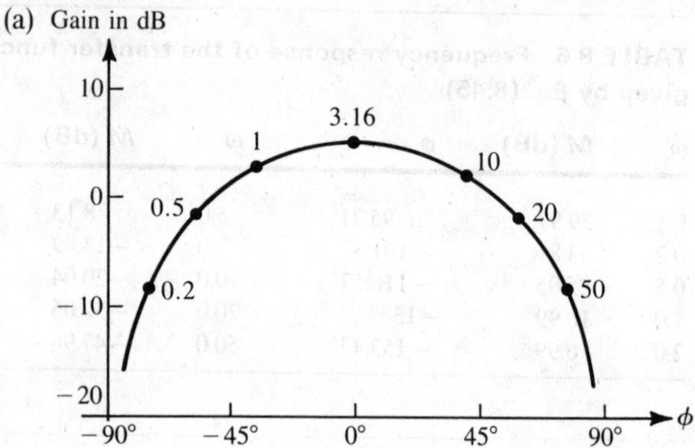

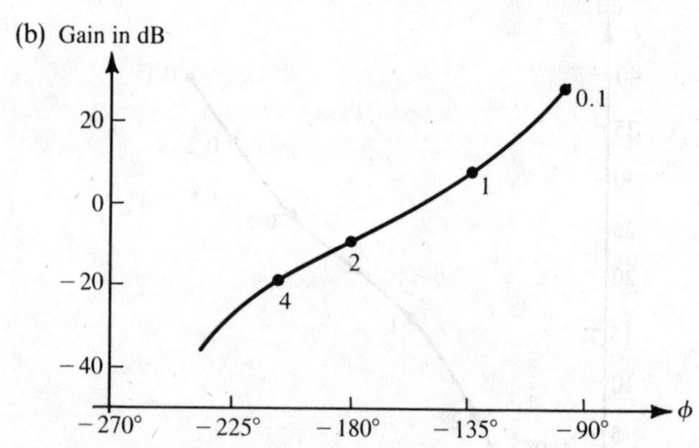

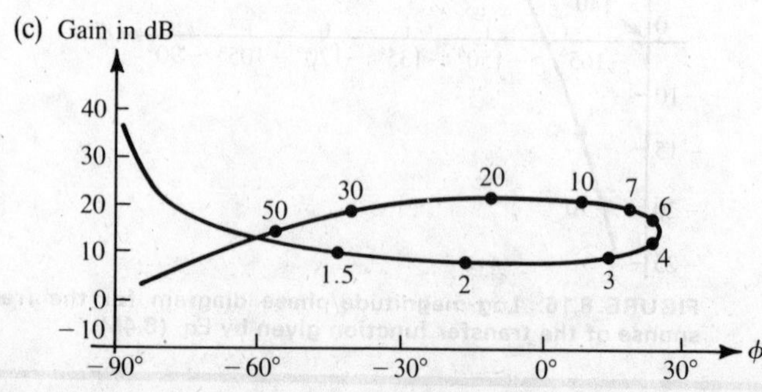

8.6 SYSTEMS WITH TRANSPORT LAG

The systems that we have considered so far have transfer functions in the form of the ratio of two polynomials of the complex frequency variable s. In many systems, pure time delays occur. Typical examples are systems with hydraulic, pneumatic, or mechanical systems. An electrical transmission line is another example. In these systems there is a time delay between the application of the input and its effect on the output. This is often called "transport lag" or "pure delay." The input and the output in such cases are related through the transfer function e^{-Ts}, where T represents the time delay in seconds.

The frequency response of this transfer function is obtained easily by replacing s by $j\omega$. As a result, we have

$$e^{-j\omega T} = Me^{j\phi} \tag{8.46}$$

where

$$M = 1 \tag{8.47}$$

$$\phi = -\omega T \tag{8.48}$$

Hence transport lag does not affect the magnitude ratio, but it introduces a phase lag that is proportional to the frequency. It should be noted that Eq. (8.48) gives the phase lag in radians (and not degrees) if ω is in radians per second and T is in seconds.

EXAMPLE 8.9

Consider the transfer function

$$G(s) = \frac{10e^{-0.1s}}{s(s+1)} \tag{8.49}$$

which is similar to Eq. (8.45), with the difference that it includes transport lag.

TABLE 8.7. Frequency response of the transfer function given by Eq. (8.49)

ω	M	ϕ	ω	M	ϕ
0	∞	$-90°$	2.0	2.24	$-164.89°$
0.1	99.5	$-96.28°$	5.0	0.39	$-197.34°$
0.2	49.03	$-102.46°$	7.0	0.202	$-211.98°$
0.5	17.89	$-119.43°$	10.0	0.100	$-231.59°$
1.0	7.07	$-140.73°$	20	0.025	$-291.73°$

The frequency response for this transfer function is shown in Table 8.7. Comparison with Table 8.4 indicates that the magnitude ratio is unchanged, but the phase lag is increased. This increase is more marked at higher frequencies.

DRILL PROBLEM 8.6

Sketch the polar plot for the frequency response of the transfer function

$$G(s) = \frac{10e^{-0.05s}}{s(s+1)(s+4)}$$

Answer:

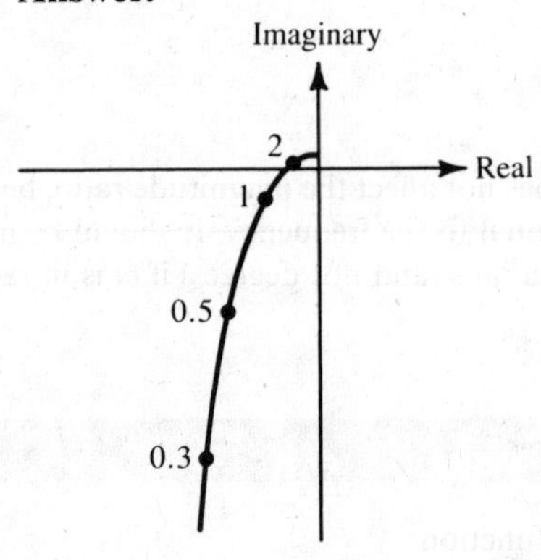

8.7 ESTIMATION OF TRANSFER FUNCTIONS FROM BODE PLOTS

It is often possible to estimate the transfer function from the Bode plots of the frequency response. Starting with the plot of the logarithmic gain against the frequency on log scale, the first step is to approximate it by straight-line segments of slope $6n$ decibels per octave, where n is an integer. The values of frequency at which the slope changes are then identified as the break frequencies, and hence the poles and zeros of the transfer function are located. At the poles the slope is reduced, whereas it is increased at zeros. If the change in slope corresponds to 12 dB/octave, the break frequency may be either a double real pole or zero, or a pair of complex poles or zeros with the undamped natural frequency given by the break frequency. For the latter case, the damping ratio may be estimated by examining the frequency response near the break frequency.

It may be noted that except for the case of minimum-phase transfer functions (i.e., stable transfer functions with all zeros in the left half of the s-plane), the transfer function cannot be uniquely determined from the gain curve alone. This follows from the fact that multiplying any transfer function by all pass functions of the form

$$\frac{s-a}{s+a} \quad \text{or} \quad \frac{s^2 - 2\zeta\omega_n s + \omega_n^2}{s^2 + 2\zeta\omega_n s + \omega_n^2}$$

or the introduction of pure delay does not alter the gain curve, but the phase shift is changed.

The following examples will illustrate the procedure.

EXAMPLE 8.10

The Bode plots of the transfer function of a dc servomotor-amplifier combination (the output is the angular velocity of the shaft and the input is the applied voltage) are shown in Fig. 8.17(a) and (b). The straight-line approximation to the gain curve is shown by the dashed lines. Hence, we get

$$G(s) = \frac{K}{\left(1 + \dfrac{s}{\alpha}\right)} = \frac{K\alpha}{s + \alpha}$$

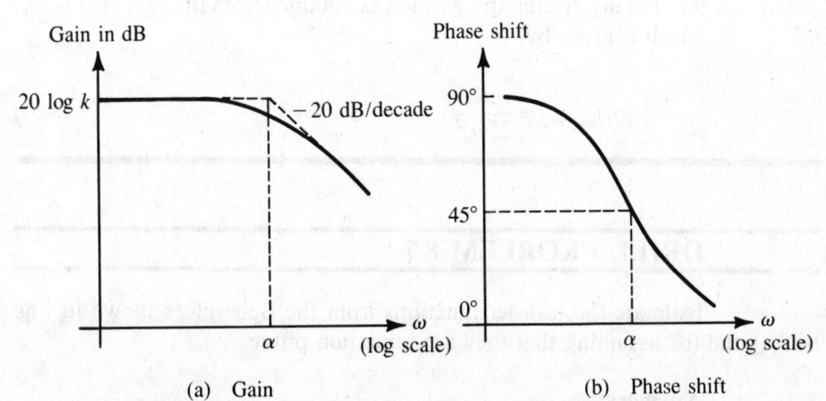

(a) Gain (b) Phase shift

FIGURE 8.17. Bode plots for a dc servomotor-amplifier combination.

This agrees with the phase-shift curve.

EXAMPLE 8.11

The Bode plots of the frequency response of a closed-loop position-control system are shown in Fig. 8.18.

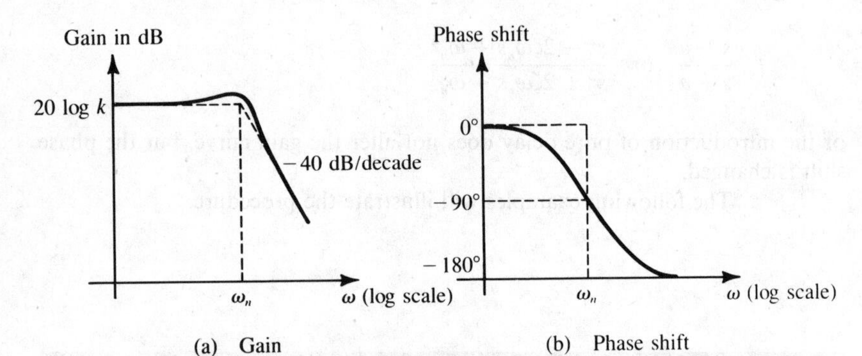

FIGURE 8.18. Bode plots of the frequency response of a closed-loop position-control system.

From the straight-line approximation of the magnitude plot, it is seen that the transfer function is of the following form.

$$G(s) = \frac{K}{s^2 + 2\zeta\omega_n s + \omega_n{}^2} \tag{8.50}$$

The value of K is determined from the low-frequency gain, and ω_n is determined as the frequency at which the phase shift is 90°. Finally, the damping ratio ζ is obtained from the gain at $\omega = \omega_n$, which is given by

$$|G(w)|_{\omega = \omega_n} = \frac{K}{2\zeta\omega_n{}^2} \tag{8.51}$$

DRILL PROBLEM 8.7

Estimate the transfer functions from the Bode plots shown in Fig. 8.19(a) and (b), assuming that they are minimum phase.

Answers:

(a) $G(s) = \dfrac{4(1 + s/2)}{s(1 + s/10)}$ (b) $\dfrac{10s}{(1 + s/2)(1 + s/10)(1 + s/30)}$

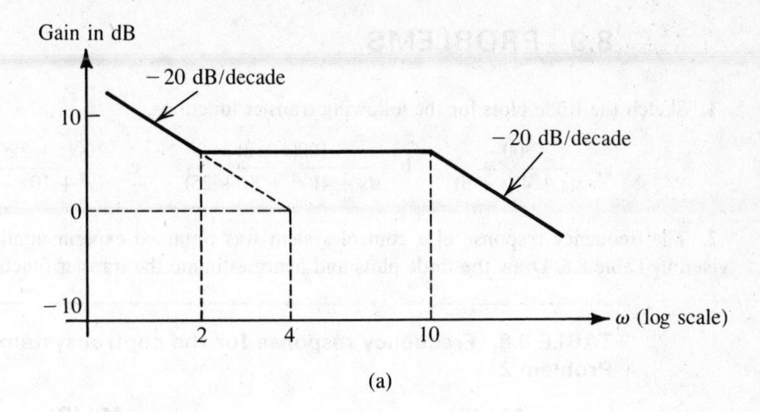

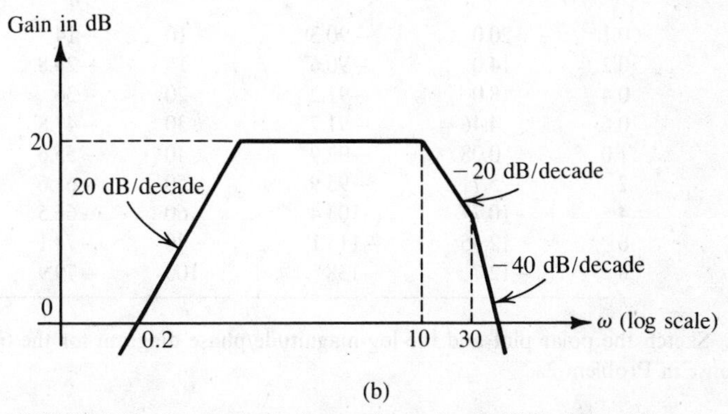

FIGURE 8.19. Bode plots of magnitude for Drill Problem 8.7.

8.8 SUMMARY

We have discussed three types of frequency response plots. These are (a) Bode plots, (b) polar plots, and (c) log-magnitude/phase diagrams. The first and the last using logarithmic gain are utilized considerably in the design of compensators for closed-loop systems. Polar plots are most commonly used for stability analysis with the Nyquist criterion of stability. Bode plots have been very popular in the past since they can be obtained without much computation. They can also be used for estimating the transfer function of a system. Computation of frequency response can be carried out conveniently using either a programmable calculator or a personal computer. These programs are discussed in Appendix D.

8.9 PROBLEMS

1. Sketch the Bode plots for the following transfer functions.

a. $\dfrac{300}{s(s+4)(s+8)}$ **b.** $\dfrac{100(s+2)}{s(s+4)(s^2+8s+25)}$ **c.** $\dfrac{100(s+2)e^{-0.2s}}{s(s^2+10s+100)}$

2. The frequency response of a control system was obtained experimentally and is given in Table 8.8. Draw the Bode plots and hence estimate the transfer function.

TABLE 8.8. Frequency response for the control system of Problem 2

ω	M (dB)	ϕ	ω	M (dB)	ϕ
0.1	20.0	−90.3°	10	−14	−180°
0.2	14.0	−90.6°	15	−26.8	−239°
0.4	8.0	−91.2°	20	−36	−252°
0.6	4.46	−91.7°	30	−47.8	−259°
1.0	0.08	−92.9°	40	−55.6	−262°
2	−5.71	−95.9°	50	−61.6	−264°
4	−10.77	−103.4°	60	−66.5	−265°
6	−12.55	−115.1°	80	−74.1	−266°
8	−12.7	−138°	100	−79.9	−267°

3. Sketch the polar plot and the log-magnitude/phase diagram for the frequency response in Problem 2.

4. Sketch the polar plot and the log-magnitude/phase diagram for each of the transfer functions given in Problem 1.

5. Estimate the transfer function for the Bode plot of magnitude shown in Fig. 8.20. Assume that it is minimum phase.

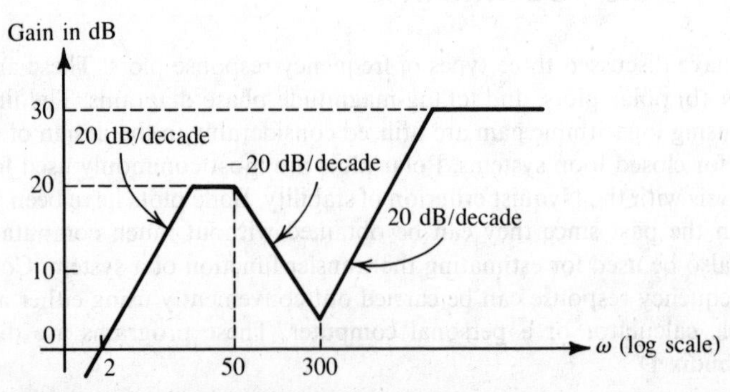

FIGURE 8.20. Bode plot of gain for Problem 5.

6. Repeat Problem 5 for the plot shown in Fig. 8.21.

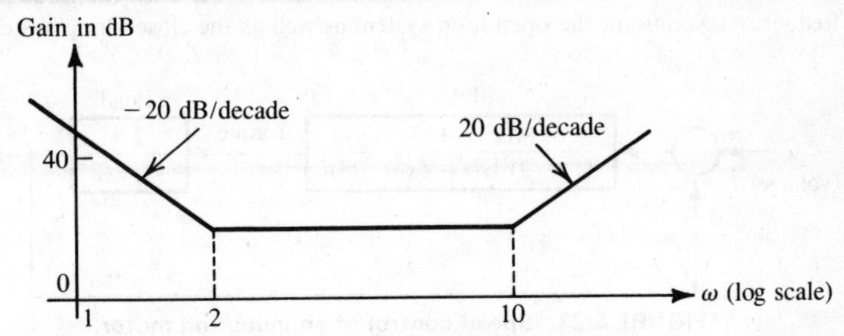

FIGURE 8.21. Bode plot of gain for Problem 6.

7. The block diagram of the pitch control system for a supersonic aircraft is shown in Fig. 7.20 (see Chapter 7, Problem 7). Draw the log-magnitude/phase plot of the frequency response of the open-loop system as well as the closed-loop system if the value of K is set at 5. [*Hint*: If the open-loop frequency response of a unity-feedback system at a certain frequency is given by $Me^{j\phi}$, then the frequency response of the closed-loop system is $Me^{j\phi}/(1 + Me^{j\phi})$.]

8. The transfer function of the forward path of a unity feedback system representing a chemical reactor is given by

$$G_p(s) = \frac{300e^{-0.05s}}{s(s+5)(s+10)}$$

Sketch the Bode plots of the frequency response of the open-loop system as well as the closed-loop system. What is the maximum value of the magnitude of the closed-loop frequency response?

9. Draw the log-magnitude/phase plots for the frequency response of the system described in Problem 8 with the loop open as well as closed.

10. The block diagram of the Otolith linear model of a human vestibular system is shown in Fig. 8.22. Draw the log-magnitude/phase plot of the frequency response of this system assuming that the gain K is set at 1.5.

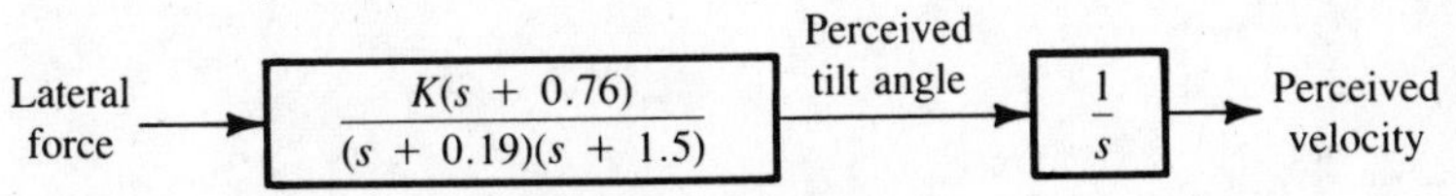

Lateral force → $\dfrac{K(s + 0.76)}{(s + 0.19)(s + 1.5)}$ → Perceived tilt angle → $\dfrac{1}{s}$ → Perceived velocity

FIGURE 8.22. Otolith linear model.

11. The transfer function relating the exciter voltage to the bucket-wheel current in a large coal dredger is given by

$$G_p(s) = \frac{K(1 + 0.613s)}{(1 + 4.42s)(1 + 0.23s)(1 + 0.66s)(1 + 0.4s)(s^2 + 0.41s + 1.5)}$$

Draw the log-magnitude/phase plot of the frequency response if K is set at 0.5.

12. The block diagram in Fig. 8.23 shows the speed-control system for an induction motor commonly used in industry. Assuming that $K = 80$, draw the Bode plots for the frequency response of the open-loop system as well as the closed-loop system.

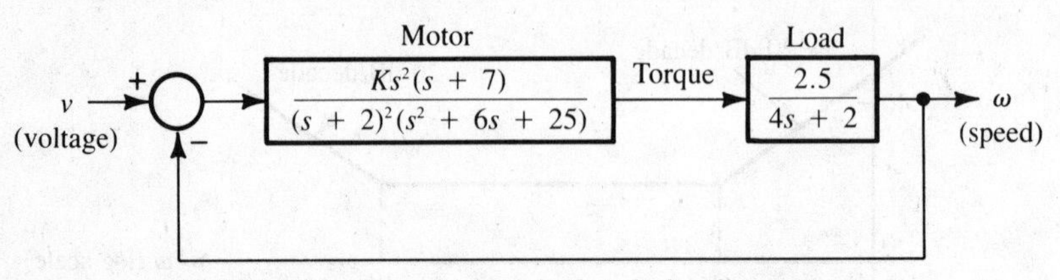

FIGURE 8.23. Speed control of an induction motor.

9 Stability from Frequency Response

9.1 INTRODUCTION

As stated earlier, in order to be useful, a control system must be stable. In Chapter 6 we discussed the Routh-Hurwitz criterion, which enables us to determine whether a system is stable by examining the characteristic polynomial, corresponding to the denominator of the closed-loop transfer function. The root locus method described in Chapter 7 enables us to determine the relative stability in terms of the damping ratio of the dominant poles. Both of these methods require the knowledge of the transfer function of the open-loop system, which must be a rational function of the complex frequency variable s—that is, a ratio of two finite-degree polynomials of s. These methods cannot be used when the system contains an ideal delay of the form e^{-Ts}, although it is possible to obtain approximate analysis by replacing e^{-Ts} by a truncated power series or by a rational function such as a Padé approximant.

In this chapter we shall study the use of frequency response methods, described in Chapter 8, for the determination of the stability of closed-loop systems. Our objective will be to utilize the frequency response of the open-loop system not only to determine the absolute stability of the closed-loop system but also to evaluate some measures of relative stability. There is no need to know the transfer function since the frequency response can be measured experimentally.

Consider the closed-loop system shown in Fig. 9.1. The main reason for possible instability of this system is that what is designed as negative feedback may turn out to be positive feedback if at some frequency a phase lag of 180° occurs in the loop. If the feedback at this frequency is sufficient to sustain oscillations, the system will act as an oscillator. The frequency response of this open-loop system transfer function can therefore be utilized for the investigation of stability of the closed-loop system.

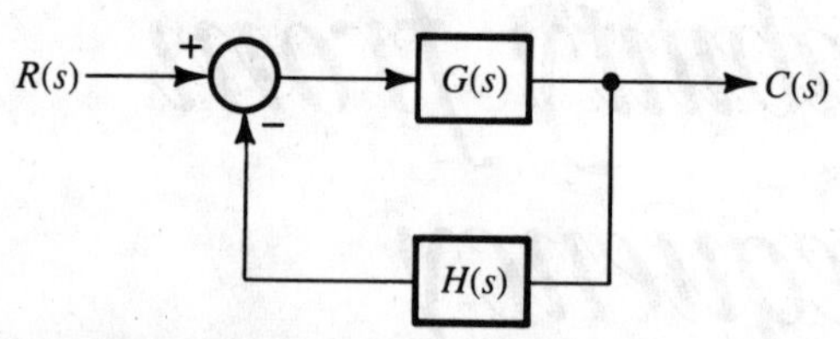

FIGURE 9.1. A closed-loop system.

From superficial considerations it will appear that the closed-loop system shown in Fig. 9.1 will be stable, provided that the open-loop transfer function $GH(j\omega)$ does not have gain of 1 or more at the frequency where the phase shift is 180°. However, this is not always true. For example, consider the polar plot of the frequency response shown in Fig. 9.2. For two values of ω the phase shift is 180° and gain more than one (at points A and B), and yet this system is stable when the loop is closed. A thorough understanding of this problem is possible by applying the criterion of stability developed by H. Nyquist

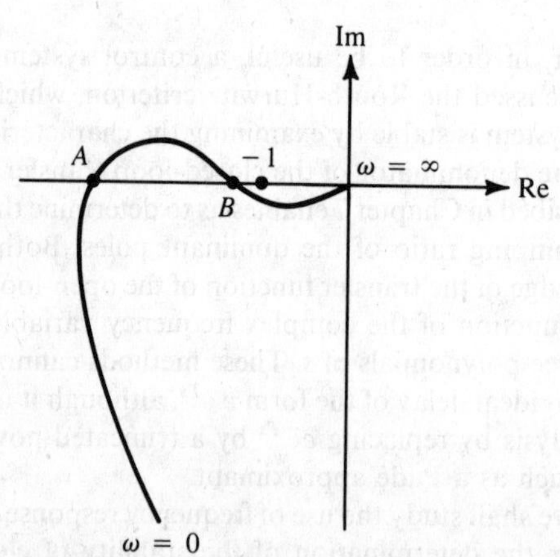

FIGURE 9.2. Polar plot of a conditionally stable system.

in 1932. It is based on a theorem in complex variable theory, due to Cauchy, called the principle of the argument, which is related to mapping of a closed path (or contour) in the s-plane for a function $F(s)$.

9.2 THE PRINCIPLE OF THE ARGUMENT

The principle of the argument is related to the theory of mapping of analytic functions of a complex variable. In order to get a thorough appreciation the theory will be reviewed briefly, and then the principle of the argument will be stated without proof. The interested reader may follow this up by reading a suitable book on complex variables [1, 2] or operational methods [3].

Let $F(s)$ be a function of the complex variable $s = \sigma + j\omega$. In our case, s has a physical meaning; it is the complex frequency variable. Since in general $F(s)$ will also be complex, we may write

$$F(s) = U(\sigma, \omega) + jV(\sigma, \omega) \tag{9.1}$$

where U and V are real functions of the variables σ and ω. Hence U is called the real part of F and V is called the imaginary part of F.

The function $F(s)$ is said to be a real function of s if $F(s)$ is real for real s. For example,

$$F(s) = 3s^2 + 2s + 4 \tag{9.2}$$

is a real function of the complex variable s, but

$$F(s) = 3s^2 + (2 + 3j)s + 4 \tag{9.3}$$

is not a real function of s.

The transfer function of a physical system is always a real function of s.

An $F(s)$, defined in a domain D in the s-plane, is said to be analytic in D if the derivative dF/ds is continuous in D. It can be shown that $F(s)$ is analytic in D if and only if the Cauchy-Riemann equations

$$\frac{\partial U}{\partial \sigma} = \frac{\partial V}{\partial \omega} \tag{9.4}$$

$$\frac{\partial U}{\partial \omega} = -\frac{\partial V}{\partial \sigma} \tag{9.5}$$

are satisfied in the domain D.

It can be proved that all rational functions of s are analytic everywhere in the s-plane except for the points of singularities. Hence all transfer functions are analytic in the s-plane except at their poles.

Just as the complex variable s is shown in a plane with the real axis σ and the imaginary axis ω, it is possible to illustrate F in a plane with real axis denoted by U and the imaginary axis by V. The former is called the s-plane and the latter the F-plane.

Any point in the s-plane will be "mapped" into the F-plane by locating the values of U and V for the given values of σ and ω. For example, consider the function

$$F(s) = \frac{2s + 3}{s + 5} \tag{9.6}$$

For $s_1 = 1 + j2$, we get

$$F(1 + j2) = \frac{2(1 + j2) + 3}{1 + j2 + 5} = 0.95 - j0.33 \tag{9.7}$$

This mapping is shown in Fig. 9.3.

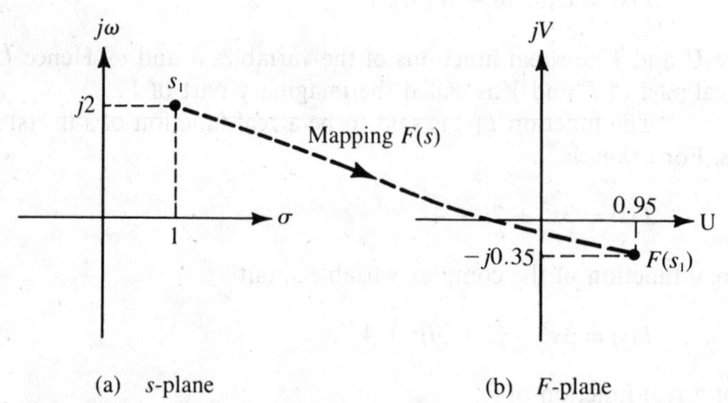

FIGURE 9.3. Mapping of a function $F(s)$.

This correspondence between points in the two planes is called a mapping or transformation. For an analytic function, every point in the s-plane maps into a unique point in the F-plane. The concept can be extended to mapping a line or curve in the s-plane to the F-plane. In particular, a smooth curve in the s-plane will map into a smooth curve in the F-plane if $F(s)$ is analytic at every point on the curve.

The polar plot of the frequency response of a transfer function $G(s)$ can be considered as the map of the positive part of the $j\omega$-axis of the s-plane into the G-plane.

We shall now consider mapping a closed curve in the s-plane. Such a curve will be called a contour, and denoted as Γ_s. Its map in the F-plane will be denoted by the contour Γ_F. The principle of the argument gives the relationship between the number of poles and zeros of $F(s)$ enclosed by Γ_s and the number of times Γ_F will encircle the origin of the F-plane. It is stated below.

Let $F(s)$ be an analytic function, except for a finite number of poles. If a contour Γ_s in the s-plane encloses Z zeros and P poles of $F(s)$ and does not pass through any pole or zero of $F(s)$, then with the transversal along Γ_s in the clockwise direction the corresponding contour Γ_F in the $F(s)$-plane encircles the origin of the F-plane N times in the clockwise direction, where $N = Z - P$. For a proof, the interested reader may refer to either Churchill [1] or Saff and Snider [2].

We shall now consider two examples that illustrate the basic idea.

EXAMPLE 9.1

Consider the function

$$F(s) = s + 2 \tag{9.8}$$

which has a zero at $s = -2$ and no pole. We shall determine the maps of two clockwise contours, Γ_1 and Γ_2, in the s-plane. The former is a circle of radius 1, with center at the origin, whereas the latter is a circle of radius 3, also centered at the origin. The pole-zero plot of $F(s)$ and the two contours are shown in the s-plane plot given in Fig. 9.4(a). The corresponding contours in the F-plane have also been labeled as Γ_1 and Γ_2, and are shown in Fig. 9.4(b). These are obtained easily in view of the translation represented by Eq. (9.8).

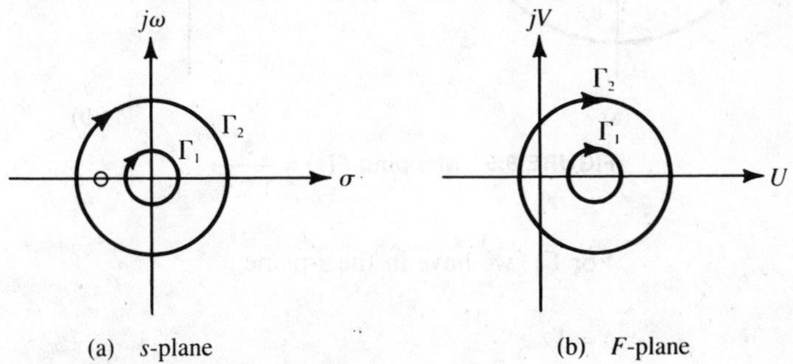

FIGURE 9.4. Mapping $F(s) = s + 2$.

It will be seen that Γ_1 does not enclose any pole or zero of $F(s)$ in the s-plane. The map of Γ_1 in the F-plane does not encircle the origin. On the other hand, Γ_2 encloses one zero of $F(s)$ in the s-plane. The map of Γ_2 in the F-plane encircles the origin once in the clockwise direction.

Both of these observations agree with the principle of the argument.

EXAMPLE 9.2

Consider the function

$$F(s) = \frac{s}{s+2} \tag{9.9}$$

which has a zero at the origin of the s-plane and a pole at $s = -2$.

We shall determine the maps of two clockwise contours: Γ_1 is a circle of radius 1 and Γ_2 is a circle of radius 3, both with centers at the origin. The pole-zero plot of $F(s)$ and the two contours are shown in Fig. 9.5(a). To obtain the maps in the F-plane, a little calculation is required.

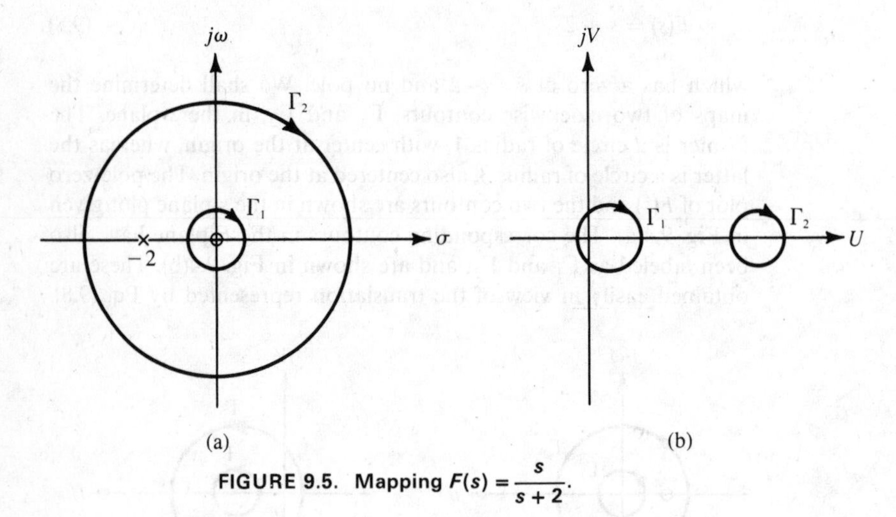

FIGURE 9.5. Mapping $F(s) = \dfrac{s}{s+2}$.

For Γ_1, we have in the s-plane

$$s = 1 \cdot e^{j\phi} \tag{9.10}$$

where ϕ varies over the entire circle in the clockwise direction. The mapping is given by

$$F(s) = \frac{1 \cdot e^{j\phi}}{1 \cdot e^{j\phi} + 2}$$

$$= \frac{1}{\sqrt{5 + 4\cos\phi}} \exp\left[j\left(\phi - \tan^{-1}\frac{\sin\phi}{2 + \cos\phi}\right)\right] \tag{9.11}$$

For the contour Γ_2, we have

$$s = 3e^{j\phi} \tag{9.12}$$

and

$$F(s) = \frac{3}{\sqrt{13 + 12\cos\phi}} \exp\left[j\left(\phi - \tan^{-1}\frac{3\sin\phi}{2 + 3\cos\phi}\right)\right] \tag{9.13}$$

The F-plane maps of Γ_1 and Γ_2 as given by Eqs. (9.11) and (9.13), respectively, are shown in Fig. 9.5(b). It will be seen that Γ_1 encloses one zero of $F(s)$, and its map in the F-plane encircles the origin once. On the other hand, Γ_2 encloses one zero and one pole of $F(s)$. Hence, in this case, $Z = 1$, $P = 1$, and $N = Z - P = 0$. As a consequence, the map of Γ_2 in the F-plane does not encircle the origin.

DRILL PROBLEM 9.1

The transfer function of a servomotor is given by

$$G(s) = \frac{10}{s(s+1)} \tag{9.14}$$

Determine the number of times the origin of the G-plane will be encircled by the map of each of the following contours in the s-plane: (a) a circle of radius 0.5 with center at the origin, (b) a circle of radius 3 with the center at the origin. Assume that the contours are taken clockwise.

Answers:
(a) -1, (b) -2. (*Note*: The negative sign implies encirclement in the opposite or counterclockwise direction.)

9.3 NYQUIST CRITERION

The overall transfer function of the system shown in Fig. 9.1 is given by

$$T(s) = \frac{G(s)}{1 + G(s)H(s)} \tag{9.15}$$

To find out if the closed-loop system will be stable, we must determine whether

$F(s) = 1 + GH(s) = 0$ has any root in the right half of the s-plane (including the $j\omega$-axis). For this purpose we must take a contour in the s-plane that encloses the entire right half plane as shown in Fig. 9.6. If $F(s)$ has a pole (or zero) at the origin of the s-plane or at some points on the $j\omega$-axis, we must make a detour along an infinitesimal semicircle, as shown in Fig. 9.6(b). This is called the Nyquist contour, and our object is to determine the number of zeros of $F(s)$ inside the Nyquist contour.

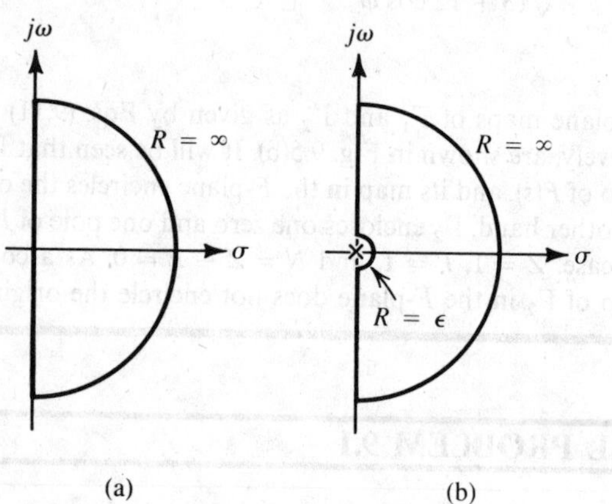

FIGURE 9.6. The Nyquist contour.

Let $F(s)$ have P poles and Z zeros within the Nyquist contour. Note that the poles of $F(s)$ are also the poles of $GH(s)$, but that the zeros of $F(s)$ are different from those of $GH(s)$ and not known. For the system to be stable, we must have $Z = 0$; that is, the characteristic equation must not have any root within the Nyquist contour.

From the principle of the argument, a map of the Nyquist contour in the F-plane will encircle the origin of the F-plane N times in the clockwise direction, where

$$N = Z - P$$

Thus the system is stable if and only if $N = -P$, so that Z will be zero. Further, note that the origin of the F-plane is the point $-1 + j0$ in the GH-plane. Hence, we get the following criterion in terms of the loop transfer function $GH(s)$.

A feedback control system is stable if and only if the number of counterclockwise encirclements of the point $-1 + j0$ by the map of the Nyquist contour in the GH-plane is equal to the number of poles of $GH(s)$ within the Nyquist contour in the s-plane.

The map of the Nyquist contour in the GH-plane is called the Nyquist plot of $GH(s)$. The polar plot of the frequency response of $GH(s)$, which is the map of the positive plot of the $j\omega$-axis, is an important part of the Nyquist plot and can be obtained either experimentally or by computation if the transfer function is known. The procedure for completing the rest of the Nyquist plot will be illustrated by a number of examples.

EXAMPLE 9.3

Consider the transfer function

$$GH(s) = \frac{60}{(s+1)(s+2)(s+5)} \tag{9.16}$$

The frequency response is readily calculated, and is given in Table 9.1. Sketches of the Nyquist contour and the Nyquist plot are shown in Fig. 9.7. The various steps of the procedure for completing the Nyquist plot are given below.

TABLE 9.1. Frequency response of the transfer function given by Eq. (9.16)

ω	*Gain*	*Phase Shift*	ω	*Gain*	*Phase Shift*
0	6	0°	3.0	0.90	−158.8°
0.25	5.77	−24.0°	4.123	0.476	−180°
0.5	5.18	−46.3°	5.0	0.309	−191.9°
1.0	3.72	−82.9°	10.0	0.052	−226.4°
2.0	1.76	−130.2°	20.0	0.007	−247.4°

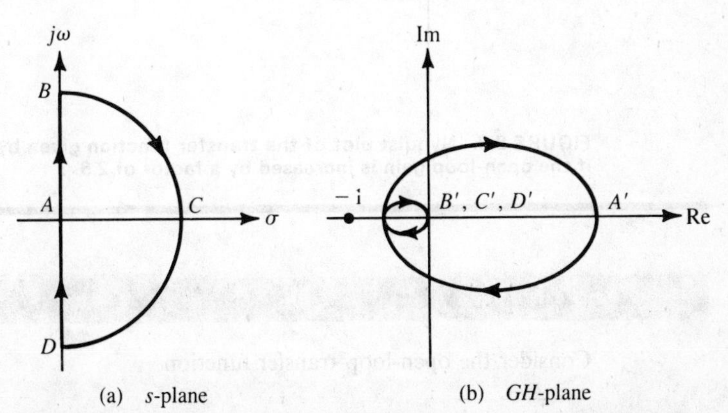

(a) *s*-plane (b) *GH*-plane

FIGURE 9.7. Nyquist contour and plot for the transfer function given by Eq. (9.16).

(i) Part AB in the s-plane is the $j\omega$-axis from $\omega = 0$ to $\omega = \infty$, and maps into the polar plot $A'B'$ in the GH-plane.

(ii) The infinite semicircle BCD maps into the origin of the GH-plane.

(iii) The part DA, which is the negative part of the $j\omega$-axis, maps into the curve $D'A'$. It may be noted that $D'A'$ is the mirror image of $A'B'$ about the real axis. This follows from the fact that $GH(-j\omega)$ is the complex conjugate of $GH(j\omega)$.

The Nyquist plot does not enclose the point $-1 + j0$ in the GH-plane; hence $N = 0$. Also, the function $GH(s)$ does not have any pole inside the Nyquist contour. This makes the system stable since $Z = N + P = 0$.

If we increase the open-loop gain of this system by a factor of 2.5, the resulting Nyquist plot will encircle the point $-1 + j0$ twice in the GH-plane, as shown in Fig. 9.8. Hence, for this case, $N = 2$. Since P is still zero, the closed-loop system is unstable, with two poles in the right half of the s-plane.

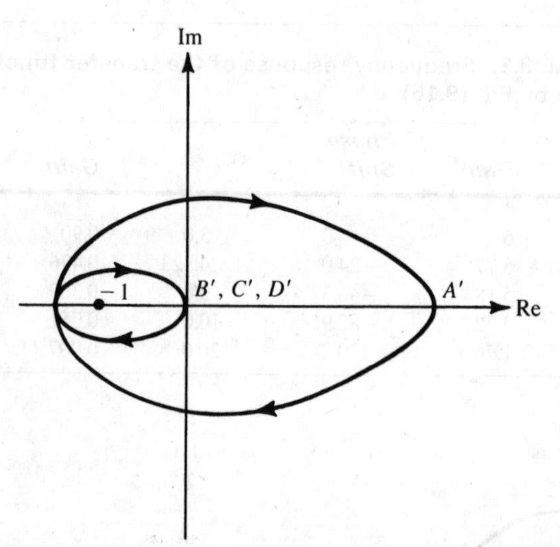

FIGURE 9.8. Nyquist plot of the transfer function given by Eq. (9.16) if the open-loop gain is increased by a factor of 2.5.

EXAMPLE 9.4

Consider the open-loop transfer function

$$GH(s) = \frac{K}{s(s + 2)(s + 6)} \tag{9.17}$$

Since this time we have a pole at the origin, the Nyquist contour must make a small detour around this pole while still attempting to enclose the entire right half of the s-plane, including the $j\omega$-axis. Hence, we get the semicircle EFA of infinitesimal radius, ε. The resulting Nyquist contour and plot are shown in Fig. 9.9. The various steps in obtaining the Nyquist plot are described below.

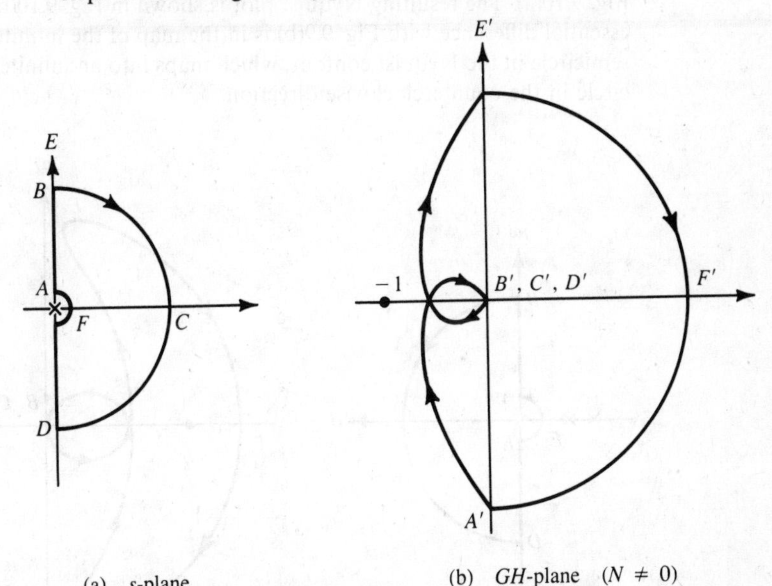

(a) s-plane (b) GH-plane ($N \neq 0$)

FIGURE 9.9. Nyquist plot for the transfer function given by Eq. (9.17).

(i) Part AB in the s-plane is the $j\omega$-axis from $j\varepsilon$ to $j\infty$ and maps into the polar plot of the frequency response $A'B'$ in the GH-plane.

(ii) The infinite semicircle BCD maps into the origin of the GH-plane.

(iii) The part DE (negative $j\omega$-axis) maps into the image $D'E'$ of the frequency response.

(iv) The infinitesimal semicircle EFA is approximately the map of $K/12s$ as $s = \varepsilon e^{j\theta}$, with θ going from $-90°$ to $90°$. It maps into an infinite semicircle, $E'F'A'$, as shown. This also follows from the fact that we have s in the denominator of the approximate expression, which will cause the infinitesimal semicircle to map into an infinite semicircle with traversal in the opposite direction due to the change in the sign of θ as it is brought to the numerator from the denominator. For the plot shown, $N = 0$ and $P = 0$, indicating that this is a stable system. Increasing the gain will cause the system to be unstable since the point -1 will be encircled twice—that is, $N = 2$ and $Z = 2$.

EXAMPLE 9.5

We shall reconsider the transfer function of the preceding example, but this time we shall take a different Nyquist contour. The detour around the pole at the origin will be taken from its left, as shown in Fig. 9.10(a). The resulting Nyquist plot is shown in Fig. 9.10(b). The essential difference with Fig. 9.9(b) is in the map of the infinitesimal semicircle in the Nyquist contour, which maps into an infinite semicircle in the counterclockwise direction.

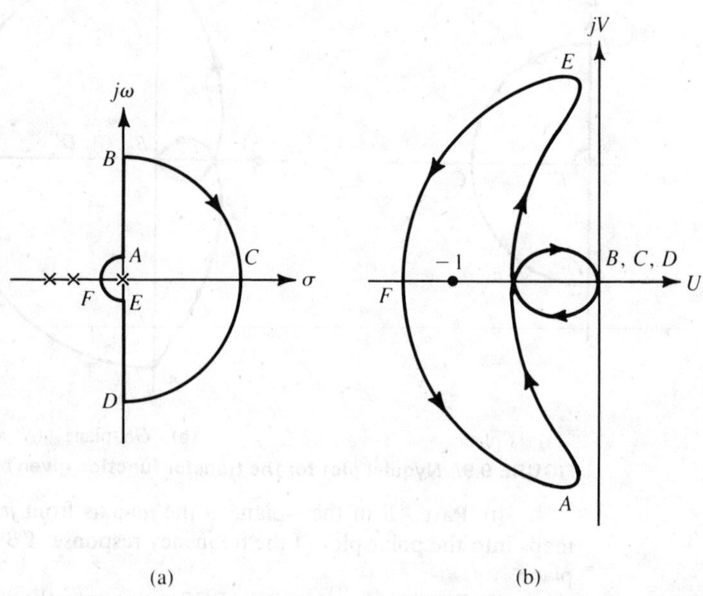

FIGURE 9.10. Nyquist plot of the transfer function given by Eq. (9.17) with the Nyquist contour enclosing the pole at the origin.

In this case, since the Nyquist contour encloses the pole at the origin, we have $P = 1$. Also, the Nyquist plot encircles the point $-1 + j0$ in the GH-plane once in the counterclockwise direction, giving us $N = -1$. Hence, the number of zeros of $1 + GH(s)$ in the right half of the s-plane (corresponding to the number of poles of the closed-loop transfer function in the right half of the s-plane) is given by

$$Z = N + P = 0 \tag{9.18}$$

Thus we have a stable system, as expected from the previous example.

Note that if we increase the loop gain, the point -1 will be inside the smaller loop in the GH-plane, giving us $N = 1$ and

$$Z = N + P = 2 \tag{9.19}$$

and hence we get an unstable system.

EXAMPLE 9.6

The polar plot of the open-loop frequency response of a system is shown in Fig. 9.11(b) by the solid curve for positive ω. We shall complete the Nyquist plot in the GH-plane to determine stability. The system is known to be open-loop stable; that is, $GH(s)$ has no pole in the right half of the s-plane.

The Nyquist contour is shown in Fig. 9.11(a) and the corresponding Nyquist plot is shown as the dotted curve in Fig. 9.11(b) with the exception of the polar plot, which is the solid part.

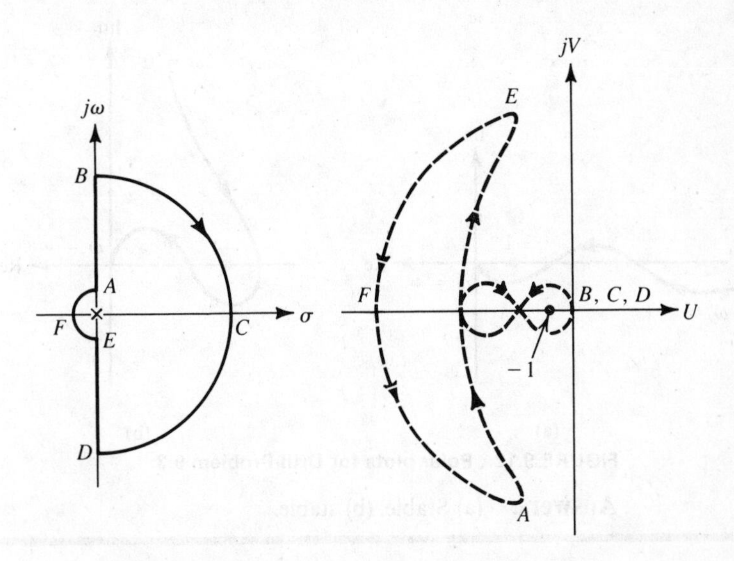

(a) Nyquist contour (b) Nyquist plot

FIGURE 9.11. Complete Nyquist plot from the polar plot of the frequency response.

From the plot, we have $N = -1$ and $P = 1$, so $Z = 0$. Hence the system is stable.

DRILL PROBLEM 9.2

For the open-loop transfer function given by

$$GH(s) = \frac{10(s+1)}{s^2(s+4)} \tag{9.20}$$

obtain the Nyquist plot and determine whether the closed-loop system will be stable.

Answer: Stable

DRILL PROBLEM 9.3

The polar plots of the open-loop frequency response of two control systems are shown in Fig. 9.12. Both of them are known to be open-loop stable. Complete the Nyquist plot for each case, and determine the stability of the closed-loop system.

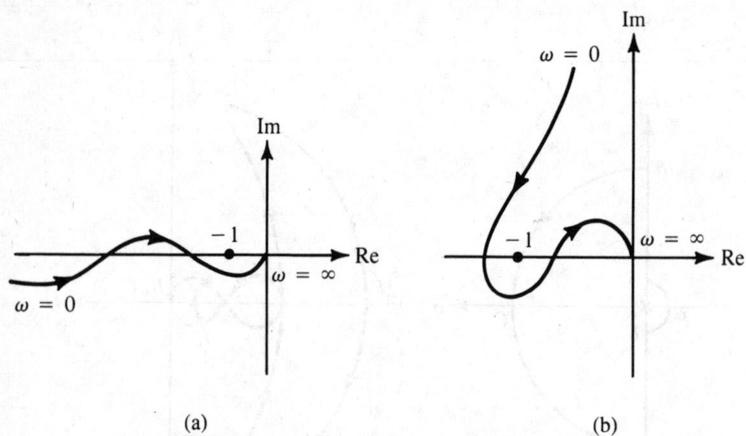

FIGURE 9.12. Polar plots for Drill Problem 9.3.

Answers: (a) Stable, (b) stable.

9.4 RELATIVE STABILITY

In most practical situations, in addition to finding out whether a closed-loop system is stable, it is also desirable to determine how close it is to instability. This information can be readily obtained from the open-loop frequency response.

The proximity of the open-loop frequency response to the point $-1 + j0$ in the GH-plane is a measure of the relative stability of a closed-loop system.

Two commonly used measures of relative instability are the gain margin and the phase margin. These are defined below.

The *gain margin* is defined as the additional gain required to make the system just unstable. It may be expressed either as a factor or in decibels.

The *phase margin* is defined as the additional phase lag required to make the system just unstable.

In the example shown in Fig. 9.13, from the location of point A where the phase lag is 180° we see that the gain must be increased by the factor 2 to make the system just unstable; that is, the gain margin $= 1/0.5 = 2$. Alternatively, we may say that the gain margin is 20 log 2 or 6 dB.

Similarly, from point B where the gain is one, the phase margin is seen to be $180° - 140° = 40°$.

The gain margin as well as the phase margin can also be determined from either the Bode plots or the log-magnitude/phase curve. The gain margin is determined from the gain at the phase crossover frequency—i.e., the frequency at which the phase shift is 180°. Similarly, the phase margin is determined from the phase shift at the gain crossover frequency—i.e., the frequency at which the gain is 0 dB. For example, the Bode plots and log-magnitude/phase plot corresponding to the polar plot of Fig. 9.13 are shown in Figs. 9.14 and 9.15, respectively. The gain margin and the phase margin are shown for each case.

In many cases, phase margin provides a better measure of relative stability. For example, consider a unity feedback system with a forward transfer function

$$G(s) = \frac{K}{s(s+2)} \tag{9.21}$$

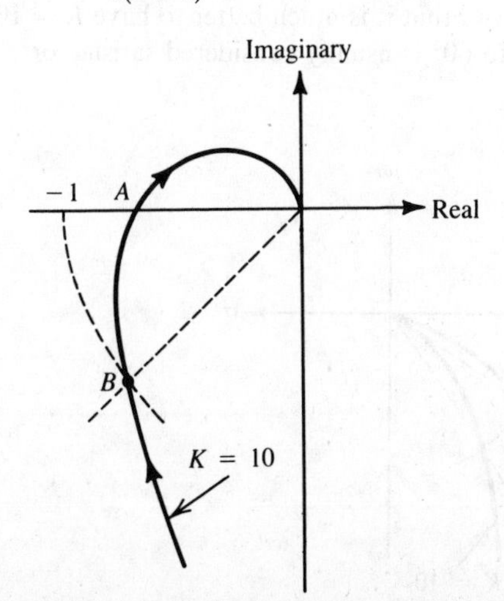

FIGURE 9.13. Gain margin and phase margin.

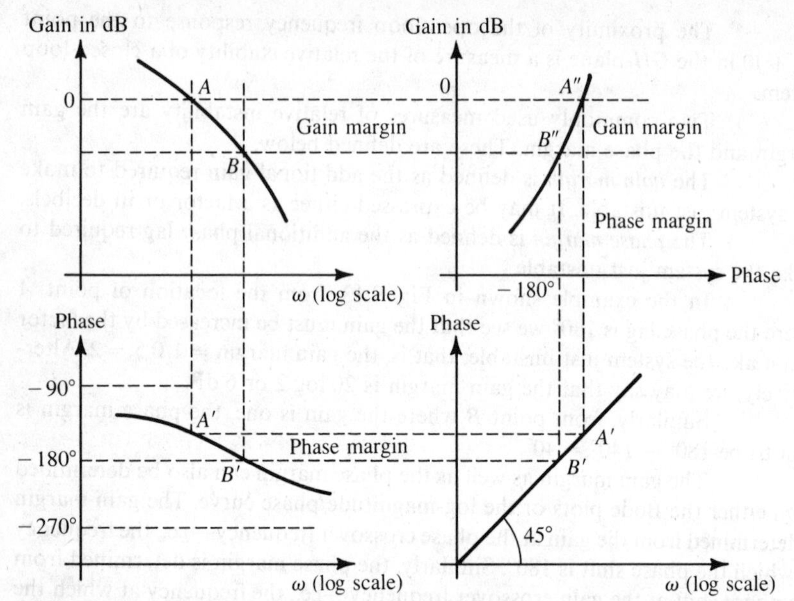

FIGURE 9.14. Bode plots for Fig. 9.13.

FIGURE 9.15. Gain-phase plot for Fig. 9.13 (obtained from Bode plots using the construction shown).

The polar plot of the open-loop transfer functions for $K = 10$ and $K = 50$ are shown in Fig. 9.16. It is evident that for both cases the gain margin is infinity since the plot never crosses the negative real axis of the GH-plane. On the other hand, the phase margins for the two cases are 35° for $K = 10$, and 16° for $K = 50$. This shows that it is much better to have $K = 10$. In practice, a phase margin of 45° to 60° is usually considered satisfactory.

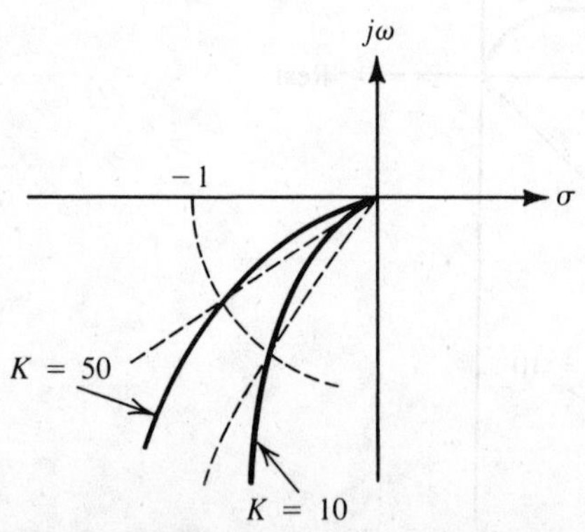

FIGURE 9.16. Gain and phase margins for a second-order system.

One can often use Bode plots for conveniently determining the value of the open-loop gain that will provide a specified gain or phase margin. This will be illustrated by means of the following example.

EXAMPLE 9.7

The open-loop transfer function of a control system is given by

$$GH(s) = \frac{K}{s(s+2)(s+10)} \tag{9.22}$$

Determine the value of K such that the system may have (a) a gain margin of 6 dB, and (b) a phase margin of 45°.

Solution

We start by assuming a suitable value of K and then drawing the Bode plots. Here we shall take $K = 20$. The resulting frequency response is shown in Table 9.2, and the Bode plots are shown in Fig. 9.17.

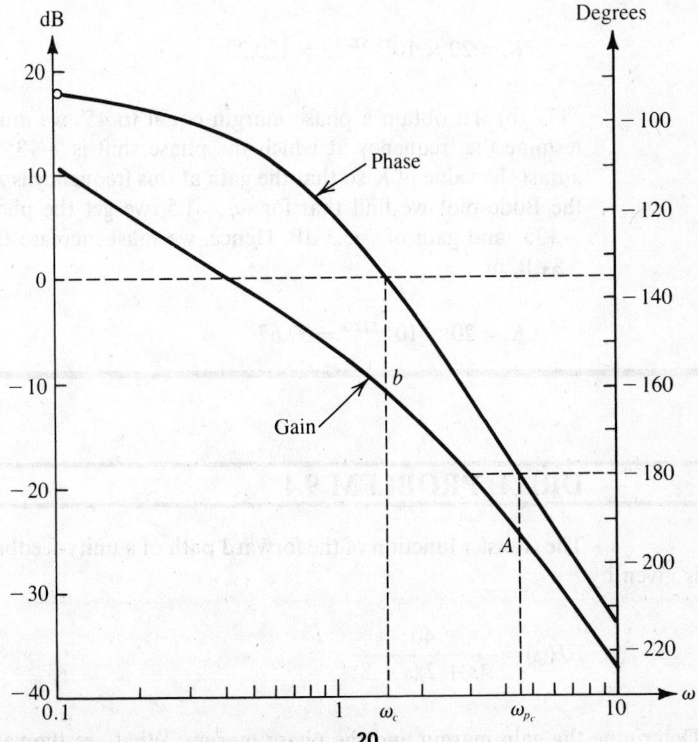

FIGURE 9.17. Bode plots of $\frac{20}{s(s+2)(s+10)}$.

TABLE 9.2. Frequency response of Eq. (9.22), with $K = 20$

ω	*Gain* (dB)	*Phase Shift*	ω	*Gain* (dB)	*Phase Shift*
0.1	20	−93.4°	3.0	−15.0	−163°
0.2	13.9	−96.9°	4.0	−19.7	−175.2°
0.5	5.75	−106.9°	5.0	−23.7	−184.8°
0.7	2.57	−113.3°	6.0	−28.7	−192.5°
1.0	−1.01	−122.3°	10.0	−37.2	−213.7°
2.0	−9.2	−146.3°			

(a) To obtain a gain margin equal to 6 dB, first we determine the phase crossover frequency—the frequency at which the phase shift is 180°. This is readily found as $\omega_{pc} = 4.47$ and the gain at this frequency is found to be -21.58 dB, as indicated by the point A on the gain curve.

Hence, we must increase the gain by 15.58 dB to get the desired gain margin. The resulting value of K is obtained as

$$K = 20 \times 10^{15.58/20} = 120.28 \tag{9.23}$$

(b) To obtain a phase margin equal to 45° we must first determine the frequency at which the phase shift is $-135°$ and then adjust the value of K so that the gain at this frequency is zero. From the Bode plot we find that for $\omega_c = 1.5$, we get the phase shift of $-135°$ and gain of -5.5 dB. Hence, we must increase the gain by 5.5 dB, or

$$K = 20 \times 10^{5.5/20} = 37.67 \tag{9.24}$$

DRILL PROBLEM 9.4

The transfer function of the forward path of a unity-feedback system is given by

$$GH(s) = \frac{40}{s(s + 2)(s + 8)} \tag{9.25}$$

Determine the gain margin and the phase margin. What are the values of the gain crossover and the phase crossover frequencies?

Answers:

Gain margin = 12 dB, phase margin = 35°, phase crossover frequency = 4 rad/s, gain crossover frequency = 1.81 rad/s.

DRILL PROBLEM 9.5

The open-loop transfer function of a unity-feedback system is given by

$$GH(s) = \frac{K}{s(s^2 + 2s + 5)} \tag{9.26}$$

What should be the value of K to obtain (a) a gain margin of 6 dB, and (b) a phase margin of 50°?

Answers: (a) $K = 5$, (b) $K = 5.6$.

9.5 THE CLOSED-LOOP FREQUENCY RESPONSE

A better measure of relative stability is the maximum magnitude of the closed-loop frequency response of the system, denoted by M_m. A large value of M_m implies the presence of dominant poles with small damping ratio and hence an undesirable transient response. On the other hand, a smaller value of M_m indicates poles farther away from the $j\omega$-axis, resulting in a well-damped system. The frequency at which the maximum gain is obtained is denoted as ω_m.

The relationship between open-loop and closed-loop frequency response can be established easily for unity-feedback systems. For such a system, we have

$$T(j\omega) = M(\omega)e^{j\phi(\omega)} = \frac{G(j\omega)}{1 + G(j\omega)} \tag{9.27}$$

From the polar plot of the open-loop transfer function, Eq. (9.27) can be given an interesting graphical interpretation, as shown in Fig. 9.18. Since for a given ω, OP represents the open-loop frequency response $G(j\omega)$, the vector AP represents $1 + G(j\omega)$. Hence, M is obtained as the ratio OP/AP.

It will now be shown that points representing a constant value of the closed-loop gain, M, lie on a circle in the GH-plane. Let

$$G(j\omega) = U(\omega) + jV(\omega) \tag{9.28}$$

Then

$$M = \left| \frac{U + jV}{1 + U + jV} \right| = \frac{(U^2 + V^2)^{1/2}}{[(1 + U)^2 + V^2)]^{1/2}} \tag{9.29}$$

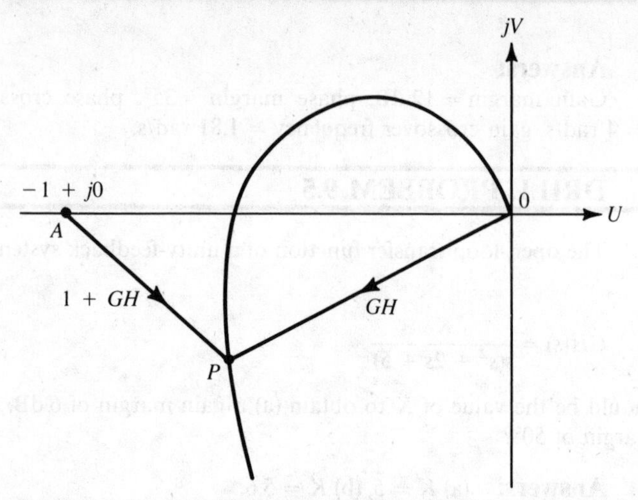

FIGURE 9.18. Graphical interpretation of the closed-loop frequency response of a unity-feedback system.

or

$$(1 - M^2)U^2 + (1 - M^2)V^2 - 2M^2U = M^2 \tag{9.30}$$

This can be rearranged as

$$\left(U - \frac{M^2}{1 - M^2}\right)^2 + V^2 = \left(\frac{M}{1 - M^2}\right)^2 \tag{9.31}$$

This is the equation of a circle with center at $U = -M^2/(M^2 - 1)$, $V = 0$, and radius $|M/(M^2 - 1)|$ for constant M.

Thus it is possible to draw a family of constant-M circles in the GH-plane, as shown in Fig. 9.19.

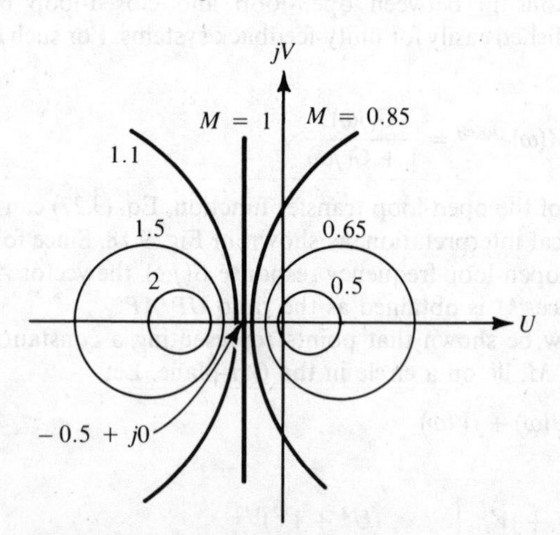

FIGURE 9.19. Constant-M circles.

With the help of these M circles, one can determine the value of M_m from the polar plot of the open-loop transfer function as the largest value of M for the circle tangent to the polar plot and satisfying Eq. (9.31). This is indicated in Fig. 9.20.

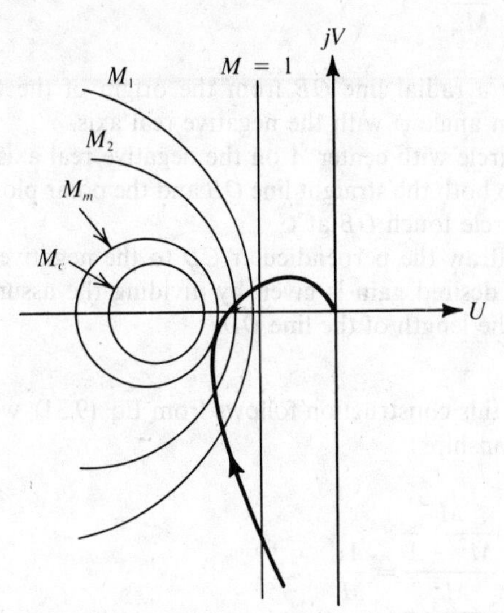

FIGURE 9.20. Determination of M_m from the family of M circles.

It is also possible to use a simple geometric construction to determine the value of K that will provide a desired M_m. The various steps are given below (see Fig. 9.21).

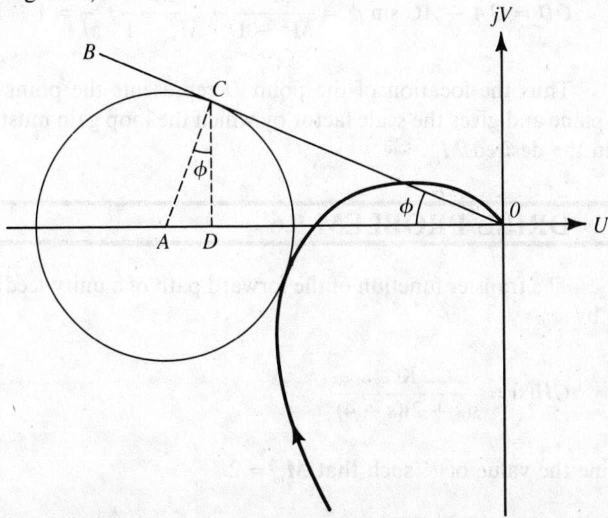

FIGURE 9.21. Construction for determining the open-loop gain for a specified M_m.

1. Draw the polar plot of $G(j\omega)$, assuming a certain value of K.
2. Calculate the angle

$$\phi = \sin^{-1}\frac{1}{M_m} \tag{9.32}$$

and draw a radial line OB from the origin of the GH-plane, making an angle ϕ with the negative real axis.
3. Draw a circle with center A on the negative real axis, which is tangent to both the straight line OB and the polar plot of $G(j\omega)$. Let the circle touch OB at C.
4. From C, draw the perpendicular CD to the negative real axis.
5. Then the desired gain is given by dividing the assumed value of K by the length of the line OD.

The proof for this construction follows from Eq. (9.31), which leads immediately to the relationships

$$\sin\phi = \frac{AC}{OA} = \frac{\dfrac{M}{M^2-1}}{\dfrac{M^2}{M^2-1}} = \frac{1}{M} \tag{9.33}$$

and

$$OD = OA - AC\sin\phi = \frac{M^2}{M^2-1} - \frac{M}{M^2-1}\cdot\frac{1}{M} = 1 \tag{9.34}$$

Thus the location of the point D represents the point $-1 + j0$ in the GH-plane and gives the scale factor by which the loop gain must be changed to obtain the desired M_m.

DRILL PROBLEM 9.6

The transfer function of the forward path of a unity-feedback system is given by

$$GH(s) = \frac{K}{s(s+2)(s+4)} \tag{9.35}$$

Determine the value of K such that $M_m = 2$.

Answer: $K = 17.3$

The constant-M circles in the polar plot transform to a family of curves in the logarithmic gain phase plots. Nichols charts (named after N. B. Nichols), which contain these constant-M curves, are available. These charts can be used very conveniently for obtaining M_m for value of the open-loop gain for a given M_m. The procedure is as follows.

For a given $KG(s)$, assume a suitable value of K, and plot the frequency response of the open-loop on the Nichols charts. Next, examine by how many decibels this response curve should be raised or lowered so that it is tangential to a specified M_m curve. This gives the gain adjustment necessary to obtain the desired M_m.

EXAMPLE 9.8

The forward transfer function of a unity-feedback system is given by

$$G(s) = \frac{K}{s(s^2 + 2s + 5)} \tag{9.36}$$

Determine the value of K so that $M_m = 3$ dB.

Solution

Assuming $K = 10$, the frequency response is calculated and is given in Table 9.3.

TABLE 9.3. Frequency response of the transfer function given by Eq. (9.36)

ω	*Gain* (dB)	*Phase Shift*	ω	*Gain* (dB)	*Phase Shift*
0.1	26	−92.3°	2.0	1.67	−166°
0.2	20	−94.6°	2.2	0.28	−177.9°
0.5	12.3	−101.9°	2.3	−0.5	183.6°
1.0	7	−116.6°	2.4	−1.34	−189.0°
1.5	4.3	−137.5°	2.6	−3.09	−198.7°
1.8	2.8	−153.9°	2.8	−4.9	−206.9°

This is plotted in the Nichols chart shown in Fig. 9.22. The plots indicate that the system is unstable, with M_m approximately infinity. To obtain $M_m = 3$ dB we must lower the plot by approximately 6 dB. Hence, the desired $K = 5$.

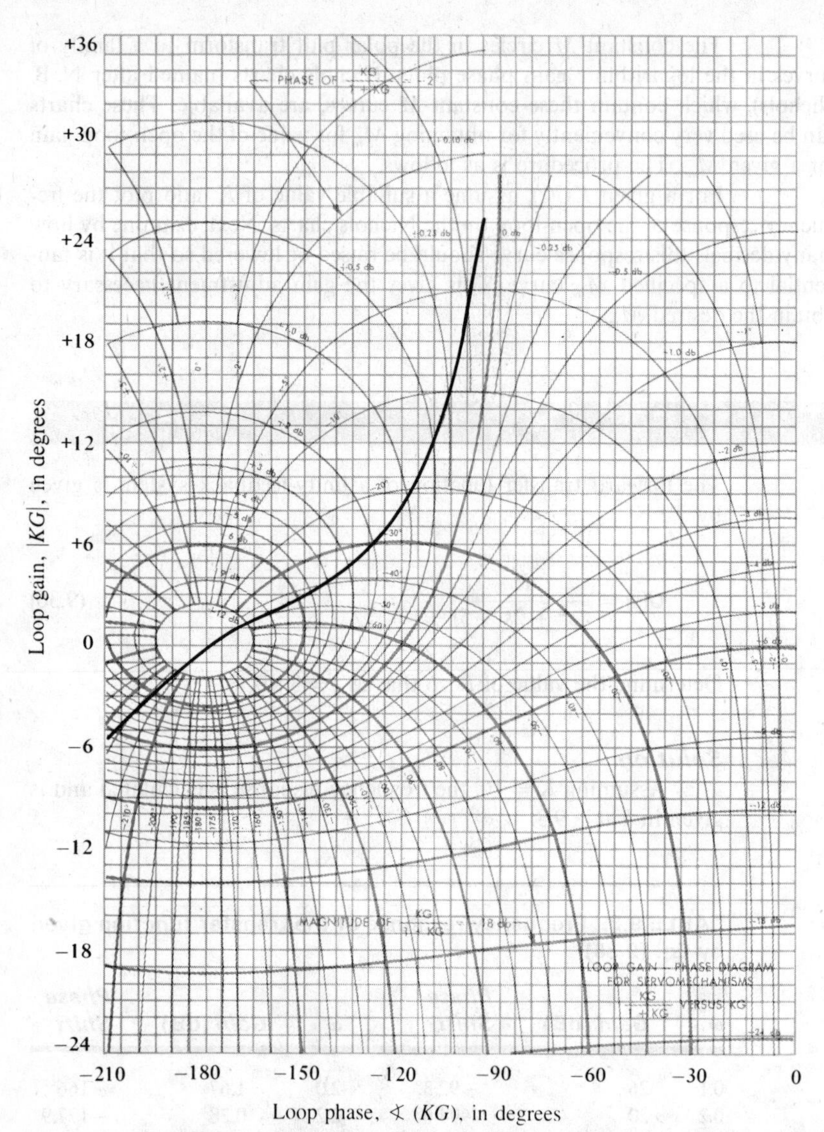

FIGURE 9.22. Nichols chart plot of the frequency response of the transfer function given by Eq. (9.36).

DRILL PROBLEM 9.7

Use the Nichols chart procedure to verify the solution to Drill Problem 9.6.

9.6 SYSTEMS WITH PURE DELAYS

In many practical situations the open-loop transfer function contains a pure delay (or transport lag) term of the form $e^{-\beta s}$. The Nyquist criterion remains valid for such cases, and we may use the frequency response curve as before. We may also use the various frequency response curves to determine relative stability, using the gain margin, the phase margin, and the maximum closed-loop gain as in the previous section. We may recall that neither the Routh criterion nor the root locus method is applicable for this case.

EXAMPLE 9.9

The transfer function of a chemical plant that is in the forward path of a unity-feedback system can be represented as

$$G(s) = \frac{Ke^{-0.05s}}{s(s+2)(s+5)} \tag{9.37}$$

We shall first use the Nyquist criterion to determine whether the closed-loop system will be stable for $K = 10$. After that, we shall use the Nichols chart to adjust K so that the system may be stable with (a) a phase margin = 45°, and (b) $M_m = 3$ dB.

The frequency response for $K = 10$ of the open-loop transfer function is given in Table 9.4. The Nyquist contour and plot are shown in Fig. 9.23. They indicate that the system is stable, with large gain and phase margins.

The Nichols chart is shown in Fig. 9.24.

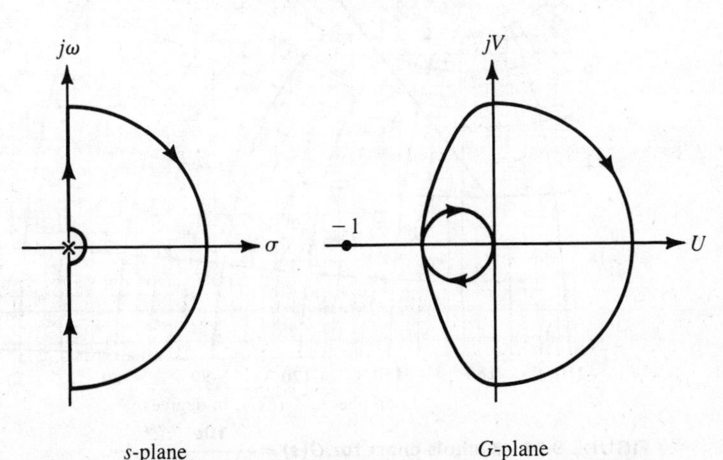

FIGURE 9.23. Nyquist plot for system represented by Eq. (9.37).

TABLE 9.4. Frequency response of Eq. (9.37), with $K = 10$

ω	*Gain*	*Phase Shift*	*Gain* (dB)	ω	*Gain*	*Phase Shift*	*Gain* (dB)
0.1	9.99	−94.3°	20	2	0.33	−162.5°	−9.7
0.2	4.97	−98.6°	13.9	3	0.16	−185.9°	−16.0
0.5	1.93	−111.2°	5.7	4	0.087	−203.6°	−21.2
0.7	1.34	−119.3°	2.5	5	0.053	−217.5°	−26
1	0.88	−130.7°	−1.14	10	0.009	−260.8°	−41.19
1.5	0.51	−147.9°	−5.83				

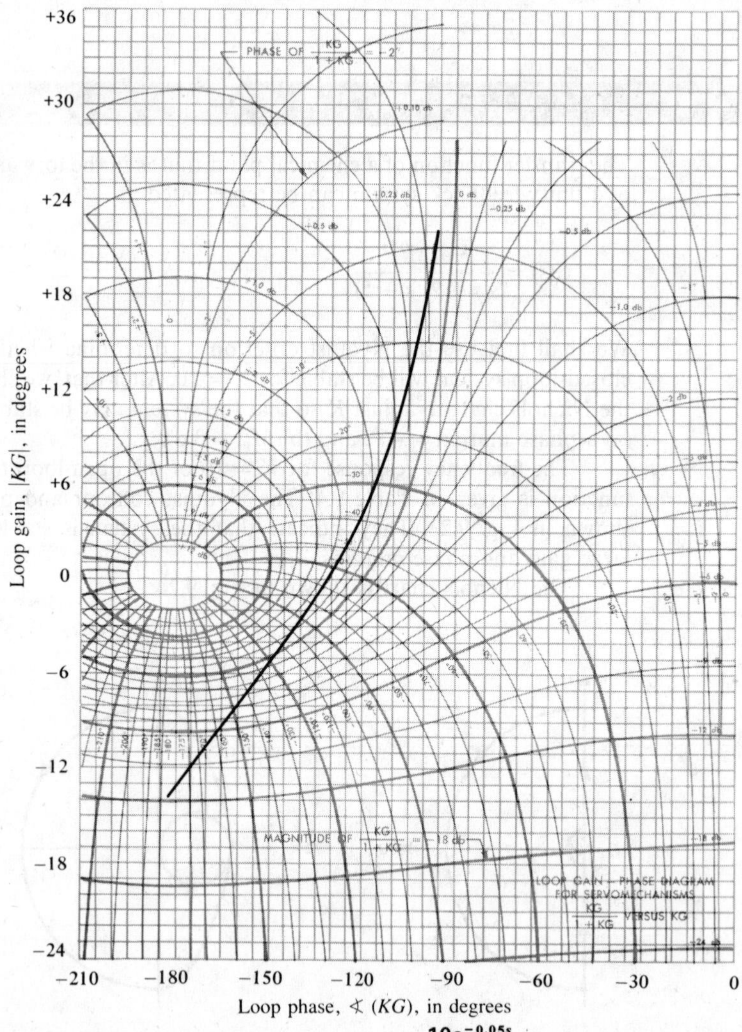

FIGURE 9.24. Nichols chart for $G(s) = \dfrac{10e^{-0.05s}}{s(s+2)(s+5)}$.

(a) From the plot, it is seen that the phase margin is 53°. To obtain a phase margin of 45°, the plot should be raised by 2 dB. Hence the desired value of the gain is obtained as

$$K = 10 \times 10^{2/20} = 12.59 \tag{9.38}$$

(b) From the plot we get $M_m = 1$ dB. To obtain $M_m = 3$ dB, the plot should be raised by 3 dB. Hence, the desired value of the gain is obtained as

$$K = 10 \times 10^{3/20} = 14.13 \tag{9.39}$$

DRILL PROBLEM 9.8

The block diagram of a satellite control system is shown in Fig. 9.25. The time required for the signals sent from the earth to reach the satellite is represented by the ideal delay in the block diagram.

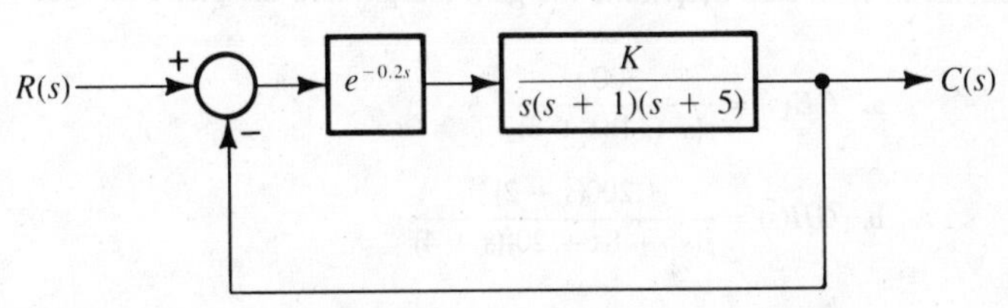

FIGURE 9.25. Block diagram of a satellite control system.

(a) Determine the value of the gain constant K so that the closed-loop system may have phase margin equal to 45°.

(b) Determine the value of the gain constant K so that M_m, the maximum gain of the closed-loop frequency response, is 3 dB. What is the value of ω_m, the frequency at which the maximum gain is obtained for this case?

Answers:

(a) $K = 3.56$; (b) $K = 3.9$, $\omega_m = 0.7$.

9.7 SUMMARY

The Nyquist criterion of stability is a powerful concept and leads to a thorough understanding of the stability of a closed-loop system through the open-loop frequency response. The ideas may be further extended to obtain measures of relative stability of the system. An important advantage of this approach is that one may be able to improve the relative stability by reshaping the frequency response by designing a compensator. This will be discussed in the next chapter.

9.8 REFERENCES

1. R. V. Churchill, *Introduction to Complex Variables and Applications*, 2nd ed., McGraw-Hill, New York, 1960.
2. E. B. Saff and A. D. Snider, *Fundamentals of Complex Analyses*, Prentice-Hall, Englewood Cliffs, N.J., 1976
3. W. Kaplan, *Operational Methods for Linear Systems*, Addison-Wesley, Reading, Mass., 1962.
4. H. B. James, N. B. Nichols, and R. S. Phillips, *Theory of Servomechanisms*, McGraw-Hill, New York, 1947.

9.9 PROBLEMS

1. Draw the Bode plots for the open-loop frequency response of the following transfer functions. In each case determine the gain margin and the phase margin.

a. $$GH(s) = \frac{300}{s(s+4)(s+8)}$$

b. $$GH(s) = \frac{200(s+2)}{s(s^2+8s+20)(s+4)}$$

2. The transfer function of the forward path of a unity feedback system is given by

$$G(s) = \frac{Ke^{-0.2s}}{s(s+2)(s+8)}$$

Determine the value of K so that the system is stable with (a) gain margin equal to 6 dB and (b) phase margin equal to 45°.

3. The polar plots of the open-loop frequency response of two unity-feedback systems for positive values of frequency are shown in Fig. 9.26. It is known that the open-loop transfer function has no pole in the right half of the s-plane. Draw the complete Nyquist plot for each case and determine whether the closed-loop system will be stable.

4. The frequency response of the forward path of a unity-feedback control system was obtained experimentally, and is given in Table 9.5.

a. Draw the Bode plots of the frequency response.

b. Increase the loop gain to make the gain margin equal to 5 dB. Determine M_m and ω_m for this value of the gain using the Nichols chart.

c. What adjustment should be made in the loop gain in order to make $M_m = 3$ dB? Determine ω_m for this case.

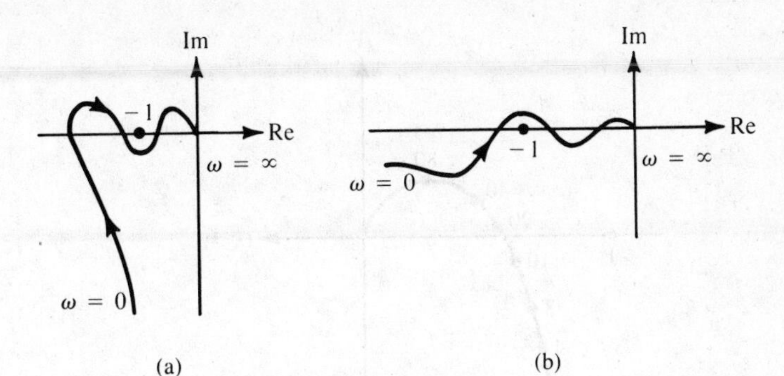

FIGURE 9.26. Polar plots for Problem 3.

TABLE 9.5. Frequency response for Problem 4

ω	M (dB)	ϕ	ω	M (dB)	ϕ
0.1	20	−90.30°	10	−13.98	−180.0°
0.2	13.98	−90.57°	15	−26.8	−239.0°
0.4	7.97	−91.15°	20	−36.0	−251.6°
0.6	4.46	−91.73°	30	−47.75	−259.4°
1.0	0.08	−92.9°	40	−55.64	−262.4°
2	−5.71	−95.9°	50	−61.63	−264.1°
4	−10.77	−103.4°	60	−66.48	−265.1°
6	−12.55	−115.1°	80	−74.04	−266.4°
8	−12.68	−138.0°	100	−79.92	−267.1°

5. The forward transfer function of a unity feedback system is given by

$$G(s) = \frac{Ke^{-0.02s}}{s(s+2)}$$

Draw the polar plot of the frequency response for $K = 10$ and obtain the value of K that will make $M_m = 2$.

6. The polar plot of the frequency of the forward path of a unity-feedback system is shown in Fig. 9.27. Determine

a. Phase margin
b. Gain margin
c. M_m
d. ω_m

7. A feedback system with two delays is shown in Fig. 9.28. It represents a typical chemical reactor.

a. Investigate the stability of this system for $K = 1$.
b. What value of K will give a phase margin of 45°?

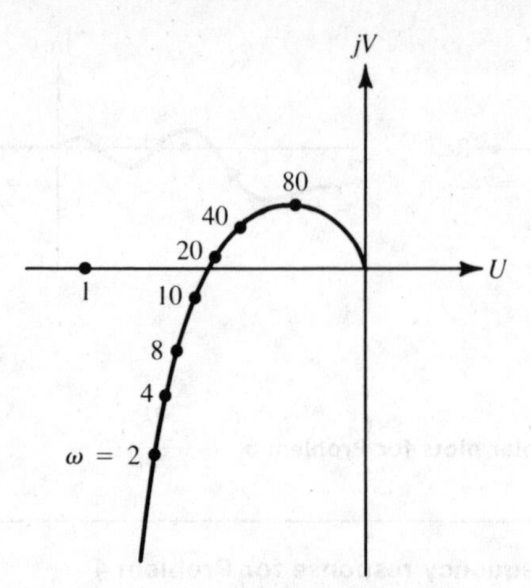

FIGURE 9.27. Polar plot for Problem 6.

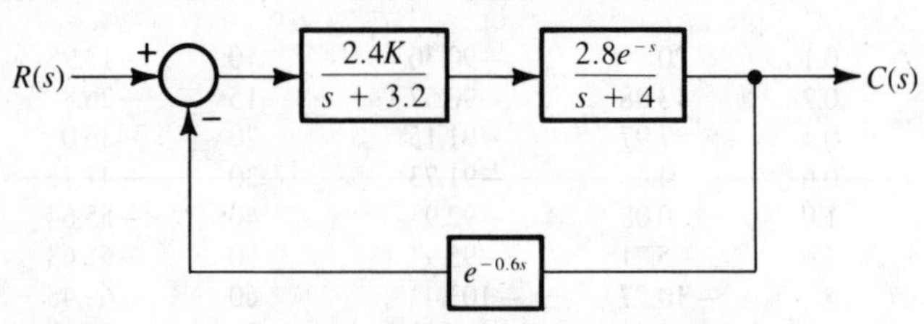

FIGURE 9.28. A feedback system with two delays.

8. The block diagram of a course-keeping control system for a ship is shown in Fig. 9.29. Determine the value of K_p so that the maximum magnitude of the closed-loop gain is 3 dB. What is the phase margin for this value of K_p?

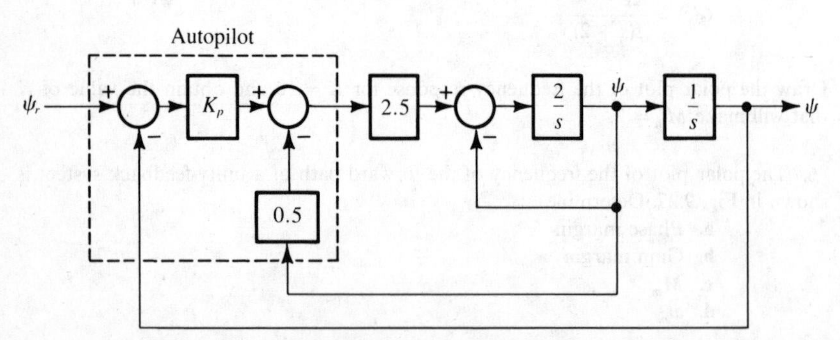

FIGURE 9.29. Course-keeping control system for a ship.

9. The linearized model for automatic gauge control in a hot-strip finishing mill is shown in Fig. 9.30.

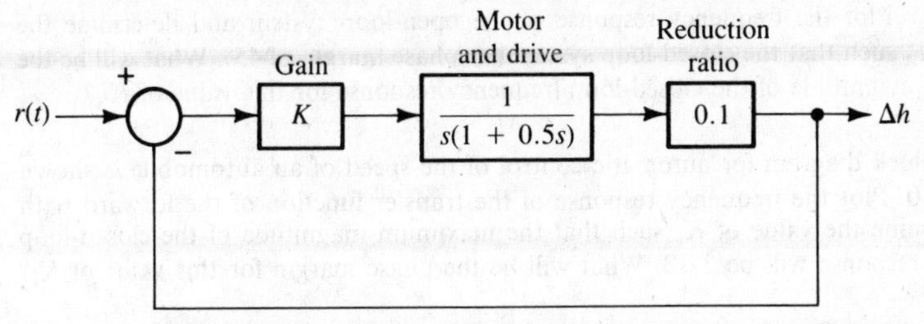

FIGURE 9.30. Gauge control system for a hot-strip finishing mill.

Draw the gain-phase plots of the open-loop transfer function and determine

a. the value of K for gain margin equal to 3 dB.
b. the value of K for phase margin equal to 45°.
c. the value of K for the maximum closed-loop gain to be 3 dB.

10. A linearized model of a driver steering-control system is shown in Fig. 9.31, where the approximate transfer function of the human driver is given by $G_1(s)$ with K_p = gain, τ = time delay, and T_n = neuromuscular system delay.

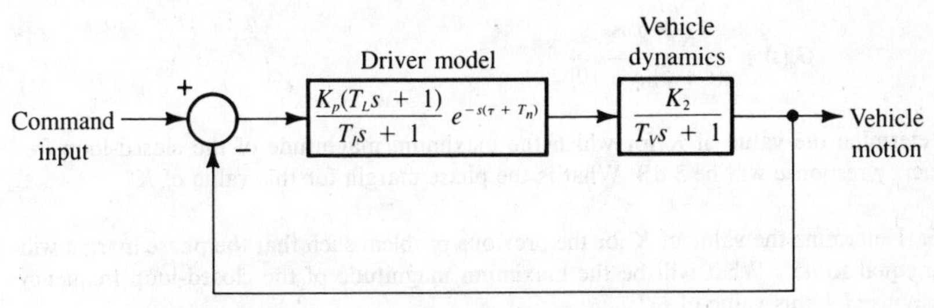

FIGURE 9.31. Driver steering-control system.

a. Investigate the stability of the closed-loop system if $K_p = 5$, $\tau + T_n = 0.3$, $K_2 = 2$, $T_L = 1$, $T_I = 2$ and $T_v = 4$.
b. How should K_p be changed to obtain phase margin of 45°?

11. The block diagram for speed control of a motor using a phase-locked loop is shown in Fig. 9.32. It has been claimed that such a system has a much higher accuracy than conventional speed control systems while requiring fewer components.

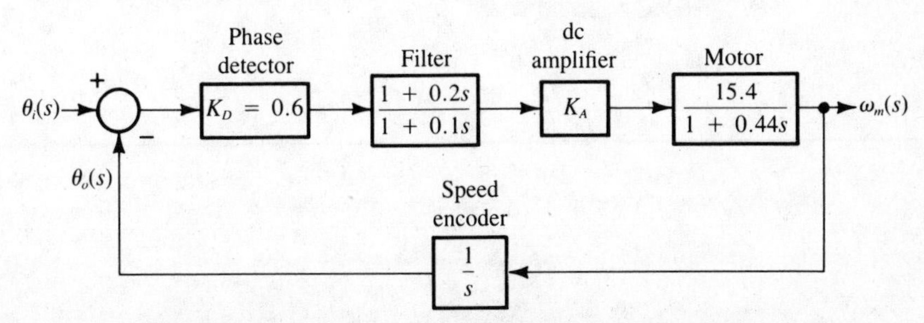

FIGURE 9.32. Speed control using a phase-locked loop.

Plot the frequency response of the open-loop system and determine the value of K_A such that the closed-loop system has phase margin of 45°. What will be the maximum magnitude of the closed-loop frequency response for this value of K_A?

12. The block diagram for automatic control of the speed of an automobile is shown in Fig. 4.10. Plot the frequency response of the transfer function of the forward path and determine the value of K_e such that the maximum magnitude of the closed-loop frequency response will be 3 dB. What will be the phase margin for this value of K_e?

13. Problem 12 of Chapter 5 described the transfer function of linearized dynamic model relating the changes in the angle of attack of a ship to changes in the angular position of its rudder. Determine the value of K such that the maximum magnitude of the closed-loop frequency response will be 3 dB. What will be the phase margin for this value of K?

14. Repeat for the type 2 system shown in Fig. 6.3 (see Chapter 6, Problem 4), where the value of α is fixed at 0.1, but the gain K is adjustable.

15. The transfer function of the forward path of a unity feedback system representing a chemical process is given by

$$G_p(s) = \frac{Ke^{-0.05s}}{s(s+5)(s+10)}$$

Determine the value of K for which the maximum magnitude of the closed-loop frequency response will be 3 dB. What is the phase margin for this value of K?

16. Determine the value of K for the previous problem such that the phase margin will be equal to 45°. What will be the maximum magnitude of the closed-loop frequency response for this value of K?

17. The transfer function of the forward path of a unity feedback system is given by

$$G(s) = \frac{10(s+5)}{s^2}$$

Sketch the polar plot of the frequency response and use the Nyquist criterion to investigate the stability of the system.

10 Design and Compensation of Control Systems

10.1 INTRODUCTION

Control systems are always designed for a specific purpose. A good control system should have the following properties.

1. It should operate with as little error as possible.
2. It should exhibit suitable damping; i.e. the controlled output should follow the changes in the reference input without unduly large oscillations or overshoots.
3. Its performance should not be appreciably affected by small changes in certain parameters.
4. It should be able to mitigate the effect of undesirable disturbances.

It has already been seen earlier that requirements 1 and 2 are contradictory, since the increase in loop gain reduces the steady-state error but causes a deterioration in the transient performance. For example, it has been shown in Chapters 3 and 4 that by increasing the loop gain, we can increase the speed of response and reduce the steady-state error to a step input, but we also tend to increase the overshoot in the response and reduce the damping ratio of the dominant poles. This was further clarified by the root locus method in Chapter 6, which showed that, in general, we reduce the damping ratio of the dominant poles as the gain is increased. Requirement 3 is necessary in order to be able to produce the actual system economically.

A control system usually requires some adjustment so that the various conflicting and demanding specifications may be met. This adjustment is called compensation, and can be done in several ways. For example, an additional component may be inserted in the forward path, as shown in Fig. 10.1. This is called series or cascade compensation. The transfer function of the compensator is denoted by $G_c(s)$, whereas that of the original process (or plant) is denoted by $G_p(s)$. Alternatively, the compensator may be inserted in the feedback path as shown in Fig. 10.2. This is called feedback compensation. A combination of these two schemes is shown in Fig. 10.3.

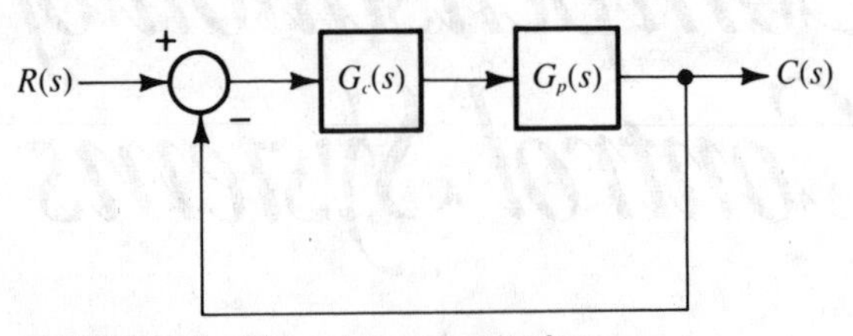

FIGURE 10.1. Cascade compensation.

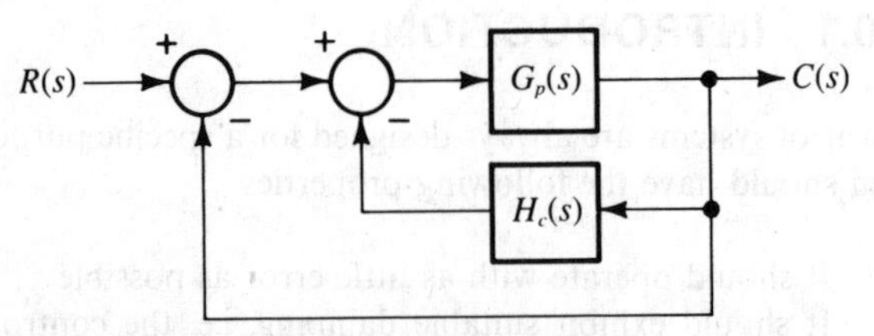

FIGURE 10.2. Feedback compensation.

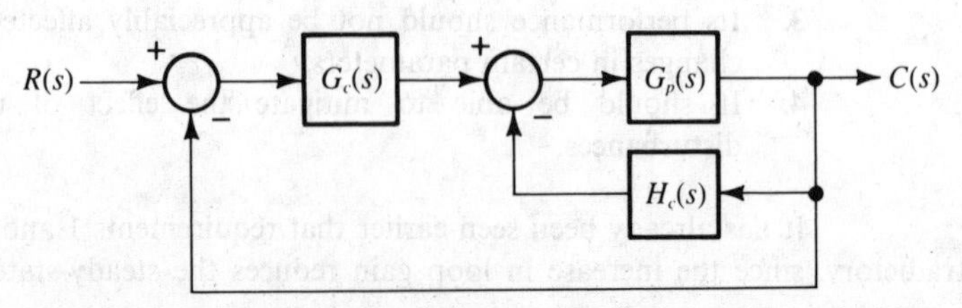

FIGURE 10.3. Combined cascade and feedback compensation.

The compensator, like the system components, may be an electrical, mechanical, hydraulic, pneumatic, or other type of device. Electric networks are often used as compensators in many control systems. The simplest among these are the lead, lag, and lag-lead networks, which will be discussed in the next section.

10.2 TYPICAL COMPENSATORS

Although many different types of compensators can be used, the most common are the three basic compensators described in this section. Each of these can be realized, using an *RC* network (with an amplifier in one case). We shall discuss the characteristics and design of each of these in detail.

10.2.1 Lead compensator

One of the simplest compensators has the first-order transfer function

$$G_c(s) = \frac{1 + s/a}{1 + s/\alpha a} = \frac{\alpha(s + a)}{s + \alpha a} \qquad \alpha > 1 \tag{10.1}$$

with pole-zero plot shown in Fig. 10.4.

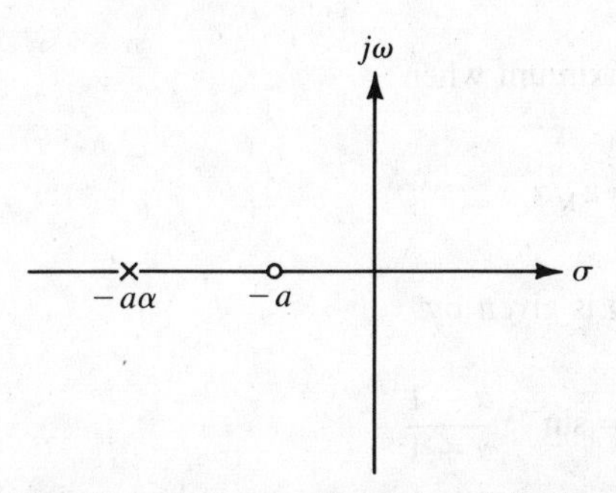

FIGURE 10.4. Pole-zero plot of lead compensator.

It is evident that this compensator provides a phase lead between the output and the input, given by

$$\phi(\omega) = \tan^{-1}\frac{\omega}{a} - \tan^{-1}\frac{\omega}{\alpha a} \tag{10.2}$$

Hence, it is called a lead compensator. The Bode plots of the gain and the phase shift of this transfer function are shown in Fig. 10.5.

It is seen that at frequencies below a, the gain is nearly zero decibels and at frequencies above αa, the gain is nearly 20 log α decibels.

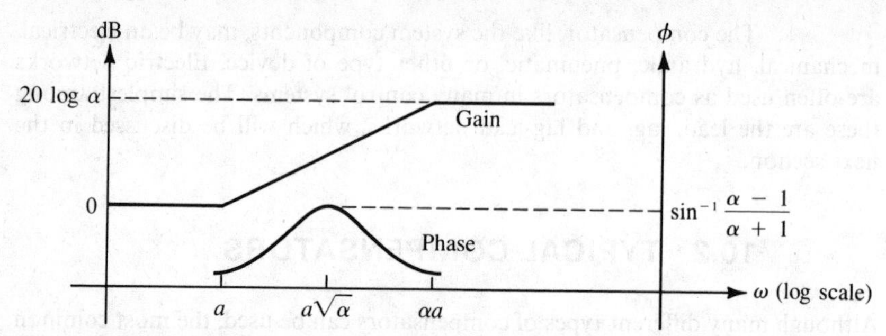

FIGURE 10.5. Gain and phase shift of a lead compensator.

Let us now determine the frequency at which the phase lead is a maximum. From Eq. (10.2), we have

$$\phi = \tan^{-1}\frac{\dfrac{\omega}{a}-\dfrac{\omega}{\alpha a}}{1+\dfrac{\omega}{a}\cdot\dfrac{\omega}{\alpha a}} = \tan^{-1}\frac{\omega a(\alpha-1)}{\omega^2+\alpha a^2}$$

$$= \tan^{-1}\frac{\alpha-1}{\dfrac{\omega}{a}+\dfrac{\alpha a}{\omega}} \tag{10.3}$$

It is easily shown that ϕ is maximum when

$$\frac{\omega}{a} = \frac{\alpha a}{\omega} \quad \text{or} \quad \omega_m = a\sqrt{\alpha} \tag{10.4a}$$

Also, the maximum value of ϕ is given by

$$\phi_m = \tan^{-1}\frac{\alpha-1}{2\sqrt{\alpha}} = \sin^{-1}\frac{\alpha-1}{\alpha+1} \tag{10.4b}$$

It may be noted that the maximum phase lead produced by the lead compensator is less than 90°. In practice, values of α greater than 20 (corresponding to $\phi_m = 64.8°$), are seldom used.

It will now be shown that given the values of a specified phase lead ϕ_c and decibel gain M_c at a particular frequency ω_c, one may always determine uniquely the transfer function of the corresponding lead compensator. Let

$$q = \tan\phi_c = \frac{\dfrac{\omega_c}{a}-\dfrac{\omega_c}{\alpha a}}{1+\dfrac{\omega_c}{a}\cdot\dfrac{\omega_c}{\alpha a}} = \frac{\dfrac{\omega_c}{a}(\alpha-1)}{\left(\dfrac{\omega_c}{a}\right)^2+\alpha} \tag{10.5}$$

and

$$c = 10^{M_c/10} = \frac{\alpha^2(\omega_c{}^2 + a^2)}{\omega_c{}^2 + \alpha^2 a^2} = \frac{\alpha^2\left(1 + \dfrac{\omega_c{}^2}{a^2}\right)}{\alpha^2 + \left(\dfrac{\omega_c}{a}\right)^2} \tag{10.6}$$

From Eq. (10.6),

$$\left(\frac{\omega_c}{a}\right)^2 = \frac{\alpha^2(c-1)}{\alpha^2 - c} \tag{10.7}$$

From Eq. (10.5),

$$q^2\left[\left(\frac{\omega_c}{a}\right)^4 + 2\alpha\left(\frac{\omega_c}{a}\right)^2 + \alpha^2\right] = \left(\frac{\omega_c}{a}\right)^2(\alpha - 1)^2 \tag{10.8}$$

Substituting for $(\omega_c/a)^2$ from (10.7) into (10.8), we get

$$q^2\left[\frac{\alpha^4(c-1)^2}{(\alpha^2 - c)^2} + 2\alpha \cdot \frac{\alpha^2(c-1)}{\alpha^2 - c} + \alpha^2\right] = \frac{\alpha^2(c-1)}{\alpha^2 - c}(\alpha - 1)^2 \tag{10.9}$$

Equation (10.9) can be simplified as

$$\begin{aligned} q^2[\alpha^4(c-1)^2 + 2\alpha^3(c-1)(\alpha^2 - c) + \alpha^2(\alpha^2 - c)^2] \\ = \alpha^2(c-1)(\alpha-1)^2(\alpha^2 - c) \end{aligned} \tag{10.10}$$

The left-hand side of Eq. (10.10) is a perfect square, and can be rearranged as

$$\begin{aligned} q^2[\alpha^4(c-1)^2 + 2\alpha^3(c-1)(\alpha^2 - c) + \alpha^2(\alpha^2 - c)^2] \\ = q^2\alpha^2(\alpha-1)^2(\alpha + c)^2 \end{aligned} \tag{10.11}$$

Since $\alpha > 1$, we can cancel α^2 and $(\alpha - 1)^2$ from the left- and right-hand sides of Eq. (10.10) to get

$$q^2(\alpha + c)^2 = (c-1)(\alpha^2 - c) \tag{10.12}$$

which can be rearranged as

$$(q^2 - c + 1)\alpha^2 + 2q^2c\alpha + (q^2c + c - 1)c = 0 \tag{10.13}$$

This is a quadratic equation that will yield one positive real solution for α, provided that

(i) $q > 0$

(ii) $c > 1$

and

$$\text{(iii)} \quad c > q^2 + 1 \tag{10.14}$$

Note that (i) corresponds to $\phi_c > 0$ and (ii) corresponds to $M_c > 0$. As these two conditions are implied from the transfer function (i), it follows that (iii) is the necessary and sufficient condition for the existence of a lead compensator that will provide the gain M_c and the phase-lead ϕ_c at a specified frequency ω_c.

Finally, from Eq. (10.7),

$$a = \frac{\omega_c}{\alpha}\left(\frac{\alpha^2 - c}{c - 1}\right)^{1/2} \tag{10.15}$$

EXAMPLE 10.1

Determine the transfer function of a lead compensator that will provide a phase lead of 45° and gain of 10 dB at $\omega = 8$ rad/s.

Solution

Here we have

$$q = \tan 45^\circ = 1 \qquad c = 10^{M_c/10} = 10$$

$$q^2 - c + 1 = 1 - 10 + 1 = -8$$

$$2q^2 c = 20;$$

and

$$(q^2 c + c - 1)c = (10 + 10 - 1)10 = 190$$

Thus, $-8\alpha^2 + 20\alpha + 190 = 0$.

This has a positive root $\alpha = 6.2812$

Hence

$$a = \frac{\omega_c}{\alpha}\left(\frac{\alpha^2 - c}{c - 1}\right)^{1/2} = 2.304$$

and the transfer function is given by

$$G_c(s) = \frac{\alpha(s + a)}{s + \alpha a} = \frac{6.2812s + 14.472}{s + 14.472}$$

DRILL PROBLEM 10.1

Determine the transfer function of a lead compensator that will provide a phase lead of 50° and gain of 8 dB at $\omega = 5$ rad/s.

Answer:

$$\frac{7.6389(s + 2.0492)}{s + 15.654}$$

DRILL PROBLEM 10.2

While designing a suitable control system for a missile, it was found necessary to introduce a lead of 35° and gain of 6.5 dB at 2.8 rad/s. What will be the transfer function of the lead compensator that will satisfy the above requirements?

Answer:

$$\frac{3.7408(s + 1.2408)}{(s + 4.6416)}$$

An electrical network that can be utilized to obtain the phase-lead characteristic is shown in Fig. 10.6. The transfer function is obtained as

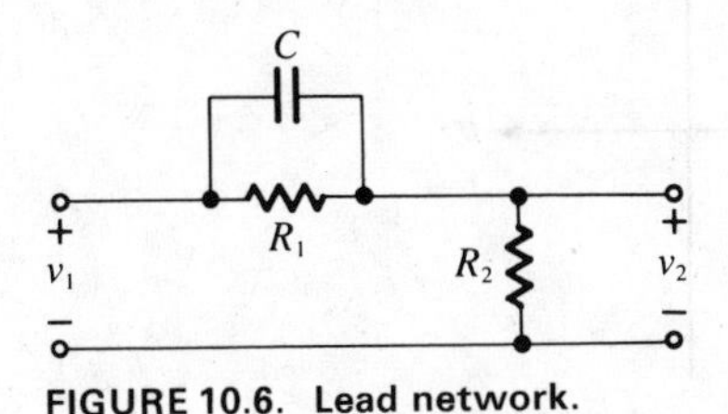

FIGURE 10.6. Lead network.

$$\frac{V_2(s)}{V_1(s)} = \frac{R_2}{R_2 + \dfrac{R_1(1/sC)}{R_1 + (1/sC)}}$$

$$= \frac{R_2\left(R_1 + \dfrac{1}{sC}\right)}{R_2\left(R_1 + \dfrac{1}{sC}\right) + R_1 \cdot \dfrac{1}{sC}} = \frac{R_2 + R_1 R_2 sC}{R_1 + R_2 + R_1 R_2 sC}$$

$$= \frac{s + \dfrac{1}{R_1 C}}{s + \dfrac{R_1 + R_2}{C(R_1 R_2)}} = \frac{s + a}{s + \alpha a} \tag{10.16}$$

where

$$a = \frac{1}{R_1 C} \quad \text{and} \quad \alpha = \frac{R_1 + R_2}{R_2}$$

It should be noted that an additional cascade gain α is required to obtain the transfer function in Eq. (10.1). This can be obtained either by increasing the gain of the main amplifier or by including an operational amplifier along with the RC network.

10.2.2 Lag compensator

The transfer function of the lag compensator is given by

$$G_c(s) = \frac{1}{\alpha}\frac{(s + \alpha a)}{s + a} \qquad \alpha > 1 \tag{10.17}$$

$$= \frac{\beta(s + b)}{s + \beta b} \qquad \beta < 1 \tag{10.18}$$

where $\beta = 1/\alpha$, $b = \alpha a$, and $\beta b = a = b/\alpha$.

The pole-zero plot is shown in Fig. 10.7.

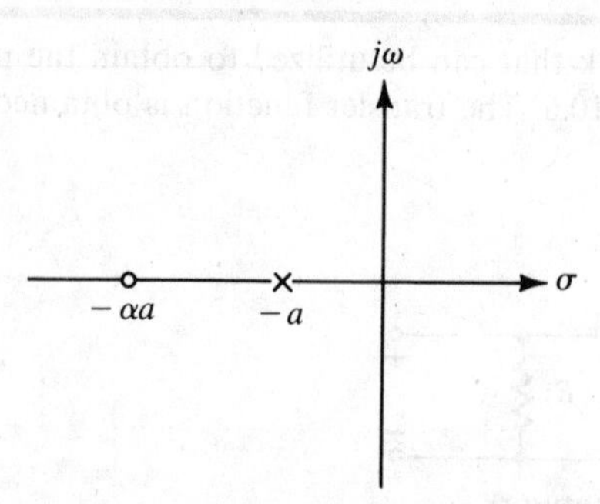

FIGURE 10.7. Pole-zero plot of lag compensator.

The Bode plots of the gain and the phase shift for the lag compensator are shown in Fig. 10.8. Since Eq. (10.18) corresponds exactly with Eq. (10.1), with α replaced by β and a replaced by b, we can use the corresponding relationships derived for the lead compensator.

For a given phase shift ϕ_c and gain M_c at a specified frequency ω_c we get an equation similar to Eq. (10.13), but with α replaced by β and a by b. Hence

$$(q^2 - c + 1)\beta^2 + 2q^2 c\beta + (q^2 c + c - 1)c = 0 \tag{10.19}$$

Note that, in this case, $0 < c < 1$ and $q < 0$. Hence $q^2 - c + 1 > 0$ and $2q^2 c > 0$. Thus a positive and real solution for β will exist if and only if

$$q^2 c + c - 1 < 0 \tag{10.20}$$

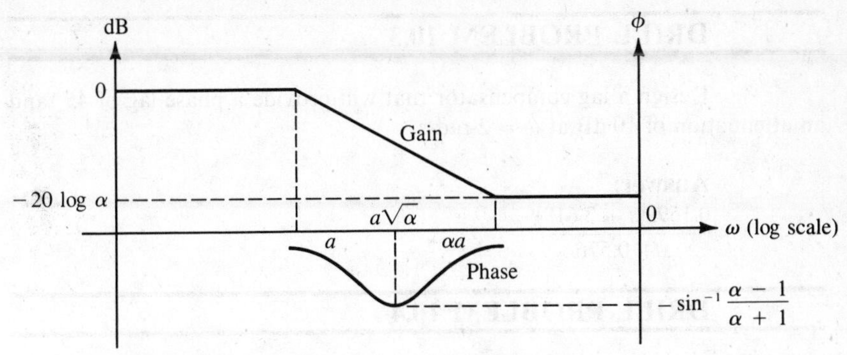

FIGURE 10.8. Bode plots for lag compensator.

Also,

$$b = \frac{\omega_c}{\beta}\left(\frac{\beta^2 - c}{c - 1}\right)^{1/2} \tag{10.21}$$

EXAMPLE 10.2

Design a lag compensator that will provide a phase lag of 50° and an attenuation of 15 dB at $\omega = 2$ rad/s.

Solution

Here we have

$$q = -\tan 50^\circ = -1.192$$

$$c = 10^{M_c/10} = 0.0316$$

$$q^2 - c + 1 = 2.3887$$

$$2q^2c = 0.0898$$

$$(q^2c + c - 1)c = -0.0292$$

Hence

$$2.3887\beta^2 + 0.0898\beta - 0.0292 = 0$$

Solving, $\beta = 0.09336$ is the positive root. Hence

$$b = \frac{\omega_c}{\beta}\left(\frac{\beta^2 - c}{c - 1}\right)^{1/2} = 3.295$$

and the transfer function is obtained as

$$G_c(s) = \frac{\beta(s + b)}{s + \beta b} = \frac{0.09336s + 0.3076}{s + 0.3076}$$

DRILL PROBLEM 10.3

Design a lag compensator that will provide a phase lag of 45° and an attenuation of 10 dB at $\omega = 2$ rad/s.

Answer:

$$\frac{0.1592(s + 3.618)}{s + 0.576}$$

DRILL PROBLEM 10.4

A lag compensator required for a position control system must provide a phase lag of 35° and attenuation of 8 dB at 1.9 rad/s. Determine the transfer function.

Answer:

$$\frac{0.2487(s + 2.5883)}{(s + 0.6438)}$$

The electrical network shown in Fig. 10.9 can be utilized to obtain the phase-lag characteristic.

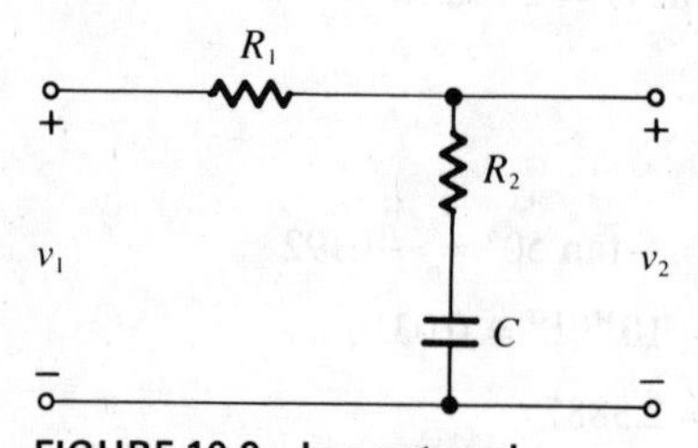

FIGURE 10.9. Lag network.

The transfer function is obtained as

$$\frac{V_2(s)}{V_1(s)} = \frac{R_2 + \dfrac{1}{sC}}{R_1 + R_2 + \dfrac{1}{sC}} = \frac{R_2}{R_1 + R_2} \cdot \frac{s + \dfrac{1}{CR_2}}{s + \dfrac{1}{C(R_1 + R_2)}}$$

$$= \beta \cdot \frac{s + b}{s + \beta b} \tag{10.22}$$

where

$$\beta = \frac{R_2}{R_1 + R_2} \quad \text{and} \quad b = \frac{1}{CR_2}$$

10.2.3 Lag-lead compensator

The transfer function of the lag-lead compensator is given by

$$G_c(s) = \frac{\alpha(s+a)}{s+\alpha a} \cdot \frac{\beta(s+b)}{s+\beta b} \qquad \text{where } \alpha = \frac{1}{\beta} \text{ and } \alpha > 1$$

$$= \frac{(s+a)(s+b)}{(s+\alpha a)\left(s+\dfrac{b}{\alpha}\right)} \tag{10.23}$$

It can be considered as a combination of lag and lead compensators. The pole-zero plot of the transfer function is shown in Fig. 10.10, which is clearly seen to consist of one pole-zero pair for the lag part much closer to the $j\omega$-axis, and another pole-zero pair for the lead part. The Bode plots of the magnitude and phase shift are shown in Fig. 10.11.

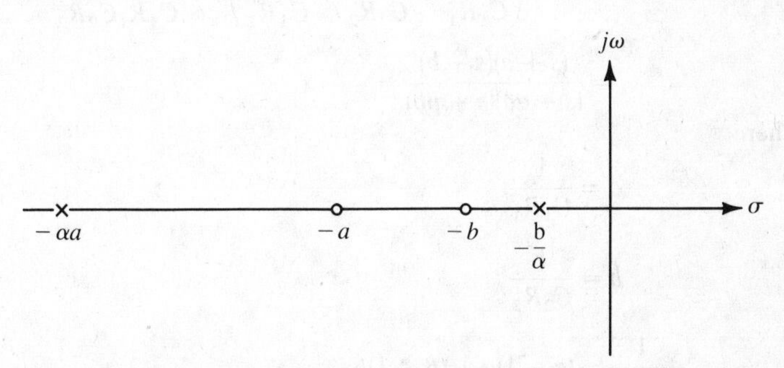

FIGURE 10.10. Pole-zero plot of lag-lead compensator.

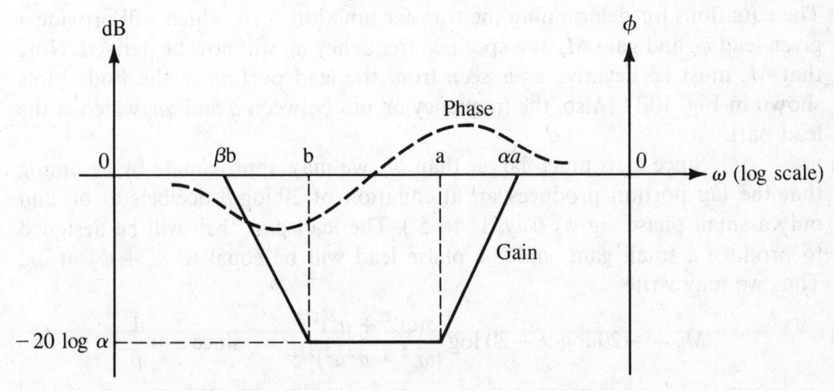

FIGURE 10.11. Bode plots for lag-lead compensator.

An electrical network that has a lag-lead characteristic is shown in Fig. 10.12.

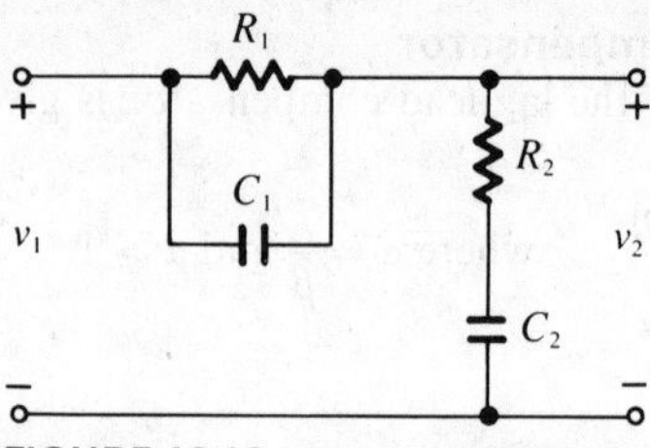

FIGURE 10.12. Lag-lead network.

The transfer function is easily derived as

$$\frac{V_2(s)}{V_1(s)} = \frac{\left(s+\dfrac{1}{C_1R_1}\right)\left(s+\dfrac{1}{C_2R_2}\right)}{s^2+\left(\dfrac{1}{C_1R_1}+\dfrac{1}{C_2R_2}+\dfrac{1}{C_1R_2}\right)s+\dfrac{1}{C_1R_1C_2R_2}}$$

$$= \frac{(s+a)(s+b)}{(s+\alpha a)(s+\beta b)} \tag{10.24}$$

where

$$a = \frac{1}{C_1R_1}$$

$$b = \frac{1}{C_2R_2}$$

$$\frac{1}{C_1R_2} = (\alpha - 1)a + (\beta - 1)b$$

The equations for determining the transfer function $G_c(s)$, which will provide a given lead ϕ_c and gain M_c at a specified frequency ω_c will now be derived. Note that M_c must be negative as is seen from the lead portion of the Bode plots shown in Fig. 10.11. Also, the frequency ω_c lies between a and αa, which is the lead part.

Since ω_c is much larger than βb, we may approximate by assuming that the lag portion produces an attenuation of 20 log β decibels at ω_c and only a small phase lag ϕ_2 (say, 1° to 5°). The lead part then will be designed to produce a small gain, and the phase lead will be equal to $\phi_c + \phi_2$ at ω_c. Thus we may write

$$M_c = -20 \log \alpha + 20 \log \frac{\alpha(\omega_c^2 + a^2)^{1/2}}{(\omega_c^2 + \alpha^2 a^2)^{1/2}} \qquad \text{since } \alpha = \frac{1}{\beta}$$

$$= 10 \log \frac{1 + \left(\dfrac{\omega_c}{a}\right)^2}{\alpha^2 + \left(\dfrac{\omega_c}{a}\right)^2} \tag{10.25}$$

or

$$\left(\frac{\omega_c}{a}\right)^2 = \frac{\alpha^2 c - 1}{1 - c} \tag{10.26}$$

where

$$c = 10^{M_c/10} < 1 \quad \text{since } M_c < 0$$

Also,

$$\phi_c + \phi_2 = \tan^{-1} \frac{\dfrac{\omega_c}{a} - \dfrac{\omega_c}{\alpha a}}{1 + \left(\dfrac{\omega_c}{a}\right)\left(\dfrac{\omega_c}{\alpha a}\right)} = \tan^{-1} \frac{\dfrac{\omega_c}{a}(\alpha - 1)}{\alpha + \left(\dfrac{\omega_c}{a}\right)^2} \tag{10.27}$$

and

$$q = \tan(\phi_c + \phi_2) = \frac{\dfrac{\omega_c}{a}(\alpha - 1)}{\alpha + \left(\dfrac{\omega_c}{a}\right)^2} \tag{10.28}$$

Eliminating ω_c/a from Eqs. (10.26) and (10.28) and canceling out the common factor $(\alpha - 1)^2$, since $\alpha \neq 1$, we get

$$(q^2 c + c - 1)c\alpha^2 + 2q^2 c\alpha + (q^2 - c + 1) = 0 \tag{10.29}$$

Note that $0 < c < 1$, $q > 0$, $2q^2 c > 0$ and $2q^2 - c + 1 > 0$. Hence, a positive root for α in Eq. (10.29) will exist if and only if

$$(q^2 c + c - 1)c < 0 \tag{10.30}$$

or, equivalently,

$$c < \frac{1}{q^2 + 1} \tag{10.31}$$

After α is determined from Eq. (10.29), the value of a is obtained from Eq. (10.26); that is,

$$a = \omega_c \left(\frac{1 - c}{\alpha^2 c - 1}\right)^{1/2} \tag{10.32}$$

The value of b must now be fixed so that the lag portion produces a phase lag of ϕ_2 at ω_c, as stipulated earlier. This is done by noting that

$$\tan(-\phi_2) = \frac{\frac{\omega_c}{b} - \frac{\omega_c}{\beta b}}{1 + \frac{\omega_c}{b}\cdot\frac{\omega_c}{\beta b}} = \frac{\frac{\omega_c}{b}(1-\alpha)}{1 + \left(\frac{\omega_c}{b}\right)^2 \alpha} \qquad \text{since } \alpha = \frac{1}{\beta} \tag{10.33}$$

This can be simplified to obtain the quadratic equation

$$\left(\frac{b}{\omega_c}\right)^2 + \frac{\alpha - 1}{\tan(-\phi_2)}\left(\frac{b}{\omega_c}\right) + \alpha = 0 \tag{10.34}$$

Equation (10.34) has two real and positive roots, and we take the smaller of these two to obtain the value of b.

EXAMPLE 10.3

Determine the transfer function of a lag-lead compensator that will provide a phase lead of 50° and attenuation of 15 dB at $\omega = 6$ rad/s.

Solution

Let the contribution of the lag part at $\omega_c = 6$ be given by $\phi_2 = 2°$. Then

$$q = \tan(50° + 2°) = 1.28$$

$$c = 10^{M_c/10} = 10^{-1.5} = 0.0316$$

$$(q^2c + c - 1)c = -0.02898$$

$$2q^2c = 0.1036$$

$$q^2 - c + 1 = 2.6066$$

Hence

$$-0.02898\alpha^2 + 0.1036\alpha + 2.6066 = 0$$

The solution gives $\alpha = 11.4376$ as the positive root. Hence

$$a = \omega_c\left(\frac{1-c}{\alpha^2 c - 1}\right)^{1/2} = 3.3337$$

and

$$\frac{\alpha - 1}{\tan(-\phi_2)} = -298.893$$

This gives the quadratic

$$\left(\frac{b}{\omega_c}\right)^2 - 298.893\left(\frac{b}{\omega_c}\right) + 11.4376 = 0$$

Solving and using the smaller root,

$$\frac{b}{\omega_c} = 0.0383 \quad \text{and} \quad b = 0.2296$$

Hence

$$G(s) = \frac{(s+a)(s+b)}{(s+\alpha a)(s+\beta b)} = \frac{(s+3.3337)(s+0.2296)}{(s+38.1295)(s+0.0201)}$$

$$= \frac{s^2+3.5633s+0.7655}{s^2+38.1496s+0.7655}$$

DRILL PROBLEM 10.5

Determine the transfer function of a lag-lead compensator that will provide a phase lead of 55° and attenuation of 20 dB at $\omega = 4$ rad/s.

Answer:

$$\frac{(s+2.1210)(s+0.1466)}{(s+45.1288)(s+0.00689)}$$

10.3 APPROACHES TO COMPENSATION

A number of different approaches for compensating a control system have been proposed in the literature. Some of these will be described here. The earliest approaches were based on modifying the frequency response of the forward path of a unity-feedback system in such a manner that the performance of the closed-loop system is satisfactory. For this method of design the steady-state accuracy is specified in terms of the error coefficients K_p, K_v, etc., and the transient performance is specified in terms of the gain margin, the phase margin, and the maximum value of the closed-loop frequency response, M_m. The speed of response is specified either in terms of the crossover frequency ω_c or the frequency ω_m at which the maximum closed-loop response is obtained. It may be noted that although these quantities are related to the transient performance in an indirect manner, they are conveniently determined from the open-loop frequency response.

Another approach to compensation is through the root locus method, where the root locus is reshaped through the use of compensators so that the dominant poles of the closed-loop system are at suitable locations. In this case we get a more direct correlation between the specifications and the desired transient response.

A more direct approach to the design of a control system is to specify the closed-loop transfer function that will give a satisfactory performance. The compensator is then designed so that the system is forced to have this transfer function. The selection of the desired closed-loop transfer function is the most important step in this procedure. One must also make the design physically realizable and economical. Other desirable properties are insensitivity to small variations in parameters, as well as disturbance rejection.

It should be evident to the reader that in all these methods there is a certain degree of arbitrariness involved in deciding what is the best control system for a particular application. Once "suitable" specifications have been obtained, there are many possible solutions that will satisfy these and one may use other criteria for selection. A unique solution is obtained when a quantitative performance index, usually called the "cost function," is minimized. This has led to the development of optimal control theory.

In the following sections we shall be considering some of the methods of compensation described earlier. These methods, and typical corresponding performance specifications, are summarized in Table 10.1.

TABLE 10.1. Typical performance specifications for different methods of compensation

Method of Compensation	*Performance Criteria*	*Type of Compensation*
Frequency response (Bode plots)	Steady-state error coefficients, phase margin, crossover frequency	Cascade lead, lag, or lag-lead compensation
Frequency response (Nichols charts)	Steady-state error coefficients, maximum closed-loop frequency response M_m and frequency ω_m	Cascade lead, lag, or lag-lead compensation
Root locus	Steady-state error coefficients, location of dominant poles of the closed-loop, root sensitivity	Cascade lead or lag compensation, feedback compensation
Pole-placement compensation	Desired closed-loop transfer function, sensitivities of the poles to parameter variations	More general compensators in cascade, feedback, or both
Time-domain Methods	Integral performance criteria in the time domain	Optimal control theory is utilized to determine the optimum input to the plant

10.4 COMPENSATION USING BODE PLOTS

Historically, this is the earliest approach to the design of control systems. Its main attraction is that one does not have to know precisely the transfer function of the plant to be controlled. The frequency response of the open-loop transfer function is measured experimentally and then reshaped, if necessary, to satisfy the performance specifications. These consist of the steady-state error coefficients, the phase margin, and the phase crossover frequency.

The first step in the design is to adjust the loop gain so that the steady-state accuracy requirement is satisfied. The frequency response of the uncompensated open-loop transfer function is then calculated. If it does not satisfy the phase margin specification, a lag, lead, or lag-lead compensator must be included in cascade in the forward path. The choice of the type of compensation depends on the gain-phase characteristic at the desired gain crossover frequency. For a lag compensator, the gain (in decibels) should be positive at the desired crossover frequency and the phase lag less than $180° - \phi_m$, where ϕ_m is the specified phase margin. For a lead compensator the gain at the desired crossover frequency should be negative and the phase lag more than $180° - \phi_m$. For each case, the magnitude of the phase shift to be provided by the compensator must be less than 90° and, preferably, less than 70°. For a lag-lead compensator, the gain at the desired crossover frequency should be positive and the phase lag more than $180° - \phi_m$.

This can be understood better by examining the polar plot shown in Fig. 10.13. Here, if we want to use a lead compensator to satisfy the design specifications, we should select ω_1, corresponding to the frequency at the point A, as the gain crossover frequency, so that the compensator will be designed

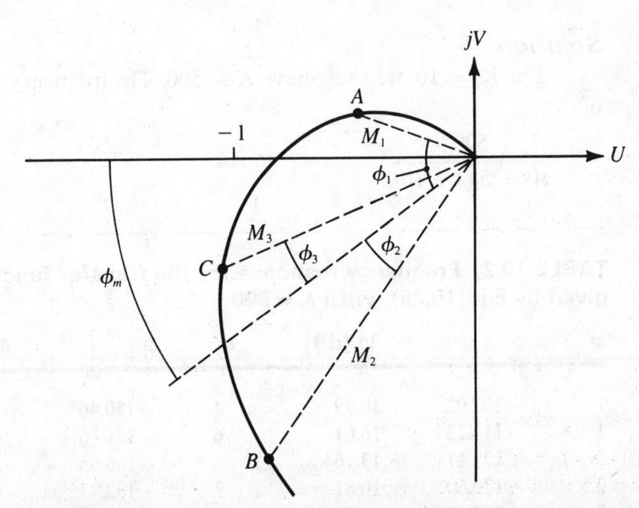

FIGURE 10.13. Polar plot of a typical open-loop transfer function requiring compensation to satisfy the phase margin specification.

to provide a lead of ϕ_1 and a gain of 20 log M_1 decibels at the frequency ω_1. Similarly, for a lag compensator to satisfy the specifications, we should select ω_2, corresponding to the frequency at the point B as the gain crossover frequency. In this case the compensator should provide a phase lag of ϕ_2 degrees and an attenuation of 20 log M_2 decibels at the frequency ω_2. Finally, for a lag-lead compensator, the crossover frequency may be selected as ω_3, corresponding to the point C. In this case the compensator should provide a phase lead of ϕ_3 degrees and attenuation of 20 log M_3 decibels at the frequency ω_3.

It will be seen from the above that with a lead compensator, we get a higher gain crossover frequency and thus a larger bandwidth than possible with a lag compensator. The larger bandwidth results in a faster response but also makes the system susceptible to more noise. The lag-lead compensator has a bandwidth intermediate between those of the lag and the lead compensators. The actual choice will depend upon what is more important for a particular application.

The procedure will be easily understood through the following examples.

EXAMPLE 10.4

The transfer function of a plant is given by

$$G_p(s) = \frac{K}{s(s+5)(s+10)} \tag{10.35}$$

Design a suitable compensator to meet the following specifications.

(a) $K_v = 10$
(b) Phase margin = 45°

Solution

For $K_v = 10$, we must have $K = 500$. The frequency response of

$$\frac{500}{s(s+5)(s+10)}$$

TABLE 10.2. Frequency response for the transfer function given by Eq. (10.35), with $K = 500$

ω	ϕ	M (dB)	ω	ϕ	M (dB)
1	−107.02°	19.79	4	−150.46°	5.166
1.5	−115.23°	16.01	6	−171.16°	−0.7723
2	−123.11°	13.164	8	−186.65°	−5.7247
2.5	−130.60°	10.81	9	−192.93°	−7.935
3	−137.66°	8.75	10	−198.43°	−10.00

is shown in Table 10.2. It is seen that the phase margin specification is not met, since the phase margin is only about 12°.

If we want to design a lead compensator, the gain crossover frequency should be more than 6 and less than 10. These limits are set by the fact that we must have a negative gain for the lead compensator, and the required phase lead should not exceed 65°. If we select $\omega_c = 8$, the lead compensator is required to have a phase lead of 186.65° − 135° = 51.65° and gain equal to 5.73 dB at $\omega_c = 8$ rad/s. Following the approach shown in Sec. 10.2.1 the transfer function for the compensator is obtained as

$$G_c(s) = \frac{12.726(s + 4.78)}{s + 60.831} \tag{10.36}$$

The frequency response of the compensated system is shown in Fig. 10.14, along with that of the uncompensated system.

It may be added that one could also have chosen $\omega_c = 7$ or $\omega_c = 9$ to design the lead compensator. These give different values of α, the additional gain required for the compensator. One may choose the value of the crossover frequency that minimizes α. Another practical criterion is the bandwidth of the system, which is related to the crossover frequency (where the gain of the open-loop transfer function crosses the zero-decibel line).

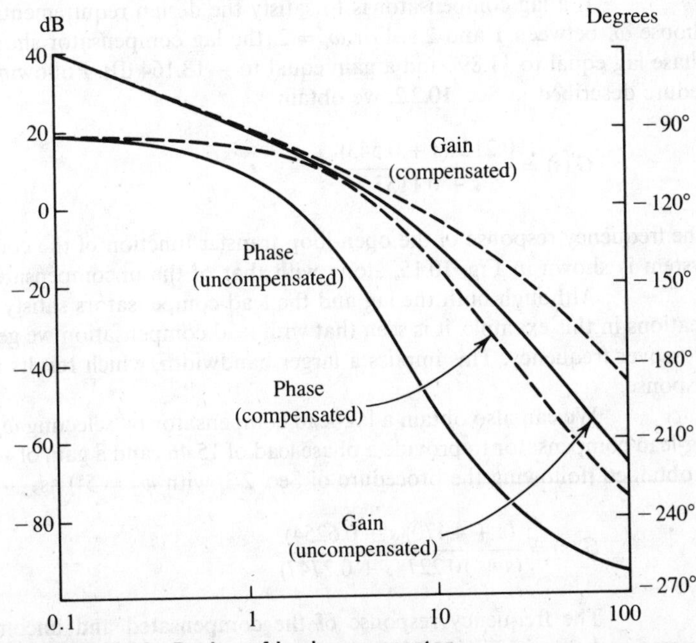

FIGURE 10.14. Design of lead compensator.

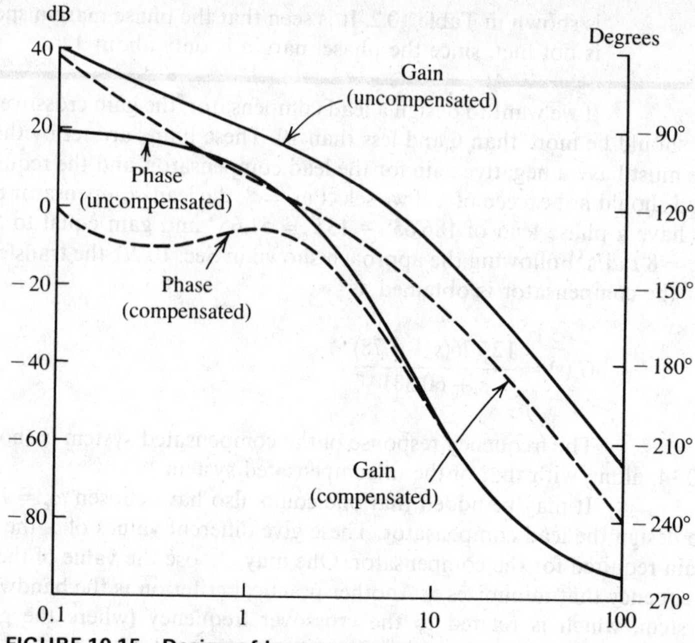

FIGURE 10.15. Design of lag compensator.

If a lag compensator is to satisfy the design requirements, we must choose ω_c between 1 and 2.5. For $\omega_c = 2$, the lag compensator should have a phase lag equal to 11.89° and a gain equal to −13.164 dB. Following the procedure described in Sec. 10.2.2, we obtain

$$G_c(s) = \frac{0.2124(s + 0.543)}{s + 0.1153} \tag{10.37}$$

The frequency response of the open-loop transfer function of the compensated system is shown in Fig. 10.15, along with that of the uncompensated system.

Although both the lag and the lead compensators satisfy the specifications in this example, it is seen that with lead compensation we get a higher crossover frequency. This implies a larger bandwidth, which results in a faster response.

We can also obtain a lag-lead compensator by selecting $\omega_c = 4$. The lag-lead compensator to provide a phase lead of 15.46° and a gain of −5.166 dB is obtained (following the procedure of Sec. 2.3, with $\phi_2 = 5°$) as

$$G_c(s) = \frac{(s + 4.470)(s + 0.6254)}{(s + 10.227)(s + 0.2747)} \tag{10.38}$$

The frequency response of the compensated and uncompensated systems is shown in Fig. 10.16.

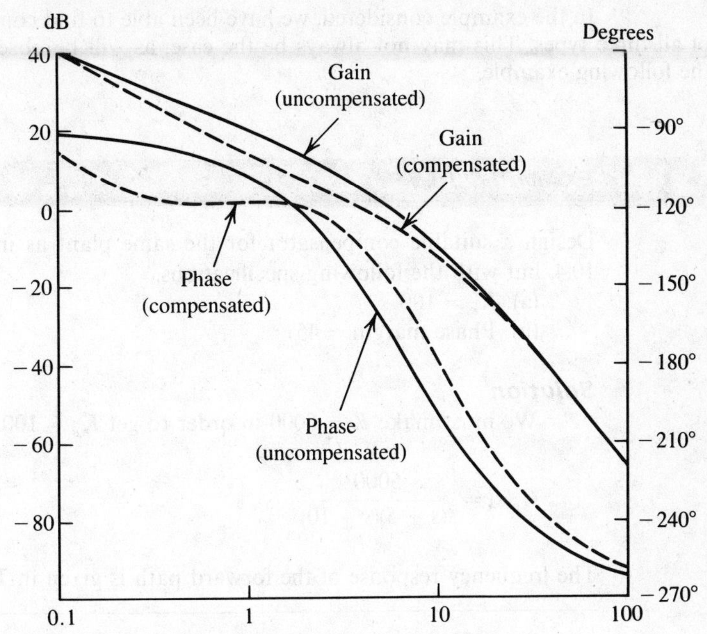

FIGURE 10.16. Design of lag-lead compensator.

The response of the compensated system to a unit step input for each of the three types of compensators is shown in Fig. 10.17. As expected, the system with the lead compensator has the fastest response, whereas that with the lag compensator has the slowest response.

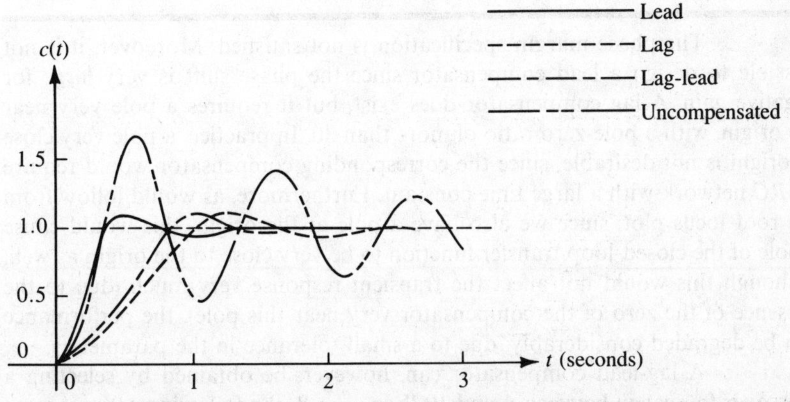

FIGURE 10.17. Responses of the compensated systems to unit step input.

In the example considered, we have been able to find compensators of all three types. This may not always be the case, as will be observed from the following example.

EXAMPLE 10.5

Design a suitable compensator for the same plant as in Example 10.4, but with the following specifications.

(a) $K_v = 100$

(b) Phase margin = 45°

Solution

We must make $K = 5000$ in order to get $K_v = 100$. Hence

$$G_p(s) = \frac{5000}{s(s+5)(s+10)} \tag{10.39}$$

The frequency response of the forward path is given in Table 10.3.

TABLE 10.3 Frequency response for the transfer function given in Eq. (10.39)

ω	ϕ	M (dB)	ω	ϕ	M (dB)
2	−123.11°	33.16	12	−198.43°	10.00
2.5	−130.6°	30.81	15	−217.87°	1.3593
4	−150.46°	25.166	16	−220.64°	−0.1046
6	−171.16°	19.228	20	−229.40°	−5.315
8	−186.65°	14.275			

The phase margin specification is not satisfied. Moreover, it is not possible to design a lead compensator since the phase shift is very large for negative gain. A lag compensator does exist, but it requires a pole very near the origin, with a pole-zero ratio of more than 30. In practice, a pole very close to origin is not desirable, since the corresponding compensator would require an RC network with a large time constant. Furthermore, as would follow from the root locus plot, since we also have a pole at the origin, this would cause a pole of the closed-loop transfer function to be very close to the origin as well. Although this would not affect the transient response very much (due to the presence of the zero of the compensator very near this pole), the performance can be degraded considerably, due to a small tolerance in the parameters.

A lag-lead compensator can, however, be obtained by selecting a crossover frequency between 4 and 10. For $\omega_c = 8$, the following compensator is obtained, with $\alpha = 13.09$.

$$G_c(s) = \frac{(s + 3.422)(s + 0.7587)}{(s + 44.499)(s + 0.05835)} \tag{10.40}$$

10.5 COMPENSATION USING NICHOLS CHARTS

The procedure discussed in the previous section is based on the assumption that a good phase margin will produce a desirable transient response for the closed-loop system. As would be evident from Example 9.4, although the lead, lag, and lag-lead compensators were designed to give the same phase margin in each case, the step responses of the resulting closed-loop systems are quite different, as shown in Fig. 10.17. Another approach is based on the fact that closed-loop poles with small damping ratios produce frequency response curves with large peaks in the magnitude. Therefore, in order to obtain a desirable transient response, one should require that M_m, the maximum magnitude ratio in the closed-loop frequency response, does not exceed a specified value, say 1.5 to 4 dB. This is done by using the Nichols chart. The procedure will be evident from the following example.

EXAMPLE 10.6

The forward transfer function of a unity-feedback system is given by

$$G_p(s) = \frac{10e^{-0.05s}}{s^2(s + 10)} \tag{10.41}$$

Design a suitable compensator so that M_m does not exceed 3.5 dB and ω_m is nearly 1.4 rad/s.

Solution

The open-loop frequency response is shown in Table 10.4. These values are plotted in the Nichols chart in Fig. 10.18. For M_m

TABLE 10.4. Frequency response for the transfer function $G_p(s)$ given in Eq. (10.41)

ω	ϕ	M (dB)	ω	ϕ	M (dB)
0.8	−186.866°	3.849	1.8	−195.361°	−10.349
1	−188.575°	−0.043	2.0	−197.04°	−12.212
1.2	−190.28°	−3.229	2.5	−201.198°	−16.181
1.4	−191.98°	−5.929	3.0	−205.294°	−19.459
1.6	−193.674°	−8.275			

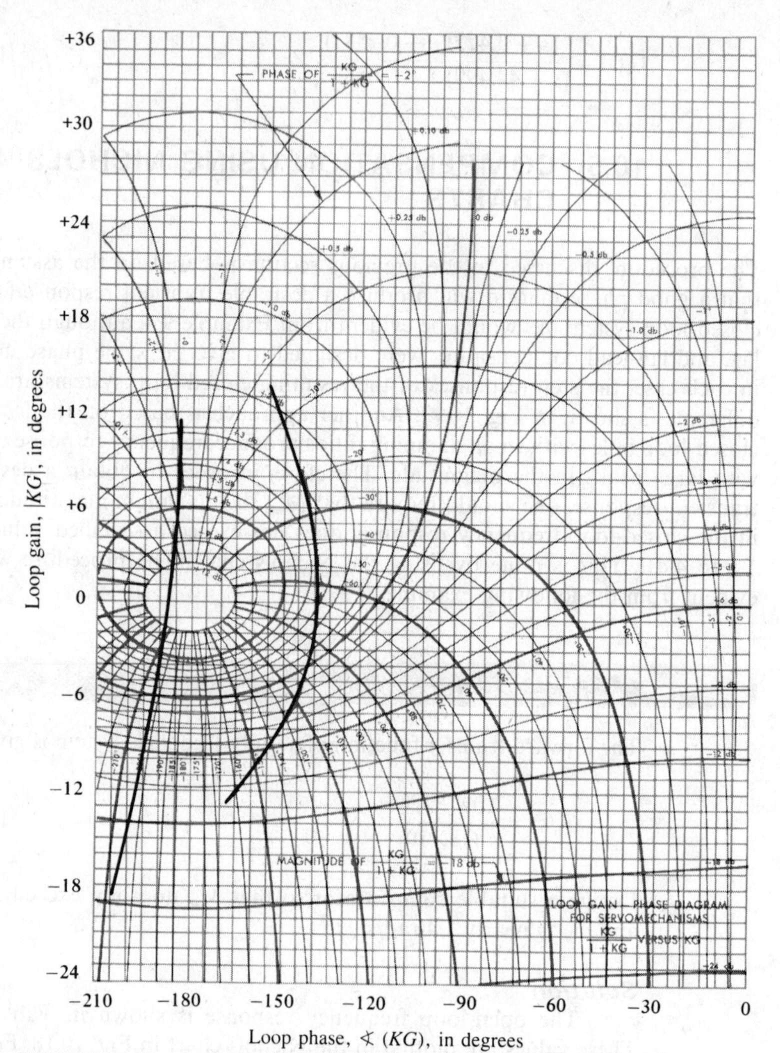

FIGURE 10.18. Nichols chart design of a lead compensation.

to be 3.5 dB at $\omega_m = 1.4$, the plot should be shifted to the right by 55° and raised by 7 dB. This was determined by first locating the point corresponding to $\omega = 1.4$ in the plot of the uncompensated system, and then deciding on the amount of lead and gain required to make this point tangential to the 3.5-dB closed-loop gain curve. (Since the Nichols chart has only 3-dB and 4-dB closed-loop gain curves, one must do a little interpolation.) Hence, the lead compensator that will provide a lead of 55° and gain of 7 dB at $\omega = 1.4$ is calculated using the procedure of Sec. 10.2.1 and the following transfer function is obtained.

$$G_c(s) = \frac{13.1224(s + 0.6887)}{s + 9.0376} \tag{10.42}$$

The frequency response of $G_p(s)G_c(s)$ is shown in Table 10.5 and has been plotted along with that of the uncompensated system on the Nichols chart in Fig. 10.18. It is seen that the M_m specification is satisfied, with ω_m nearly equal to 1.4.

TABLE 10.5. Frequency response for $G_p(s)G_c(s)$ as given by Eqs. (10.41) and (10.42)

ω	ϕ	M (dB)	ω	ϕ	M (dB)
0.5	−151.5°	13.86	1.8	−137.6°	−1.58
0.7	−145°	9.2	2.0	−138.5°	−2.67
1.0	−139.4°	4.83	2.5	−142.06°	−5.0
1.2	−137.7°	2.75	3.0	−146.6°	−6.9
1.4	−137°	1.07	3.5	−167.7°	−12.8

10.6 COMPENSATION USING ROOT LOCUS

The compensation methods described in the two preceding sections specify the transient performance in an indirect manner. A more direct approach is to specify the location of the dominant poles of the closed-loop system. This may usually be done by specifying the damping ratio of these poles and requiring that they be to the left of a certain vertical line in the s-plane in order that the settling time will be less than a specified value.

10.6.1 Cascade compensation using root locus

The root locus method can be utilized to design a cascade compensator to meet these specifications. From the root locus plot of the uncompensated system, one determines the value of the loop gain K to obtain the dominant poles with the desired damping ratio. If these poles are suitably located, but the steady-state specification is not met, one may add a lag compensator with pole and zero sufficiently close to the origin so that the root locus is not altered appreciably but the low-frequency gain is suitably increased. On the other hand, if the dominant poles must be moved to the left, a lead compensator must be designed. The procedure for the two cases will be evident from the following examples.

EXAMPLE 10.7

The forward transfer function of a unity-feedback system is given by

$$G_p(s) = \frac{K}{s(s+3)(s+6)} \tag{10.43}$$

It is specified that the dominant poles of the closed-loop system have a damping ratio equal to 0.5 and that the magnitude of the real part of the poles be not less than 1. Also, the steady-state error to a unit ramp input must not exceed 0.1. Design a suitable compensator.

Solution

The root locus plot for open-loop function is shown in Fig. 10.19. From this plot, it is determined that for $K = 28$, the characteristic polynomial of the closed-loop system has roots with damping ratio of 0.5 and is given by $(s + 7)(s^2 + 2s + 4)$. The real part of the complex poles is -1, which just meets the specifications.

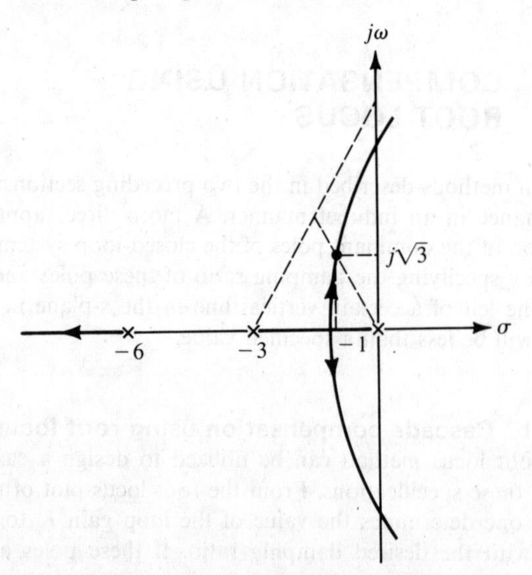

FIGURE 10.19. Root locus plot for $G_p(s)$.

For $K = 28$, we get

$$K_v = \frac{28}{18} = \frac{14}{9}$$

which is less than the desired value of 10.

Hence, a lag compensator must be included, which does not appreciably affect the closed-loop poles and has a dc gain equal to 45/7 in order to provide the desired K_v. The following transfer function for the compensator will be satisfactory.

$$G_c(s) = \frac{s + (0.01)(45/7)}{s + 0.01} = \frac{s + 0.0643}{s + 0.01} \tag{10.44}$$

where the distances from the origin of the compensated pole and zero are chosen to be small compared with the distance of the dominant poles from the origin so that the compensator will not contribute significant phase lag near the crossover frequency. A plot of the root locus of the system with the lag compensator is shown in Fig. 10.20.

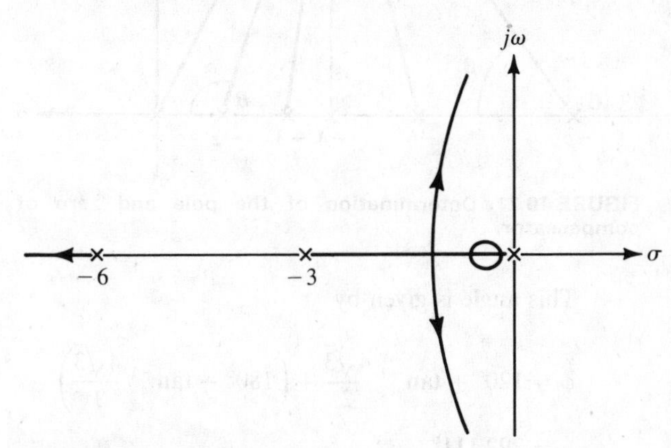

FIGURE 10.20. Root locus plot for $G_p(s)G_c(s)$.

It will be seen that the root locus plot for $G_p(s)G_c(s)$ is not significantly different. An additional pole is added to the closed-loop system, but since it is very near a zero, the contribution to the transient response is negligible.

EXAMPLE 10.8

Consider the same system as in Example 10.7, but with the specification that the magnitude of the real part of the dominant poles must not be less than 4.

Solution

In this case we must use a lead compensator to shift the root locus to the left. The first step in the procedure is to determine the total angle subtended by the poles and zeros of $G_p(s)$ at the desired root location, which in this case will be taken as $-4 + j4\sqrt{3}$, as shown in Fig. 10.21.

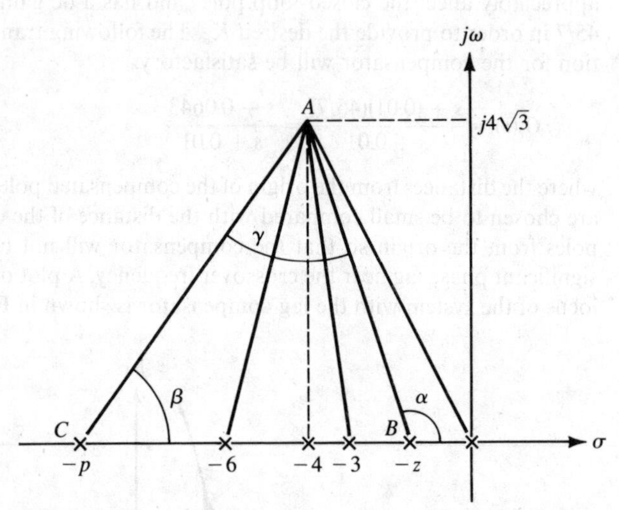

FIGURE 10.21. Determination of the pole and zero of the lead compensator.

This angle is given by

$$\phi = 120^\circ + \tan^{-1}\frac{4\sqrt{3}}{2} + \left(180^\circ - \tan^{-1}\frac{4\sqrt{3}}{1}\right)$$

$$= 292.111^\circ$$

Now if the zero of the compensator is located at $-z$, and the pole at $-p$, these must be selected in such a manner that

$$\phi + \beta - \alpha = 180^\circ$$

Hence $\gamma = \alpha - \beta = \phi - 180^\circ = 112.111^\circ$.

This suggests a simple construction for obtaining p for a given choice of z. For example, we can arbitrarily select the location of the zero at the point B, and then determine location of the pole by drawing the line AC making an angle $\gamma = 112.111^\circ$, as shown in Fig. 10.21. A solution is possible as long as the point C lies on the negative real axis, the limiting case will occur when AC is parallel to the real axis. Hence, a solution will not exist if z is greater than 4 −

$4\sqrt{3}\tan 22.1111°$—that is, 1.18. Some possible solutions are given below.

z	0.2	0.4	0.6	0.7	0.8	0.9	1.0	1.15
p	63.579	78.04	102.38	122.08	152.0	202.9	308.78	1195.6
p/z	317.9	195.1	170.6	174.4	190.09	225.44	308.78	1387.5

If we select $z = 0.6$, which requires the smallest p/z ratio in this table, we get the compensated open-loop transfer function

$$G_p(s)G_c(s) = \frac{K(s + 0.6)}{s(s + 3)(s + 6)(s + 102.38)} \tag{10.45}$$

and the value of K to obtain the complex poles at $s = -4 + j4\sqrt{3}$ is determined from Fig. 10.21 as

$$K = \frac{8 \times 7 \times 7.211 \times 98.624}{7.7175} = 5160.5$$

For this value of K we get

$$K_v = \frac{5160.5 \times 0.6}{3 \times 6 \times 102.38} = 1.68$$

Hence, in order to get $K_v = 10$, we need an additional lag compensator that will provide a dc gain of 10/1.68 or 5.95. The following will be adequate

$$G_c'(s) = \frac{s + 0.06}{s + 0.01} \tag{10.46}$$

We shall now consider an alternative procedure in which we start by determining the location of the poles of the closed-loop transfer function that will satisfy the specifications. The transfer function of the open-loop system is then altered, by the choice of suitable compensators, so that these poles are obtained for the closed-loop transfer function. This may be done most conveniently by designing compensators the zeros of which cancel the poles of the open-loop transfer function. Thus, in effect, the poles of the open-loop transfer function are moved so that the transfer function of the closed loop system is forced to have the specified poles. For example, the specifications for this problem can be met if the dominant poles of the closed-loop transfer function are located at

$s = -8 \pm j8\sqrt{3}$. This can be done by using the compensator

$$G_s(c) = \frac{s+3}{s+p_1} \cdot \frac{s+6}{s+p_2} \tag{10.47}$$

where the values p_1 and p_2 can be selected by using the geometric construction discussed earlier. As will be seen from Fig. 10.22, we must select the locations of p_1 and p_2 such that the angles α and β add up to 60° in order that the point A will satisfy the angle of the root locus. Many solutions are possible. The most convenient solution is to make $p_1 = p_2$, so that $\alpha = \beta = 30°$ and the poles of the compensators are located at $s = -32$. From Fig. 10.22, by using the rules of root locus, it is easily found that for this case $K = 12{,}288$, and the resulting $K_v = 12$.

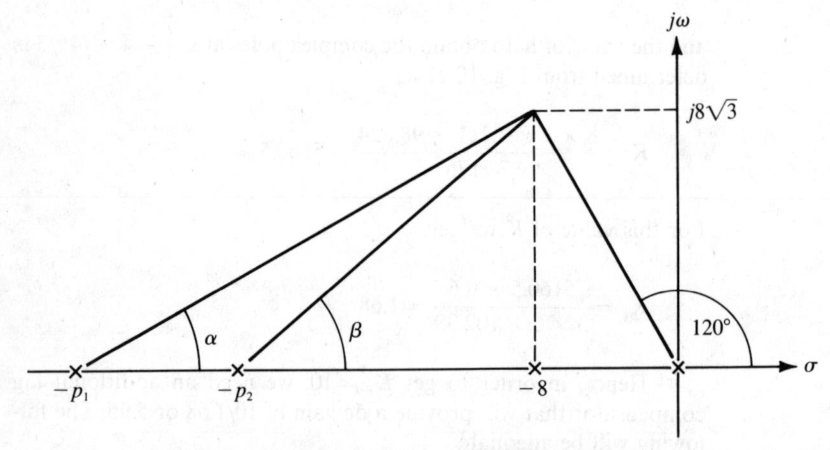

FIGURE 10.22. Construction for locating the poles of compensator while canceling the poles of the plant transfer function.

Although the pole-zero cancellation approach discussed above appears simple as well as quite elegant, a word of caution is necessary. In most practical situations, we must allow some tolerances in the values of the components. As a result, exact pole-zero cancellation may not always be obtained. Hence, to be realistic, one must take this fact into consideration and plot the root locus for the case when the poles of the plant transfer function are not exactly canceled by the zeros of the compensator. In such cases, we may get some poles of the closed-loop system transfer function that are much closer to the $j\omega$-axis and will lead to larger time constants. However, since these poles are very close to the zeros of the compensator, their contribution to the transient response may still be quite small. This would need careful investigation for a practical design.

10.6.2 Feedback compensation using root locus

The root locus method can also be used for the design of feedback compensation. The idea, again, is to shape the locus so that the dominant poles have desired locations. It will be illustrated through the following example.

EXAMPLE 10.9

The forward transfer function of a unity-feedback position-control system is given by

$$G_p(s) = \frac{60}{s(s+2)} \tag{10.48}$$

It is desired to add velocity feedback as shown in Fig. 10.23, so that the damping ratio of the closed-loop poles is 0.5. Determine the value of α for this case.

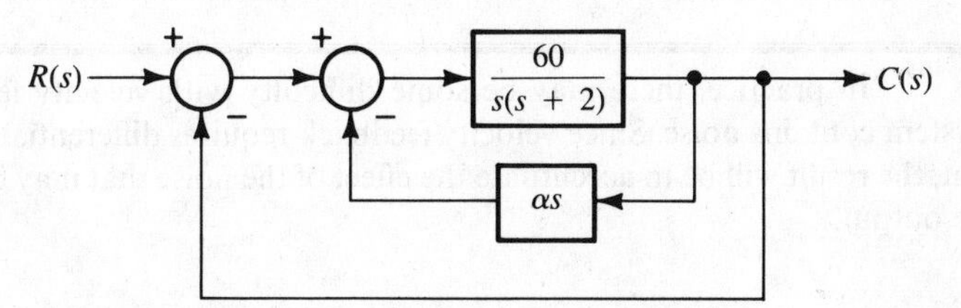

FIGURE 10.23. System with velocity feedback.

Solution

The characteristic polynomial of the system with velocity feedback is obtained as

$$\begin{aligned} P(s) &= s^2 + (60\alpha + 2)s + 60 \\ &= (s^2 + 2s + 60) + 60\alpha s \end{aligned} \tag{10.49}$$

Thus we may draw the root locus for

$$\frac{60\alpha s}{s^2 + 2s + 60}$$

and determine the value of α for 0.5 damping ratio. The root locus plot is shown in Fig. 10.24. The value of α for the desired root location is found to be 0.0958.

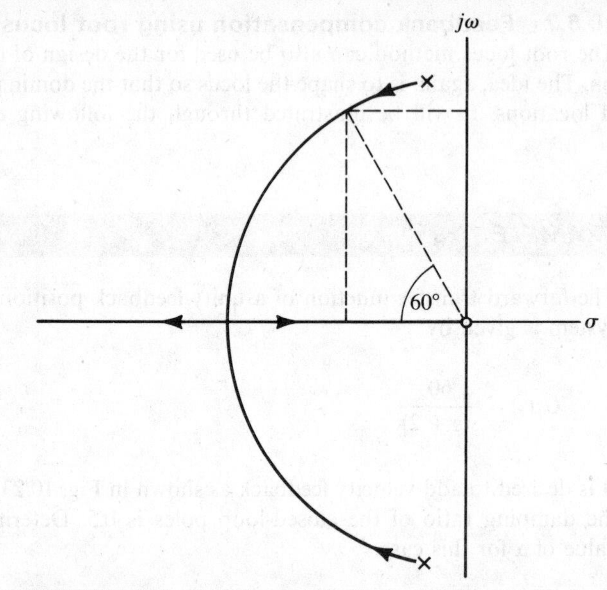

FIGURE 10.24. Root locus plot for the system with velocity feedback.

In practice, there may be some difficulty with velocity feedback if the system contains noise. Since velocity feedback requires differentiation of the output, the result will be to accentuate the effect of the noise that may be present in the output.

10.7 POLE-PLACEMENT COMPENSATION

The philosophy of design in this approach is to select the poles of the closed-loop system in such a manner that the specifications for steady-state accuracy as well as good transient response are satisfied. A compensator is then designed that forces the closed-loop system to have this transfer function. The general scheme for compensation that will achieve this is shown in Fig. 10.25. As will be seen, two compensators are required, both of which have the same denominator. Let

$$KG_p(s) = \frac{N_p(s)}{D_p(s)} \tag{10.50}$$

$$G_u(s) = \frac{N_u(s)}{\Delta(s)} \tag{10.51}$$

and

$$G_c(s) = \frac{N_c(s)}{\Delta(s)} \tag{10.52}$$

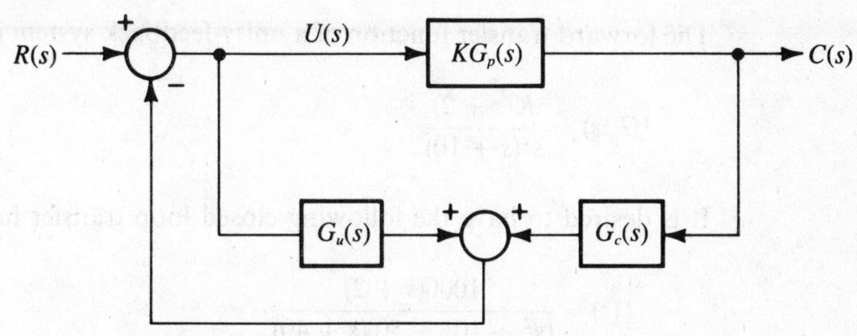

FIGURE 10.25. Compensators for pole placement.

Then, applying Mason's rule, the overall transfer function is obtained as

$$\frac{C(s)}{R(s)} = \frac{KG_p(s)}{1 + G_u(s) + KG_p(s)G_c(s)} \tag{10.53}$$

Let the desired closed-loop transfer function be $N_p(s)/D(s)$. Then, from Eq. (10.53),

$$\frac{N_p(s)}{D(s)} = \frac{\dfrac{N_p(s)}{D_p(s)}}{1 + \dfrac{N_u(s)}{\Delta(s)} + \dfrac{N_p(s)}{D_p(s)} \cdot \dfrac{N_c(s)}{\Delta(s)}}$$

$$= \frac{N_p(s)\Delta(s)}{\Delta(s)D_p(s) + N_u(s)D_p(s) + N_p(s)N_c(s)} \tag{10.54}$$

Equation (10.54) can be rearranged to give

$$[D(s) - D_p(s)]\Delta(s) = N_c(s)N_p(s) + N_u(s)D_p(s) \tag{10.55}$$

It may be noted that $D(s)$ and $D_p(s)$ are monic polynomials of the same degree, say n. Hence, $D(s) - D_p(s)$ is a polynomial of degree $n - 1$. If we assume $\Delta(s)$ to be a stable monic polynomial of degree $n - 1$, then $N_c(s)$ and $N_u(s)$ can be obtained as polynomials of degree $n - 1$, if and only if $N_p(s)$ and $D_p(s)$ are relatively prime—that is, if there are no cancellations in the plant transfer function $G_p(s)$. As will be shown in Chapter 11, this is equivalent to the requirement that the system be both controllable and observable. A formal proof of this result is given in Kailath [1]. We shall illustrate the procedure by means of an example.

EXAMPLE 10.10

The forward transfer function of a unity-feedback system is given by

$$G_p(s) = \frac{K(s+2)}{s^2(s+10)} \tag{10.56}$$

It is desired to have the following closed-loop transfer function.

$$T(s) = \frac{1000(s+2)}{(s^2+10s+50)(s+40)} \tag{10.57}$$

Note that the closed-loop transfer function is designed to make the damping ratio of the dominant poles equal to 0.707, and the settling less than 1 second. Also, the steady-state error to a step input is zero, since $T(0) = 1$. Here we have

$$N_p(s) = 1000s + 2000$$
$$D_p(s) = s^3 + 10s^2$$
$$D(s) = s^3 + 50s^2 + 450s + 2000$$
$$D(s) - D_p(s) = 40s^2 + 450s + 2000$$

Let us make $\Delta(s) = (s+10)(s+20) = s^2 + 30s + 200$. Then

$$[D(s) - D_p(s)]\Delta(s) = 40s^4 + 1650s^3 + 23{,}500s^2 + 150{,}000s + 400{,}000 \tag{10.58}$$

and, according to Eq. (10.55), this should be equal to $N_c(s)N_p(s) + N_u(s)D_p(s)$.

Let

$$N_c(s) = b_2s^2 + b_1s + b_0$$
$$N_u(s) = a_1s + a_0$$

Note in the above that the degree of $N_c(s)$ has been made equal to that of $\Delta(s)$ to make the transfer function of the compensator a proper rational function. Also, the degree of $N_u(s)$ has been made one, so the product $N_u(s)D_p(s)$ has the same degree as $[D(s) - D_p(s)]\Delta(s)$.

Then

$$N_c(s)N_p(s) = 1000b_2s^3 + (1000b_1 + 2000b_2)s^2 + (1000b_0 + 2000b_1)s + 2000b_0$$

and

$$N_u(s)D_p(s) = a_1s^4 + (10a_1 + a_0)s^3 + 10a_0s^2$$

The coefficients b_2, b_1, b_0, a_1, and a_0 must be selected in such a manner that they match the coefficients in Eq. (10.58). The equations for the coefficients can be arranged in the matrix form, as follows:

$$\begin{bmatrix} 1 & 0 & 0 & 0 & 0 \\ 10 & 1 & 1000 & 0 & 0 \\ 0 & 10 & 2000 & 1000 & 0 \\ 0 & 0 & 0 & 2000 & 1000 \\ 0 & 0 & 0 & 0 & 2000 \end{bmatrix} \begin{bmatrix} a_1 \\ a_0 \\ b_2 \\ b_1 \\ b_0 \end{bmatrix} = \begin{bmatrix} 40 \\ 1{,}650 \\ 23{,}500 \\ 150{,}000 \\ 400{,}000 \end{bmatrix} \tag{10.59}$$

The solution of these equations gives $a_1 = 40$, $a_0 = 5750$, $b_2 = -4.5$, $b_1 = -25$, and $b_0 = 200$. From these coefficients we get the transfer functions $G_u(s)$ and $G_c(s)$ of the compensators. Since these are proper rational functions with poles on the negative real axis [due to our choice of $\Delta(s)$], they can be realized as *RC* networks—with amplifiers in cascade, if necessary.

It is important to note that a solution will exist only if the matrix in the left-hand side of Eq. (10.59) is nonsingular. Furthermore, the first two columns of the matrix are made up from the coefficients of $D_p(s)$ and the remaining three columns from the coefficients of $N_p(s)$. It turns out that this is the well-known Sylvester matrix, and will be nonsingular if and only if $D_p(s)$ and $N_p(s)$ are relatively prime. For details, the interested reader may see section 4.5 of Kailath [1]. It will be seen that the transfer functions of the two compensators are proper with poles on the negative real axis of the s-plane. They can, therefore, be designed as *RC* networks, with amplifiers in cascade, if necessary.

In general, the approach presented in this section is better than the pole-zero cancellation method described in Sec. 10.6.1 since we do not attempt to cancel the poles of the plant transfer function. However, it should be noted that the overall function has hidden modes, the roots of $\Delta(s)$. Although these are canceled, since $\Delta(s)$ is present in both the numerator and the denominator of Eq. (10.54), there may be some difficulty due to imperfect cancellation caused by the tolerances in the parameters. Hence $\Delta(s)$ must be selected carefully. Nevertheless, it is much easier to design the compensators with greater precision than to cancel the poles of the transfer function of the plant (which may actually be nonlinear) by the zeros of the transfer function of the compensator.

10.8 PID CONTROLLERS

In our deliberations so far we have considered that the plant to be controlled is completely known to us. In practice this is not always the case. It may still be possible to obtain good performance of the closed-loop system by introducing a PID (proportional, integral, and derivative) controller as shown in the block diagram of Fig. 10.26.

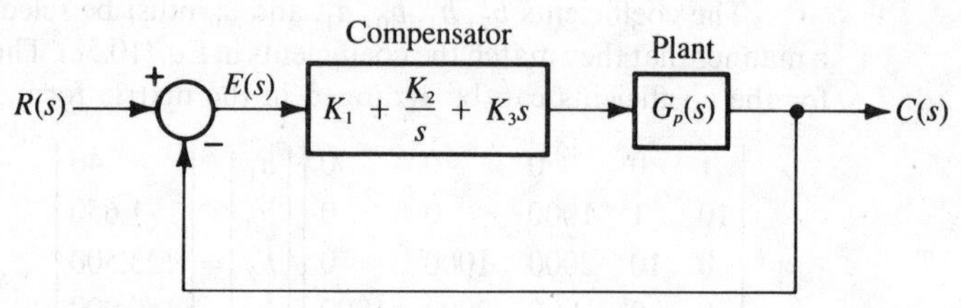

FIGURE 10.26. A PID controller.

The input to the plant consists of three components: (1) K_1E, which is proportional to the error; (2) K_2E/s, which is proportional to the integral of the error; and (3) K_3sE, which is proportional to the derivative of the error. The first component increases the loop gain of the system and thereby reduces its sensitivity to plant parameter variations. The second component increases the order of the system and reduces the steady-state error. The last component may tend to stabilize the system by introducing the derivative term.

The values of the gain constants K_1, K_2, and K_3 can often be determined by trial and error if $G_p(s)$ is not known exactly. If the parameters of the plant are subject to large variations, the gain constants can be adjusted to improve the performance. This scheme is suitable for adaptive control of systems subject to such variations.

In practical implementation, sometimes a problem may be caused by noise, which is invariably present in all physical systems. Differentiation of noise is seldom desirable, since it has the effect of accentuating it. Hence the derivative component of the PID controller is difficult to implement except in the case of discrete-time systems. This will be discussed in Chapter 11.

EXAMPLE 10.11

Consider the case when

$$G_p(s) = \frac{10}{s(s+2)}$$

With a PID controller, the order of the closed-loop system will be increased to 3. It is desired to have the closed-loop poles at -10

and $-2 \pm j2$. We shall determine the values of K_1, K_2, and K_3 to accomplish this, and calculate the response of the compensated system to a unit step.

$$T(s) = \frac{C(s)}{R(s)}$$

$$= \frac{\left(K_1 + \frac{K_2}{s} + K_3 s\right) G_p(s)}{1 + \left(K_1 + \frac{K_2}{s} + K_3 s\right) G_p(s)}$$

$$= \frac{10(K_3 s^2 + K_1 s + K_2)}{s^3 + 2s^2 + 10K_1 s + 10K_3 s^2 + 10K_2}$$

Hence

$$s^3 + (10K_3 + 2)s^2 + 10K_1 s + 10K_2 = (s + 10)(s^2 + 4s + 8)$$
$$= s^3 + 14s^2 + 48s + 80$$

This gives $K_1 = 4.8$, $K_2 = 8$, and $K_3 = 1.2$.

Since $R(s) = 1/s$,

$$C(s) = \frac{10(1.2s^2 + 4.8s + 8)}{s(s + 10)(s^2 + 4s + 8)}$$

$$c(t) = 1 - 1.765e^{-10t} + 0.1765e^{-2t} \cos 2t + 0.1471e^{-2t} \sin 2t$$

Note that this system will have zero steady-state error to a step input. It can be verified that the steady-state error to a ramp input will also be zero. This is caused by the addition of an extra pole at the origin by the controller.

10.9 SUMMARY

We have discussed several methods for compensating a control system so that its response is desirable. The approach used for design depends to some degree on the form of the specifications and in some cases may appear arbitrary. Moreover, we have not considered the effect of random disturbances or sensitivity to parameter variations.

In the design of a control system, we may wish to consider criteria based on optimizing a cost function. This is a subject for advanced study and is the main theme for optimal control theory.

10.10 REFERENCES

1. T. Kailath, *Linear Systems*, Prentice-Hall, Englewood Cliffs, N.J., 1980 (section 4.5).
2. J. J. Dazzo and C. H. Houpis, *Feedback Control System Analysis and Synthesis*, 2d ed., McGraw-Hill, New York, 1966.
3. J. G. Truxal, *Automatic Feedback Control System Synthesis*, McGraw-Hill, New York, 1955.

10.11 PROBLEMS

1. For each of the following open-loop transfer functions, determine the value of K so that the magnitude of the closed-loop frequency response does not exceed 3 dB. For each case, also determine ω_m, the frequency at which the maximum closed-loop response occurs.

a. $\dfrac{K}{s(s+2)(s+8)}$

b. $\dfrac{K(s+2)}{s(s^2+16s+400)(s+4)}$

2. For Problem 1 design a suitable compensator for each case so that steady-state error of the closed-loop system to a unit ramp input does not exceed 0.1 in addition to meeting the other specifications.

3. The transfer function of the forward path of a unity feedback system representing a chemical process given by

$$G_p(s) = \frac{Ke^{-0.05s}}{s(s+5)(s+10)}$$

It is desired that the closed-loop system should have $K_v = 5$ and a phase margin of 45°.

a. Design a lead compensator that will achieve gain crossover at 5 rad/s.
b. Design a lag compensator that will achieve gain crossover at 2 rad/s.

4. The forward transfer function of a unity feedback system is given by

$$G_p(s) = \frac{1}{s^2(1+0.1s)}$$

Design a suitable compensator so that the closed-loop system is stable with $M_m \leq 1.5$ and ω_m equal to approximately 1.4 rad/s.

5. The forward transfer function of a unity-feedback system is given by $KG_p(s)G_c(s)$. It is known that

$$G_p(s) = \frac{1}{s(s+1)(s+12)} \quad \text{and} \quad G_c(s) = \frac{\alpha(s+1)}{s+\alpha}$$

a. Determine K and α so that the closed-loop system has dominant poles with the real part equal to -4 and a damping ratio of 0.5.
b. What will be the response of the closed-loop system to a unit step?

6. Use the Nichols chart to determine the gain margin, the phase margin, M_m and ω_m for the system in Problem 5 with and without compensation. [For the latter case take $G_c(s) = 1$ and $K = 224/3$.]

7. Design a suitable compensator for Problem 3 if the closed-loop system is required to have $K_v = 5$ and $M_m = 3$ dB, with (a) $\omega_m = 5$ rad/s and (b) $\omega_m = 2$ rad/s, approximately.

8. The transfer function of the forward path of a unity-feedback system is given by

$$G(s) = \frac{K}{s(s+3)(s+5)}$$

It is desired that the closed-loop system should satisfy the following specifications.
a. The steady-state error for a unit step input should be zero.
b. The steady-state error for unit ramp input should not exceed 0.1.
c. The complex poles of the closed-loop system should have a damping ratio of 0.5.

Design a suitable compensator and determine the sensitivity of the complex poles of the closed-loop system transfer function to changes in K.

9. For Problem 3, determine the range of gain crossover frequencies over which it is possible to design (a) a lead compensator, (b) a lag compensator, and (c) a lag-lead compensator to satisfy the same specifications.

10. The linearized transfer function of a helicopter near hover can be approximated by

$$G(s) = \frac{K(s+0.3)}{(s+0.65)(s^2-0.2s+0.1)}$$

It is desired to stabilize the system by adding compensation so that the poles of the closed-loop transfer function are located at $s = -2$, $-1 + j1$, and $-1 - j1$. Also the system is required to give zero steady-state error to a step input. Use the method of Sec. 10.7 and locate the poles of the compensators at $s = -4$ and -5.

11. The control system for a tracking antenna used in radar can be represented by a unity-feedback system with the forward transfer function

$$G(s) = \frac{K}{s(s+4)(s+8)}$$

The closed-loop system should satisfy the following specifications.
a. The steady-state error to a unit ramp should be 0.05.
b. The damping ratio of the complex poles should be 0.707.

Design a suitable compensator using the procedure described in Sec. 10.7. (*Hint*: First determine a closed-loop transfer function that will satisfy these specifications.)

12. Consider the feedback system with two delays shown in Fig. 9.28 (see Chapter 9, Problem 8), which represents a typical chemical reactor. Design a suitable compensator to meet the following specifications.

- **a.** The steady-state error to a unit step input should be 0.05.
- **b.** The phase margin should be 45°.

13. Consider the block diagram of an automatic gauge control system for a hot-strip finishing mill shown in Fig. 9.30 (see Chapter 9, Problem 10). It is desired to design a compensator to satisfy the following specifications.

- **a.** The steady-state error to a unit ramp input should not exceed 0.05.
- **b.** The phase margin should be at least 45°.

Discuss the type of compensator that will be most suitable if the bandwidth of the system is to be kept as small as possible. What should be the parameters of this compensator? In addition, determine the response of the compensated system to a unit step.

14. Repeat Problem 13 if the following specifications are to be satisfied.

- **a.** The steady-state error to a unit ramp input should not exceed 0.05.
- **b.** The maximum magnitude of the closed-loop frequency response should not exceed 3 dB.

Determine the response of the compensated system to a unit step.

15. Repeat Problem 13 if the following specifications are to be met.

- **a.** The steady-state error to a unit ramp input should not exceed 0.05.
- **b.** The damping ratio of the complex poles should be 0.5.
- **c.** The undamped natural frequency of the system should not be less than 15 rad/s.

16. The block diagram of a driver steering control system is shown in Fig. 9.31 (see Chapter 9, Problem 11). Design a compensator so that the following specifications are satisfied.

- **a.** The steady-state error to a unit step input should be 0.05.
- **b.** The phase margin should be 45°.

Determine the response of the compensated system to a unit step.

17. The block diagram of a speed-control system using a phase-locked loop is shown in Fig. 9.32 (see Chapter 9, Problem 12). It is desired to replace the filter by a lead compensator and adjust the value of K_A such that the following specifications are satisfied.

- **a.** The steady-state error to a unit step input does not exceed 0.01.
- **b.** The phase margin is not less than 45°.

If it is possible, then design such a compensator. What other alternative is available if a lead compensator that will satisfy these specifications does not exist?

18. The simplified block diagram for the pitch control system of a supersonic aircraft is shown in Fig. 7.20 (see Chapter 7, Problem 7). Design a compensator that will cause the steady-state error to a unit step to be less than 0.02 and the damping ratio of the dominant poles to be 0.6.

19. For Problem 18 use the method of Sec. 10.7 to obtain a closed-loop system transfer function that has dominant poles located at $-5 \pm j5$ and a steady-state error of zero to a ramp input. The poles of the compensator should be placed at -10. (*Hint*: Start with a closed-loop system transfer function that will satisfy these specifications.)

20. The linearized transfer function of a helicopter near hover can be approximated by

$$G(s) = \frac{K(s + 0.3)}{(s + 0.65)(s^2 - 0.2s + 0.1)}$$

It is desired to design a PID controller to stabilize the system as well as to reduce the steady-state error to step inputs to zero. This controller will be connected in cascade with $G(s)$ in a unity-feedback system. The transfer function of the PID controller was selected as

$$G_c(s) = \frac{K(s + 0.5)(s + 1)}{s}$$

a. Determine the value of K such that the complex poles of the closed-loop system transfer function will have damping ratio of 0.707.
b. Calculate the response of the compensated system to a unit step.
c. What will be the steady-state error of this system to a unit ramp input?

21. Consider Problem 20 with the PID controller of the form

$$G_c(s) = \frac{K(s + \alpha)(s + \beta)}{s}$$

In this case we have three adjustable parameters: K, α, and β. It is therefore possible to select them in such a way as to specify the location of the complex poles of the transfer function of the closed-loop system. For example, one may select β equal to 0.65, canceling one of the poles of $G(s)$.

a. Determine the values of K and α such that the complex poles of the closed-loop system transfer function are located at $-1 \pm j1$.
b. Calculate the response of the compensated system to a unit step.
c. What will be the steady-state error of this system to a unit ramp input?

22. A video tracking system (see Fig. 10.27) is used to monitor the automatic docking of a satellite in a space rendezvous. A video camera is geared to a motor. The postion of the camera with respect to an inertial reference is θ_1. The position of the point to be tracked with respect to the same reference is θ_c. It has been determined that a satisfactory performance can be obtained if, for a step input,

s_1 the settling time for the closed-loop system is less than 0.5

s_2 the maximum overshoot for the closed loop system is less than 16%

and for a ramp input,

s_3 the steady-state error is less than 0.01.

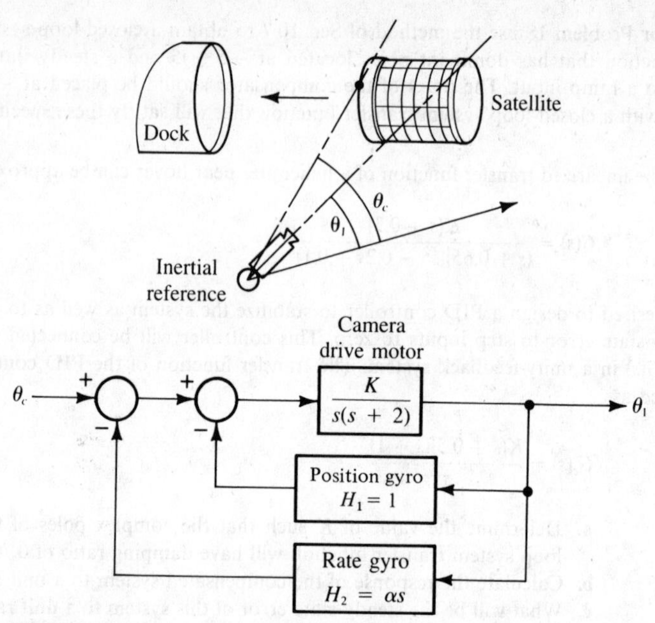

FIGURE 10.27. A video tracking system.

a. Show that K *and* α can be chosen to meet the first two specifications (s_1, s_2).

b. In order to meet the last specification, design a suitable compensator.

11

Digital Control

11.1 INTRODUCTION

The use of a digital computer as a part of a control system has many advantages. Some of these are listed below.

1. The need for amplification of analog signals is eliminated, and this results in the reduction of noise in the system.
2. Higher resolution and accuracy can be obtained with digital transducers. These are also less affected by noise.
3. Digitally coded data can be stored for any length of time and transmitted or used, as many times as necessary, without any loss in accuracy.
4. Overall costs are reduced, due to the possibility of time-sharing.
5. Implementation of optimal and adaptive control is much more convenient with a digital computer as part of the control loop.

With the rapid advances in the area of microprocessors in recent years, digital control has become even more attractive and economical in a large number of applications. One effect of the microelectronics revolution has been a significant reduction in the size and weight of microcomputers to the extent that now it is quite practical to incorporate them as components of space vehicles, automobiles, ships, cameras, and other devices without appreciable increase in physical dimensions. This has provided a great impetus to the design of control systems with one or more microprocessors as integral components.

The block diagram of a typical computer control system is shown in Fig. 11.1. In this diagram, the analog error signal, $e(t)$, is transformed into a sequence of numbers, $m(kT)$, by an analog-to-digital (A/D) converter. This requires sampling the analog signal at the instants $t = kT$, where k is an integer and replacing $e(kT)$ by a sequence of numbers, $m(kT)$. The sequence $m(kT)$ is then processed by the computer to obtain another sequence of numbers, $u(kT)$. This is then transformed into an analog signal $u(t)$ by a digital-to-analog (D/A) converter before it is applied to the process.

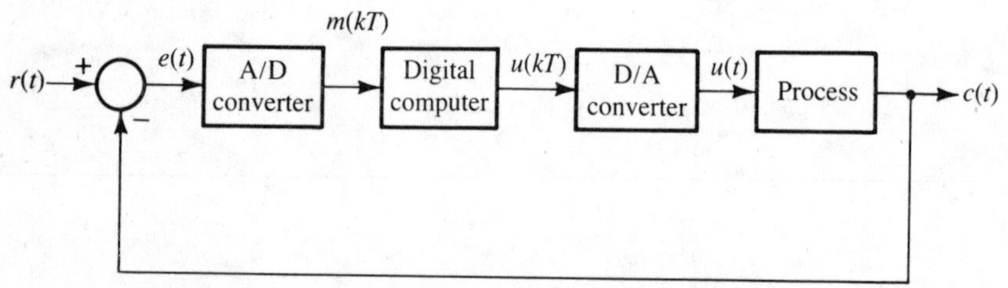

FIGURE 11.1. A typical computer control system.

Although, in practice, the sampling period T is not necessarily constant, we shall make the simplifying assumption that it is. Typically, as a part of the computer hardware there is a clock that sends a pulse every T seconds and then the A/D converter sends a number to the computer. Thus, the input sequence $m(kT)$ supplied to the computer can change only at discrete intervals of time—that is, at $t = kT$, $k = 0, 1, 2, \ldots$. In addition to the process of discretizing, the A/D converter also quantizes the discrete signal to obtain a digital output, or a sequence of numbers suitable for use by the computer. It is common practice to call the resulting system a discrete-time or sampled-data system.

The actual input to the process, denoted as $u(t)$, is usually a continuous-time function, reconstructed from the output $u(kT)$ of the digital computer.

The mathematical analysis of discrete-time systems can be carried out conveniently by many methods. The two most outstanding are (1) methods based on using z transforms and (2) the state-variable method. The former is a modification of the Laplace-transform method used for continuous-time systems, whereas the latter follows directly from the state equations for continuous-time systems. In this chapter we shall study the z-transform method. The state-variable method will be discussed in Chapter 12.

11.2 DATA EXTRAPOLATORS

A data extrapolator is a device that reconstructs a sampled function into a continuous-time signal based on a knowledge of the past samples. This device, which is also called a data hold, follows the digital computer in the system and is a part of the D/A converter.

Data extrapolators are classified according to the number of prior samples required for predicting the sampled function during waiting intervals. The simplest case is that of the "zero-order hold," in which the value of the reconstructed function during any waiting period is simply equal to the value of the sampled function at the beginning of the interval. This type of hold, therefore, leads to the "staircase" approximation of the continuous-time function, as shown in Fig. 11.2. Its operation is similar to an electronic clamping circuit, which keeps its output level equal to the magnitude of an input pulse and then resets itself when the next pulse arrives.

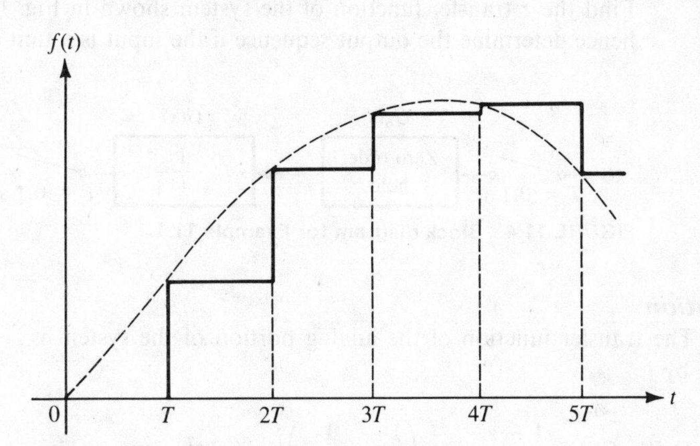

FIGURE 11.2. Staircase approximation.

The transfer function of a zero-order hold can be obtained easily from its impulse response, which is a rectangular pulse of unit height and duration T, as shown in Fig. 11.3. It can be regarded as the result of subtracting

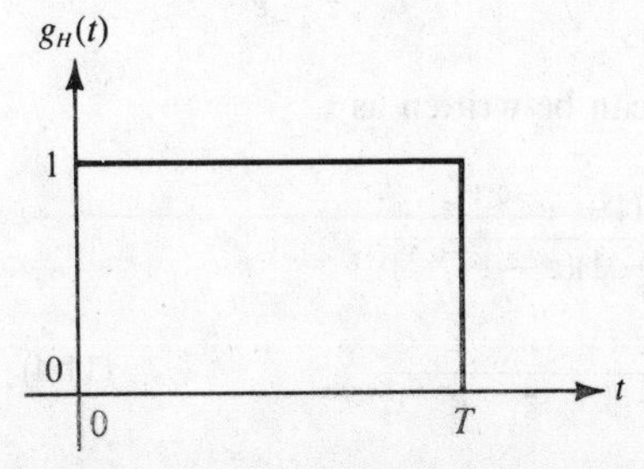

FIGURE 11.3. Impulse response of a zero-order hold.

from the unit step function occurring at $t = 0$ another step function delayed by time T. Hence, taking the Laplace transform we get

$$G_H(s) = \frac{1}{s}(1 - e^{-sT}) \tag{11.1}$$

An extrapolator that depends upon two prior samples is known as the "first-order hold." In this case, during the waiting interval, the function is extrapolated as a straight line determined by the last two samples. In the same way, one may envisage higher-order holds. These are seldom used in feedback control systems.

EXAMPLE 11.1

Find the z transfer function of the system shown in Fig. 11.4, and hence determine the output sequence if the input is a unit step.

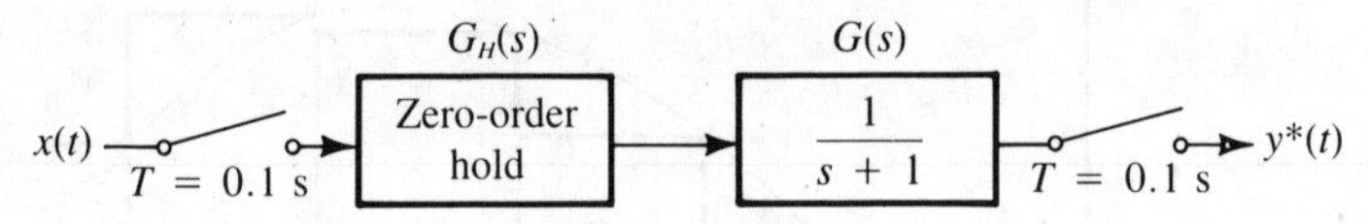

FIGURE 11.4. Block diagram for Example 11.1.

Solution

The transfer function of the analog portion of the system is given by

$$G_H(s) \cdot G(s) = \frac{1 - e^{-sT}}{s(s+1)} = \left(\frac{1}{s} - \frac{1}{s+1}\right)(1 - e^{-sT}) \tag{11.2}$$

Hence, the z transfer function is obtained as

$$G_HG(z) = \left(\frac{z}{z-1} - \frac{z}{z - e^{-0.1}}\right)(1 - z^{-1}) = \frac{1 - e^{-0.1}}{z - e^{-0.1}} \tag{11.3}$$

The z transform of the output can be written as

$$Y(z) = G_HG(z) \cdot X(z) = \frac{(1 - e^{-0.1})z}{(z-1)(z - e^{-0.1})}$$

$$= \frac{z}{z-1} - \frac{z}{z - e^{-0.1}} \tag{11.4}$$

and $y(nT) = 1 - e^{-0.1n}$, as expected.

It may be noted that the zero-order hold followed by the sampler reconstructs the unit step accurately, so the input to $G(s)$, given by $u(t)$, is actually the unit step. Hence $y(nT)$ is identical to the nth sample of the step response of $1/(s+1)$.

EXAMPLE 11.2

An error-sampled closed-loop system is shown in Fig. 11.5. Determine the output sequence for a unit step input.

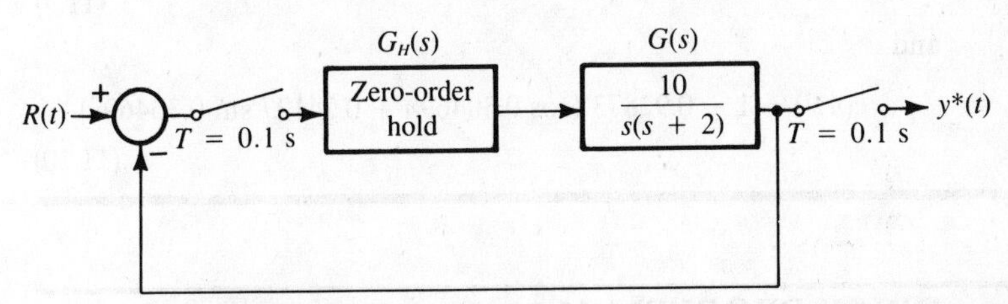

FIGURE 11.5. Block diagram of a closed-loop sampled-data system.

Solution

The transfer function of the forward path is obtained as

$$G_H(s)\cdot G(s) = \frac{1-e^{-sT}}{s}\cdot\frac{10}{s(s+2)}$$

$$= 10\left(\frac{1/2}{s^2} - \frac{1/4}{s} + \frac{1/4}{s+2}\right)(1-e^{-sT}) \tag{11.5}$$

and

$$G_HG(z) = \left[\frac{0.5z}{(z-1)^2} - \frac{2.5z}{z-1} + \frac{2.5z}{z-e^{-0.2}}\right](1-z^{-1})$$

$$= \frac{(2.5e^{-0.2}-2)z + 2.5 - 3e^{-0.2}}{(z-1)(z-e^{-0.2})} \tag{11.6}$$

The closed-loop system transfer function is given by

$$\frac{C(z)}{R(z)} = \frac{G_HG(z)}{1+G_HG(z)}$$

$$= \frac{(2.5e^{-0.2}-0.2)z + 2.5 - 3e^{-0.2}}{(z-1)(z-e^{-0.2}) + (2.5e^{-0.2}-0.2)z + 2.5 - 3e^{-0.2}}$$

$$= \frac{0.046872z + 0.043808}{z^2 - 1.771904z + 0.8625385} \tag{11.7}$$

and the z transform of the input is

$$R(z) = \frac{z}{z-1} \tag{11.8}$$

Hence

$$\frac{1}{z}C(z) = \frac{0.046827z + 0.043808}{(z-1)(z^2 - 1.771904z + 0.8625385)}$$

$$= \frac{1}{z-1} - \frac{z - 0.81873}{z^2 - 2(0.92873) \times (0.95394z)z + 0.92873^2} \tag{11.9}$$

and

$$c(nT) = 1 - 0.92873^n(\cos 0.30469n + 0.24127 \sin 0.30469n) \tag{11.10}$$

DRILL PROBLEM 11.1

An error-sampled control system is shown in Fig. 11.6. For $K = 2$, find the output if the input is a unit step. Determine the values of the output for t equal to 0.5, 1, 2, 3, and ∞ seconds.

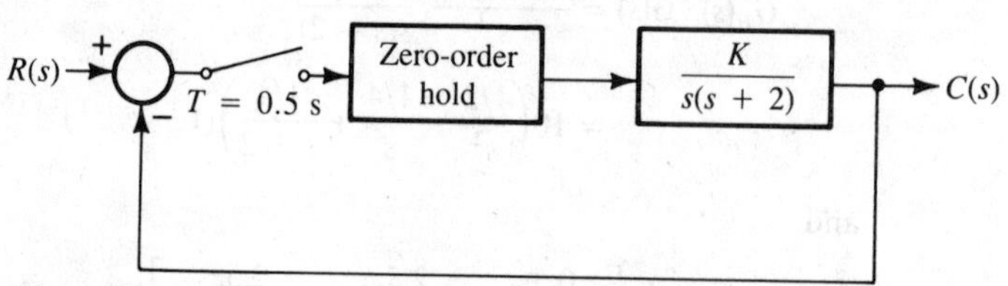

FIGURE 11.6. An error-sampled closed-loop system.

Answers:

$c(nT) = 1 - 0.7072^n(\cos 0.5787n + 0.5794 \sin 0.5787n)$

$c(0.5) = 0.1838$, $c(1) = 0.5337$, $c(2) = 1.0628$, $c(3) = 1.1419$, $c(\infty) = 1$

11.3 STABILITY OF CLOSED-LOOP DISCRETE-TIME SYSTEMS

The necessary and sufficient condition for the stability of a discrete-time system is that all the poles of its z transfer function lie inside the unit circle in the z-plane. This fact follows from the transformation $z = e^{sT}$, which maps the left half of the s-plane inside the unit circle.

Keeping this in mind, the three basic techniques—(a) the Routh-Hurwitz criterion, (b) the Nyquist criterion, and (c) the root locus method—can be used with slight modifications for discrete-time systems.

11.3.1 The Routh-Hurwitz criterion

The conventional Routh test was devised to test the presence of roots of a polynomial in the right half plane, whereas we now need to determine the presence of roots outside the unit circle. This can be done by using the bilinear transformation

$$w = u + jv = \frac{z+1}{z-1} \quad \text{or} \quad z = \frac{w+1}{w-1} = \frac{u+1+jv}{u-1+jv} \tag{11.11}$$

which maps the unit circle of the z-plane into the imaginary axis of the w-plane and the interior of the unit circle into the left half of the w-plane.

EXAMPLE 11.3

The characteristic polynomial of a system is given by

$$Q(z) = z^3 + 5z^2 + 3z + 2 = 0 \tag{11.12}$$

Then

$$Q(w) = \left(\frac{w+1}{w-1}\right)^3 + 5\left(\frac{w+1}{w-1}\right)^2 + 3\left(\frac{w+1}{w-1}\right) + 2 = 0$$

or

$$(w+1)^3 + 5(w+1)^2(w-1) + 3(w+1)(w-1)^2 + 2(w-1)^3 = 0 \tag{11.13}$$

This is simplified to obtain the polynomial

$$11w^3 - w^2 + w - 3 = 0 \tag{11.14}$$

The Routh table is shown below.

w^3	11	1
w^2	-1	-3
w^1	-32	0
w^0	-3	

There is one sign change in the first column of the Routh table. This indicates one pole in the right half of the w-plane, and hence an unstable system.

DRILL PROBLEM 11.2

Use the Routh criterion to determine the maximum value of K for which the system shown in Fig. 11.13 will be stable.

Answer: $K = 9.5686$.

11.3.2 The Nyquist criterion

The Nyquist criterion is directly applicable to the open-loop transfer function $GH(z)$, with the essential difference that this time we map the unit circle of the z-plane into the GH-plane.

To apply the Nyquist criterion, therefore, one would plot $GH(z)$ as the z-plane is traversed around the unit circle in the counterclockwise* direction and determine N, the number of counterclockwise encirclements of $GH(z)$ about the point $-1 + j0$ in the $GH(z)$ plane. For a stable system, we must have $N = -P$, where P is the number of poles of $GH(z)$ outside the unit circle.

Often the open-loop transfer function $GH(z)$ has one or more poles at $z = 1$, corresponding to the origin of the s-plane in a continuous system. In such a case, the contour in the z-plane is modified to exclude the point $1 + j0$ by taking a detour, as shown in Fig. 11.7.

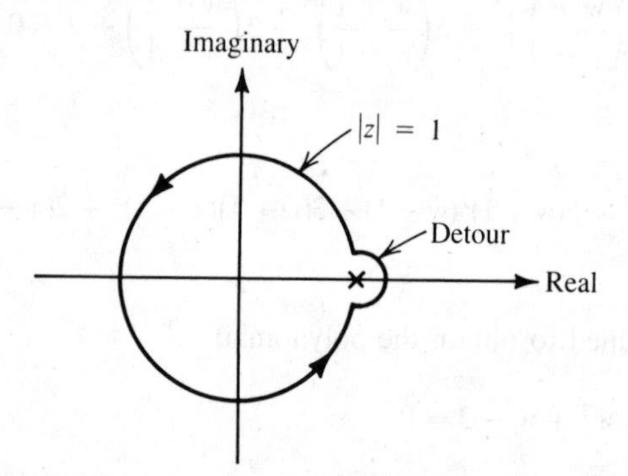

FIGURE 11.7. The Nyquist contour.

* It may be noted that we traverse around the unit circle in the z-plane in the counterclockwise direction, whereas we had traversed around the right half of the s-plane in the clockwise direction (see Sec. 9.3). This can be understood by appreciating that the right half of the s-plane maps outside the unit circle in the z-plane, and we enclose this region if it is to our right while we traverse along the Nyquist contour.

An alternative approach is to use the w transformation and then map $GH(jv)$, using the conventional Nyquist plot. This method has the advantage that Bode plots, as well as Nichols charts, can also be used.

We shall use a slightly different form of the bilinear transformation that introduces a simplification. Define

$$w = u + jv = \frac{2}{T}\frac{z-1}{z+1} \tag{11.15}$$

Since

$$z = e^{sT} \tag{11.16}$$

for $s = j\omega$, we have

$$jv = \frac{2}{T}\frac{e^{j\omega T}-1}{e^{j\omega T}+1} = j\frac{2}{T}\tan\frac{\omega T}{2} \tag{11.17}$$

From Eq. (11.17) we see that whereas z goes around the unit circle, the w-plane frequency v stays real and goes from 0 to ∞. The purpose of selecting the scale factor $2/T$ in Eq. (11.15) is to ensure that the w-plane transfer function would approach that in the s-plane as T is made smaller, and the error constants, K_p, K_v, etc., are correct. This will be illustrated by means of an example.

EXAMPLE 11.4

Consider the closed-loop system shown in Fig. 11.5, for which the step response was calculated in Example 11.2. For $T = 0.1$ second the forward-path transfer function (including the zero-order hold) was obtained as

$$G(z) = \frac{(2.5e^{-0.2} - 2)z + (2.5 - 3e^{-0.2})}{(z-1)(z-e^{-0.2})}$$

$$= \frac{0.04683(z+0.9355)}{(z-1)(z-0.8187)} \tag{11.18}$$

To transform to the w-plane, we first solve Eq. (11.15) to obtain z in terms of w. Hence

$$z = \frac{1+(wT/2)}{1-(wT/2)} \tag{11.19}$$

Substituting this into Eq. (11.18), we obtain after some simplification

$$G(w) = \frac{-0.00083(w-20)(w+600.4)}{w(w+1.99336)} \tag{11.20}$$

It can be verified easily that the value of K_v for $G(w)$ is identical to that for $G(s)$. Furthermore, the denominator of Eq. (11.20) would be very nearly the same as the denominator of $G(s)$ if s were replaced by w, and the two numerators would also be equal as w approaches zero.

The frequency response $G(jv)$ can be calculated easily from Eq. (11.20) and is shown in Table 11.1. The polar plot of the frequency response, shown in Fig. 11.8, can be used for Nyquist analysis as well as for information in relative stability, as discussed in Chapter 9.

TABLE 11.1. Frequency response $G(jv)$ for Example 11.4

v	Gain	Phase	v	Gain	Phase
0.1	49.94	−93.15°	2.0	1.774	−140.6°
0.2	24.87	−96.28°	5.0	0.382	−171.8°
0.5	9.70	−105.47°	10.0	0.109	−194.4°
1.0	4.475	−119.4°	20.0	0.035	−217.4°

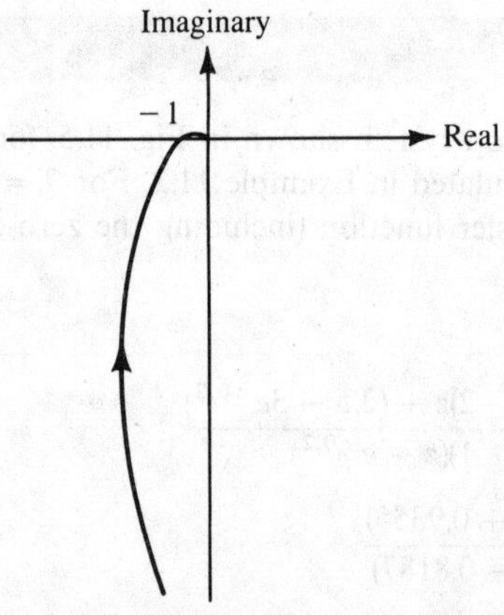

FIGURE 11.8. Polar plot of $G(jv)$.

The main advantage of this approach is that one can design a compensator on the basis of the frequency response. This will be discussed further in Sec. 11.5.

11.3.3 The root locus method

Since the response of the closed-loop system depends on its poles, which are the roots of

$$1 + KGH(z) = 0 \tag{11.21}$$

the root locus technique can be used with the same rules of construction as for the continuous-time case. The only difference arises in the manner in which the resulting root locations are related to stability and time response.

The closed-loop system will be unstable when the root locus crosses the unit circle in the z-plane. Furthermore, lines of constant damping in the s-plane are parallel to the $j\omega$-axis, and these transform into circles of radii smaller than one. For example, the line $\sigma = -a$ in the s-plane will map in the z-plane into a circle with its center at the origin and a radius e^{-aT}. On the other hand, a line of constant damping ratio, which is a radial line in the s-plane, maps into a spiral in the z-plane. These mappings are shown in Fig. 11.9.

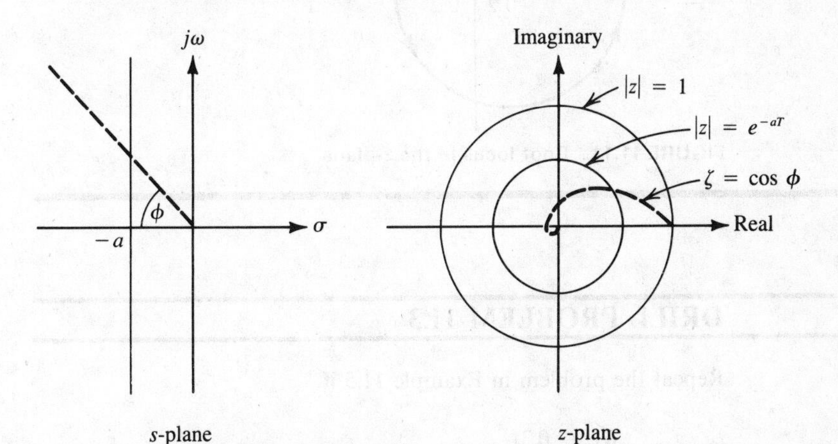

FIGURE 11.9. Mapping from the s-plane into the z-plane, of lines with constant damping and constant damping ratio.

EXAMPLE 11.5

The block diagram of a unity-feedback sampled-data system is shown in Fig. 11.10. Determine the value of the gain K so that all the closed-loop poles are inside a circle of radius 0.7071.

Solution

A sketch of the root locus is shown in Fig. 11.11. The largest value of K that will satisfy the above condition is 1.5, with poles at $-0.15 \pm j0.691$.

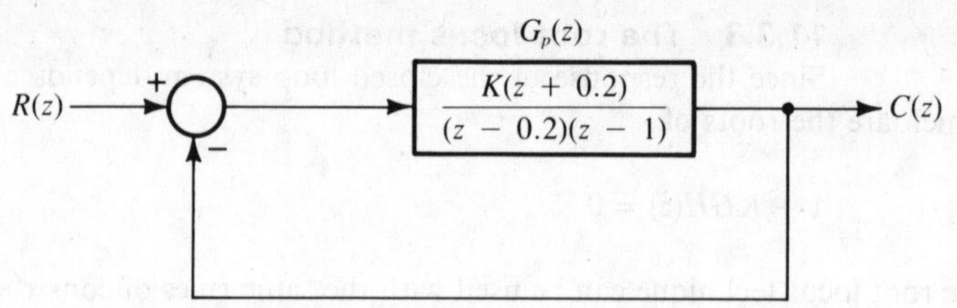

FIGURE 11.10. Block diagram of a sampled-data system.

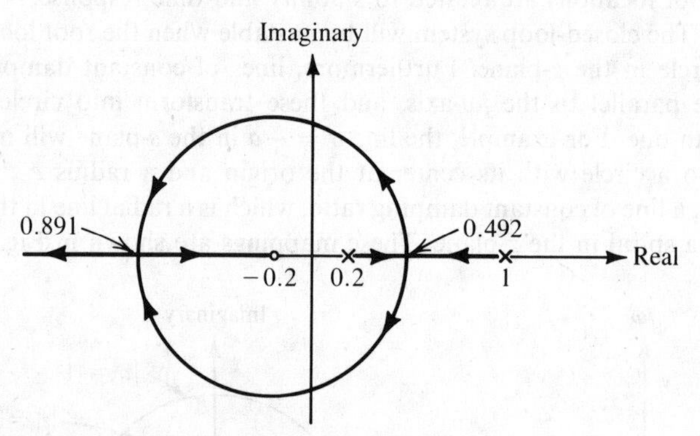

FIGURE 11.11. Root locus in the z-plane.

DRILL PROBLEM 11.3

Repeat the problem in Example 11.5 if

$$G_p(z) = \frac{K(z + 0.3)}{(z - 0.3)(z - 1)}$$

Answer:
$K = 0.66$ gives poles at $0.32 \pm j0.629$.

11.4 DESIGN OF DIGITAL CONTROLLERS

In a large number of practical situations the sampled and quantized error signal must be digitally processed before it is reconstructed and applied to the process. The block diagram of such a system is shown in Fig. 11.12.

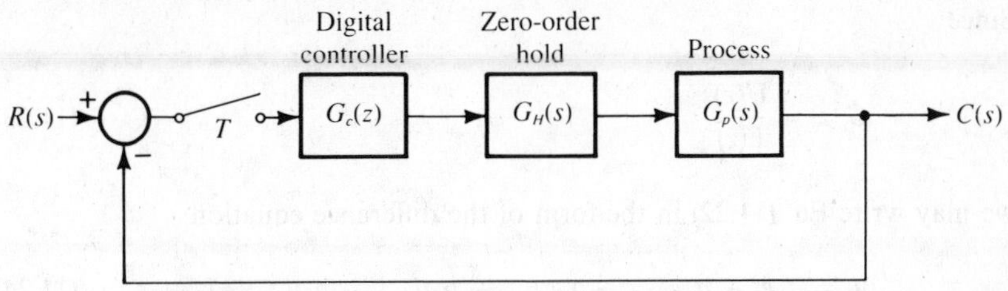

FIGURE 11.12. Block diagram of system with digitally processed error signal.

The transfer function $G_c(z)$ of the digital controller is selected in such a way as to improve the system performance in order to satisfy the specifications. This can be carried out in several ways, in a manner similar to the case of continuous-time systems.

For example, we may apply the frequency response method, using Bode plots or Nichols charts, if we first carry out the bilinear transformation from the z-plane to the w-plane. An additional advantage of this transformation is related to computation. Since the entire left half of the s-plane maps inside the unit circle of the z-plane, we must include many significant figures to distinguish the various poles and zeros. The transformation to the w-plane avoids this difficulty. The procedure is outlined below.

1. Determine the z transfer function of the combination of the zero-order hold and the process. Let this be denoted by $G_HG_p(z)$.
2. Apply the bilinear transformation $z = (1 + wT/2)/(1 - wT/2)$ to obtain $G_HG_p(w)$.
3. Design the controller (or compensator) $G_c(w)$ that will satisfy the desired characteristics specified in terms of the gain margin, phase margin, bandwidth, and maximum magnitude of the closed-loop frequency response and resonant frequency, using the frequency response methods discussed earlier for continuous-time systems. For physical realizability and stability of the controller it is necessary and sufficient that $G_c(w)$ be stable (with all poles in the left half of the w-plane) and that it be a proper rational function of w; that is, the number of finite zeros must not exceed the number of finite poles.
4. From $G_c(w)$, determine the digital controller $G_c(z)$ by substituting $w = 2(z - 1)/T(z + 1)$ in $G_c(w)$.

We shall now discuss the implementation of the digital controller. Considering the block diagram of Fig. 11.12, suppose that we find that a suitable transfer function for the controller is given by

$$G_c(z) = \frac{a_0z^2 + a_1z + a_2}{z^2 + b_1z + b_2} = \frac{a_0 + a_1z^{-1} + a_2z^{-2}}{1 + b_1z^{-1} + b_2z^{-2}} \tag{11.22}$$

Since

$$G_c(z) = \frac{U(z)}{E(z)} \tag{11.23}$$

we may write Eq. (11.22) in the form of the difference equation

$$u_k = a_0 e_k + a_1 e_{k-1} + a_2 e_{k-2} - b_1 u_{k-1} - b_2 u_{k-2} \tag{11.24}$$

where

$$u_i \triangleq u(iT) \tag{11.25}$$
$$e_i \triangleq e(iT) \tag{11.26}$$

Equation (11.24) tells us that the input $u(kT)$ at the kth sampling instant is obtained as a linear combination of the present and two previous values of the error, $e(kT)$, $e(kT - T)$, and $e(kT - 2T)$, as well as two previous values of the input, $u(kT - T)$ and $u(kT - 2T)$. This simple numerical algorithm is easily carried out on any microprocessor. An important feature that must be noted is the ease with which the parameters of the controller can be changed in that it merely implies altering the values of the coefficients a_i and b_i in the program. On the other hand, the controller in a continuous-time system is not easily altered because the values of physical components must now be changed. In practice, however, we shall have to take into consideration the limitations of a microcomputer caused by finite word lengths and quantization errors.

EXAMPLE 11.6

As an example, consider the block diagram of the attitude control system for a satellite shown in Fig. 11.13. The transfer function $G_p(z)$ includes a zero-order hold.

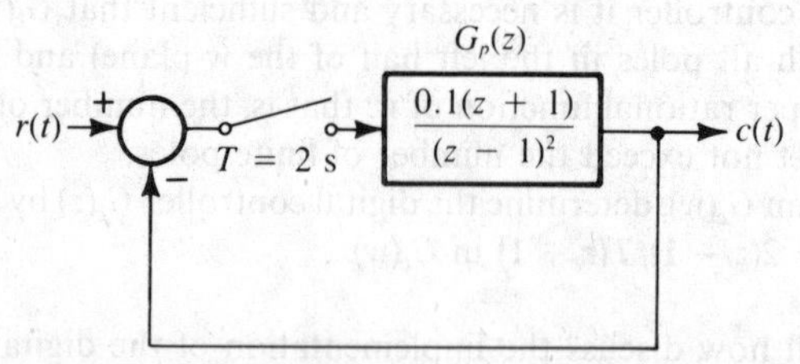

FIGURE 11.13. Block diagram of an uncompensated attitude control system of a satellite.

It is desired to design a digital compensator for the forward path so that the closed-loop system may have phase margin of 45°. The step response of the final design will also be calculated.

Using the bilinear transformation

$$z = \frac{1 + (wT/2)}{1 - (wT/2)} \tag{11.27}$$

we get

$$G_p(w) = \frac{-0.05w + 0.05}{w^2} \tag{11.28}$$

The frequency response of this transfer function is easily calculated by replacing w by jv and is given in Table 11.2. The frequency response is shown in Fig. 11.14. The system is clearly unstable.

We shall design a lead compensator to provide a phase margin of 45°. Let us select

$$v_c = 0.4 \qquad \phi_c = 66.8^\circ \qquad M_c = 9.46 \text{ dB} \tag{11.29}$$

Following the procedure described in Chapter 10, the transfer function of the compensator is found as

$$G_c(w) = \frac{44.884361(w + 0.1426267)}{w + 6.401741} \tag{11.30}$$

The frequency response of the compensated system is given in Table 11.3. Plotting it on the Nichols chart, as in Fig. 11.15, shows that the maximum magnitude of the closed-loop frequency response is 3 dB.

We shall now reverse the transformation to obtain the transfer function of the digital compensator. Recalling that

$$w = \frac{2}{T}\left(\frac{z-1}{z+1}\right) \tag{11.31}$$

we get

$$G_c(z) = \frac{6.92892(z - 0.75035)}{z + 0.729793} \tag{11.32}$$

and the closed-loop system transfer function is evaluated as

$$\frac{C(z)}{R(z)} = \frac{0.692892(z+1)(z-0.75035)}{(z+0.5704)(z-0.5738+j0.1967)(z-0.5738-j0.1967)} \tag{11.33}$$

TABLE 11.2. Frequency response of the transfer function given by Eq. (11.28)

v	*Gain* (dB)	*Phase*	*v*	*Gain* (dB)	*Phase*
0.1	14.02	−185.7°	0.5	−13.01	−206.6°
0.2	2.11	−191.31°	0.7	−18.09	−215.0°
0.3	−4.73	−196.7°	1.0	−23.01	−225.0°
0.4	−9.46	−201.8°	2.0	−31.07	−243.4°

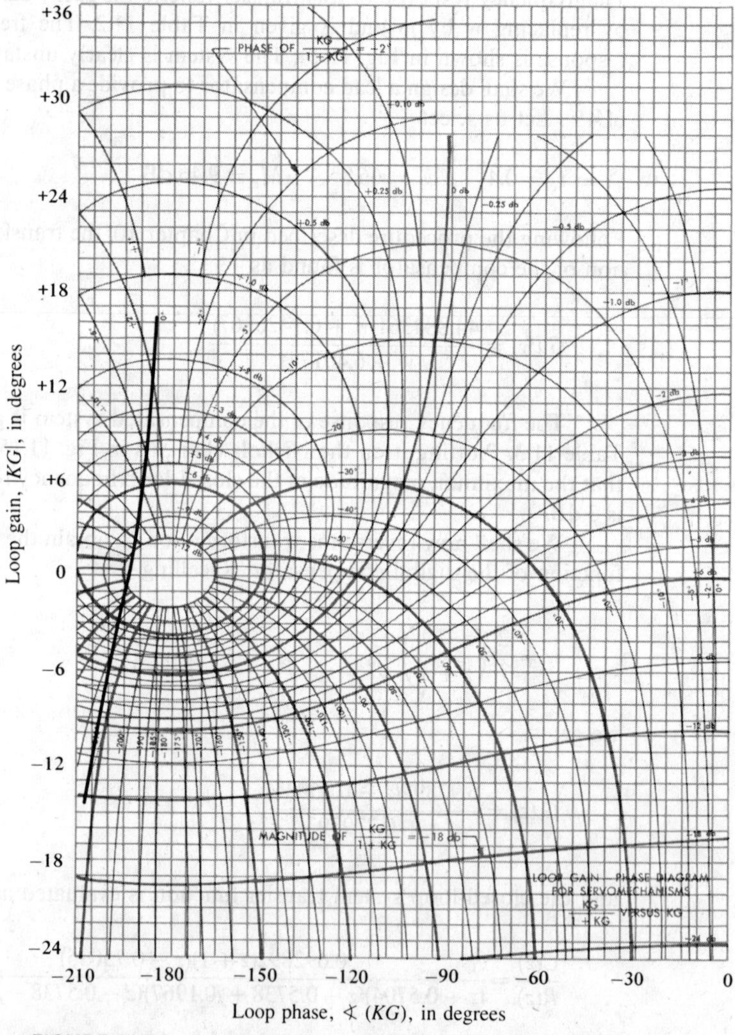

FIGURE 11.14. Frequency response for $G_p(w)$ in Eq. (11.28).

TABLE 11.3. Frequency response of the compensated system given by $G_c(w)G_p(w)$

v	*Gain* (dB)	*Phase*	v	*Gain* (dB)	*Phase*
0.1	15.76	−151.6°	0.6	−3.13	−139.7°
0.2	6.83	−138.6°	0.7	−4.15	−142.7°
0.3	2.60	−134.7°	0.8	−4.95	−145.9°
0.4	0	−135°	1.0	−6.11	
0.5	−1.8	−137°	2.0	−8.52	−174.9°

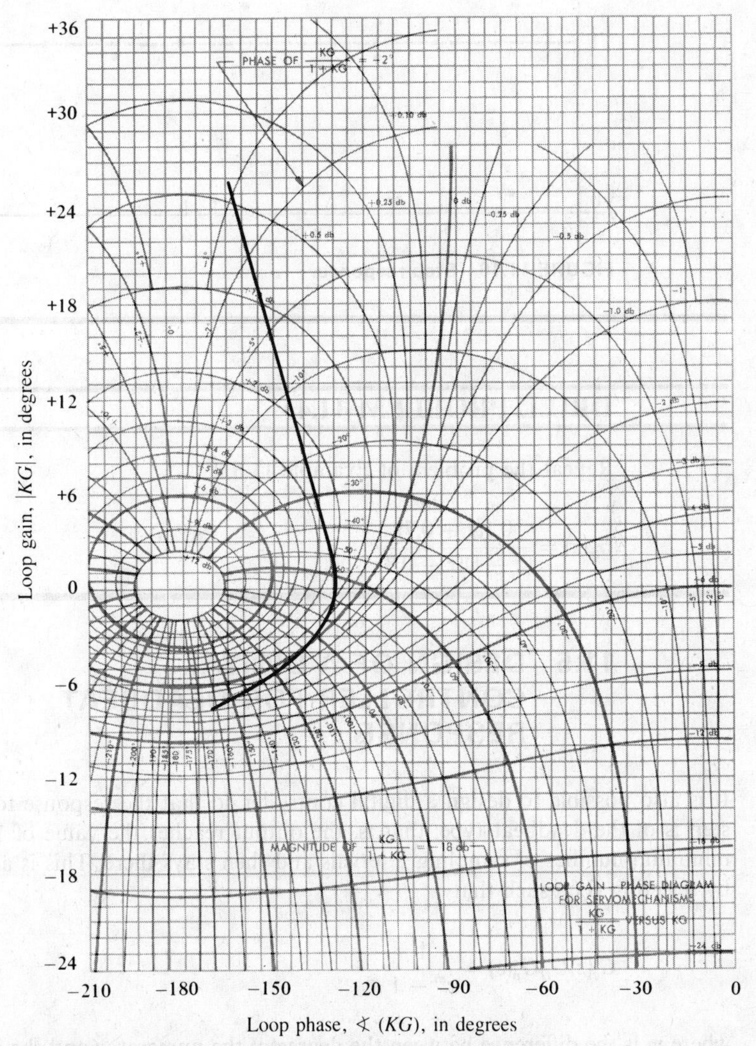

FIGURE 11.15. Frequency response of $G_c(w)G_p(w)$.

The poles of the closed-loop transfer function are well within the unit circle. The response of the system to a unit step input is shown in Fig. 11.16.

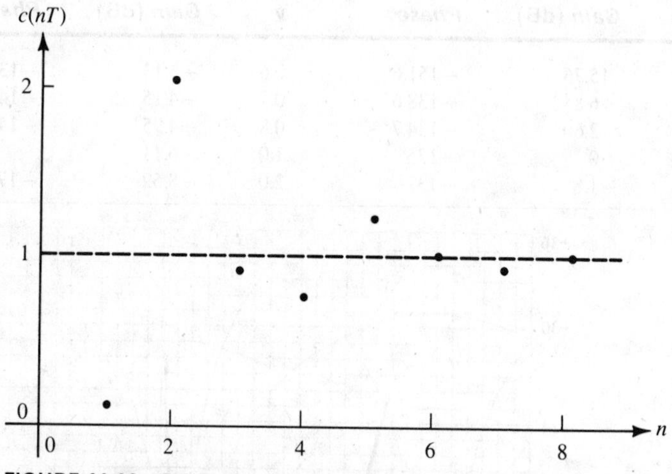

FIGURE 11.16. Step response.

DRILL PROBLEM 11.4

Repeat the problem of Example 11.6 if

$$G_p(z) = \frac{0.1(z + 0.7)}{(z - 1)(z - 0.8)}$$

11.5 DESIGN OF A DIGITAL CONTROLLER FOR DEADBEAT RESPONSE

It is also possible to design a digital controller so that the response to a unit step is of the deadbeat type; that is, the output reaches the value of 1 in the minimum number of sampling intervals and then stays there. This is achieved by selecting $G_c(z)$ such that

$$G_c(z)G_H G_p(z) = \frac{1}{z^m - 1} \tag{11.34}$$

where m is the difference between the degree of the numerator and the denominator of $G_H G_p(z)$. The closed-loop transfer function with unity feedback is then obtained as z^{-m}, which implies that the response will be the unit step delayed by m sampling intervals. This will be evident from Example 11.7.

It should be emphasized that the response of the system reaches the steady state exactly after m sampling intervals. This is an important property of discrete-time systems and should be contrasted with the fact that theoretically it requires infinite time for a stable continuous-time system to reach the steady state in response to a step input.

EXAMPLE 11.7

Consider the system shown in Fig. 11.17, where $G_H(z)$ is a zero-order hold and

$$G_p(s) = \frac{10}{s(s+2)} \tag{11.35}$$

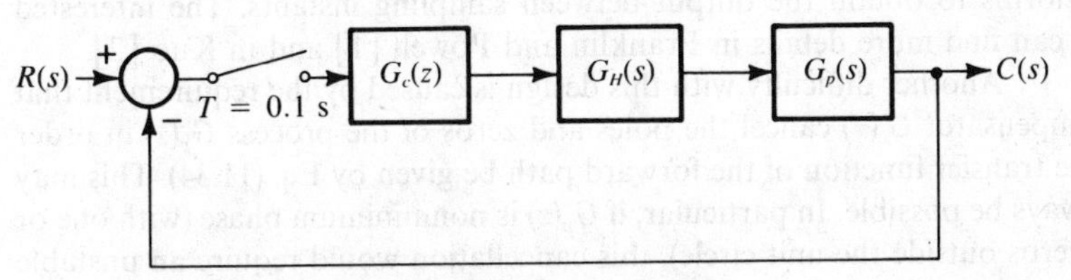

FIGURE 11.17. A closed-loop system.

Then, as in Example 11.2,

$$G_H G_p(z) = \frac{0.046827z + 0.043808}{(z-1)(z-e^{-0.2})}$$

$$= \frac{0.046827(z + 0.93553)}{(z-1)(z-0.8187)} \tag{11.36}$$

Since here $m = 1$, we shall make

$$G_c(z) = \frac{z - 0.8187}{0.046827z + 0.043808}$$

$$= \frac{21.3552 - 17.8414z^{-1}}{1 + 0.93553z^{-1}} \tag{11.37}$$

so that

$$G_c(z)G_H G_p(z) = \frac{1}{z-1} \tag{11.38}$$

Hence

$$\frac{C(z)}{R(z)} = \frac{1}{z} \tag{11.39}$$

and

$$C(z) = \frac{1}{z-1} = z^{-1} + z^{-2} + z^{-3} + \cdots \tag{11.40}$$

It should be noted that, although the response is exactly one at the sampling instants, there is no guarantee that $c(t)$ has no oscillation between the sampling instants. In this example the sampling period of 0.1 second is much smaller than the system time constant, and hence we may expect that $c(nT)$ gives a reasonably accurate description of $c(t)$. One may use the method of modified z transforms to obtain the output between sampling instants. The interested reader can find more details in Franklin and Powell [1] and in Kuo [2].

Another difficulty with this design is caused by the requirement that the compensator $G_c(\ )$ cancel the poles and zeros of the process $G_p(z)$ in order that the transfer function of the forward path be given by Eq. (11.34). This may not always be possible. In particular, if $G_p(z)$ is nonminimum phase (with one or more zeros outside the unit circle), this cancellation would require an unstable compensator, with a serious problem in case of imperfect cancellation due to tolerances.

A more practical approach to the design of a deadbeat control system will be discussed in Chapter 12.

11.6 SUMMARY

The theory of digital control systems has been discussed here from the frequency response point of view. The use of z transforms and w transforms makes it possible to apply all the analysis and design techniques developed for continuous-time systems.

Often it is more convenient to utilize time-domain methods for the analysis and design of discrete-time systems, where a digital computer can be used together with a state-variable formulation. This approach will be discussed in Chapter 13.

In this chapter we have only been able to discuss some of the basic concepts of digital control. Many details, especially those of interfacing between the analog and digital components, have been left out. The interested reader is referred to the books mentioned in the next section.

11.7 REFERENCES

1. G. R. Franklin and J. D. Powell, *Digital Control of Dynamic Systems*, Addison-Wesley, Reading, Mass., 1980.
2. B. C. Kuo, *Digital Control Systems*, Holt, Rinehart and Winston, New York, 1980.
3. R. Isermann, *Digital Control Systems*, Springer, New York, 1981.
4. Paul Katz, *Digital Control Using Microprocessors*, Prentice-Hall, Englewood Cliffs, N.J., 1981.

11.8 PROBLEMS

1. An error-sampled control system has a block diagram as shown in Fig. 11.5, with $T = 0.2$ second and

$$G(s) = \frac{K}{s(s+1)}$$

a. Determine the maximum value of K for which this system will be stable.
b. Determine the response of this system to a unit step with gain margin set to 2.

2. Design a digital compensator for the system in the previous problem that will give a deadbeat response to a unit step (with K set for a gain margin of 2).

3. The block diagram of the control system for a tracking antenna is shown in Fig. 11.18.

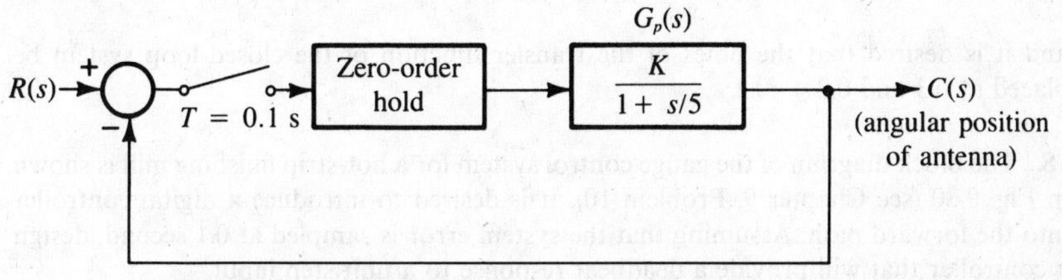

FIGURE 11.18. Block diagram of the control system for a tracking antenna.

a. Design a suitable digital compensator so that for $K = 10$ the phase margin is not less than 45°.
b. Determine the response of the compensated system to a unit step input.

4. The block diagram for the cruise control system for an automobile is shown in Fig. 11.19.

a. For $G_c(z) = K$, determine the range of values of K for which the system will be stable.
b. Design a suitable digital compensator so that the steady-state error to a unit step input may not exceed 0.02 and the phase margin is 45°.

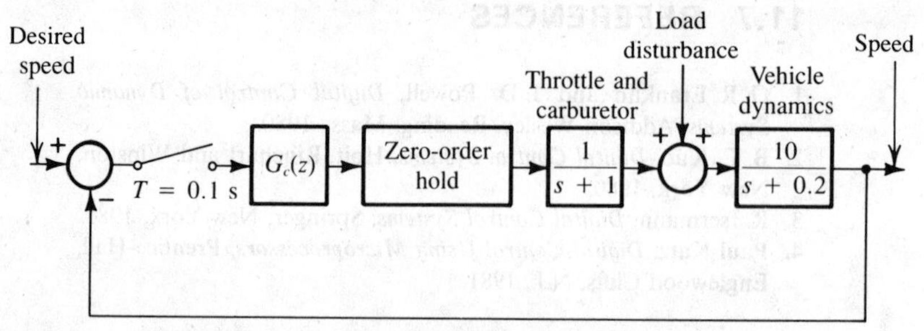

FIGURE 11.19. A digital cruise control system.

5. For Problem 4 design a digital compensator for deadbeat response to a step-load disturbance.

6. Sketch the root locus for the system described in Problem 4, where $G_c(z) = K$ varies from zero to infinity. Use both the z-plane and the w-plane to suggest a suitable value for K.

7. The procedure for designing compensators that will place all the poles of the transfer function of the closed-loop system at specified locations, provided that there are no common factors between the numerator and denominator of the transfer function of the forward path, as discussed in Sec. 10.7, can also be applied to discrete-time systems. The only important difference is that the z transfer function is used and the poles should lie inside the unit circle for a stable system.

Use this procedure to design compensators with all poles at 0.1 in the z-plane if the forward-path transfer function is

$$G_p(z) = \frac{12z(z - 0.6)}{(z - 1)(z - 0.95)(z - 0.9)}$$

and it is desired that the poles of the transfer function of the closed-loop system be placed at 0.3 and $0.2 \pm j0.2$.

8. The block diagram of the gauge control system for a hot-strip finishing mill is shown in Fig. 9.30 (see Chapter 9, Problem 10). It is desired to introduce a digital controller into the forward path. Assuming that the system error is sampled at 0.1 second, design a controller that will provide a deadbeat response to a unit step input.

9. For the gauge control system discussed in the Problem 8 design a digital controller so that the steady-state error to a unit ramp does not exceed 0.1 and the phase margin is 45°. Determine the response of this system to a unit step input.

10. The block diagram of a speed-control system using a phase-locked loop is shown in Fig. 9.32 (see Chapter 9, Problem 12). It is desired to introduce a digital controller in the forward path. Assuming that the system error is sampled at 0.05 second, design a controller to satisfy the following specifications.

a. The steady-state error to a unit step input will not exceed 0.05.
b. The compensated system will have a phase margin of 45°.
Determine the response of this system to a unit step input.

11. The block diagram of a discrete-time control system is shown in Fig. 11.20, where the A/D converter includes a sampler with a period of 0.1 second and the D/A converter includes a zero-order hold. The diagram includes the compensators, $G_u(z)$ and $G_y(z)$. It is desired that the poles of the transfer function of the closed-loop system be located at $0.2 \pm j0.2$ in the z-plane. The poles of the compensators should be placed at 0.15. Design suitable compensators if the transfer function $G_p(s)$, which represents a dc servomotor, is given by

$$G_p(s) = \frac{4}{s(s+1)}$$

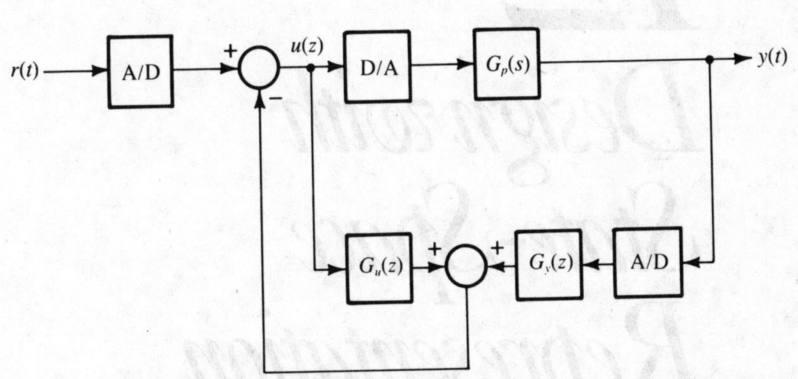

FIGURE 11.20. A discrete-time system with compensators for pole placement.

12. Consider the block diagram of a digital cruise control system shown in Fig. 11.19. It is desired that the poles of the transfer function of the closed-loop system be placed at $0.2 \pm j0.2$ and the steady-state error to a unit step input should not exceed 0.05. Design compensators using the configuration shown in Fig. 11.20, placing the poles of the compensator at 0.15. Determine the response of the compensated system to a unit step input.

13. The transfer function of the forward path of a discrete-time system is given by

$$G_p(z) = \frac{Kz(z-0.6)}{(z-1)(z-0.95)(z-0.9)}$$

It is to be used with unity feedback. What value of K will make the system stable with phase margin equal to 45°?

14. For the system described in Problem 13 design a digital compensator to satisfy the following specifications.

- **a.** The steady-state error to unit ramp input should not exceed 0.1.
- **b.** The maximum magnitude of the closed-loop frequency response should not exceed 3 dB.

Determine the response of the compensated system to a unit step input.

12 Design with State-Space Representation

12.1 INTRODUCTION

The transfer function representation of physical systems has led to the development of the root locus and other *s*-plane methods for analysis and design. Furthermore, it relates directly to the frequency response approach, which is also very useful for the analysis and design of closed-loop systems. Both of these approaches, however, are limited to linear time-invariant systems.

On the other hand, time-domain methods, based on the state-space representation are more powerful. In addition to being applicable to nonlinear and time-varying systems, they are readily extended to multivariable systems—i.e., systems with several inputs and outputs. Their main attraction is that they are natural and convenient for computer solutions. In particular, design and analysis of discrete-time systems can be carried out more easily with digital computers when the state-space representation is employed.

In this chapter we shall study the application of state-space methods to the design of control systems. These are based on the key concepts of controllability and observability. To explain the idea of controllability we shall start with the computation of the input sequence that will transfer the state of a discrete-time system from a given initial value to a specified final value.

12.2 DETERMINATION OF THE INPUT SEQUENCE FOR TRANSFERRING THE STATE OF A DISCRETE-TIME SYSTEM

If a linear time-invariant single-input discrete-time system of order n is controllable, then it is possible to determine uniquely the input sequence that will transfer the system from any given initial state to any specified final state in no more than n sampling intervals. The concept of controllability will be discussed in the next section. Here we shall see how to obtain the required sequence of inputs.

Consider a single-input system with state-transition equation

$$x(kT + T) = Fx(kT) + Gu(kT) \tag{12.1}$$

where x is an n-dimensional state vector and T is the sampling interval. As discussed in Sec. 3.2, the matrices F and G are obtained from the state-space description of the corresponding continuous-time system. That is, if

$$\dot{x} = Ax + Bu \tag{12.2}$$

then

$$F = e^{AT} \tag{12.3}$$

and

$$G = \int_0^T e^{At}\,dt\,B \tag{12.4}$$

If we assume that the initial state of the system is given as $x(0)$, then

$$x(T) = Fx(0) + Gu(0) \tag{12.5}$$

$$\begin{aligned} x(2T) &= Fx(T) + Gu(T) \\ &= F^2x(0) + FGu(0) + Gu(T) \end{aligned} \tag{12.6}$$

Continuing in this manner, it is easily shown that

$$\begin{aligned} x(nT) = F^n x(0) &+ F^{n-1}Gu(0) + F^{n-2}Gu(T) \\ &+ \cdots + FGu(nT - 2T) + Gu(nT - T) \end{aligned} \tag{12.7}$$

Since the final state $x(nT)$ and the initial state $x(0)$ are known, we may rearrange Eq. (12.7) to obtain the following matrix equation.

$$[F^{n-1}G \quad F^{n-2}G \quad \cdots \quad FG \quad G]\begin{bmatrix} u(0) \\ u(T) \\ \vdots \\ u(nT) \end{bmatrix} = x(nT) - F^{n}x(0) \qquad (12.8)$$

A solution to this equation will exist if and only if the matrix on the left-hand side is nonsingular, and in that case we shall get a unique solution for the desired input sequence. It will be seen in the next section that this condition implies the controllability of the system.

EXAMPLE 12.1

A control system with a sampler and zero-order hold is shown in Fig. 12.1. Determine the input sequence that will transfer the system from the origin of the state space to the point $\begin{bmatrix} 1 \\ 0 \end{bmatrix}$ in two steps.

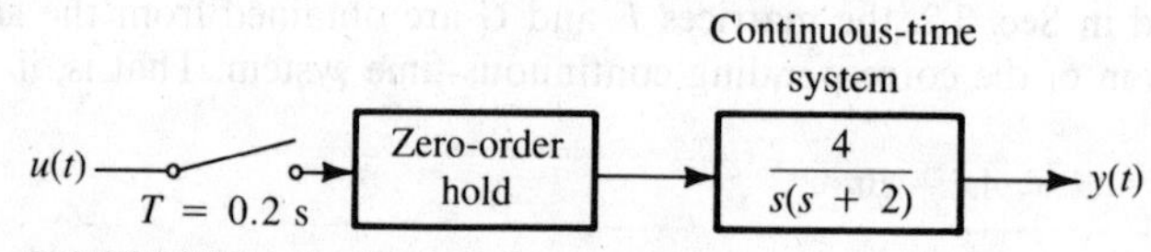

FIGURE 12.1. A discrete-time system.

Solution

A state-space representation for the continuous-time portion of the system is obtained as

$$\begin{bmatrix} \dot{x}_1 \\ \dot{x}_2 \end{bmatrix} = \begin{bmatrix} 0 & 1 \\ 0 & -2 \end{bmatrix}\begin{bmatrix} x_1 \\ x_2 \end{bmatrix} + \begin{bmatrix} 0 \\ 1 \end{bmatrix} u(T) \qquad (12.9)$$

$$y = [4 \quad 0]\begin{bmatrix} x_1 \\ x_2 \end{bmatrix} \qquad (12.10)$$

The matrices F and G for this system were calculated in Sec. 3.2 in Example 3.9. These are given below.

$$F = \begin{bmatrix} 1 & 0.16484 \\ 0 & 0.67032 \end{bmatrix} \qquad (12.11)$$

$$G = \begin{bmatrix} 0.01758 \\ 0.16484 \end{bmatrix} \qquad (12.12)$$

Here we need two sampling intervals for the transfer. Hence,

$$x(2T) - F^2x(0) = \begin{bmatrix} 1 \\ 0 \end{bmatrix} \tag{12.13}$$

Also,

$$[FG \quad G] = \begin{bmatrix} 0.04475 & 0.01758 \\ 0.11050 & 0.16484 \end{bmatrix} \tag{12.14}$$

and

$$\begin{bmatrix} u(0) \\ u(T) \end{bmatrix} = \begin{bmatrix} 0.04475 & 0.01758 \\ 0.11050 & 0.16484 \end{bmatrix}^{-1} \begin{bmatrix} 1 \\ 0 \end{bmatrix} = \begin{bmatrix} 30.33245 \\ -20.33245 \end{bmatrix} \tag{12.15}$$

It can be easily verified that with the input sequence $u(0) =$ 30.33245 and $u(T) = -20.33245$, the state $x(2T) = \begin{bmatrix} 1 \\ 0 \end{bmatrix}$.

The response of the system is said to be "deadbeat" and "ripple-free" in this case, because after two sampling intervals the desired state has been reached exactly, without any overshoot or ripples. Note that in the case of a continuous-time system, the steady state is reached only after an infinite amount of time has elapsed (at least theoretically, although in the case of a stable system, "for all practical purposes" the steady-state is reached after an elapsed time equal to 5 times the largest time constant).

EXAMPLE 12.2

Instead of generating the input sequence, it is also possible to close the loop with a digital compensator, as shown in Fig. 12.2. The transfer function of the digital controller, $D(z)$, can then be calculated such that, with a step input of magnitude 4, the error sequence $e(kT)$ is transformed into the desired input sequence $u(kT)$.*

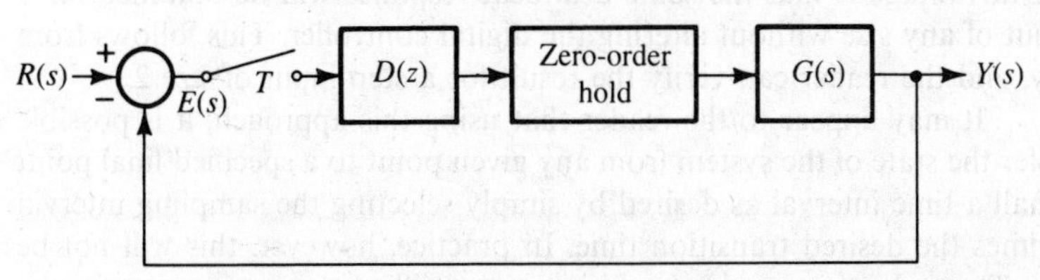

FIGURE 12.2. A discrete-time system with deadbeat response.

---

* This follows from Eq. (12.10), so we get $y = 4x_1 = 4$ when $x_1 = 1$.

For this example, we first calculate the samples of the output for the input sequence calculated in Example 12.1. Thus

$$x(T) = \begin{bmatrix} 1 & 0.16484 \\ 0 & 0.67032 \end{bmatrix} \begin{bmatrix} 0 \\ 0 \end{bmatrix} + \begin{bmatrix} 0.01758 \\ 0.16484 \end{bmatrix} 30.3345$$

$$= \begin{bmatrix} 0.53324 \\ 5 \end{bmatrix} \tag{12.16}$$

and

$$x(2T) = \begin{bmatrix} 1 & 0.16484 \\ 0 & 0.67032 \end{bmatrix} \begin{bmatrix} 0.53324 \\ 5 \end{bmatrix} + \begin{bmatrix} 0.01758 \\ 0.16484 \end{bmatrix} (-20.33245)$$

$$= \begin{bmatrix} 1 \\ 0 \end{bmatrix} \tag{12.17}$$

Hence $y(0) = 0$, $y(T) = 4x_1(T) = 2.13298$, and $y(2T) = 4$. Thus the following error sequence is obtained.

$$\begin{aligned} e(0) &= 4 - y(0) = 4 \\ e(T) &= 4 - y(T) = 1.86702 \end{aligned} \tag{12.18}$$

The z transfer function of the digital controller is obtained by dividing the z transform of the $u(t)$ by the z transform of $e(t)$. This is easy since both $e(t)$ and $u(t)$ will be zero for $t > T$, and we have already calculated the values of $e(0)$, $e(T)$, $u(0)$, and $u(T)$. Hence

$$D(z) = \frac{u(0) + u(T)z^{-1}}{e(0) + e(T)z^{-1}} = \frac{30.3345 - 20.33245z^{-1}}{4 + 1.86702z^{-1}} \tag{12.19}$$

An advantage of this closed-loop implementation is that $u(kT) = 0$ for $k > 1$, so the system is kept in the state $\begin{bmatrix} 1 \\ 0 \end{bmatrix}$ after being brought there. Another advantage is that the same deadbeat response will be obtained for a step input of any size without altering the digital controller. This follows from linearity, and the reader can verify the result for a step input of size 2.

It may appear to the reader that using this approach, it is possible to transfer the state of the system from any given point to a specified final point in as small a time interval as desired by simply selecting the sampling interval as $1/n$ times the desired transition time. In practice, however, this will not be possible. The reduction in the transition time will cause an increase in the magnitude of the input signal applied to the process. Hence there will be a physical limitation imposed by the allowable magnitude of the input to the process. For example, there is always a limit to the magnitude of the voltage that we can safely apply to a servomotor.

DRILL PROBLEM 12.1

The state equations for a third-order system, corresponding to the transfer function

$$G(s) = \frac{1}{s(s+1)(s+2)}$$

are given by

$$\begin{bmatrix} \dot{x}_1 \\ \dot{x}_2 \\ \dot{x}_3 \end{bmatrix} = \begin{bmatrix} 0 & 1 & 0 \\ 0 & 0 & 1 \\ 0 & -2 & -3 \end{bmatrix} \begin{bmatrix} x_1 \\ x_2 \\ x_3 \end{bmatrix} + \begin{bmatrix} 0 \\ 0 \\ 1 \end{bmatrix} u(t)$$

$$y(t) = x_1(t)$$

(a) Determine the input sequence which will transfer the system from the origin of the state space to the state

$$\begin{bmatrix} 1 \\ 0 \\ 0 \end{bmatrix}$$

in three sampling periods if the sampling interval is 0.5 second.

(b) If the system is connected in the closed-loop form as shown in Fig. 12.2, determine the transfer function of the digital controller that will provide a deadbeat ripple-free response to a step input.

Answers:

(a) $u(0) = 16.082354$, $u(T) = -15.670808$, $u(2T) = 3.588458$

(b) $$D(z) = \frac{16.082354 - 15.670808z^{-1} + 3.588458z^{-2}}{1 + 0.765828z^{-1} + 0.1107z^{-2}}$$

DRILL PROBLEM 12.2

Repeat Drill Problem 12.1 if the sampling interval is reduced to 0.1 second.

Answers:

(a) $u(0) = 1159.417$, $u(T) = -1998.334$, $u(2T) = 858.917$

(b) $$D(z) = \frac{1159.417 - 1998.334z^{-1} + 858.917z^{-2}}{1 + 0.8206z^{-1} + 0.5144z^{-2}}$$

12.3 CONTROLLABILITY AND OBSERVABILITY

We are now ready to study the very important concepts of controllability and observability. These tell us whether it is at all possible to control all the states of a system completely by a suitable choice of an input and whether it is at all possible to reconstruct the states of a system from its input and output. A precise knowledge of the state equations of the system is assumed.

12.3.1 Controllability

A system is said to be completely controllable if it is possible to find an input $u(t)$ that will transfer the system from any given initial state $x(t_0)$ to any given final state $x(t_f)$ over a specified interval of time $t_f - t_0$. (*Note*: For the sake of brevity we shall be using the term "uncontrollable" for a system that is not completely controllable.)

THEOREM 12.1

A linear time-invariant continuous-time system described by the equations

$$\dot{x} = Ax + Bu \qquad x \in R^n,\ y \in R^p$$

$$y = Cx \qquad u \in R^m$$

is completely controllable if and only if the rank of the controllability matrix, defined as

$$U \triangleq [B \quad AB \quad A^2B \quad \cdots \quad A^{n-1}B] \tag{12.20}$$

is equal to n.

To appreciate this fact, consider the discrete-time system described by

$$x(kT + T) = F(T)x(kT) + G(T)u(kT) \tag{12.21}$$

It was seen that an input sequence $u(0), u(T), \ldots, u((n-1)T)$ for transferring the system from a given initial state $x(0)$ to a given final state $x(nT)$ could be determined uniquely by inverting the following matrix (for the scalar-input case)

$$U_d = [F^{n-1}G \quad F^{n-2}G \quad \cdots \quad FG \quad G] \tag{12.22}$$

It can be shown (by recognizing that $F = e^{AT}$, and by using the Cayley-Hamilton theorem), that the inverse of U_d will exist if and only if the rank of U is n (that is, if U is nonsingular for the case when B is a column vector).

If a system is not completely controllable, it implies that it has one or more natural modes that cannot be affected by the input, directly or indirectly. This is illustrated in the following example.

EXAMPLE 12.3

Consider the system described by

$$\begin{bmatrix} \dot{x}_1 \\ \dot{x}_2 \end{bmatrix} = \begin{bmatrix} -0.5 & 0 \\ 0 & -2 \end{bmatrix} \begin{bmatrix} x_1 \\ x_2 \end{bmatrix} + \begin{bmatrix} 0 \\ 1 \end{bmatrix} u(t) \tag{12.23}$$

Here,

$$U = \begin{bmatrix} 0 & 0 \\ 1 & -2 \end{bmatrix} \tag{12.24}$$

and is singular. Hence the system is uncontrollable. This is more obvious if we write the two differential equations separately; that is,

$$\begin{aligned} \dot{x}_1 &= -0.5x_1 \\ \dot{x}_2 &= -2x_2 + u(t) \end{aligned} \tag{12.25}$$

It is evident that whereas x_2 can be changed by $u(t)$, the state x_1 is not coupled either directly to the input or to the state x_2. Hence this state (or the mode $x_1(0)e^{-0.5t}$) is uncontrollable. On the other hand, if we had $\dot{x}_1 = -0.5x_1 + x_2$, we would have had

$$A = \begin{bmatrix} -0.5 & 1 \\ 0 & -2 \end{bmatrix} \quad B = \begin{bmatrix} 0 \\ 1 \end{bmatrix} \quad U = \begin{bmatrix} 0 & 1 \\ 1 & -2 \end{bmatrix} \tag{12.26}$$

giving us a controllable system since we can now control x_1 indirectly through x_2.

12.3.2 Observability

A system is said to be completely observable if the state can be determined from a knowledge of the input $u(t)$ and the output $y(t)$ over a finite interval of time.

Since, by definition, if we know the initial state of the system we can determine all future outputs and states for a specified input, it follows that a system is observable if and only if the initial state can be determined from the input/output records for a finite interval. (*Note*: For the sake of brevity we shall be loosely using the term "unobservable" for a system that is not completely observable.)

THEOREM 12.2

A linear system is completely observable if and only if the rank of the observability matrix, defined as

$$V \triangleq \begin{bmatrix} C \\ CA \\ \vdots \\ CA^{n-1} \end{bmatrix} \tag{12.27}$$

is equal to n.

To appreciate this, let us again consider the discrete-time single-input single-output system

$$\left.\begin{aligned} x(kT+T) &= F(T)x(kT) + G(T)u(kT) \qquad x \in R^n \\ y(kT) &= Cx(kT) \end{aligned}\right\} \tag{12.28}$$

We shall assume that the initial state $x(0) = x_0$ is not known, and we want to determine it from observations of the output $y(kT)$, $k = 0, 1, \cdots, n-1$. To simplify matters, let us make $u(kT) = 0$ for all k. Then

$$\begin{aligned} y(0) &= Cx(0) \\ y(T) &= Cx(T) = CF(T)x(0) \\ y(2T) &= CF^2(T)x(0) \\ y(nT-T) &= CF^{n-1}(T)x(0) \end{aligned} \tag{12.29}$$

or

$$\begin{bmatrix} C \\ CF \\ CF^2 \\ \vdots \\ CF^{n-1} \end{bmatrix} x(0) = \begin{bmatrix} y(0) \\ y(T) \\ \vdots \\ y(nT-T) \end{bmatrix} \tag{12.30}$$

Hence the state vector $x(0)$ can be uniquely determined if and only if the matrix to the left is nonsingular.*

---

* Since

$$F = e^{AT} = I + AT + \frac{1}{2!}(AT)^2 + \cdots$$

one may invoke the Cayley-Hamilton theorem to establish, from the above, the main theorem for observability.

EXAMPLE 12.4

Consider the system described by

$$\begin{bmatrix} \dot{x}_1 \\ \dot{x}_2 \end{bmatrix} = \begin{bmatrix} -0.5 & 0 \\ 0 & -2 \end{bmatrix} \begin{bmatrix} x_1 \\ x_2 \end{bmatrix} + \begin{bmatrix} 0 \\ 1 \end{bmatrix} u(T) \tag{12.31}$$

$$y = [0 \quad 1] \begin{bmatrix} x_1 \\ x_2 \end{bmatrix} \tag{12.32}$$

We have already seen in Eq. (12.24) that this system is uncontrollable, since the state x_1 is not affected either by u or by x_2. We now form the observability matrix

$$V = \begin{bmatrix} C \\ CA \end{bmatrix} = \begin{bmatrix} 0 & 1 \\ 0 & -2 \end{bmatrix} \tag{12.33}$$

which is singular. The reason this system is unobservable is that the state x_1 does not affect the output, nor does it affect x_2 (which is coupled to the output).

EXAMPLE 12.5

Consider the system described by

$$\begin{bmatrix} \dot{x}_1 \\ \dot{x}_2 \end{bmatrix} = \begin{bmatrix} 0 & 1 \\ -2 & -3 \end{bmatrix} \begin{bmatrix} x_1 \\ x_2 \end{bmatrix} + \begin{bmatrix} 0 \\ 1 \end{bmatrix} u \tag{12.34}$$

$$y = [1 \quad 1] \begin{bmatrix} x_1 \\ x_2 \end{bmatrix} \tag{12.35}$$

This system has

$$U = \begin{bmatrix} 0 & 1 \\ 1 & -3 \end{bmatrix} \qquad \det U = -1 \tag{12.36}$$

Hence it is controllable.
Also,

$$V = \begin{bmatrix} 1 & 1 \\ -2 & -2 \end{bmatrix} \qquad \det V = 0 \tag{12.37}$$

Hence it is unobservable.

This is explained easily from the transfer function

$$\frac{Y(s)}{U(s)} = \frac{s+1}{s^2+3s+2} = \frac{s+1}{(s+1)(s+2)} = \frac{1}{s+2} \tag{12.38}$$

Since there is a cancellation of the factor $(s+1)$, the mode at $s = -1$, giving rise to a term of the form Ke^{-t} does not appear at the output, even though it is present inside the system.

For example, in many electronic circuits, we have some "parasitic" oscillations that do not appear at the output.

It may be noted that in the particular case when

$$A = \begin{bmatrix} 0 & 1 & 0 & \cdots & 0 \\ 0 & 0 & 1 & \cdots & 0 \\ \vdots & \vdots & \vdots & & \vdots \\ 0 & 0 & 0 & \cdots & 1 \\ -a_0 & -a_1 & -a_2 & \cdots & -a_{n-1} \end{bmatrix} \qquad B = \begin{bmatrix} 0 \\ 0 \\ \vdots \\ 0 \\ 1 \end{bmatrix} \tag{12.39}$$

the controllability matrix is

$$U = \begin{bmatrix} 0 & & & & 1 \\ 0 & & & \cdot & x \\ \vdots & & \cdot & & x \\ 0 & 1 & & & \\ 1 & -a_{n-1} & & \cdots & x \end{bmatrix} \quad \text{and} \quad \det U = -1 \tag{12.40}$$

where the x's indicate elements that are, in general, different from 0 or 1.

Because of the triangular nature of U, with 1's on the diagonal, U is nonsingular and the system is always controllable. It is customary to call this the controllable canonical form.

For any given transfer function

$$\frac{Y(s)}{R(s)} = \frac{b_{n-1}s^{n-1} + b_{n-2}s^{n-2} + \cdots + b_1 s + b_0}{s^n + a_{n-1}s^{n-1} + \cdots + a_1 s + a_0} \tag{12.41}$$

one may write the state equations in the controllable canonical form directly. We shall have A and B as above, and

$$C = [b_0 \quad b_1 \quad b_2 \quad \cdots \quad b_{n-1}] \tag{12.42}$$

It is interesting to note that the inverse of U can also be written by inspection. It is given by

$$U^{-1} = \begin{bmatrix} a_1 & a_2 & \cdots & a_{n-1} & 1 \\ a_2 & a_3 & & 1 & 0 \\ \vdots & \vdots & \cdot & \vdots & \vdots \\ a_{n-1} & 1 & & 0 & 0 \\ 1 & 0 & \cdots & 0 & 0 \end{bmatrix} \tag{12.43}$$

which is an upper triangular matrix.

This can easily be verified for $n = 2$ and 3 and then proved by induction.

In the same way we can write the state equations for any given transfer function in the observable canonical form by inspection. In this case the observability matrix will have the same form as in Eq. (12.40) and the inverse of the observability matrix will be of the same form as Eq. (12.43).

COROLLARY

A linear time-invariant single-input single-output system is controllable and observable if and only if the numerator and denominator polynomials of its transfer function are coprime (that is, they do not have a common factor, with the exception of a constant).

This can be proved by writing the state equations for the given transfer function in the diagonal canonical form (or the Jordan form in the case of multiple poles) by performing a partial fraction expansion, as shown in Example 3.8. In the case of a common factor, the residue at that particular pole (which is canceled by a zero) will be zero. The state equations that will thus be obtained will have the form shown in Example 12.4, with a zero in the corresponding column of C. This will make that state unobservable.

In general, if the numerator and the denominator of the transfer function have common factors, we can write the state equations either in the controllable canonical form or in the observable canonical form by inspection. However, the state equations in the controllable canonical form will not be completely observable. Similarly, the state equations in the observable canonical form will not be completely controllable. The state equations will be both controllable and observable if and only if there are no pole-zero cancellations. In this case, the state equations are said to be a minimal realization of the transfer function.

DRILL PROBLEM 12.3

The state equations of a system are given below. Determine whether the system is controllable and observable.

$$\dot{x} = \begin{bmatrix} -1 & 0 & 3 \\ 2 & -1 & -1 \\ -3 & 1 & -2 \end{bmatrix} x + \begin{bmatrix} 1 \\ 0 \\ 0 \end{bmatrix} u$$

$$y = [1 \quad 2 \quad 1]x$$

Answer:
The system is both observable and controllable.

DRILL PROBLEM 12.4

Repeat for the state equations given below

$$\dot{x} = \begin{bmatrix} -6 & 2 & -4 \\ -18 & 3 & -8 \\ -6 & 1 & -3 \end{bmatrix} x + \begin{bmatrix} 1 \\ 3 \\ 1 \end{bmatrix} u$$

$$y = [1 \quad -1 \quad 2]\, x$$

Answer:
Controllable but unobservable.

DRILL PROBLEM 12.5

Repeat for the following state equations

$$\dot{x} = \begin{bmatrix} -6 & -18 & -6 \\ 2 & 3 & 1 \\ -4 & -8 & -3 \end{bmatrix} x + \begin{bmatrix} 2 \\ -3 \\ 7 \end{bmatrix} u$$

$$y = [1 \quad 3 \quad 1]\, x$$

Answer:
Observable but uncontrollable.

12.4 STATE-VARIABLE FEEDBACK

Consider the block diagram shown in Fig. 12.3. We have a linear system described by the state equations

$$\dot{x} = Ax + Bu \tag{12.44}$$

$$y = Cx \tag{12.45}$$

The system is assumed to be of the single-input single-output type.

The state vector is fed back, and thus

$$u = K[r - k^T x] \tag{12.46}$$

where

$$k^T = [k_1 \quad k_2 \quad \cdots \quad k_n], \tag{12.47}$$

a row vector of the same dimension as x.

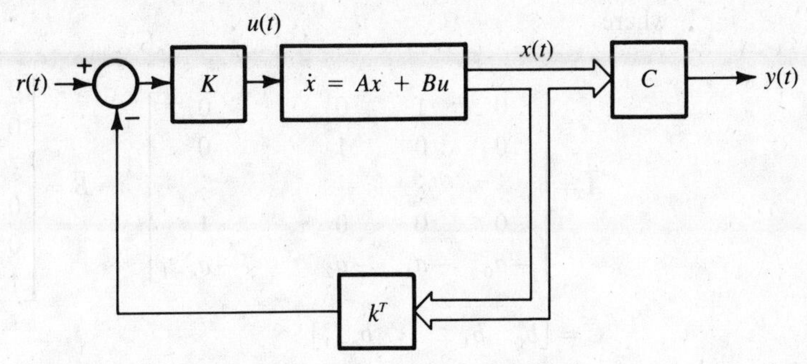

FIGURE 12.3. System with state-variable feedback.

Substitution of Eq. (12.46) into Eqs. (12.44) and (12.45) leads to

$$\dot{x} = A_f x + KBr \tag{12.48}$$

and

$$y = Cx \tag{12.49}$$

where

$$A_f = A - BKk^T \tag{12.50}$$

Equations (12.48) and (12.49) are the state equations for the closed-loop system. Note that C remains unchanged.

THEOREM 12.3

If the system is completely controllable, one can always determine the product Kk^T so that the eigenvalues of A_f (corresponding to the poles of the closed-loop system) will be placed at arbitrarily specified locations.

Proof

If the system is completely controllable, one can always find a nonsingular transformation matrix P such that $\bar{x} = P^{-1}x$, and the resulting matrices $\bar{A} = P^{-1}AP$, $\bar{B} = P^{-1}B$ and $\bar{C} = CP$ are in the controllable canonical form. The method for determining P will be described later in Sec. 12.4.1.

Thus we get

$$\dot{\bar{x}} = Ax + Bu \tag{12.51}$$

$$y = \bar{C}\bar{x} \tag{12.52}$$

where

$$\bar{A} = \begin{bmatrix} 0 & 1 & 0 & \cdots & 0 \\ 0 & 0 & 1 & \cdots & 0 \\ \vdots & \vdots & \vdots & & \vdots \\ 0 & 0 & 0 & & 1 \\ -a_0 & -a_1 & -a_2 & & -a_{n-1} \end{bmatrix} \qquad \bar{B} = \begin{bmatrix} 0 \\ 0 \\ \vdots \\ 0 \\ 0 \\ 1 \end{bmatrix} \tag{12.53}$$

$$\bar{C} = [b_0 \quad b_1 \quad \cdots \quad b_{n-1}]$$

and

$$\det(sI - A) = s^n + a_{n-1}s^{n-1} + \cdots + a_1 s + a_0 \tag{12.54}$$

Also,

$$\bar{B}K = \begin{bmatrix} 0 \\ 0 \\ \vdots \\ 0 \\ K \end{bmatrix} \quad \text{and} \quad \bar{B}K\bar{k}^T = \begin{bmatrix} 0 & 0 & \cdots & 0 \\ \vdots & \vdots & & \vdots \\ K\bar{k}_1 & K\bar{k}_2 & \cdots & K\bar{k}_n \end{bmatrix} \tag{12.55}$$

Hence,

$$\bar{A}_f = \bar{A} - \bar{B}K\bar{k}^T$$

$$= \begin{bmatrix} 0 & 1 & \cdots & 0 \\ \vdots & \vdots & & \vdots \\ 0 & 0 & \cdots & 1 \\ -a_0 - K\bar{k}_1 & -a_1 - K\bar{k}_2 & \cdots & -a_{n-1} - K\bar{k}_n \end{bmatrix} \tag{12.56}$$

From the desired pole locations, we can get the characteristic polynomial for $\bar{A}_f$ as

$$s^n + \alpha_{n-1}s^{n-1} + \cdots + \alpha_1 s + \alpha_0 = \det(sI - \bar{A}_f) \tag{12.57}$$

A comparison with the last row of $\mathbf{A}_f$ gives

$$K\bar{k}_1 = \alpha_0 - a_0 \quad K\bar{k}_2 = \alpha_1 - a_1 \quad \cdots \quad K\bar{k}_n = \alpha_{n-1} - a_{n-1} \tag{12.58}$$

If we select the gain constant K so that the dc gain of the closed-loop transfer function is unity, then we get a unique value for $\bar{k}$. Also, since

$$\bar{k}^T\bar{x} = \bar{k}^T P^{-1} x = k^T x \tag{12.59}$$

we get $k^T = \bar{k}^T P^{-1}$ as the feedback matrix for the system before transformation.

EXAMPLE 12.6

Consider a system described by the state equations

$$\dot{x} = \begin{bmatrix} 0 & 1 & 0 \\ 0 & 0 & 1 \\ -20 & -6 & 0 \end{bmatrix} x + \begin{bmatrix} 0 \\ 0 \\ 1 \end{bmatrix} u \tag{12.60}$$

$$y = [3 \quad 1 \quad 0]\, x \tag{12.61}$$

Since these are in the controllable canonical form, the open-loop transfer function is seen to be

$$\frac{Y(s)}{U(s)} = \frac{s+3}{s^3 + 0s^2 + 6s + 20}$$

$$= \frac{s+3}{(s+2)(s^2 - 2s + 10)} \tag{12.62}$$

Evidently this system is unstable. The desired closed-loop transfer function is given by

$$\frac{Y(s)}{R(s)} = \frac{40(s+3)}{(s+4)(s^2+8s+30)} = \frac{40(s+3)}{s^3 + 12s^2 + 62s + 120} \tag{12.63}$$

From the denominator of Eq. (12.63),

$$A_f = \begin{bmatrix} 0 & 1 & 0 \\ 0 & 0 & 1 \\ -120 & -62 & -12 \end{bmatrix} \tag{12.64}$$

Hence, subtracting A_f from A, we get

$$BKk^T = A - A_f = \begin{bmatrix} 0 & 0 & 0 \\ 0 & 0 & 0 \\ 100 & 56 & 12 \end{bmatrix} \tag{12.65}$$

A comparison of the numerators gives

$$K = 40 \tag{12.66}$$

Hence,

$$k^T = [2.5 \quad 1.4 \quad 0.3] \tag{12.67}$$

EXAMPLE 12.7

Consider the state equations

$$\dot{x} = \begin{bmatrix} -1 & 1 & 0 \\ 0 & -4 & 2 \\ 0 & 0 & -10 \end{bmatrix} x + \begin{bmatrix} 1 \\ 0 \\ -1 \end{bmatrix} u \tag{12.68}$$

$$y = [1 \quad 0 \quad 1] x \tag{12.69}$$

In this case, the system can be transformed into the controllable canonical form using $\bar{x} = P^{-1}x$, where

$$P = \begin{bmatrix} 38 & 14 & 1 \\ -2 & -2 & 0 \\ -4 & -5 & -1 \end{bmatrix} \tag{12.70}$$

The procedure for determining P is shown in the next section (see Example 12.8). Hence

$$\bar{A} = P^{-1}AP = \begin{bmatrix} 0 & 1 & 0 \\ 0 & 0 & 1 \\ -40 & -54 & -15 \end{bmatrix}$$

$$\bar{B} = P^{-1}B = \begin{bmatrix} 0 \\ 0 \\ 1 \end{bmatrix} \tag{12.71}$$

$$\bar{C} = CP = [34 \quad 9 \quad 0]$$

and

$$\frac{Y(s)}{U(s)} = \frac{9s + 34}{s^3 + 15s^2 + 54s + 40} = \frac{9s + 34}{(s + 1)(s^2 + 14s + 40)} \tag{12.72}$$

It is desired that

$$\frac{Y(s)}{R(s)} = \frac{32(9s + 34)}{(s + 8)(s^2 + 12s + 136)} \tag{12.73}$$

Since

$$(s+8)(s^2+12s+136) = s^3+20s^2+232s+1088 \tag{12.74}$$

we have

$$\bar{A}_f = \begin{bmatrix} 0 & 1 & 0 \\ 0 & 0 & 1 \\ -1088 & -232 & -20 \end{bmatrix}$$

$$\bar{B}K\bar{k}^T = \bar{A} - \bar{A}_f = \begin{bmatrix} 0 & 0 & 0 \\ 0 & 0 & 0 \\ 1048 & 178 & 5 \end{bmatrix} \tag{12.75}$$

A comparison of the numerators makes

$$K = 32 \tag{12.76}$$

Hence

$$\bar{k}^T = [32.75 \quad 5.5625 \quad 0.15625] \tag{12.77}$$

and

$$k^T = \bar{k}^T P^{-1} = [1.09375 \quad 2.53125 \quad 0.9375] \tag{12.78}$$

It should be noted that state feedback does not affect the C matrix. Hence, the numerator of the closed-loop transfer function remains the same as that of the open-loop function, except for the multiplier K. The latter is usually chosen to make the dc gain equal to one—i.e., zero steady-state error to step input.

12.4.1 Transformation to controllable canonical form

We shall now consider the problem of transformation to the controllable canonical form. First, we note that the controllability matrix of the transformed equations is given by

$$\begin{aligned} \bar{U} &= [\bar{B} \quad \bar{A}\bar{B} \quad \bar{A}^2\bar{B} \quad \cdots \quad \bar{A}^{n-1}\bar{B}] \\ &= [P^{-1}B(P^{-1}AP)P^{-1}B(P^{-1}AP)(P^{-1}AP)P^{-1}B \cdots (P^{-1}AP) \cdots \\ &\qquad (P^{-1}AP)P^{-1}B] \\ &= P^{-1}[B \quad AB \quad \cdots \quad A^{n-1}B] \\ &= P^{-1}U \end{aligned} \tag{12.79}$$

Hence

$$P = U\bar{U}^{-1} \tag{12.80}$$

Given A and B, we can easily calculate U. $\bar{U}^{-1}$ is also determined directly from the determinant of $(sI - A)$ as shown in Eq. (12.43).

Let

$$\det(sI - A) = s^n + a_{n-1}s^{n-1} + \cdots + a_1 s + a_0 \tag{12.81}$$

Then, we know that

$$\bar{A} = \begin{bmatrix} 0 & 1 & 0 & \cdots & 0 \\ 0 & 0 & 1 & \cdots & 0 \\ \vdots & & & & \\ -a_0 & -a_1 & -a_2 & \cdots & -a_{n-1} \end{bmatrix}$$

$$\bar{B} = \begin{bmatrix} 0 \\ 0 \\ \vdots \\ 0 \\ 1 \end{bmatrix} \tag{12.82}$$

and

$$\bar{U}^{-1} = \begin{bmatrix} a_1 & a_2 & \cdots & a_{n-1} & 1 \\ a_2 & a_3 & \cdots & 1 & 0 \\ \vdots & \vdots & & \vdots & \vdots \\ a_{n-1} & 1 & & 0 & 0 \\ 1 & 0 & \cdots & 0 & 0 \end{bmatrix} \tag{12.83}$$

This enables us to determine P from Eq. (12.80).

EXAMPLE 12.8

Consider the system in Example 12.7. Here

$$A = \begin{bmatrix} -1 & 1 & 0 \\ 0 & -4 & 2 \\ 0 & 0 & -10 \end{bmatrix} \quad B = \begin{bmatrix} 1 \\ 0 \\ -1 \end{bmatrix} \quad C = [1 \;\; 0 \;\; 1] \tag{12.84}$$

Hence,

$$U = [B \quad AB \quad A^2B] = \begin{bmatrix} 1 & -1 & -1 \\ 0 & -2 & 28 \\ -1 & 10 & -100 \end{bmatrix}$$

$$\det(sI - A) = \begin{vmatrix} s+1 & -1 & 0 \\ 0 & s+4 & -2 \\ 0 & 0 & s+10 \end{vmatrix} = (s+1)(s+4)(s+10) \qquad (12.85)$$

$$= s^3 + 15s^2 + 54s + 40$$

This gives

$$\bar{U}^{-1} = \begin{bmatrix} 54 & 15 & 1 \\ 15 & 1 & 0 \\ 1 & 0 & 0 \end{bmatrix} \qquad (12.86)$$

and

$$P = U\bar{U}^{-1} = \begin{bmatrix} 38 & 14 & 1 \\ -2 & -2 & 0 \\ -4 & -5 & -1 \end{bmatrix} \qquad (12.87)$$

Finally,

$$\bar{A} = \begin{bmatrix} 0 & 1 & 0 \\ 0 & 0 & 1 \\ -40 & -54 & -15 \end{bmatrix} \quad \bar{B} = \begin{bmatrix} 0 \\ 0 \\ 1 \end{bmatrix}; \quad \bar{C} = CP = [34 \quad 9 \quad 0] \qquad (12.88)$$

DRILL PROBLEM 12.6

The system described by the state equations in Drill Problem (12.3) is to be used as part of a closed-loop system with state-variable feedback, as shown earlier in the block diagram of Fig. 12.3. Determine the state-variable feedback so that the transfer function of the closed-loop system is given by

$$\frac{5(s^2 + 4s + 16)}{(s + 5)(s^2 + 6s + 16)}$$

Answers:
$K = 5$, $k_1 = 1.4$, $k_2 = 149/95$, $k_3 = 36/95$.

12.5 STATE-VARIABLE FEEDBACK: A TRANSFER FUNCTION APPROACH

The method for calculating the state feedback vector described in the previous section is very convenient if the state equations of the system are given in the controllable canonical form. If this is not the case, then the equations must first be transformed to the canonical form. In this section we shall study an alternative approach based on the use of transfer functions. In addition to providing a simpler method for calculating the state feedback vector, it also gives better insight into the problem.

Consider the block diagram shown in Fig. 12.4, where the effect of state-variable feedback is to return

$$f(t) = k^T x(t) \tag{12.89}$$

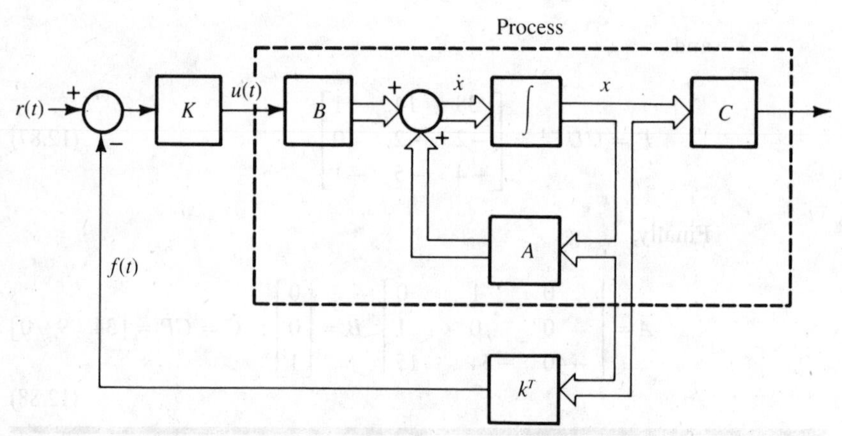

FIGURE 12.4. System with state-variable feedback.

It also follows that the input to the process is

$$u(t) = K[r(t) - f(t)] \tag{12.90}$$

Taking the Laplace transform of both sides of Eq. (12.89), we have

$$F(s) = k^T X(s) = k^T (sI - A)^{-1} B U(s) \tag{12.91}$$

so we have

$$\frac{F(s)}{U(s)} = k^T (sI - A)^{-1} B \triangleq \frac{f(s)}{a(s)} \tag{12.92}$$

where

$$a(s) = \det(sI - A) \tag{12.93}$$

$$f(s) = k^T \operatorname{adj}(sI - A)B \tag{12.94}$$

Similarly, for the process, we have

$$\frac{Y(s)}{U(s)} = C(sI - A)^{-1}B \triangleq \frac{b(s)}{a(s)} \tag{12.95}$$

where

$$b(s) = C \operatorname{adj}(sI - A)B \tag{12.96}$$

From the block diagram, we also have

$$U(s) = K[R(s) - F(s)] \tag{12.97}$$

Thus

$$\frac{F(s)}{U(s)} = \frac{F(s)}{K[R(s) - F(s)]} = \frac{f(s)}{a(s)}. \tag{12.98}$$

which can be solved to obtain

$$\frac{F(s)}{R(s)} = \frac{Kf(s)}{a(s) + Kf(s)} \tag{12.99}$$

Hence

$$\frac{Y(s)}{R(s)} = \frac{Y(s)}{U(s)} \cdot \frac{U(s)}{F(s)} \cdot \frac{F(s)}{R(s)} = \frac{Kb(s)}{a(s) + Kf(s)} \tag{12.100}$$

It follows that the numerator of the closed-loop transfer function can only be a multiple by K of the numerator of the open-loop transfer function; i.e., the zeros of the transfer functions are identical. The denominator of the closed-loop transfer function can be altered arbitrarily by the choice of $Kf(s)$. Let the desired closed-loop transfer function be

$$\frac{Y(s)}{R(s)} = \frac{Kb(s)}{\alpha(s)} \tag{12.101}$$

Comparing with Eq. (12.100), we get

$$f(s) = \frac{1}{K}[\alpha(s) - a(s)] \tag{12.102}$$

Hence $f(s)$ can be determined from the denominators of the open-loop and the desired closed-loop transfer functions.

Returning to Eq. (12.92) and utilizing (12.102), we obtain

$$\frac{f(s)}{a(s)} = \frac{\alpha(s) - a(s)}{Ka(s)} = k^T(sI - A)^{-1}B$$
$$= k^T B s^{-1} + k^T A B s^{-2} + k^T A^2 B s^{-3} + \cdots \tag{12.103}$$

The last equation is obtained from a power series expansion of $(sI - A)^{-1}$ and has been derived in Appendix B.

This provides a simple procedure for obtaining the elements of k, through the following steps.

1. From the specified locations of the poles of the closed-loop transfer function, determine $\alpha(s)$. Also determine K by comparing the numerators of the open- and closed-loop transfer functions.
2. Determine $f(s)$ by subtracting $a(s)$ from $\alpha(s)$ and dividing by K.
3. Divide $f(s)$ by $a(s)$ to obtain a power series in negative powers of s. Only n terms are needed, where n is the order of the system.
4. Calculate $k^T B, k^T AB, k^T A^2 B, \ldots, k^T A^{n-1} B$. Note that these are obtained easily by first premultiplying B by A as many times as necessary, and then premultiplying by k^T.
5. By comparing the coefficients, as in Eq. (12.103), we get n linear simultaneous equations for the n unknown elements of the feedback vector k^T.

To illustrate the procedure we shall solve the transfer function of Example (12.7) by this method.

EXAMPLE 12.9

From Example 12.7 the open-loop transfer function is given by

$$\frac{Y(s)}{U(s)} = \frac{9s + 34}{s^3 + 15s^2 + 54s + 40} \tag{12.104}$$

as in Eq. (12.72). Also, from Eq. (12.73), the desired closed-loop transfer function is

$$\frac{Y(s)}{R(s)} = \frac{32(9s + 34)}{s^3 + 20s^2 + 232s + 108} \tag{12.105}$$

Hence

$$K = 32 \tag{12.106}$$

and

$$f(s) = \frac{1}{K}[\alpha(s) - a(s)] = \frac{5}{32}s^2 + \frac{89}{16}s + \frac{131}{4} \tag{12.107}$$

By long division,

$$\frac{f(s)}{a(s)} = \frac{\frac{5}{32}s^2 + \frac{89}{16}s + \frac{131}{4}}{s^3 + 15s^2 + 54s + 40}$$

$$= \frac{5}{32}s^{-1} + \frac{103}{32}s^{-2} - \frac{767}{32}s^{-3} + \cdots \tag{12.108}$$

Let

$$k^T = [k_1 \quad k_2 \quad k_3] \tag{12.109}$$

Then

$$k^T B = k_1 - k_3 = \frac{5}{32} \tag{12.110}$$

$$k^T AB = -k_1 - 2k_2 + 10k_3 = \frac{103}{32} \tag{12.111}$$

and

$$k^T A^2 B = -k_1 + 28k_2 - 100k_3 = -\frac{767}{32} \tag{12.112}$$

solving the last three equations, we get

$$k^T = [1.09375 \quad 2.53125 \quad 0.9375] \tag{12.113}$$

which agrees with the answer in Eq. (12.78).

DRILL PROBLEM 12.7

Repeat Drill Problem 12.6 using the transfer function approach discussed in this section.

12.6 ASYMPTOTIC STATE OBSERVERS

The concept of state feedback discussed in Sec. 12.4 is very powerful since it allows us to place the poles of the closed-loop system at desired locations provided that the system is completely controllable. In practice, however, the states of a system are seldom available for feedback. Hence, in order to implement state feedback we must somehow obtain (or reconstruct) the states from the input and the output of the system. This will be possible if and only if the system is completely observable.

One simple (but impractical) solution to the problem of state reconstruction is to differentiate the output of the system $n - 1$ times, where n is the order of the system. This would then provide us with a state vector consisting of the output and its $n - 1$ derivatives. This is seldom desirable due to the inevitable presence of noise, which gets accentuated by differentiation. A better alternative is to use an analog or digital computer model of the system, and obtain the states by applying the same input to the model and the system. Thus, if the system is described the Eqs. (12.44) and (12.45), we may set up a model given by the equations

$$\dot{\hat{x}} = A\hat{x} + Bu \tag{12.114}$$

$$\hat{y} = C\hat{x} \tag{12.115}$$

where $\hat{x}(t)$ is the state vector for this model.

In general, $x(t)$ and $\hat{x}(t)$ will not be equal unless the initial conditions $x(0)$ and $\hat{x}(0)$ are also equal. Since $x(0)$ is not known, this is not possible. However, we may improve the situation by utilizing a correction factor based on our knowledge of the output $y(t)$. Consider changing the differential equation (12.114) to

$$\dot{\hat{x}} = A\hat{x} + Bu + L(y - \hat{y}) \tag{12.116}$$

where L is a suitably chosen gain vector of the same dimension as x.

If we define the error in state reconstruction as

$$\tilde{x}(t) \triangleq x(t) - \hat{x}(t) \tag{12.117}$$

then differentiating both sides of Eq. (12.117) and substituting for $\dot{x}$ and $\dot{\hat{x}}$ from Eqs. (12.44) and (12.116), respectively, we have

$$\dot{\tilde{x}} = A(x - \hat{x}) + L(y - \hat{y}) \tag{12.118}$$

Expressing y and $\hat{y}$ in terms of x and $\hat{x}$ through Eqs. (12.45) and (12.115), and using Eq. (12.117), we may rewrite Eq. (12.118) as

$$\dot{\tilde{x}} = (A - LC)\tilde{x} \tag{12.119}$$

Equation (12.119) can be solved to give the state reconstruction error as

$$\tilde{x}(t) = e^{(A-LC)t}\tilde{x}(0) \tag{12.120}$$

where

$$\tilde{x}(0) = x(0) - \hat{x}(0) \tag{12.121}$$

It follows from Eq. (12.120) that although initially the state reconstruction error will not be zero, we can make it decrease asymptotically by proper choice of the vector L so that all the eigenvalues of the matrix $A - LC$ have negative real parts.

THEOREM 12.4

If the system is completely observable, one can always determine a gain vector L so that the eigenvalues of $(A - LC)$ will be placed at arbitrarily specified locations.

Proof

If the system is completely observable, one can always find a nonsingular transformation matrix P such that $\bar{x} = P^{-1}x$, and the resulting matrices $\bar{A} = P^{-1}AP$, $\bar{B} = P^{-1}B$, and $\bar{C} = CP$ are in the observable canonical form. The method for determining P will be described in Sec. 12.6.1.

Thus we get

$$\dot{\bar{x}} = \bar{A}\bar{x} + \bar{B}u \tag{12.122}$$

and

$$y = \bar{C}\bar{x} \tag{12.123}$$

where

$$\bar{A} = \begin{bmatrix} 0 & 0 & \cdots & 0 & -a_0 \\ 1 & 0 & \cdots & 0 & -a_1 \\ 0 & 1 & \cdots & 0 & -a_2 \\ \vdots & \vdots & \cdots & \vdots & \vdots \\ 0 & 0 & \cdots & 0 & -a_{n-2} \\ 0 & 0 & \cdots & 0 & -a_{n-1} \end{bmatrix} \tag{12.124}$$

$$\bar{C} = [0 \quad 0 \quad \cdots \quad 0 \quad 1] \tag{12.125}$$

and

$$\det(sI - \bar{A}) = \det(sI - A) = s^n + a_{n-1}s^{n-1} + \cdots + a_1 s + a_0 \tag{12.126}$$

It follows that if

$$\bar{L} = \begin{bmatrix} l_0 \\ l_1 \\ \vdots \\ l_{n-1} \end{bmatrix} \tag{12.127}$$

then

$$\bar{A} - \bar{L}\bar{C} = \begin{bmatrix} 0 & 0 & \cdots & 0 & -(a_0 + l_0) \\ 1 & 0 & & 0 & -(a_1 + l_1) \\ 0 & 1 & & 0 & -(a_2 + l_2) \\ \vdots & \vdots & & \vdots & \vdots \\ 0 & 0 & \cdots & 0 & -(a_{n-2} + l_{n-2}) \\ 0 & 0 & \cdots & 1 & -(a_{n-1} + l_{n-1}) \end{bmatrix} \tag{12.128}$$

Hence

$$\det(\bar{A} - \bar{L}\bar{C}) = s^n + (a_{n-1} + l_{n-1})s^{n-1} + \cdots + (a_1 + l_1)s + (a_0 + l_0) \tag{12.129}$$

Thus, by a proper choice of the elements of $\bar{L}$, we can place the eigenvalues of $(\bar{A} - \bar{L}\bar{C})$, at desired locations. Also, if we make

$$L = P\bar{L} \tag{12.130}$$

or

$$\bar{L} = P^{-1}L \tag{12.131}$$

then

$$\begin{aligned} \bar{A} - \bar{L}\bar{C} &= P^{-1}AP - (P^{-1}L)(CP) \\ &= P^{-1}(A - LC)P \end{aligned} \tag{12.132}$$

which shows that $(\bar{A} - \bar{L}\bar{C})$ and $(A - LC)$ have the same eigenvalues.

Note that the proof for Theorem 12.4 is very similar to that for Theorem 12.3. This is because of the duality between controllability and observability, and is the result of the fact that $\bar{A}$ and $\bar{B}$ for the controllable canonical form are obtained by transposing $\bar{A}$ and $\bar{C}$ for the observable canonical form. The block diagram for the implementation of the observer is shown in Fig. 12.5.

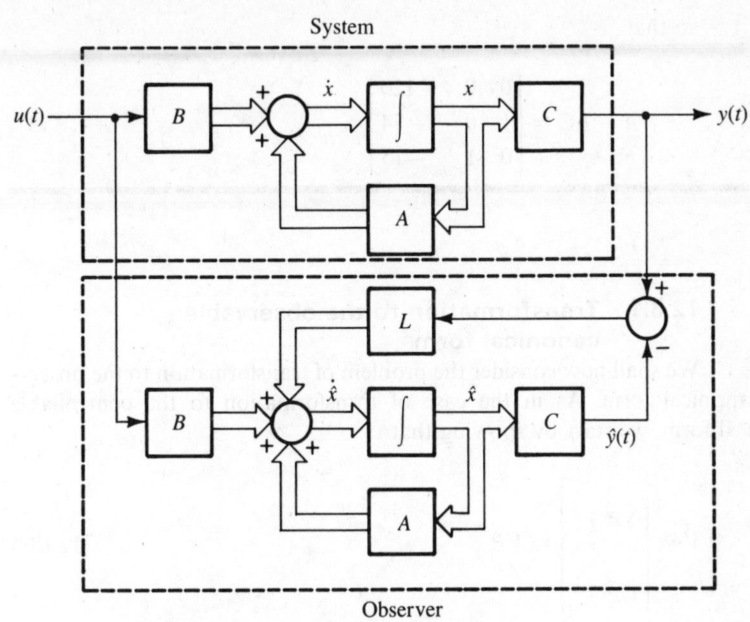

FIGURE 12.5. Asymptotic observer.

EXAMPLE 12.10

Consider a system described by the state equations

$$\dot{x} = \begin{bmatrix} 0 & 0 & -1 \\ 1 & 0 & -2 \\ 0 & 1 & -2 \end{bmatrix} x + \begin{bmatrix} 1 \\ 2 \\ 0 \end{bmatrix} u \tag{12.133}$$

$$y = [0 \quad 0 \quad 1]x \tag{12.134}$$

We shall design an observer that has eigenvalues at $s = -4, -5$, and -6. Hence the characteristic polynomial of the observer is given by

$$\Delta(s) = (s + 4)(s + 5)(s + 6) = s^3 + 15s^2 + 74s + 120 \tag{12.135}$$

Since the equations are already in the observable canonical form, the output gain vector is given by

$$L = \begin{bmatrix} 119 \\ 72 \\ 13 \end{bmatrix} \tag{12.136}$$

so that

$$A - LC = \begin{bmatrix} 0 & 0 & -120 \\ 1 & 0 & -74 \\ 0 & 1 & -15 \end{bmatrix} \tag{12.137}$$

12.6.1 Transformation to the observable canonical form

We shall now consider the problem of transformation to the observable canonical form. As in the case of transformation to the controllable canonical form, we start by showing that

$$\bar{V} = \begin{bmatrix} \bar{C} \\ \bar{C}\bar{A} \\ \vdots \\ \bar{C}\bar{A}^{n-1} \end{bmatrix} = VP \tag{12.138}$$

where P is the matrix required for the transformation; that is, $\bar{A} = P^{-1}AB$, $\bar{B} = P^{-1}B$, and $\bar{C} = CP$.

Hence

$$P = V^{-1}\bar{V} \tag{12.139}$$

and

$$P^{-1} = \bar{V}^{-1}V \tag{12.140}$$

The latter is more suitable since $\bar{V}^{-1}$ can be obtained directly from the characteristic polynomial for A. Let

$$\det(sI - A) = s^n + a_{n-1}s^{n-1} + \cdots + a_1 s + a_0 \tag{12.141}$$

Then it can be shown that

$$\bar{V}^{-1} = \begin{bmatrix} a_1 & a_2 & \cdots & a_{n-1} & 1 \\ a_2 & a_3 & \cdots & 1 & 0 \\ \vdots & \vdots & \cdots & \vdots & \vdots \\ a_{n-1} & 1 & \cdots & 0 & 0 \\ 1 & 0 & \cdots & 0 & 0 \end{bmatrix} \tag{12.142}$$

which can therefore be written by inspection if $\Delta(s)$ is known.

EXAMPLE 12.11

Consider the system described in Example 12.7, for which

$$A = \begin{bmatrix} -1 & 1 & 0 \\ 0 & -4 & 2 \\ 0 & 0 & -10 \end{bmatrix} \qquad C = [1 \quad 0 \quad 1] \tag{12.143}$$

The observability matrix is given by

$$V = \begin{bmatrix} 1 & 0 & 1 \\ -1 & 1 & -10 \\ 1 & -5 & 102 \end{bmatrix} \tag{12.144}$$

Also, as shown in Example 12.8,

$$\Delta(s) = \det(sI - A) = s^3 + 15s^2 + 54s + 40 \tag{12.145}$$

This gives

$$\bar{V}^{-1} = \begin{bmatrix} 54 & 15 & 1 \\ 15 & 1 & 0 \\ 1 & 0 & 0 \end{bmatrix} \tag{12.146}$$

so that

$$P^{-1} = \bar{V}^{-1}V = \begin{bmatrix} 40 & 10 & 6 \\ 14 & 1 & 5 \\ 1 & 0 & 1 \end{bmatrix} \tag{12.147}$$

and

$$P = \begin{bmatrix} -0.017857 & 0.178571 & -0.785714 \\ 0.160714 & -0.607143 & 2.071429 \\ 0.017857 & -0.1785715 & 1.785714 \end{bmatrix} \tag{12.148}$$

It can easily be verified that $P^{-1}AP = \bar{A}$ and $CP = \bar{C}$.

DRILL PROBLEM 12.8

Design an observer for the system described in Drill Problem 12.3 so that the eigenvalues of $(A - LC)$ are at -4 and $-4 \pm j2$.

Answer:

$$L = \begin{bmatrix} -3 \\ 16 \\ -21 \end{bmatrix}$$

12.7 COMBINED OBSERVER-CONTROLLER COMPENSATOR

Our main objective in designing an observer for the system was to reconstruct the states so that they may be utilized for state-variable feedback. The block diagram of a compensator combining the observer with state feedback is shown in Fig. 12.6.

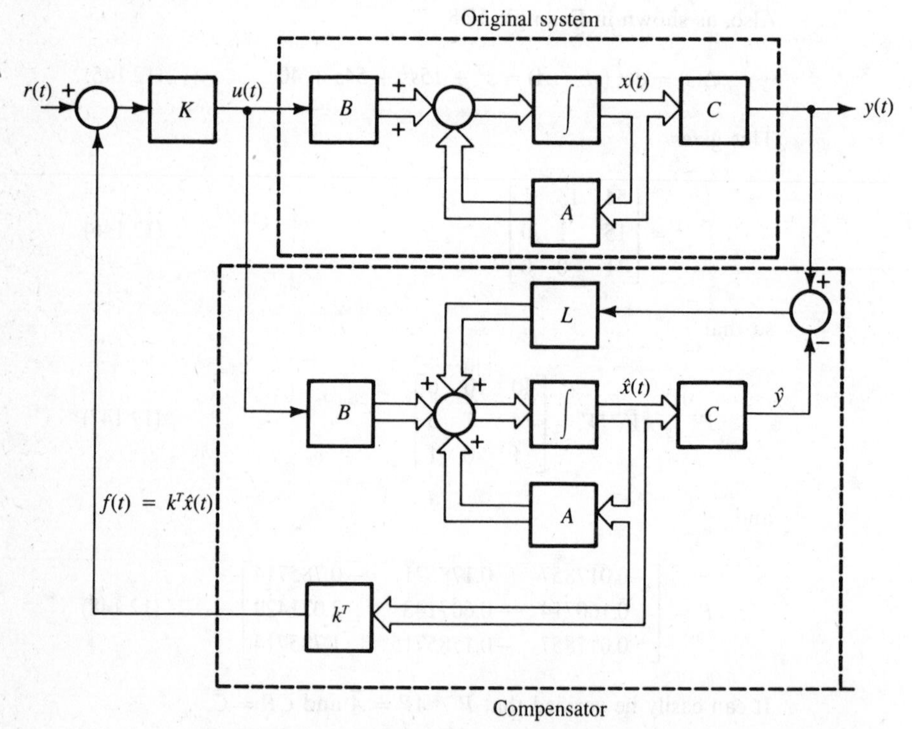

FIGURE 12.6. Compensator combining observer with state feedback.

It will be seen that the observer, driven by the system input $u(t)$ and the output $y(t)$, reconstructs the state $\hat{x}(t)$, which is utilized for state feedback so that

$$u(t) = K[r(t) - k^T\hat{x}(t)] \tag{12.149}$$

A basic question that arises here is whether the use of $\hat{x}(t)$ for state-variable feedback, instead of the actual state $x(t)$ will affect the stability of the system. We shall now attempt to answer this question.

In our design of the observer, we had seen that $\hat{x}(t)$ approaches $x(t)$ asymptotically if the eigenvalues of $(A - LC)$ have negative real parts. Hence, in the steady-state, $\hat{x}(t)$ and $x(t)$ will be identical for all practical purposes. It can be shown that the transfer function $Y(s)/R(s)$ of the system shown in Fig. 12.6 is identical to the transfer function of the system shown in Fig. 12.5 (for example, see section 4.2 of Kailath [1]). This follows intuitively from the fact that transfer functions are based on zero initial conditions, with the result that $\hat{x}(t)$ and $x(t)$ are identical for this case. The transfer function of the overall system is given by

$$\frac{Y(s)}{R(s)} = \frac{Kb(s)\Delta(s)}{\alpha(s)\Delta(s)} \tag{12.150}$$

where $b(s)$ is the numerator of the transfer function of the original system, $\alpha(s)$ the desired characteristic polynomial, and $\Delta(s)$ the characteristic polynomial of the observer; that is,

$$\Delta(s) = \det(sI - A + LC) \tag{12.151}$$

This shows that, as long as the roots of $\Delta(s)$ have negative real parts, we can design the observer and the state feedback separately, and combine them as shown in Fig. 12.6.

The actual implementation of the combined observer-controller compensator requires the use of a model for the system. This is not a problem, since one can easily construct an electronic analog from the given state equations and generate the states using integrators. Alternatively, one may implement the scheme through a microprocessor designed specially to integrate the state equations for the observer and hence calculate $k^T\hat{x}$.

It is very revealing to redraw the block diagram of Fig. 12.6 as in Fig. 12.7. This follows from the fact that the input to the compensator consists

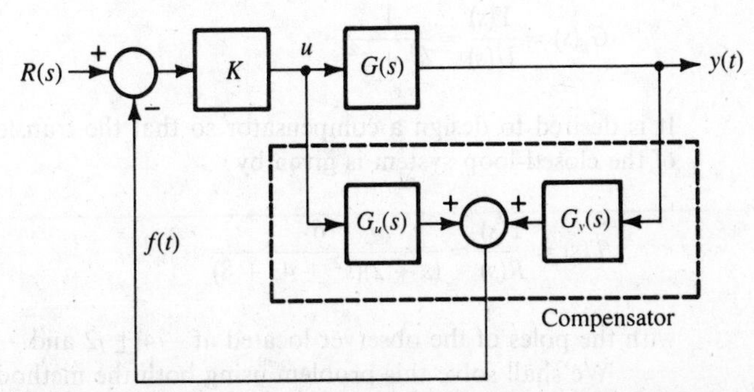

FIGURE 12.7. Compensator combining observer with state feedback.

of $u(t)$ and $y(t)$ and its output is $f(t) = k^T\hat{x}(t)$. Hence, the block marked compensator in Fig. 12.6 can be replaced by the two transfer functions shown in Fig. 12.7 so that

$$F(s) = G_u(s)U(s) + G_y(s)Y(s) \tag{12.152}$$

Therefore, one may also solve the problem of compensator design

$$G_u(s) = \frac{n_u(s)}{\Delta(s)}$$

and

$$G_y(s) = \frac{n_y(s)}{\Delta(s)}$$

so that

$$\frac{Y(s)}{R(s)} = \frac{Kb(s)\Delta(s)}{\alpha(s)\Delta(s)} \tag{12.153}$$

This is precisely the problem that was discussed in Chapter 10 (Sec. 10.7). Note that although there is no essential difference between the two schemes for compensator design, the transfer function approach requires less computation in addition to the fact that it is intuitively more appealing. It should be appreciated, however, that the block diagram shown in Fig. 12.7 was motivated by the state-variable approach shown in Fig. 12.6.

EXAMPLE 12.12

Consider a linear system described by the transfer function

$$G_p(s) = \frac{Y(s)}{U(s)} = \frac{1}{s^3 + s^2} \tag{12.154}$$

It is desired to design a compensator so that the transfer function of the closed-loop system is given by

$$T(s) = \frac{Y(s)}{R(s)} = \frac{16}{(s + 2)(s^2 + 4s + 8)} \tag{12.155}$$

with the poles of the observer located at $-4 \pm j2$ and -5.

We shall solve this problem using both the methods.

Method I: Using State Equations

First we shall design the observer. From the given transfer function, we can write the following state equations in the observable canonical form

$$\dot{x} = \begin{bmatrix} 0 & 0 & 0 \\ 1 & 0 & 0 \\ 0 & 1 & -1 \end{bmatrix} x + \begin{bmatrix} 1 \\ 0 \\ 0 \end{bmatrix} u \tag{12.156}$$

$$y = [0 \quad 0 \quad 1] x \tag{12.157}$$

Since the desired characteristic polynomial for the observer is

$$\begin{aligned} \Delta(s) &= (s + 5)(s^2 + 8s + 20) \\ &= s^3 + 13s^2 + 60s + 100 \end{aligned} \tag{12.158}$$

we immediately get

$$L = \begin{bmatrix} 100 \\ 60 \\ 12 \end{bmatrix} \tag{12.159}$$

To find the state feedback vector we shall first transform Eqs. (12.156) and (12.157) to the controllable canonical form. Following the approach discussed in Sec. 12.4, we obtain the controllability matrix

$$U = [B \quad AB \quad A^2B] = \begin{bmatrix} 1 & 0 & 0 \\ 0 & 1 & 0 \\ 0 & 0 & 1 \end{bmatrix} \tag{12.160}$$

and

$$\bar{U}^{-1} = \begin{bmatrix} 0 & 1 & 1 \\ 1 & 1 & 0 \\ 1 & 0 & 0 \end{bmatrix} \tag{12.161}$$

so that

$$P = U\bar{U}^{-1} = \begin{bmatrix} 0 & 1 & 1 \\ 1 & 1 & 0 \\ 1 & 0 & 0 \end{bmatrix} \tag{12.162}$$

and

$$P^{-1} = \begin{bmatrix} 0 & 0 & 1 \\ 0 & 1 & -1 \\ 1 & -1 & 1 \end{bmatrix} \quad (12.163)$$

It is easily verified that

$$\bar{A} = P^{-1}AP = \begin{bmatrix} 0 & 1 & 0 \\ 0 & 0 & 1 \\ 0 & 0 & -1 \end{bmatrix} \quad (12.164)$$

and

$$\bar{B} = P^{-1}B = \begin{bmatrix} 0 \\ 0 \\ 1 \end{bmatrix} \quad (12.165)$$

Since the desired characteristic polynomial is given by

$$\begin{aligned} \alpha(s) &= (s+2)(s^2+4s+8) \\ &= s^3 + 6s^2 + 16s + 16 \end{aligned} \quad (12.166)$$

we immediately get

$$K\bar{k}^T = [16 \quad 16 \quad 5] \quad (12.167)$$

By comparing the numerators of the transfer functions $G_p(s)$ and $T(s)$, we get

$$K = 16 \quad (12.168)$$

and hence

$$\bar{k}^T = \begin{bmatrix} 1 & 1 & \frac{5}{16} \end{bmatrix} \quad (12.169)$$

Finally, since

$$\bar{k}^T \bar{x} = \bar{k}^T P^{-1} x = k^T x \quad (12.170)$$

we get

$$k^T = \bar{k}^T P^{-1} = \begin{bmatrix} \frac{5}{16} & \frac{11}{16} & \frac{5}{16} \end{bmatrix} \quad (12.171)$$

This completes the design, with the parameters as shown in Fig. 12.6.

Method II: Using Transfer Functions

Following the development in Sec. 10.7, we note that

$$b(s) = N_p(s) = 16 \tag{12.172}$$

$$a(s) = D_p(s) = s^3 + s^2 \tag{12.173}$$

$$\alpha(s) = D(s) = s^3 + 6s^2 + 16s + 16 \tag{12.174}$$

and

$$\Delta(s) = s^3 + 13s^2 + 60s + 100 \tag{12.175}$$

Hence

$$\begin{aligned}[][D(s) - d_p(s)]\Delta(s) \\ &= (5s^2 + 16s + 16)(s^3 + 13s^2 + 60s + 100) \\ &= (5s^5 + 81s^4 + 524s^3 + 1668s^2 + 2560s + 1600)\end{aligned} \tag{12.176}$$

Let

$$N_y(s) = q_2 s^2 + q_1 s + q_0 \tag{12.177}$$

$$N_u(s) = r_2 s^2 + r_1 s + r_0 \tag{12.178}$$

Then

$$N_y(s)N_p(s) = 16q_2 s^2 + 16q_1 s + 16q_0 \tag{12.179}$$

$$N_u(s)D_p(s) = r_2 s^5 + (r_2 + r_1)s^4 + (r_1 + r_0)s^3 + r_0 s^2 \tag{12.180}$$

Hence, equating the corresponding coefficients of both sides of

$$N_u(s)N_p(s) + N_u(s)D_p(s) = [D(s) - D_p(s)]\,\Delta(s) \tag{12.181}$$

we get

$$\begin{matrix} r_2 = 5 & r_1 = 76 & r_0 = 448 \\ q_2 = 75 & q_0 = 160 & q_1 = 100 \end{matrix} \tag{12.182}$$

It may be noted that it is possible to reduce the order of the observer to $(n - 1)$. This follows from the fact that if the system is in the observable canonical form, the last state is equal to the output. Hence, we need to reconstruct only the remaining $(n - 1)$ states. As the design of the reduced-state observer using the state-variable approach is more involved, we shall not discuss it here. The interested reader can find more details in the references listed in Sec. 12.9.

The transfer function approach allows the design of the reduced-order observer-compensator without any additional complication. As a matter of fact, the discussion in Sec. 10.7 of Chapter 10 assumes that the degree of $\Delta(s)$ is $(n - 1)$, and then calculates the compensator transfer functions.

DRILL PROBLEM 12.9

Design a reduced-order compensator for the system described in Example 12.11. The poles of the observer are to be located as $-4 \pm j2$. (*Hint*: Use the approach discussed in Sec. 10.7.)

Answers:

$$\Delta(s) = s^2 + 8s + 20$$

$$N_u(s) = 5s + 51$$

$$N_y(s) = \frac{193}{16}s^2 + 28s + 20$$

12.8 SUMMARY

The method of state-variable feedback is very powerful since it allows the poles of the closed-loop system to be placed anywhere in the s-plane. For example, consider the simple position-control system with two open-loop poles, one at the origin and the other at $-1/\tau_m$, where τ_m = motor time constant. Then, by adjusting the forward gain only, we can locate the closed-loop poles at any point on the corresponding root locus plot.

On the other hand, if we can feed back the velocity as well as the position, this will enable us to place the poles anywhere we like.

Note, however, that we have to differentiate to get the velocity. Although it is not much of a problem in this case, for a higher-order system many differentiations may be required. This is not desirable, due to the fact that the noise, which is invariably present, will cause problems.

There is another way to generate the state of the system without requiring differentiation. If the system is observable, this is possible by including the so-called "asymptotic observer." Its use, however, requires that we know the system state equations exactly. The same assumption was also made in designing the state-variable feedback. One important point to consider, then, is the effect on this control scheme of small changes in parameters. This is the so-called parameter sensitivity problem. In particular, we would like to know how the eigenvalues of the closed-loop system move in the s-plane with small variations in parameters. This is a topic for further study, and is usually referred to as the problem of robust control.

12.9 REFERENCES

1. T. Kailath, *Linear Systems*, Prentice-Hall, Englewood Cliffs, N.J., 1980.
2. C. T. Chen, *Introduction to Linear System Theory*, Holt, Rinehart and Winston, New York, 1970.
3. D. G. Luenberger, "Observing the state of a linear system," *IEEE Transactions on Military Electronics*, vol. MIL-8, pp. 74–80, 1964.

12.10 PROBLEMS

1. The state equations of a system are as follows:

$$\dot{x} = \begin{bmatrix} 0 & 0 & -8 \\ 1 & 0 & -14 \\ 0 & 1 & -7 \end{bmatrix} x + \begin{bmatrix} 2 \\ 1 \\ 0 \end{bmatrix} u \quad \text{and} \quad y = [0 \quad 0 \quad 1]x$$

Determine whether the system is (a) controllable, (b) observable, and (c) stable.

2. For the error-sampled control system shown in Fig. 12.8, the response to a step input is required to be deadbeat and ripple-free, in addition to being as fast as possible. Determine the transfer function $D(z)$ of a suitable digital compensator.

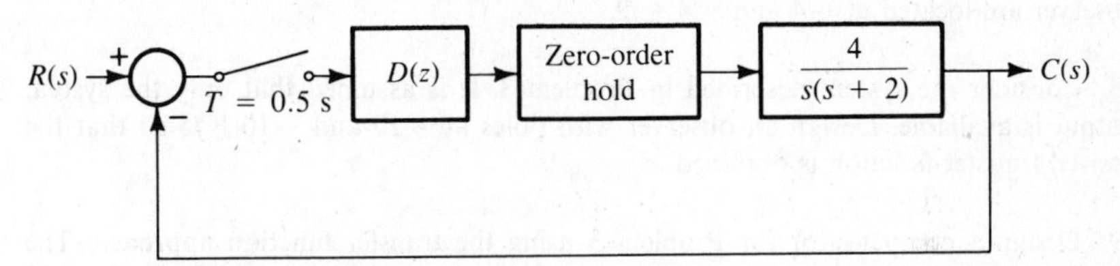

FIGURE 12.8. Block diagram of a closed-loop discrete-time system.

3. The state equations of a system are given by

$$\dot{x} = \begin{bmatrix} -2 & 0 & 1 \\ 1 & -3 & 0 \\ 1 & 1 & -1 \end{bmatrix} x + \begin{bmatrix} 1 \\ 0 \\ 1 \end{bmatrix} u$$

$$y = [1 \quad -1 \quad 2]\, x$$

Determine the state feedback law $u = K(r - k^T x)$ such that the closed-loop system has the following transfer function.

$$\frac{Y(s)}{R(s)} = \frac{12.5(3s^2 + 16s + 24)}{(s + 5)(s^2 + 10s + 60)}$$

4. If only the output is available for feedback in the previous problem, what compensator in the feedback path will give the same closed-loop transfer function?

5. The state equations for a linear system are given below.

a. Transform them to the controllable canonical form, and hence, determine the transfer function of the system.

$$\dot{x} = \begin{bmatrix} 0 & 1 & -6 \\ 1 & 0 & 5 \\ 0 & -1 & 2 \end{bmatrix} x + \begin{bmatrix} 1 \\ 1 \\ 0 \end{bmatrix} u$$

$$y = [0 \quad 1 \quad 2]x$$

b. Determine the state feedback vector that will move the poles to -2 and $-1 \pm j1$.

6. The state equations for a helicopter near hover are as follows:

$$\dot{x} = \begin{bmatrix} -0.002 & -1.4 & 9.8 \\ -0.01 & -0.4 & 0 \\ 0 & 1 & 2 \end{bmatrix} x + \begin{bmatrix} 9.8 \\ 6.3 \\ 0 \end{bmatrix} u$$

$$y = [0 \quad 0 \quad 1]x$$

Determine the state feedback vector that will move the poles of the closed-loop system to -3 and $-2 \pm j2$.

7. Design an asymptotic observer for the previous problem so that the poles of the observer are located at -4 and $-4 \pm j2$.

8. Consider the system described in Problem 3. It is assumed that only the system output is available. Design an observer with poles at -10 and $-10 \pm j5$ so that the desired transfer function is obtained.

9. Design a compensator for Problem 3 using the transfer function approach. The poles of the observer should be at -10 and $-10 \pm j5$.

10. Repeat Problem 9 with a second-order observer, having poles at $-10 \pm j5$.

11. Consider the state equations for a helicopter at hover, as given in Problem 6. We shall now investigate the case when the input to the system is sampled and held constant between the sampling instants. The sampling interval is given by $T = 0.1$ second.

a. Calculate the input sequence that will transfer the state of the system from

$$\begin{bmatrix} 0 \\ 0 \\ 0 \end{bmatrix} \quad \text{to} \quad \begin{bmatrix} 1 \\ 0 \\ 0 \end{bmatrix}$$

in three sampling intervals.

b. Design a digital controller so that the response of this system to a step input is deadbeat and ripple-free, using the state-variable approach presented in Sec. 12.2.

12. For Problem 11 determine the state feedback vector that will place the eigenvalues at 0.25 and $0.2 \pm j0.2$ in the z-plane. What will be the response of this system to a unit step input?

13. Design an observer for the discrete-time system described in Problem 11 that will reconstruct the states from the input and the output. The eigenvalues of the observer are to be located at 0.1, 0.12, and 0.15 in the z-plane.

14. Design an observer-compensator of the minimum order for the discrete-time system of Problem 11 using the method described in Sec. 10.7. The poles of the system are to be located as in Problem 12 and those of the observer are to be at 0.1 and 0.12 in the z-plane.

15. The inverted-pendulum problem shown in Fig. 12.9 has been the subject of considerable study. Although the system is inherently unstable, it is possible to keep the stick from falling by applying a suitable input $u(t)$ to the motorized cart. The linearized state equations are of the form

$$\dot{x} = Ax + Bu$$

where

$$A = \begin{bmatrix} 0 & 1 & 0 & 0 \\ 0 & 0 & -1 & 0 \\ 0 & 0 & 0 & 1 \\ 0 & 0 & 10 & 0 \end{bmatrix} \quad \text{and} \quad B = \begin{bmatrix} 0 \\ 1 \\ 0 \\ -1 \end{bmatrix}$$

The state variables are as follows:

x_1 = the horizontal displacement, $z(t)$
x_2 = the horizontal velocity, $\dot{z}(t)$
x_3 = the angular rotation, $\theta(t)$
x_4 = the angular velocity, $\dot{\theta}(t)$

Determine the state feedback vector that will stabilize the system and place its poles at -1, -2, and $-1 \pm j1$.

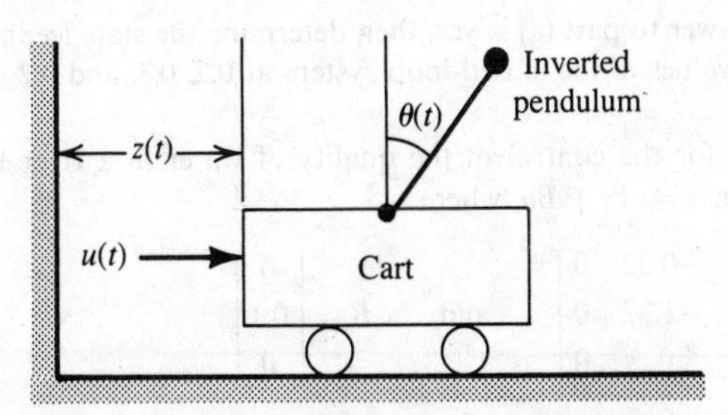

FIGURE 12.9. Cart with inverted pendulum.

16. Design an observer to reconstruct the states of the system described in Problem 15 assuming that only the output $z(t)$ is available for feedback. The eigenvalues of the observer are to be placed at -3, -3, -4, and -4.

17. The linearized equations for a satellite in a circular equatorial orbit are given by $\dot{x} = Ax + Bu$, where

$$A = \begin{bmatrix} 0 & 1 & 0 & 0 \\ 3 & 0 & 0 & 2 \\ 0 & 0 & 0 & 1 \\ 0 & -2 & -3 & 0 \end{bmatrix} \quad \text{and} \quad B = \begin{bmatrix} 0 \\ 1 \\ 0 \\ 0 \end{bmatrix}$$

with the following state variables.

x_1 = the distance from the center of the earth
x_2 = the rate of change of x_1
x_3 = angular displacement in the equatorial plane
x_4 = the rate of change of x_3

The input $u(t)$ is the thrust produced by a rocket engine.

a. Verify that this is a controllable system.
b. Determine the state variable feedback that will place the poles of the system at -1, -2, and $-1 \pm j1$.

18. Determine the state transition equation for the system described in the previous problem assuming a sampling interval of 0.1 second and that the input is held constant between sampling instants. Calculate the state-variable feedback that will cause the poles of the resulting discrete-time system to be placed at 0.15, 0.2, and $0.1 \pm j0.12$ in the z-plane.

19. Design an observer-compensator of the minimal order for Problem 18 if only the state x_3 is available. Use the transfer function approach of Sec. 10.7 in the z-plane, placing the poles of the observer at 0.1 and $0.1 \pm j0.1$.

20. The discrete-time model of the mixing system of a rotary cement kiln is as follows:

$$x(k+1) = \begin{bmatrix} 0 & 1 & 0 & 0 \\ 0.24 & 0 & 0 & 0.76 \\ 0 & 0.07 & 0.93 & 0 \\ 0 & 0 & 0 & 1 \end{bmatrix} x(k) + \begin{bmatrix} 1 \\ 0.76 \\ 0 \\ 0 \end{bmatrix} u(k)$$

a. Is this system controllable?
b. If the answer to part (a) is yes, then determine the state feedback vector that will place the eigenvalues of the closed-loop system at 0.2, 0.3, and $0.2 \pm j\,0.2$.

21. The state equations for the control of the quality of water in a river by effluent treatment are of the form $\dot{x} = Ax + Bu$, where

$$A = \begin{bmatrix} -1.2 & -0.32 & 0 \\ 0 & -1.32 & 0 \\ -1 & 0 & 0 \end{bmatrix} \quad \text{and} \quad B = \begin{bmatrix} 0 \\ 0.1 \\ 0 \end{bmatrix}$$

The state variables are

x_1 = dissolved oxygen
x_2 = biochemical oxygen demand
x_3 = integral of error between the specified reference level and dissolved oxygen level

The input $u(t)$ is to be obtained from state feedback; that is,

$$u(t) = -k^T x$$

where k is the state feedback vector.

Determine the value of k so that the characteristic polynomial of the resulting system is given by

$$P(s) = s^3 + 7.2s^2 + 21.6s + 27$$

22. Repeat Problem 21 if all the eigenvalues of the closed-loop system are to be located at $s = -3$.

23. The matrices A, B, and C for a single-input single-output system are as follows:

$$A = \begin{bmatrix} -1 & -2 & 0 & 0 & 0 & 0 \\ 2 & -1 & 0 & 0 & 0 & 0 \\ 0 & 0 & -5 & 0 & 0 & 0 \\ 0 & 0 & 0 & -2 & 1 & 0 \\ 0 & 0 & 0 & 0 & -2 & 1 \\ 0 & 0 & 0 & 0 & 0 & -2 \end{bmatrix} \qquad B = \begin{bmatrix} 1 \\ 0 \\ 0 \\ 0 \\ 0 \\ 1 \end{bmatrix}$$

$$C = [0 \quad 11 \quad -1 \quad 2 \quad 0 \quad 7]$$

a. Draw an analog computer simulation diagram for the system.
b. Which of the states of the system are uncontrollable?
c. Which of the states are unobservable?

24. A discrete-time system is described by the transfer function

$$G(z) = \frac{z^2 + \alpha z}{z^3 - 0.8z^2 - 0.21z + 0.01}$$

a. Derive a set of state equations in the controllable canonical form and draw a simulation block diagram.
b. For what values of the parameter α will this representation be unobservable? What happens to the transfer function in this case?
c. Derive a set of state equations in the observable canonical form and draw a simulation block diagram.
d. For what values of α will the observable representation of the system become uncontrollable? What happens to the system transfer function for these values of α?

25. Sometimes it is desirable to "decouple" multiple-input multiple-output systems with equal number of inputs and outputs so that the resulting transfer function matrix is of the diagonal form. This is of considerable practical importance in systems where we want to adjust the value of one output without affecting the others. It can be accomplished by state feedback in systems that satisfy certain conditions. The input for these systems is obtained as

$$u = Fx + Gv$$

The matrices A, B, and C for a two-input two-output systems are

$$A = \begin{bmatrix} 0 & 1 & 0 & 0 & 0 \\ 0 & 0 & 1 & 0 & 0 \\ -6 & 5 & 2 & -5 & -6 \\ 0 & 0 & 0 & 0 & 1 \\ 0 & 0 & 0 & -2 & 3 \end{bmatrix} \quad B = \begin{bmatrix} 0 & 0 \\ 0 & 0 \\ 1 & 0 \\ 0 & 0 \\ 0 & 1 \end{bmatrix}$$

$$C = \begin{bmatrix} 3 & 1 & 0 & 0 & 0 \\ 0 & 1 & 2 & 0 & 1 \end{bmatrix}$$

a. Determine the transfer function matrix relating $Y(s)$ to $U(s)$.
b. Determine the transfer function matrix relating $Y(s)$ to $V(s)$ if the matrices F and G are as given below.

$$F = \begin{bmatrix} 4 & -10 & -6 & 5 & 6 \\ 3 & 7 & 5 & 1 & -4 \end{bmatrix} \quad G = \begin{bmatrix} 1 & 0 \\ -2 & 1 \end{bmatrix}$$

26. The regulation of the glucose level in the blood can be expressed by the following linearized differential equations.

$$\dot{x}_1 = -ax_1 - bx_2$$

$$\dot{x}_2 = -cx_1 - dx_2 + u$$

where x_1 is the deviation of the extracellular insulin level from the mean, x_2 is the deviation of the extracellular glucose level from the mean, and u is the rate of glucose intravenous injection. It is desired to stabilize the glucose level by feeding back the states—that is, by making u a linear combination of x_1 and x_2. Determine the feedback vector that will locate the eigenvalues of closed-loop system at $-1 \pm j1$ if the parameters are given by $a = 0.78$, $b = 0.208$, $c = 4.34$, and $d = 2.92$.

13 Nonlinear Systems

13.1 INTRODUCTION

In our discussions so far we have considered only linear systems. In practice, however, all physical systems have some kind of nonlinearities. Sometimes it may even be desirable to introduce a nonlinearity deliberately in order to improve the performance of a system and make its operation safer. This may also result in making the system more economical than is possible with linear components alone.

One of the simplest examples of a system with an intentionally introduced nonlinearity is a relay-controlled or on/off system. For instance, in a typical home-heating system, a furnace is turned on when the temperature falls below a certain specified value and off when the temperature exceeds another given value. Another example is a nonlinear controller designed to realize a damping ratio that varies with the magnitude of the actuating signal.

Nonlinear systems differ from the usual linear systems in several aspects. Perhaps the most significant of these is the fact that the principle of superposition is not applicable to nonlinear systems. As a result, the output of a nonlinear system, with zero initial conditions, to step function inputs of different magnitudes is shown in Fig. 13.1. Altering the size of the input does not change the shape of the response of the linear system, whereas for the nonlinear system there is considerable change in both the percentage overshoot and the frequency of oscillation.

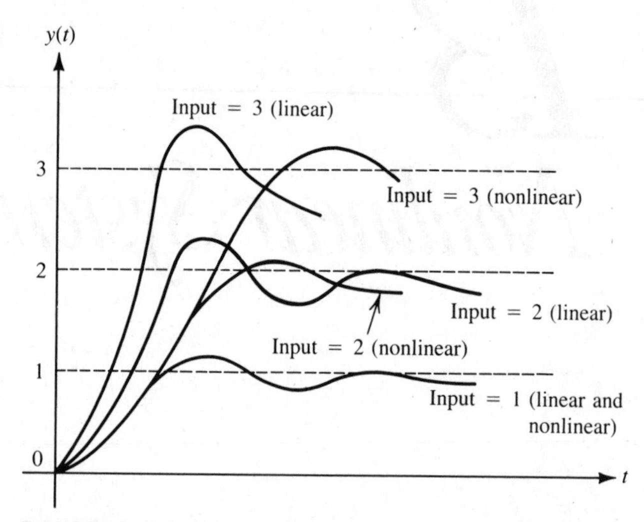

FIGURE 13.1. Step responses of linear and nonlinear systems.

Similar observations may be made about stability. In linear systems, stability is a characteristic of the system, independent of the magnitude or nature of the input or of the initial conditions. In the case of nonlinear systems, stability may depend on the magnitude of the input as well as on the initial conditions. Furthermore, application of a sinusoidal input to a stable linear system causes the steady-state output to be a sinusoid of the same frequency, which will, in general, differ from the input in phase and magnitude. In nonlinear systems, on the other hand, the steady-state output may contain harmonics of the input, and in some cases even subharmonics may arise.

Other unusual features of nonlinear systems include such phenomena as limit cycles and jump resonance. The former means that independently of the magnitude of the input or the initial conditions, the system may produce oscillations of a certain frequency and amplitude, which may not be sinusoidal. The latter imply jumps in magnitude and phase as the frequency is changed near resonance.

Although our list of the special features of nonlinear systems is far from complete, it shows that the world of nonlinear systems is much richer than that of linear systems. Since, in practice, all physical systems have some nonlinearities, we need some understanding of the basic methods for the analysis of nonlinear systems. This chapter represents an attempt to do so.

13.2 COMMON NONLINEARITIES

In most control systems we cannot avoid the presence of certain types of nonlinearities. These can be classified as static or dynamic. A device for which there is a nonlinear relationship between the input, $x(t)$, and the output, $y(t)$, that does not involve a differential equation is called a static nonlinearity. On the other hand, the input and the output may be related through a nonlinear differential equation. Such a device is called a dynamic nonlinearity. In this section we shall discuss briefly the basic features of some common nonlinearities.

13.2.1 Saturation

This is one of the most common static nonlinearities. A simple example is an amplifier for which the output is proportional to the input only for a limited range of the input. As the magnitude of the input exceeds the range, the output approaches a constant, as shown in Fig. 13.2. Although the change from one range to the other is usually gradual, it is often sufficiently accurate in most cases to approximate the curve by a set of straight lines, as shown.

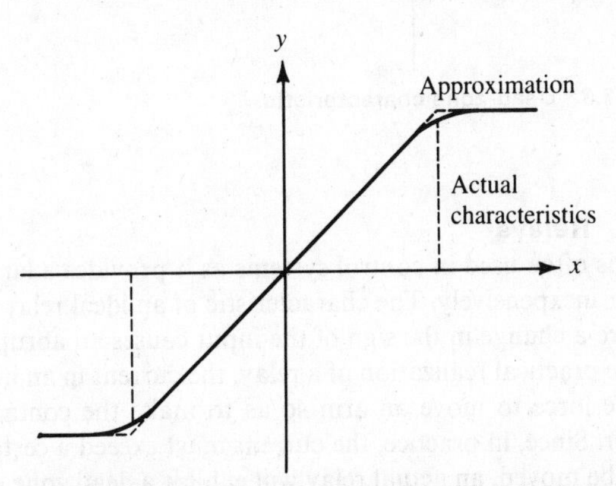

FIGURE 13.2 Saturation characteristic and piecewise linear approximation.

Besides amplifiers, many other physical devices exhibit saturation. Another well-known example of saturation is the relationship between the magnetic flux and current in an iron-cored coil.

13.2.2 Dead zones

In many physical devices the output is zero until the magnitude of the input exceeds a certain value. For example, while developing the mathematical model for a dc servomotor, we had assumed that any voltage applied to the armature windings will cause the armature to rotate, if the field current is maintained constant. In practice, rotation will result only if the torque produced by the motor is sufficient to overcome the static friction. As a result, if we plot the relationship between the steady-state angular velocity and the applied voltage, we get the characteristic shown in Fig. 13.3, which exhibits the dead-zone phenomenon. Many other devices exhibit similar characteristics.

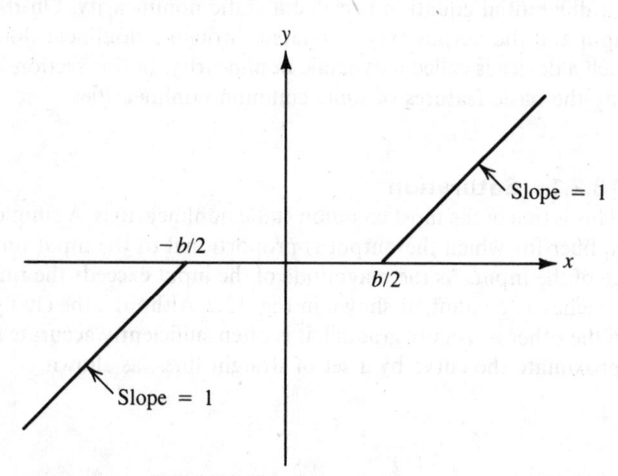

FIGURE 13.3. Dead-zone characteristic.

13.2.3 Relays

A relay is often used in control systems as it provides a large power amplification rather inexpensively. The characteristic of an ideal relay is shown in Fig. 13.4(a), where a change in the sign of the input causes an abrupt change in the output. In the practical realization of a relay, the current in an iron-cored coil exerts magnetic force to move an arm so as to make the contact in one direction or another. Since, in practice, the current must exceed a certain value before the arm can be moved, an actual relay will exhibit a dead zone as shown in Fig. 13.4(b). Furthermore, due to the phenomenon of magnetic hysteresis, a larger current is needed to close the relay than the current at which the contacts open. Hence, in practice, most relays exhibit a dead-zone characteristic as well as hysteresis, as shown in Fig. 13.4(c).

13.2.4 Friction

Frictional forces oppose motion whenever there is a sliding contact between mechanical surfaces. The predominant part of the frictional force is called viscous friction, which is proportional to the relative velocity between the moving surfaces. This is linear in nature, as shown in Fig. 13.5(a). In addition to this viscous friction, there are two components of the total frictional force that are nonlinear. One of them is the coulomb friction, which produces a constant force opposing the motion. The other is called stiction, which is the force required to initiate motion, and is always greater than the force of coulomb friction. The characteristic relating the total frictional force to velocity is shown in Fig. 13.5(b).

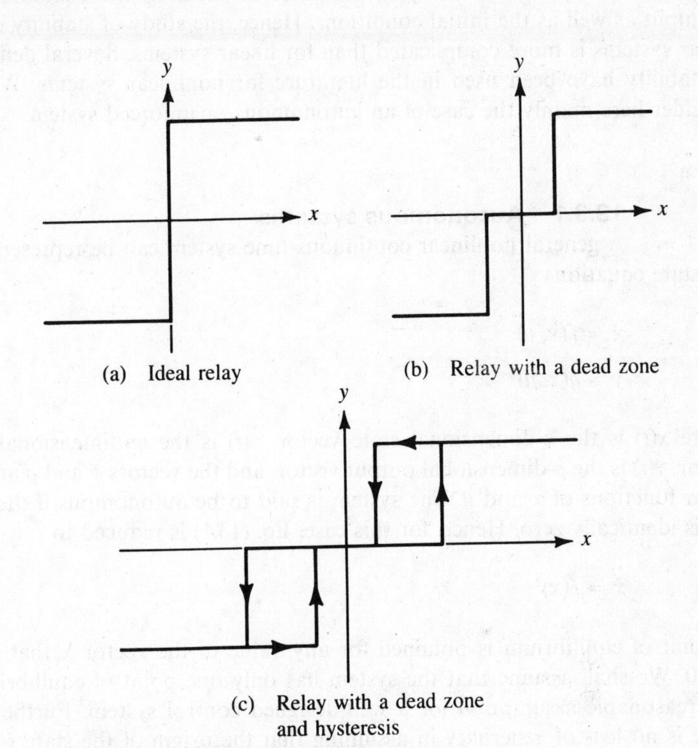

FIGURE 13.4. Characteristics of a relay.

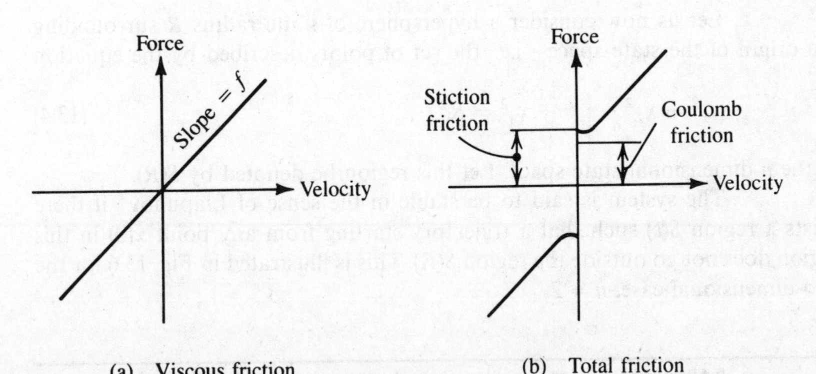

FIGURE 13.5. Characteristics of frictional forces.

13.3 STABILITY OF NONLINEAR SYSTEMS

As stated in Sec. 13.1, the stability of a nonlinear system depends not only on the physical properties of the system but also on the magnitude and nature of the input as well as the initial conditions. Hence, the study of stability of nonlinear systems is more complicated than for linear systems. Several definitions of stability have been used in the literature for nonlinear systems. We shall consider here mainly the case of an autonomous or unforced system.

13.3.1 Autonomous systems

A general nonlinear continuous-time system can be represented by the state equations

$$\dot{x} = f(x, u) \tag{13.1}$$

$$y = g(x, u) \tag{13.2}$$

where $x(t)$ is the n-dimensional state vector, $u(t)$ is the m-dimensional input vector, $y(t)$ is the p-dimensional output vector, and the vectors f and g are nonlinear functions of x and u. The system is said to be autonomous if the input $u(t)$ is identically zero. Hence, for this case, Eq. (13.1) is reduced to

$$\dot{x} = f(x) \tag{13.3}$$

A point of equilibrium is obtained for any value of the vector x that makes $\dot{x} = 0$. We shall assume that the system has only one point of equilibrium. It is a reasonable assumption for a well-designed control system. Furthermore, there is no loss of generality in assuming that the origin of the state space is this point of equilibrium, since this can always be the case with suitable transformation of the coordinates.

Let us now consider a hypersphere of finite radius R surrounding the origin of the state space—i.e., the set of points described by the equation

$$x_1{}^2 + x_2{}^2 + \cdots + x_n{}^2 = R^2 \tag{13.4}$$

in the n-dimensional state space. Let this region be denoted by $S(R)$.

The system is said to be stable in the sense of Liapunov* if there exists a region $S(\varepsilon)$ such that a trajectory starting from any point $x(0)$ in this region does not go outside the region $S(R)$. This is illustrated in Fig. 13.6 for the two-dimensional case, $n = 2$.

---

* After A. M. Liapunov, a Russian mathematician who did pioneering work in this area during the nineteenth century.

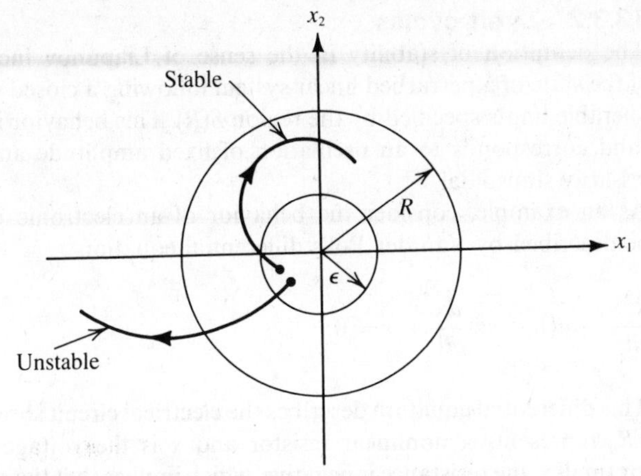

FIGURE 13.6. Stability in the sense of Liapunov.

The system is said to be asymptotically stable if there exists a $\delta > 0$ such that the trajectory starting from any $x(0)$ within $S(\delta)$ does not leave $S(R)$ at any time and finally returns to the point of equilibrium (the origin of the state space). The trajectory in Fig. 13.7(a) shows asymptotic stability.

The system is said to be monotonically stable if it is asymptotically stable and the distance of the state from the origin decreases monotonically with time. The trajectory in Fig. 13.7(b) shows monotonic stability.

A system is said to be globally stable if the regions $S(\delta)$ and $S(R)$ extend to infinity. A system is said to be locally stable if the region $S(\delta)$ is small and when subjected to small perturbations the state remains within the small specified region $S(R)$.

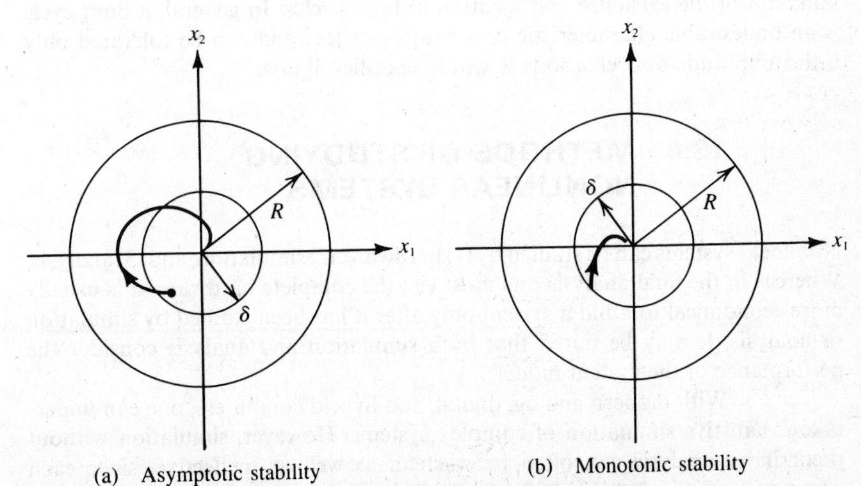

(a) Asymptotic stability (b) Monotonic stability

FIGURE 13.7. Asymptotic and monotonic stability.

13.3.2 Limit cycles

The definition of stability in the sense of Liapunov includes the possibility of the state of a perturbed linear system following a closed trajectory within the tolerable limits specified by the region $S(R)$. This behavior is called a limit cycle, and corresponds to an oscillation of fixed amplitude and period, but not necessarily sinusoidal.

As an example, consider the behavior of an electronic oscillator, which can be described by Van der Pol's differential equation

$$\frac{d^2x}{dt^2} - \mu(1 - x^2)\frac{dx}{dt} + x = 0 \tag{13.5}$$

This differential equation describes the electrical circuit shown in Fig. 13.8, where R_n represents a nonlinear resistor and x is the voltage across a capacitor. For small x, the resistance is negative, which implies that the amplitude of oscillations will increase with an exponential envelope. As x increases, the resistance increases, and is positive for $x > 1$. For an intermediate value of x, the oscillations are stable. The waveform of x is not sinusoidal.

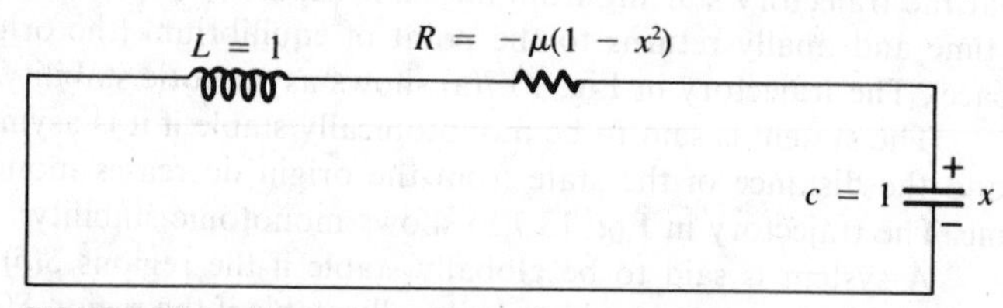

FIGURE 13.8. Equivalent circuit for Eq. (13.5).

An important aim in the study of nonlinear systems is the determination of the existence and location of limit cycles. In general, a limit cycle is an undesirable characteristic of a control system and can be tolerated only if the amplitude of oscillations is within specified limits.

13.4 METHODS OF STUDYING NONLINEAR SYSTEMS

Nonlinear systems can be studied by (1) hardware, (2) simulation, and (3) analysis. Whereas in the final analysis one must test the complete hardware, it is usually more economical to build a system only after it has been studied by simulation or analysis. It may be noted that both simulation and analysis consider the performance of the system model.

With modern analog, digital, and hybrid computers, one can undertake exhaustive simulation of complex systems. However, simulation without preliminary analysis can often be wasteful as well as ineffective, since each computer run merely provides a set of outputs for given parameters and inputs.

Theoretical analysis, on the other hand, provides at least in a qualitative manner considerable insight into the behavior of the system for different values of the parameters as well as various types of disturbances. The study can then be completed by simulation on a computer in an intelligent manner, since the choice of the type of computer and the simulation program will be based on the preliminary analysis.

Although a number of methods have been employed by control engineers for the study of nonlinear systems, we shall consider here the two most well-known methods. The first is the describing-function method, which is an attempt to extend the familiar frequency response approach to nonlinear systems. Although this method is based on a simplifying approximation and is hence inexact, its main advantage is its basic simplicity. It often gives the designer an appreciation of the system behavior and indicates the modifications that should be made to the system for satisfactory performance. Another popular approach is the phase-plane method, which provides a graphical technique for obtaining the solution of nonlinear differential equations of the second order. Although this method is not applicable to higher-order systems, quite often such systems can be represented by approximate second-order models.

A more general method for the study of the stability of nonlinear systems is based on the classical work of A. M. Liapunov. This will be presented briefly in a later section of this chapter.

13.5 DESCRIBING-FUNCTION ANALYSIS

If we apply a sine wave of period T to a nonlinear device, the steady-state output will, in general, have the same period but will not necessarily be sinusoidal. Fourier analysis of the nonsinusoidal waveform will lead to a fundamental component and several harmonics. Note that it is assumed that subharmonics do not exist. If we further assume that the remainder of the feedback system shown in Fig. 13.9 is of the low-pass nature, then the harmonics will be attenuated and may be neglected to obtain a reasonable approximation to the performance of the system. Furthermore, if at some frequency, the loop gain is one, with a phase shift of 180°, possibilities of oscillations exist.

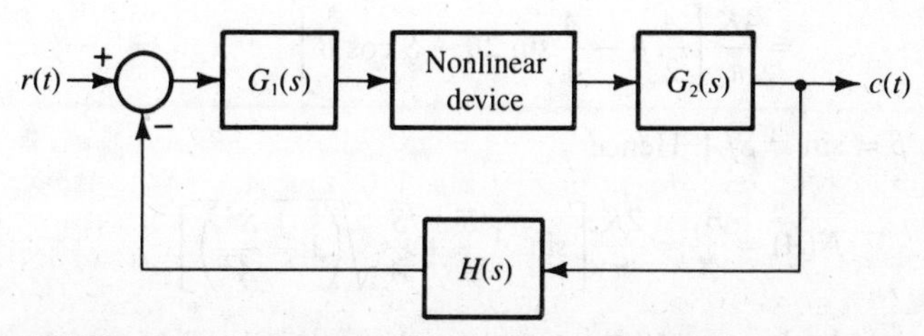

FIGURE 13.9. A typical nonlinear feedback system.

The describing function of a nonlinear device is defined as the ratio of the complex number representing the fundamental component of its output to the complex number representing the sinusoidal input; that is,

$$N(j\omega) = \frac{Y_1(j\omega)}{X(j\omega)} \tag{13.6}$$

In general, $N(j\omega)$ will be a function of the amplitude of the input as well as its frequency. For a static nonlinearity, $N(j\omega)$ is a function only of the amplitude X.

The describing functions of some typical nonlinearities will now be derived.

13.5.1 Saturation

The saturation characteristic is shown in Fig. 13.10, simplified so that it can be represented by straight-line segments. Let $x = A \sin \omega t$, where $A > S$. The waveform of the output is shown in Fig. 13.11.

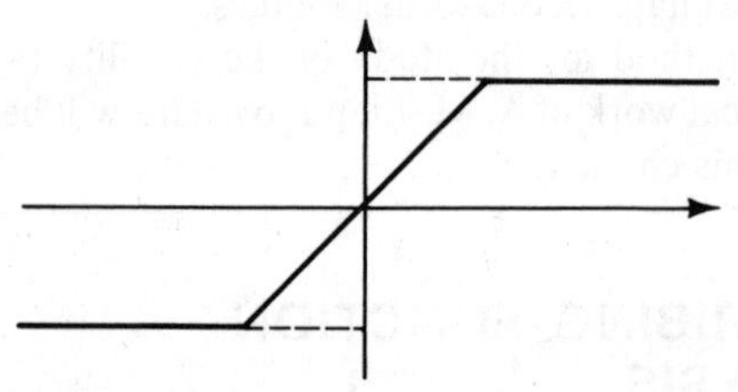

FIGURE 13.10. Saturation characteristic (piecewise linear approximation).

Since the output in Fig. 13.11 has odd symmetry, its fundamental component will not have a cosine term. It will, therefore, be of the form $B_1 \sin \theta$, where $\theta = \omega t$. Taking advantage of the half-wave symmetry, it is seen that

$$\begin{aligned} B_1 &= \frac{4}{\pi} \int_0^{\pi/2} y \sin \theta \\ &= \frac{4}{\pi} \left[\int_0^{B} KA \sin^2 \theta \, d\theta + \int_{\beta}^{\pi/2} KS \sin \theta \, d\theta \right] \\ &= \frac{4K}{\pi} \left[\frac{A}{2} \beta - \frac{A}{4} \sin 2\beta + S \cos \beta \right] \end{aligned} \tag{13.7}$$

where $\beta = \sin^{-1} S/A$. Hence

$$N(A) = \frac{B_1}{A} = \frac{2K}{\pi} \left[\sin^{-1} \frac{S}{A} + \frac{S}{A} \sqrt{\left(1 - \frac{S^2}{A^2}\right)} \right] \tag{13.8}$$

In this case $N(A)$ is a real number and independent of the frequency.

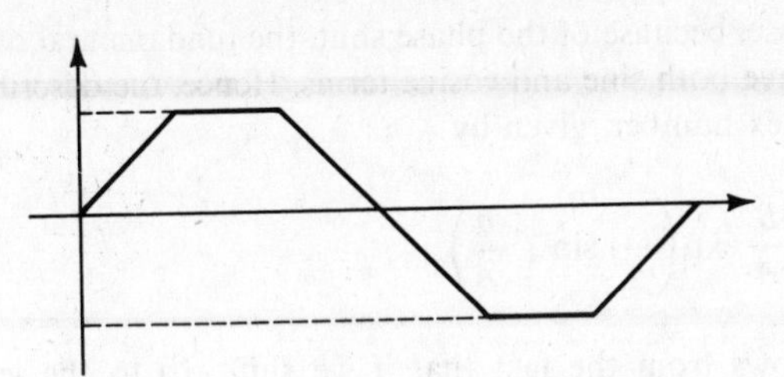

FIGURE 13.11. The output for a sine-wave input.

13.5.2 Two-position relay

The characteristic of a two-position relay is shown in Fig. 13.12. If the input is $x = A \sin \omega t$, then the output is equal to $-B$ for $0 \leq \omega t \leq \sin^{-1} b/A$, and equal to $+B$ for $\sin^{-1} b/A \leq \omega t \leq \pi + \sin^{-1} b/A$. In this case, the resulting output is easily seen as a square wave lagging behind $x(t)$ by $\sin^{-1} b/A$, as shown in Fig. 13.13.

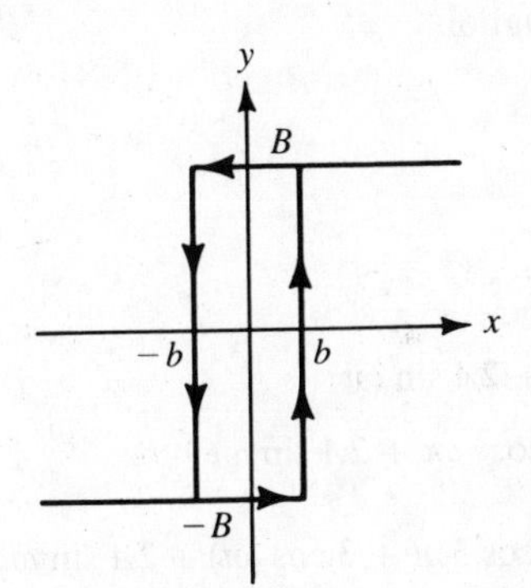

FIGURE 13.12. A typical two-position relay.

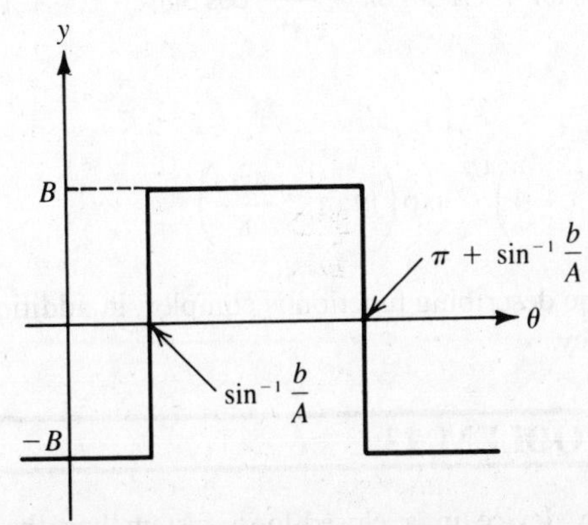

FIGURE 13.13. Output of a two-position relay to the sinusoidal input $x = A \sin \theta$.

In this case, because of the phase shift, the fundamental component of the output will have both sine and cosine terms. Hence, the describing function will be a complex number, given by

$$N(A) = \frac{4B}{\pi A} \exp\left(-j \sin^{-1} \frac{b}{A}\right) \tag{13.9}$$

This follows from the fact that if we shift $y(t)$ to the left by the amount with $\sin^{-1} b/A$, the resulting square wave will contain only sine terms, with the fundamental given by

$$B_1 = \frac{4B}{\pi A} \tag{13.10}$$

13.5.3 A nonlinear differential equation

Consider a nonlinear system where the input and output are related through the nonlinear differential equation

$$y(t) = x^2 \frac{dx}{dt} + 2x \tag{13.11}$$

Since $x = A \sin \omega t$, we get

$$\begin{aligned} y(t) &= A^3\omega \sin^2 \omega t \cos \omega t + 2A \sin \omega t \\ &= A^3\omega \cos \omega t - A^3\omega \cos^3 \omega t + 2A \sin \omega t \\ &= A^3\omega \cos \omega t - \frac{A^3\omega}{4}(\cos 3\omega t + 3 \cos \omega t) + 2A \sin \omega t \\ &= \frac{A^3\omega}{4} \cos \omega t + 2A \sin \omega t - \frac{A^3\omega}{4} \cos 3\omega t \end{aligned} \tag{13.12}$$

Hence

$$N(A) = \left(\frac{A^4\omega^2}{16} + 4\right)^{1/2} \exp\left(j \tan^{-1} \frac{A^2\omega}{8}\right) \tag{13.13}$$

In this case, the describing function is complex, in addition to being a function of the frequency.

DRILL PROBLEM 13.1

An amplifying device in a closed-loop system has the nonlinear characteristic shown in Figure 13.14. Derive the describing function.

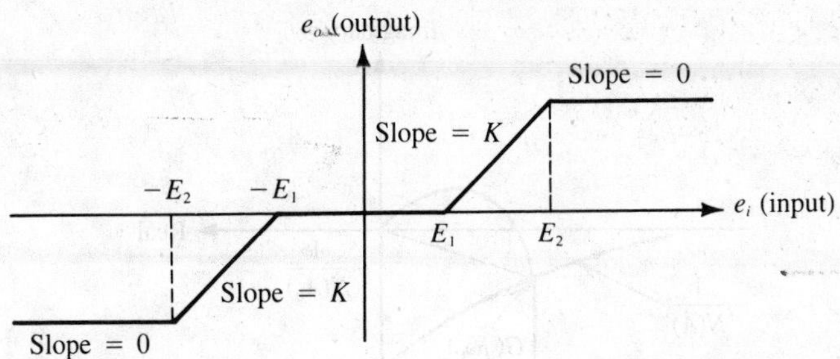

FIGURE 13.14. Characteristic of a nonlinear amplifying device.

Answer:

$$\frac{4K}{\pi}\left[\sqrt{1-\left(\frac{E_1}{A}\right)^2}-\sqrt{1-\left(\frac{E_2}{A}\right)^2}+\frac{\pi}{2A}(E_1-E_2)\right.$$
$$\left.-\frac{E_2}{A}\exp\left(j\sin^{-1}\frac{E_2}{A}\right)+\frac{E_1}{A}\exp\left(j\sin^{-1}\frac{E_1}{A}\right)\right]$$

DRILL PROBLEM 13.2

The output, $y(t)$, of a nonlinear device is related to the input, $x(t)$, through the following differential equation

$$y = 4\left(\frac{dx}{dt}\right)^2 + 6x + 3x^2\frac{dx}{dt}$$

Determine the describing function for the device.

Answer:

$$\left(\frac{9}{16}A^4\omega^2 + 36\right)^{1/2}\exp\left(j\tan^{-1}\frac{A^2\omega}{8}\right)$$

13.5.4 Stability analysis using the describing function

The basic purpose of describing-function analysis is to determine the stability of a closed-loop system with a nonlinear element. This is done by plotting on the G-H plane the quantity $-1/N(A)$, where $N(A)$ is the describing function, along with the polar plot of the transfer function $GH(s)$ of the linear portion. If these two curves intersect for some value of ω and A, as shown in Fig. 13.15, this indicates the possibility of limit cycle oscillations of amplitude A at this frequency due to a loop gain of one with a 180° phase shift.

It should be noted, however, that the describing function method is an approximation, since the higher harmonics have been neglected. Thus the

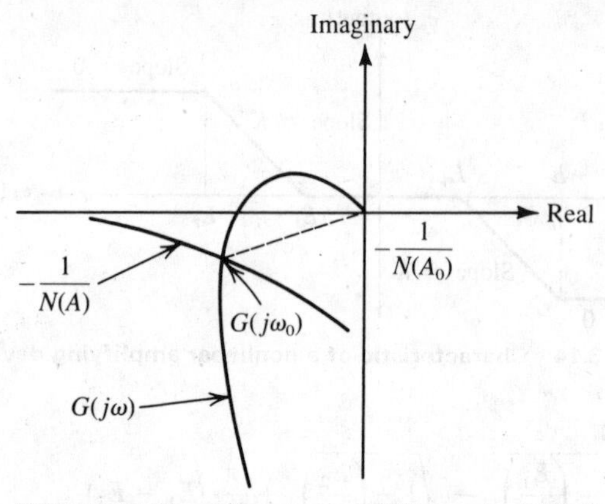

FIGURE 13.15. Polar plots of $GH(s)$ and $-1/N(A)$.

accuracy is higher in those cases where the linear portion effectively filters out the harmonics. If this is not the case, results may be misleading and should be verified by computer simulation.

EXAMPLE 13.1

Consider the control system shown in Fig. 13.16, where the nonlinear element (NL) is a two-position relay of the type show in Fig. 13.12.

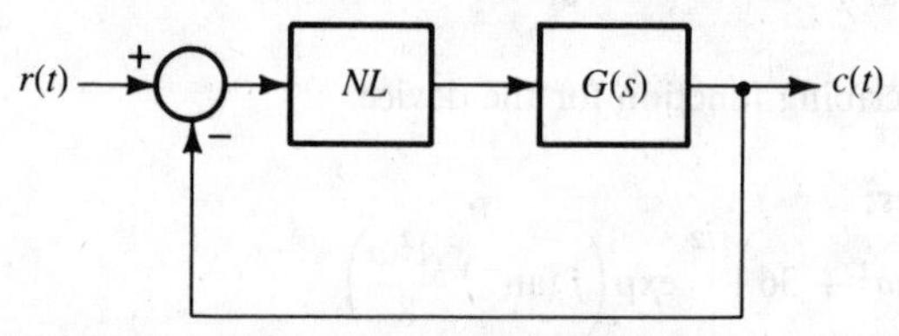

FIGURE 13.16. A relay-controlled system.

It will be assumed that the describing function of the relay is given by

$$N(A) = \frac{5}{A}\exp\left(-j\sin^{-1}\frac{0.5}{A}\right) \tag{13.14}$$

and the transfer function of the linear part of the system is given by

$$G(s) = \frac{4}{s(s+1)} \tag{13.15}$$

The polar plots of $G(j\omega)$ and $-1/N(A)$ are shown in Fig. 13.17. From the plot it will be seen that the two curves intersect for ω approximately equal to 4 rad/s and A approximately equal to 1.11. This

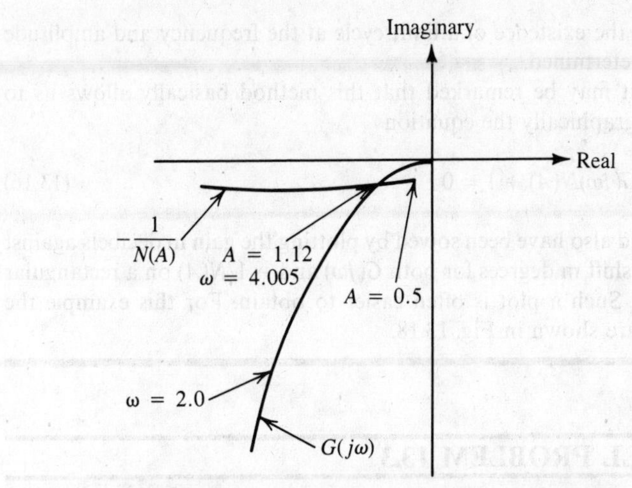

FIGURE 13.17. Polar plots of $G(j\omega)$ and $-1/N(A)$ for Example 13.1.

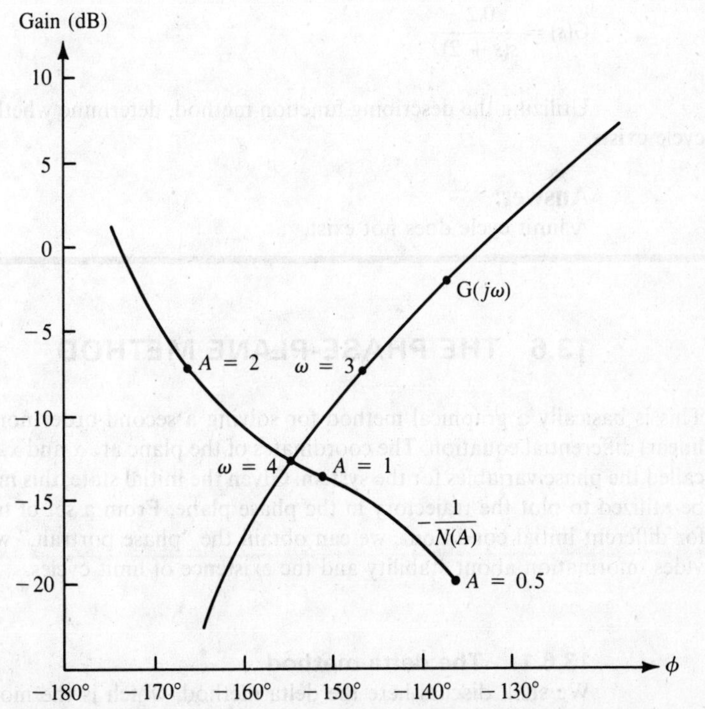

FIGURE 13.18. Log-magnitude/phase plots for Example 13.1.

shows the existence of a limit cycle at the frequency and amplitude thus determined.

It may be remarked that this method basically allows us to solve graphically the equation

$$G(j\omega)N(A) + 1 = 0 \qquad (13.16)$$

It could also have been solved by plotting the gain in decibels against phase shift in degrees for both $G(j\omega)$ and $-1/N(A)$ on a rectangular graph. Such a plot is often easier to obtain. For this example the plots are shown in Fig. 13.18.

DRILL PROBLEM 13.3

A unity-feedback instrument servo is driven by an amplifier that saturates at 70 percent of the rated voltage of the motor. Assume that the gain of the unsaturated amplifier is 50. The transfer function of the linear portion of the system, excluding the amplifier, is given by

$$G(s) = \frac{0.2}{s(s+2)}$$

Utilizing the describing function method, determine whether a limit cycle exists.

Answer:
A limit cycle does not exist.

13.6 THE PHASE-PLANE METHOD

This is basically a graphical method for solving a second-order nonlinear (or linear) differential equation. The coordinates of the plane are x and $\dot{x}$, which are called the phase variables for the system. Given the initial state, this method can be utilized to plot the trajectory in the phase plane. From a set of trajectories for different initial conditions, we can obtain the "phase portrait," which provides information about stability and the existence of limit cycles.

13.6.1. The delta method

We shall discuss here the delta method, which is the most general of all the methods for plotting the trajectory in the phase plane. Consider the general second-order differential equation of the form

$$\frac{d^2x}{dt^2} + f(x, \dot{x}, t) = 0 \tag{13.17}$$

The delta method is based on rearranging the above equation in the form

$$\frac{d^2x}{dt^2} + \omega_0{}^2(x + \delta) = 0 \tag{13.18}$$

where δ is, in general, a function of x, $\dot{x}$, and t. Normalizing the time by defining

$$\tau = \omega_0 t \tag{13.19}$$

we get

$$\frac{d^2x}{d\tau^2} + x + \delta(x_1, x_2, \tau) = 0 \tag{13.20}$$

where $x_1 = x$ and $x_2 = dx/d\tau$. Equation (13.20) may also be written as

$$\frac{d^2x}{d\tau^2} = \frac{dx_2}{d\tau} = \frac{dx_2}{dx_1} \cdot \frac{dx_1}{d\tau} = -x_1 - \delta \tag{13.21}$$

or

$$\frac{dx_2}{dx_1} = \frac{-x_1 - \delta}{x_2} \tag{13.22}$$

A graphical interpretation is given in Fig. 13.19. For a given initial point P_0, the value of δ is calculated from the known values of $x_1(0)$ and $y_2(0)$. This locates

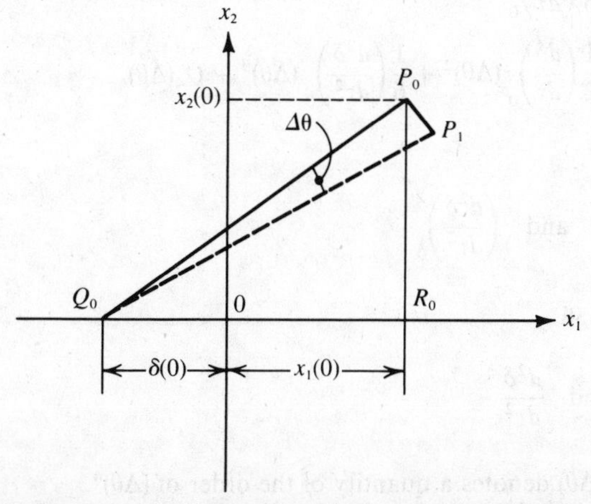

FIGURE 13.19. Graphical interpretation of the delta method.

the point Q_0. The slope of the straight line P_0Q_0 is then obtained as

$$\frac{P_0R_0}{Q_0R_0} = \frac{x_2(0)}{x_1(0) + \delta(0)} \tag{13.23}$$

Hence the trajectory at the point P_0 is perpendicular to Q_0P_0 and can be approximated by a small circular arc drawn with center at Q_0 and radius Q_0P_0. For positive time, the trajectory will be in the clockwise direction. This follows by noting that the slope

$$\frac{dx_2}{dx_1} = \frac{-x_1 - \delta}{x_2} \tag{13.24}$$

is negative for positive x_1 and x_2; that is, x_2 decreases as x_1 is increased.

By drawing the circular arc we obtain the next point P_1 on the trajectory, where the angle $P_0Q_0P_1$, denoted as $\Delta\theta$, is small. At the point P_1 we can again calculate the value of δ and, after locating the new center, Q_1, draw another circular arc.

An interesting feature of the delta method is that the time along the trajectory can be evaluated directly. It is easily shown that the time for the displacement P_0P_1 in Fig. 13.18 is given by

$$\Delta t = \frac{\Delta\theta}{\omega_0} \text{ seconds} \tag{13.25}$$

From the Taylor series expansions for the increments of x_1 and x_2, the expressions for the errors in the approximation are obtained as follows:

$$\varepsilon_{x_1} = \frac{1}{6}\left(\frac{d\delta}{d\tau}\right)_0 (\Delta\theta)^3 + O_4(\Delta\theta) \tag{13.26}$$

$$\varepsilon_{x_2} = \frac{1}{2}\left(\frac{d\delta}{d\tau}\right)_0 (\Delta\theta)^2 + \frac{1}{6}\left(\frac{d^2\delta}{d\tau^2}\right)_0 (\Delta\theta)^3 + O_4(\Delta\theta) \tag{13.27}$$

where

$$\left(\frac{d\delta}{d\tau}\right)_0 \quad \text{and} \quad \left(\frac{d^2\delta}{d\tau^2}\right)_0$$

denote the values of

$$\frac{d\delta}{d\tau} \quad \text{and} \quad \frac{d^2\delta}{d\tau^2}$$

at $\tau = 0$, and $O_4(\Delta\theta)$ denotes a quantity of the order of $(\Delta\theta)^4$.

The approximation may be improved if, instead of using $\delta(\tau_0)$, one uses the average value of δ over the interval Δt. This is equivalent to using the modified Euler method, and the error at each step is given by

$$\varepsilon_{x_1} = -\frac{1}{12}\left(\frac{d\delta}{d\tau}\right)_0 (\Delta\theta)^3 + O_4(\Delta\theta) \tag{13.28}$$

$$\varepsilon_{x_2} = \frac{1}{24}\left(\frac{d^2\delta}{d\tau^2}\right)_0 (\Delta\theta)^3 + O_4(\Delta\theta) \tag{13.29}$$

The procedure is described below, and the graphical construction is shown in Fig. 13.20.

1. Locate the initial point P_0 for $x_1(0)$, $x_2(0)$ at $\tau = \tau_0$.
2. Calculate the initial value of δ and locate Q_0 on the x_1 axis.
3. Draw the circular arc P_0P_M with its center at Q_0, the incremental angle being chosen equal to $\Delta\theta/2$.
4. Calculate the value of δ by using the intermediate values of x_1 and x_2 at P_M where $\tau = \tau_0 + \Delta\theta/2$. Locate Q_M $(-\delta_M, 0)$ on the x_1-axis.
5. Draw the circular arc P_0P_1 with its center at Q_M, the incremental angle being $\Delta\theta$. The arc P_0P_1 represents a portion of the trajectory.

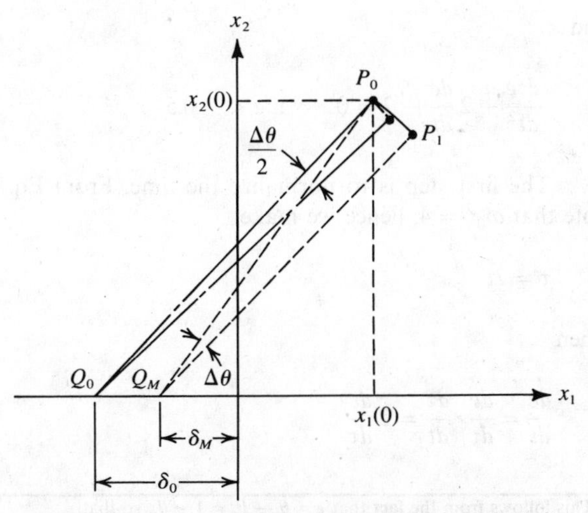

FIGURE 13.20. Construction using the delta method.

EXAMPLE 13.2

Consider the position-control system with a saturating amplifier, approximated by straight-line segments, as shown in Fig. 13.21. It is desired to obtain $\theta_c(t)$ if θ_r is a unit step for $a = 2$ and $K_m = 1$, assuming zero initial conditions.

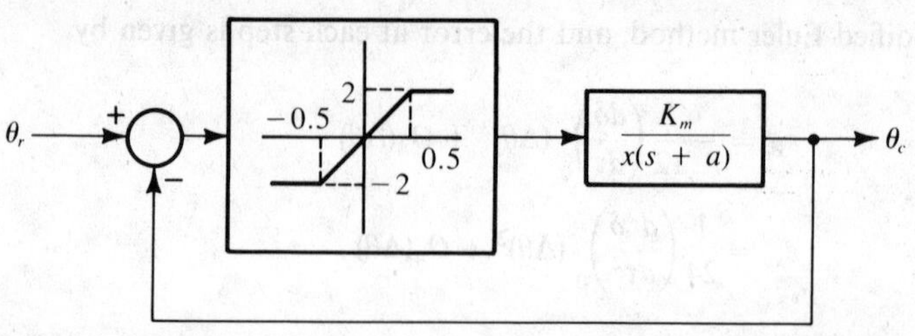

FIGURE 13.21. Position-control system with saturating amplifier.

Solution

In the linear region, when the error is less than 0.5, we obtain the following differential equation for the error e:*

$$\frac{d^2e}{dt^2} + 2\frac{de}{dt} + 4e = 0 \qquad -0.5 \leq e \leq 0.5 \tag{13.30}$$

In the saturation region we get

$$\frac{d^2e}{dt^2} + 2\frac{de}{dt} + 2 = 0 \qquad \text{if } e > 0.5 \tag{13.31}$$

and

$$\frac{d^2e}{dt^2} + 2\frac{de}{dt} - 2 = 0 \qquad \text{if } e < -0.5 \tag{13.32}$$

The first step is to normalize the time. From Eq. (13.30) we note that $\omega_0{}^2 = 4$; hence we make

$$\tau = 2t \tag{13.33}$$

Then

$$\frac{de}{dt} = \frac{de}{d\tau} \cdot \frac{d\tau}{dt} = 2\frac{de}{d\tau} \tag{13.34}$$

---

* This follows from the fact that $e = \theta_r - \theta_c = 1 - \theta_c$, so that

$$\frac{de}{dt} = -\frac{d\theta_c}{dt} \qquad \text{and} \qquad \frac{d^2e}{dt^2} = -\frac{d^2\theta_c}{dt^2}$$

Hence, the system equation

$$\frac{d^2\theta_c}{dt^2} + 2\frac{d\theta_c}{dt} = 4e$$

can be written as Eq. (13.30).

and

$$\frac{d^2e}{dt^2} = \frac{d}{d\tau} \cdot \frac{d\tau}{dt}\left(\frac{de}{d\tau}\right) = 4\frac{d^2e}{d\tau^2} \tag{13.35}$$

Hence, Eqs. (13.30), (13.31), and (13.32) are changed as follows:

$$\frac{d^2e}{d\tau^2} + \frac{de}{d\tau} + e = 0 \qquad \text{if } -0.5 \leq e \leq 0.5 \tag{13.36}$$

$$\frac{d^2e}{d\tau^2} + \frac{de}{d\tau} + 0.5 = 0 \qquad \text{if } e > 0.5 \tag{13.37}$$

and

$$\frac{d^2e}{d\tau^2} + \frac{de}{d\tau} - 0.5 = 0 \qquad \text{if } e < -0.5 \tag{13.38}$$

Making $x_1 = e$ and $x_2 = de/d\tau$, for the linear region, expressed by Eq. (13.36), we get $\delta = x_2$; from Eq. (13.37), for $e > 0.5$, we get $\delta = x_2 - x_1 + 0.5$; and from Eq. (13.38), for the region $e < -0.5$, we get $\delta = x_2 - x_1 - 0.5$.

We can now plot the trajectory in the normalized phase plane starting from the point $x_1 = 1$, $x_2 = 0$, which is in the saturation region. Note how the boundaries between the linear and nonlinear regions are clearly indicated in the phase plane. The completed trajectory is shown in Fig. 13.22, where $\Delta\theta$ has been taken as 10°.

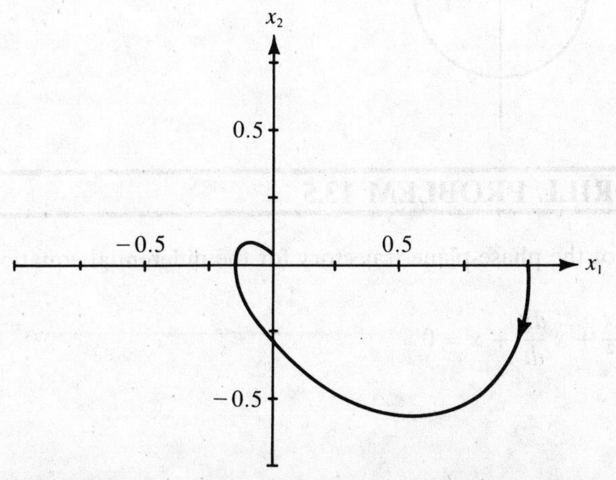

FIGURE 13.22. Trajectory for system with saturation.

The choice of $\Delta\theta$ is often quite critical to the accuracy of the solution. This is similar to the problem of selection of the time interval Δt in the numerical solution of a differential equation. The method shown in Fig. 13.20 overcomes this difficulty somewhat, and one should select $\Delta\theta$ consistent with the graphical accuracy that is possible. With a normalized differential equation a value of $\Delta\theta = 5°$ to $10°$ may be appropriate in most cases.

DRILL PROBLEM 13.4

Draw the phase-plane trajectory corresponding to the differential equation

$$\frac{d^2x}{dt^2} + x = 0$$

with

$$x(0) = 1 \quad \text{and} \quad \frac{dx}{dt}(0) = 0$$

Answer:
The trajectory is a circle of unit radius, as shown.

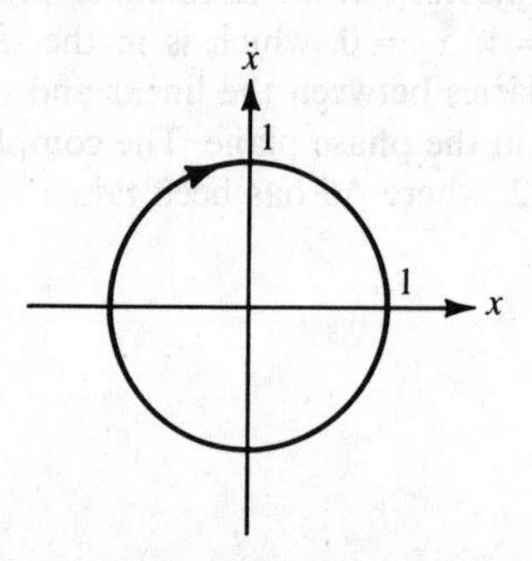

DRILL PROBLEM 13.5

Plot the phase-plane trajectory for the differential equation.

$$\frac{d^2x}{dt^2} + x\frac{dx}{dt} + x = 0$$

if

$$x(0) = 1 \quad \text{and} \quad \frac{dx}{dt}(0) = 0$$

Answer:

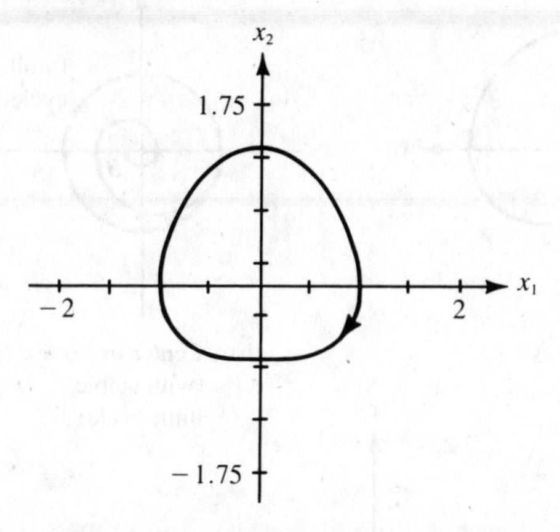

13.6.2 Stability from the phase plane

The stability of the system can be determined by examining the phase-plane trajectory near each of the singular points. A singular point in the phase plane is defined as the point where the slope of the trajectory is indeterminate; that is,

$$\frac{dx_2}{dx_1} \text{ takes the form } \frac{0}{0}$$

Some of the common singular points are the focus, the center (leading to limit cycles), and the node, and these are sketched in Fig. 13.23.

The nature of each singularity can be examined by plotting a few trajectories in the region.

DRILL PROBLEM 13.6

Locate and identify all singularities of the nonlinear differential equation

$$\frac{d^2x}{dt^2} - \left[2 - \left(\frac{dx}{dt}\right)^2\right]x + x^2 = 0$$

Answer:
Singularities are at (0, 0) and (2, 0) in the phase plane.

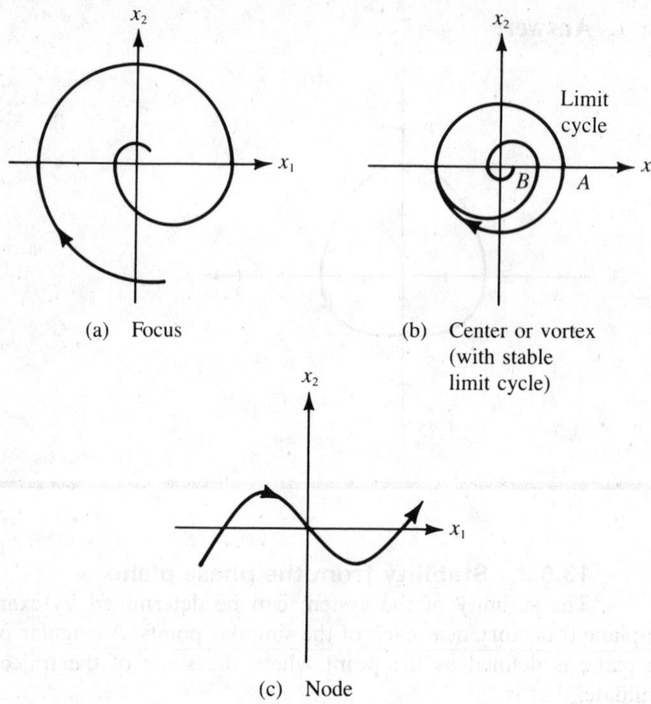

FIGURE 13.23. Common singularities.

13.7 STABILITY OF NONLINEAR SYSTEMS USING LIAPUNOV'S DIRECT METHOD

The simple stability criteria developed for linear systems are not applicable to nonlinear systems since the concept of the roots of a characteristic polynomial is no longer valid. As a result, many different classes of stability have been defined for nonlinear systems. We shall first discuss stability in the sense of Liapunov.

Consider a region ε in the state space enclosing an equilibrium point x_0. Then this is a point of stable equilibrium provided that there is a region $\delta(\varepsilon)$ contained within ε such that any trajectory starting in the region δ does not leave the region ε.

Note that with this definition it is not necessary for the trajectory to approach the point of equilibrium. It is only required that the trajectory stay within the region ε. This permits the existence of oscillations of limited amplitude. Such oscillations are indicated by a closed trajectory in the state space and are called limit cycles. A limit cycle is said to be stable if trajectories on either side of it will approach it.

Liapunov's direct method provides a means for determining the stability of a system without actually solving for the trajectories in the state space. It is based on the simple concept that the energy stored in a stable system cannot increase with time. Given a set of nonlinear state equations, one first defines a scalar function $V(x)$ that has properties similar to those of energy and then examines its derivatives with respect to time.

THEOREM 13.1

A system defined by $\dot{x} = f(x)$ is asymptotically stable in the vicinity of the equilibrium point at the origin if there exists a scalar function V such that

1. $V(x)$ is continuous and has continuous first partial derivatives at the origin.
2. $V(x) > 0$ for $x \neq 0$, and $V(0) = 0$.
3. $\dot{V}(x) < 0$ for all $x \neq 0$.

Note that the above conditions are sufficient but not necessary for stability.

THEOREM 13.2

A system defined by $\dot{x} = f(x)$ is unstable in a region Ω about the equilibrium point at the origin if there exists a scalar function $V(x)$ such that

1. $V(x)$ is continuous and has continuous first partial derivatives in the region Ω.
2. $V(x) \geq 0$ for $x \neq 0$ and $V(0) = 0$.
3. $\dot{V}(x) > 0$ in the region Ω.

Again, it should be noted that these conditions are sufficient but not necessary.

EXAMPLE 13.3

Consider the system described by the equations

$$\begin{aligned} \dot{x}_1 &= x_2 \\ \dot{x}_2 &= -x_1 - x_2^3 \end{aligned} \tag{13.39}$$

If we assume

$$V = x_1^2 + x_2^2 \tag{13.40}$$

which satisfies the conditions 1 and 2, then we get

$$\dot{V} = 2x_1\dot{x}_1 + 2x_2\dot{x}_2 = 2x_1x_2 + 2x_2(-x_1 - x_2^3) = -2x_2^4 \tag{13.41}$$

It will be seen that $\dot{V} < 0$ for all nonzero values of x_2, and hence the system is asymptotically stable.

EXAMPLE 13.4

Consider the system described by

$$\begin{aligned} \dot{x}_1 &= -x_1 + 2x_1^2x_2 \\ \dot{x}_2 &= -x_2 \end{aligned} \tag{13.42}$$

Let

$$V = \frac{1}{2}x_1^2 + x_2^2 \tag{13.43}$$

which is positive definite. Then

$$\dot{V} = x_1\dot{x}_1 + 2x_2\dot{x}_2 = -x_1^2(1 - 2x_1x_2) - 2x_2^2 \tag{13.44}$$

Clearly $\dot{V}$ is negative if $1 - 2x_1x_2 > 0$. This defines a region of stability in the state space.

It is evident that the main problem with this approach is the selection of a suitable positive definite function $V(x)$ such that its derivative is either positive definite or negative definite. Unfortunately, there is no general method for obtaining a suitable Liapunov function for every nonlinear system.

Gibson [1] has described the variable-gradient method for generating Liapunov functions. It will not be described here but the interested reader may refer to section 8.13 of the Gibson text.

DRILL PROBLEM 13.7

Consider a linear system described by the second-order differential equation

$$\frac{d^2x}{dt^2} + \alpha\frac{dx}{dt} + Kx = 0$$

Determine (a) a suitable Liapunov function for this system, and hence (b) the range of values of α and K for the system to be stable.

Answers:

(a) $V = \left(K + \frac{1}{4}\alpha^2\right)x_1{}^2 + \left(\frac{1}{2}\alpha x_1 + x_2\right)^2$ where $x_1 = x$ and $x_2 = \frac{dx}{dt}$.

(b) Stable for $\alpha > 0$ and $K > 0$.

DRILL PROBLEM 13.8

Determine a suitable Liapunov function for the differential equation for a nonlinear position control system

$$\frac{d^2x}{dt^2} + \alpha\frac{dx}{dt} + K\left(\frac{dx}{dt}\right)^3 + x = 0$$

where $\alpha > 0$.

Answer:

$V = x_1{}^2 + x_2{}^2$, where $x_1 = x$ and $x_2 = \frac{dx}{dt}$.

13.8 SUMMARY

We have discussed briefly two approaches to the study of nonlinear systems. The describing-function method is attractive since it attempts to extend the familiar frequency response concepts to nonlinear systems, but it is only approximate because of the assumption that the effect of the harmonics can be neglected. The phase-plane approach gives us the approximate solution of a nonlinear differential equation in a graphical form and allows complete stability analysis by examining the points of singularity. It is possible to implement this method on a microcomputer. However, it is limited to second-order systems only. The stability of a nonlinear system can be determined effectively using Liapunov's theorems, but practical application is limited due to the lack of a general rule for obtaining Liapunov functions.

13.9 REFERENCES

1. D. P. Atherton, *Nonlinear Control Engineering*, Van Nostrand Reinhold, New York, 1975.
2. J. E. Gibson, *Nonlinear Automatic Control*, McGraw-Hill, New York, 1963.

3. N. Minorsky, *Theory of Nonlinear Control Systems*, McGraw-Hill, New York, 1969.
4. G. J. Thaler and M. P. Pastel, *Analysis and Design of Nonlinear Feedback Control Systems*, McGraw-Hill, New York, 1962.

13.10 PROBLEMS

1. The characteristic curve for a nonlinear spring is shown in Fig. 13.24. Derive an expression for its describing function.

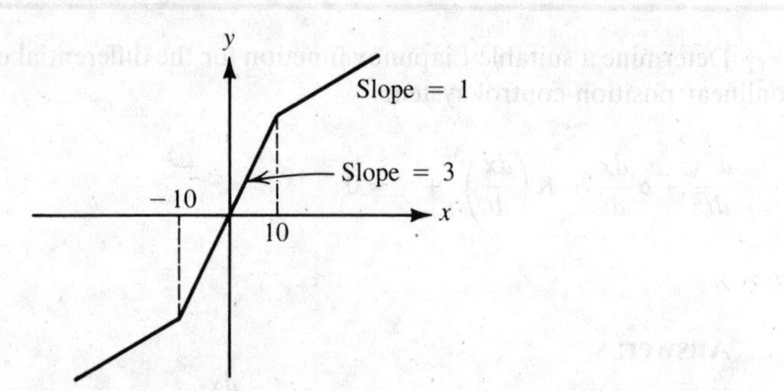

FIGURE 13.24. Characteristic curve for a nonlinear spring.

2. The block diagram of a position-control system is shown in Fig. 13.25. The transfer function of the servomotor is given by

$$G(s) = \frac{Ke^{-0.1s}}{s(s+10)}$$

It is driven by an amplifier with a gain of 20 in the linear range, which saturates when its input exceeds 1.

a. Investigate the stability of the system when $K = 10$.
b. What is the largest value of K for no limit cycle to exist?

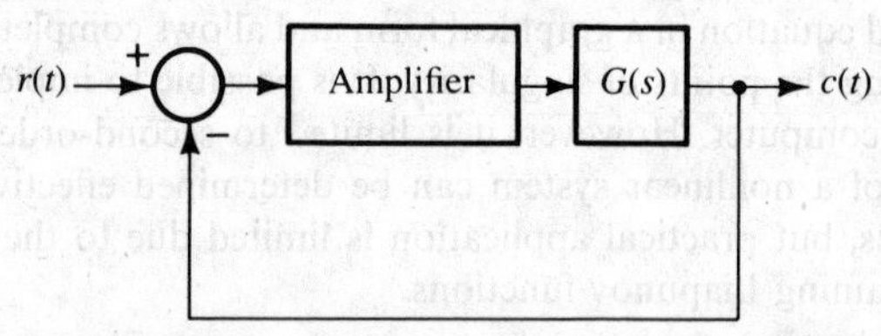

FIGURE 13.25. Position-control system with saturating amplifier.

3. The input $x(t)$ and the output $y(t)$ of a nonlinear device are related through the differential equation

$$y(t) = \left(\frac{dx}{dt}\right)^3 + x^2\frac{dx}{dt}$$

Determine the describing function for this device.

4. Determine the describing function of an amplifier that has the nonlinear gain characteristic shown in Fig. 13.26.

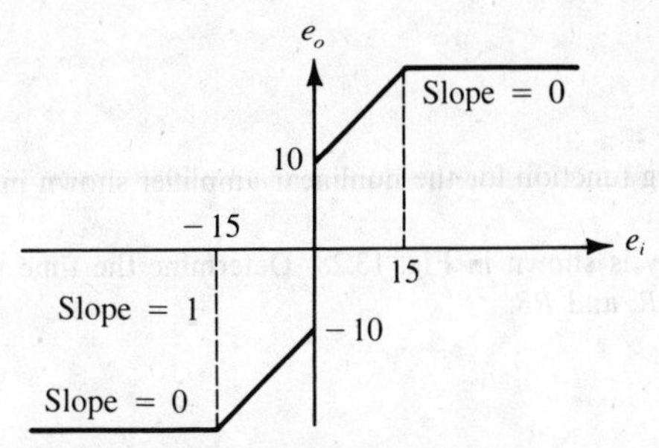

FIGURE 13.26. A nonlinear amplifier.

5. The amplifier in the position control system shown in Fig. 13.25 has the characteristic shown in Fig. 13.27.

a. Investigate the stability of the system.
b. What is the maximum value of K for no limit cycle to exist?

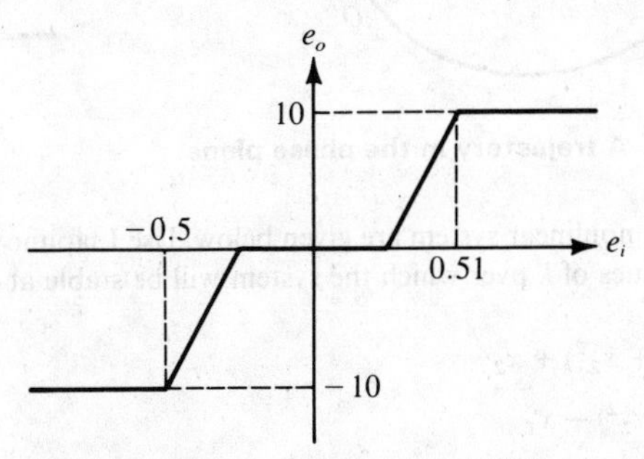

FIGURE 13.27. A nonlinear amplifier.

6. Draw the phase-plane trajectory of the response of the position-control system considered in Problem 2 if the input is a step of size 2.

7. Draw the phase-plane trajectory of the response of the position control system considered in Problem 5 if the input is a step of size 1.5.

8. A unity-feedback position-control system has an effective motorload inertia $J = 1$ and an effective friction coefficient $f = 10$. The torque developed by the motor is 100 times the square of the error. Plot the phase-plane trajectory in response to a step input of 1 rad.

9. The differential equation of a nonlinear device is given below. Use Liapunov's method to determine whether the system is stable.

$$\frac{d^2x}{dt^2} + 2x^2\frac{dx}{dt} + x = 0$$

10. Verify that either (a) $V_1 = 0.5x_1^4 + x_2^2$ or (b) $V_2 = x_1^4 + 2x_2^2 + 2x_1x_2 + x_1^2$ is a Liapunov function for the system described by the equations

$$\dot{x}_1 = x_2$$
$$\dot{x}_2 = -x_2 - x_1^3$$

11. Determine the describing function for the nonlinear amplifier shown in Fig. 13.27.

12. A phase-plane trajectory is shown in Fig. 13.28. Determine the time required to traverse the segments PQ, QR, and RS.

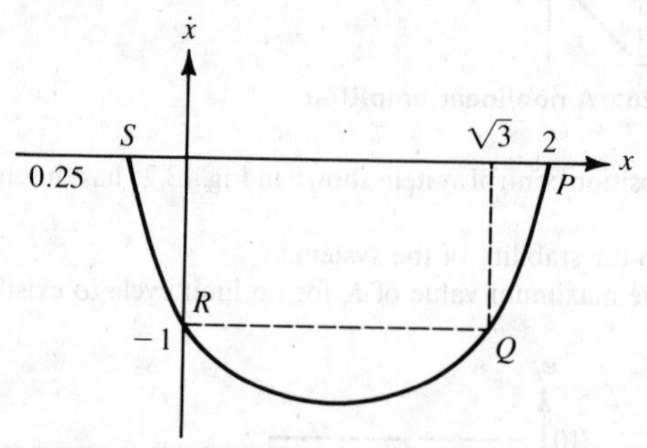

FIGURE 13.28. A trajectory in the phase plane.

13. The state equations for a nonlinear system are given below. Use Liapunov's method to determine the range of values of k over which the system will be stable at the origin.

$$\dot{x}_1 = -kx_1(x_1^2 + x_2^2) + x_2$$
$$\dot{x}_2 = kx_1(x_1^2 + x_2^2) - x_1$$

(*Hint*: Try a Liapunov function of the type $V = ax_1^2 + bx_2^2$, where the constants a and b are adjusted to make $\dot{V}$ negative definite.)

14. The Liapunov function for testing the stability of a linear system, described by the equation $\dot{x} = Ax$ can be obtained by defining

$$V = x^T A x$$

so that

$$\dot{V} = x^T(A^T P + PA)x = -x^T Q x$$

We can then set $Q = I$ and solve for the symmetric matrix P which satisfies the equation

$$A^T P + PA = -I$$

If the resulting P is positive definite, then the system is stable.

Use this method to determine P and check stability if

$$A = \begin{bmatrix} -5 & 2 \\ 1 & -3 \end{bmatrix}$$

15. Repeat if

$$A = \begin{bmatrix} -5 & 1 & -2 \\ 1 & 0 & -1 \\ 3 & 2 & 4 \end{bmatrix}$$

16. Using Liapunov's method, determine the stability of the nonlinear differential equation

$$\ddot{x} + 10(\dot{x})^5 + 4\dot{x} + x = 0.$$

17. An instrument servo uses a motor and an ideal relay in a unity-feedback system. The transfer function of the motor is given by

$$G(s) = \frac{4.5}{s(s + 1.5)}$$

whereas the relay switches between -4 and $+4$. If the block diagram of the system is as shown in Fig. 13.25, in which the amplifier is replaced by the ideal relay, draw the phase-plane trajectory of the system in response to a unit step. Assume zero initial conditions.

18. Repeat Problem 17 for the case when the relay has dead zone of 0.1 on each side of zero.

19. Use the describing function for the relay with dead zone, as in Problem 18, to investigate the existence of limit cycles in the instrument servo described in the Problem 17.

14

Epilogue

In the previous chapters of this book an attempt was made to present the basic concepts of control theory as developed over the years. Since this work is intended as a textbook for a one-semester undergraduate course, it was not possible to describe all aspects of the subject. There are many important areas that have been left out. Some of these will now be described briefly. It is hoped that this will arouse the interest of the reader for further studies in this fascinating area.

Most practical control systems are subject to random disturbances. For example, an antenna in a radar-controlled system may be subjected to gusts of wind, or a navigation control system may be affected by waves. Noise may also be introduced by sensors or generated internally in amplifiers. To get the best results in such cases we must try to filter the noise as much as possible. This is the subject of stochastic control and requires a thorough understanding of the theory of random processes. An interesting development in this area has been the theory of optimal filtering and prediction, proposed independently by Wiener in the United States and Kolmogorov in the U.S.S.R. in the 1940s. It was followed in the 1960s by the well-known Kalman filter theory, which allows us to obtain the optimal estimates of the states of a system from the noise-contaminated measurements of its inputs and outputs.

In a large number of control systems the parameters of the dynamic equations may vary considerably. For example, the differential equations for the motion of an airplane near sea level will be quite different from those for the same airplane flying at the height of 10 km above sea leval. As another example, the dynamic equations of a nuclear power plant change with the amount of power developed. For the best performance, the controller for such a system must adapt itself to changes in parameters. This has led to the development of the theory of adaptive control. As a result of the microelectronics revolution, it is now quite practical to use microprocessors for implementing adaptive control in these cases.

In all of our deliberations it was assumed that the model of the plant to be controlled is known precisely. In practice, this is seldom the case. Often one has to determine the model of the system from its response to test data. This is called the problem of system identification, and has been a subject for considerable research. Complications are introduced by the presence of noise in the measurements, with the result that there is always some uncertainty in the values of the parameters of the model. Furthermore, most physical systems have some inherent nonlinearities, with the result that exact analysis is often very involved. One of the topics of considerable amount of current research is the theory of the design of robust control systems, the performance of which will not be affected by small variations in parameters.

In Chapter 11 we had a glimpse of the theory of digital control. This subject has attracted considerable attention due to the microelectronics revolution and the consequent availability of inexpensive microprocessors. Several books have appeared on the subject during the past five years, and many engineering schools have graduate courses on digital control. In fact, some schools now offer such course at the undergraduate level.

In Chapter 12 we presented a rather brief discussion of the state-space analysis and design of control systems. This is a very powerful approach, particularly in the case of systems with many inputs and outputs. Recently it has been found that a combination of state-space and transfer function methods leads to a greater insight into the behavior of such systems. One example of this new approach was seen in Chapter 10, where the poles of the closed-loop transfer function were placed at specified locations by the design of suitable compensators instead of the more commonly used approach based on state-variable feedback with an asymptotic state observer. It has also been found that the design of compensators for multivariable systems can be carried out with greater ease if they are represented in the form of polynomial matrices.

Some other areas of great interest to control engineers are optimal control theory and the general theory of nonlinear systems. Both of these subjects require a better background in some areas of mathematics than expected in a typical undergraduate program in engineering. On the other hand, one cannot ignore the fact that most real systems have some nonlineariities.

There are a number of excellent books that deal with these advanced topics in sufficient detail. For the enthusiastic and interested reader, some of these reference books are listed below.

SUGGESTIONS FOR FURTHER READING

1. K. J. Astrom, *Introduction to Stochastic Control Theory*, Academic Press, New York, 1970.
2. D. P. Atherton, *Nonlinear Control*, Van Nostrand Reinhold, New York, 1975.
3. S. Barnett, *Polynomials and Linear Control Systems*, Marcel Dekker, New York, 1983.
4. G. F. Franklin and J. D. Powell, *Digital Control*, Addison-Wesley, Reading, Mass., 1980.
5. R. Isermann, *Digital Control Systems*, Springer, New York, 1981.
6. T. Kailath, *Linear Control Systems*, Prentice-Hall, Englewood Cliffs, N.J., 1980.
7. P. Katz, *Digital Control Using Microprocessors*, Prentice-Hall, Englewood Cliffs, N.J., 1981.
8. B. C. Kuo, *Digital Control Systems*, Holt, Rinehart and Winston, New York, 1980.
9. J. S. Meditch, *Stochastic Optimal Linear Estimation and Control*, McGraw-Hill, New York, 1969.
10. A. P. Sage and C. C. White III, *Optimum Systems Control*, 2d ed., Prentice-Hall, Englewood Cliffs, N.J., 1977.
11. G. N. Saridis, *Self-Organizing Control of Stochastic Systems*, Marcel Dekker, New York, 1977.
12. N. K. Sinha and B. Kuzsta, *Modeling and Identification of Dynamic Systems*, Van Nostrand Reinhold, New York, 1983.

A Review of Laplace Transforms

A.1 DEFINITION

The Laplace transform of a time function $f(t)$ is defined as

$$F(s) = \mathscr{L}[f(t)] \triangleq \int_0^{\infty} f(t)e^{-st}\, dt \tag{A.1}$$

where $s = \sigma + j\omega \triangleq$ complex frequency.

This integral will exist if

$$\lim_{t\to\infty} e^{-\sigma t} f(t) = 0$$

for some finite σ.

A.2 SHORT TABLE OF LAPLACE TRANSFORMS

f(t)	*F(s)*	*Pole-Zero Plot of F(s)*
Unit impulse $\delta(t)$	1	
Unit step u_t	$\frac{1}{s}$	
e^{-at}	$\frac{1}{s+a}$	
$\cos \omega_0 t$	$\frac{s}{s^2+{\omega_0}^2}$	
$\sin \omega_0 t$	$\frac{\omega_0}{s^2+{\omega_0}^2}$	
$e^{-at} \cos \omega_0 t$	$\frac{s+a}{(s+a)^2+{\omega_0}^2}$	
$e^{-at} \sin \omega_0 t$	$\frac{\omega_0}{(s+a)^2+\omega^2}$	

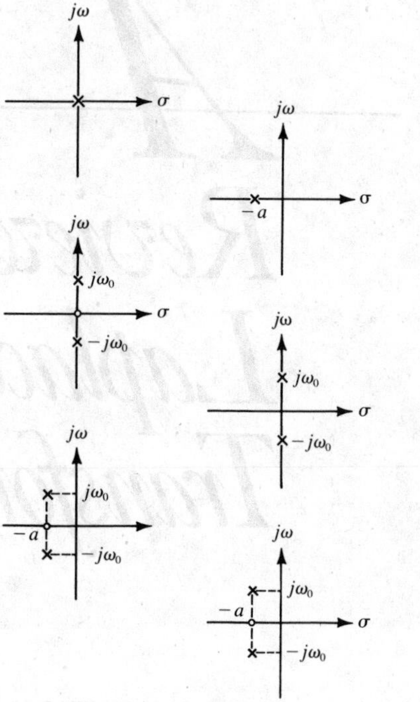

A.3 THEOREMS

1. $\mathscr{L}[e^{-at}f(t)] = F(s+a)$ — Translation in the s domain (A.2)

2. $\mathscr{L}[f(t-T)u_{t-T} = e^{-sT}F(s)$ — Translation in the time domain (A.3)

3. $\mathscr{L}[tf(t)] = -\frac{dF}{ds}$ — Multiplication by t (A.4)

4. $\mathscr{L}\left[\frac{df}{dt}\right] = sF(s) - f(0)$ — Differentiation in the time domain (A.5)

5. $\lim_{t \to 0} f(t) = \lim_{s \to \infty} [sF(s)]$ — Initial-value theorem (A.6)

6. $\lim_{t \to \infty} f(t) = \lim_{s \to 0} [sF(s)]$ Final-value theorem [valid only if all the poles of $F(s)$ have negative real part with the exception of one pole at the origin] (A.7)

A.4 INVERSE LAPLACE TRANSFORMATION

The inverse Laplace tranform of $F(s)$ is defined as

$$f(t) = \frac{1}{2\pi j} \int_{c-j\infty}^{c+j\infty} F(s)e^{st}\, ds \tag{A.8}$$

In practice, if $F(s)$ is a rational function of s (that is, the ratio of two polynomials in s), it is easier to use partial fraction expansion. If $F(s)$ is strictly proper—that is, the degree of the numerator is less than that of the denominator—we can obtain (for the case of distinct poles)

$$f(s) = \frac{K(s-z_1)(s-z_2)\cdots(s-z_m)}{(s-p_1)(s-p_2)\cdots(s-p_n)} \qquad n > m$$

$$= \frac{A_1}{s-p_1} + \frac{A_2}{s-p_2} + \cdots + \frac{A_k}{s-p_k} + \cdots + \frac{A_n}{s-p_n} \tag{A.9}$$

The residue

$$A_k = [(s-p_k)F(s)]_{s=p_k}$$

$$= \frac{K(p_k-z_1)(p_k-z_2)\cdots(p_k-z_m)}{(p_k-p_1)(p_k-p_2)\cdots(p_k-p_{k-1})(p_k-p_{k-2})\cdots(p_k-p_n)}$$

$$= K \cdot \frac{\text{product of directed distances from each zero to pole } p_k}{\text{product of directed distances from all other poles to } p_k} \tag{A.10}$$

After the residues are evaluated,

$$f(t) = \sum_{k=1}^{n} A_k e^{p_k t} \tag{A.11}$$

Note that some of the poles may be complex. In that case, they will occur in conjugate pairs. The residues at real poles will be real, and the residues at complex conjugate pairs of poles must also be conjugates of each other.

EXAMPLE A.1

$$F(s) = \frac{10(s^2 + 2s + 5)}{(s+1)(s^2 + 6s + 25)} = \frac{10(s + 1 - j2)(s + 1 + j2)}{(s+1)(s + 3 - j4)(s + 3 + j4)}$$

$$= \frac{A_1}{s+1} + \frac{A_2}{s + 3 - j4} + \frac{A_3}{s + 3 + j4} \tag{A.12}$$

$$A_1 = \frac{10(j2)(-j2)}{(2 - j4)(2 + j4)} = 2 \qquad A_2 = \frac{10(-2 + j2)(-2 + j6)}{j8(-2 + j4)}$$

$$A_3 = A_2^* \qquad\qquad = 4 + j3$$

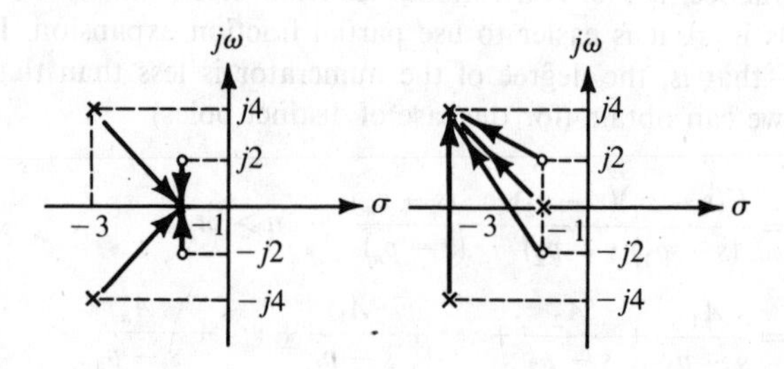

Hence

$$f(t) = 2e^{-t} + (4 + j3)e^{(-3 + j4)t} + (4 - j3)e^{(-3 - j4)t}$$

$$= 2e^{-t} + e^{-3t}(8 \cos 4t - 6 \sin 4t) \tag{A.13}$$

Note that for

$$F(s) = \frac{C + jD}{s + \alpha - j\beta} + \frac{C - jD}{s + \alpha + j\beta}$$

we have

$$f(t) = e^{-\alpha t}(2C \cos \beta t - 2D \sin \beta t) \tag{A.14}$$

For the case of multiple poles, the evaluation of the residues is slightly more complicated. The procedure will be illustrated by means of an example.

EXAMPLE A.2

Consider

$$F(s) = \frac{N(s)}{(s-p_1)^3(s-p_2)} = \frac{A_{11}}{s-p_1} + \frac{A_{12}}{(s-p_1)^2} + \frac{A_{13}}{(s-p_1)^3} + \frac{A_2}{s-p_2} \tag{A.15}$$

Then

$$A_{13} = [(s-p_1)^3 F(s)]_{s=p_1} \tag{A.16}$$

$$A_{12} = \left\{\frac{d}{ds}[(s-p_1)^3 F(s)]\right\}_{s=p_1} \tag{A.17}$$

$$A_{11} = \frac{1}{2}\left\{\frac{d^2}{ds^2}[(s-p_1)^3 F(s)]\right\}_{s=p_1} \tag{A.18}$$

and A_2 is found as before. The resulting inverse transform is

$$f(t) = A_{11}e^{p_1 t} + A_{12}te^{p_1 t} + \frac{1}{2}A_{13}t^2 e^{p_1 t} + A_2 e^{p_2 t} \tag{A.19}$$

In general, if the pole at p_i has multiplicity m, then

$$A_{ir} = \frac{1}{(m-r)!}\frac{d^{m-r}}{ds^{m-r}}[(s-p_i)^m F(s)]_{s=p_i} \qquad r = 1, 2, \ldots, m \tag{A.20}$$

A.5 THE TRANSFER FUNCTION

The transfer function of a system is defined as the ratio of Laplace transforms of the output and the input assuming zero initial conditions.

EXAMPLE A.3

For the *RC* network shown,

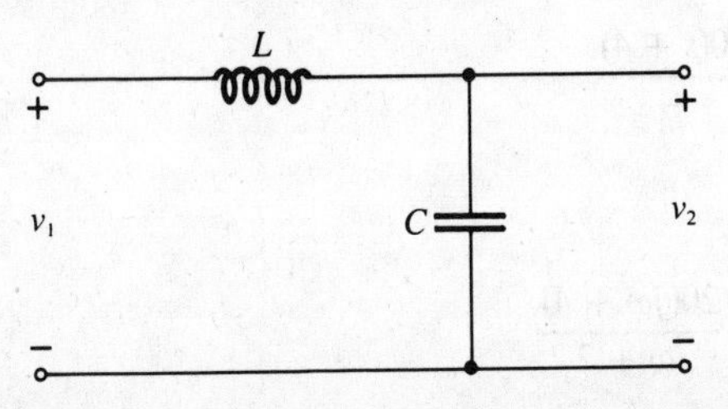

$$G(s) = \frac{V_2(s)}{V_1(s)} = \frac{\dfrac{1}{sC}}{sL + \dfrac{1}{sC}} = \frac{1}{s^2LC + 1} \tag{A.21}$$

A.6 FREQUENCY RESPONSE FROM THE TRANSFER FUNCTION

Replacing s by $j\omega$ in the transfer function gives us $G(j\omega) = M(\omega)e^{j\phi(\omega)}$, which is a complex number for a given ω. Thus if the input is a sine wave of frequency ω_1, the steady-state output will be a sine wave of the same frequency, with magnitude $M(\omega_1)$ times that of the input and leading it by the angle $\phi(\omega_1)$. The function $G(j\omega)$ is called the frequency response of the system and can be given the same graphical interpretation as for the calculation of residues. Thus, if

$$G(s) = \frac{K(s - z_1)(s - z_2) \cdots (s - z_m)}{(s - p_1)(s - p_2) \cdots (s - p_n)} \tag{A.22}$$

then

$$G(j\omega_1) = \frac{K(j\omega_1 - z_1)(j\omega_1 - z_2) \cdots (j\omega_1 - z_m)}{(j\omega_1 - p_1)(j\omega_1 - p_2) \cdots (j\omega_1 - p_n)}$$

$$= K \cdot \frac{\text{product of directed distances from each zero to } s = j\omega_1}{\text{product of directed distances from each pole to } s = j\omega_1} \tag{A.23}$$

EXAMPLE A.4

Let

$$G(s) = \frac{20(s + 4)}{s + 2} \tag{A.24}$$

then

$$G(j\omega) = \frac{20(j\omega + 4)}{j\omega + 2} \tag{A.25}$$

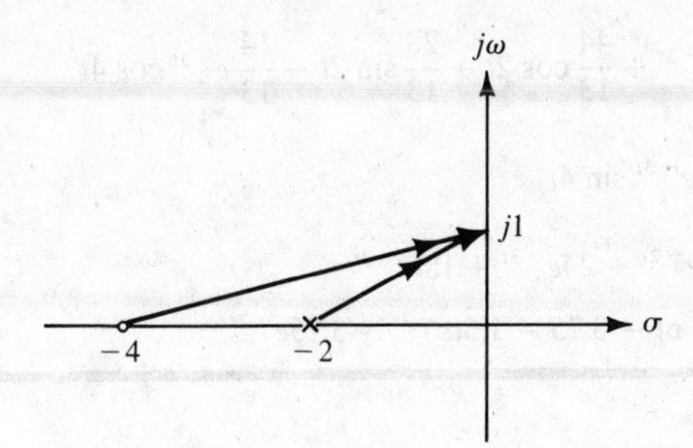

and

$$G(j1) = \frac{20(j1+4)}{j1+2} = 36 - j8 = 36.878e^{-j0.219} \qquad \text{(A.26)}$$

DRILL PROBLEM A.1

Determine the Laplace transforms of the following.

(a) $4e^{-3t}\cos 2t$ (b) te^{-2t} (c) $\frac{d}{dt}(3e^{-2t}\cos 4t)$

Answers:

(a) $\frac{4(s+3)}{(s+3)^2+2^2}$ (b) $\frac{1}{(s+2)^2}$

(c) $\frac{3s(s+2)}{(s+2)^2+4^2} - 3 = \frac{-6s-60}{s^2+4s+20}$.

DRILL PROBLEM A.2

Determine the inverse Laplace transforms of the following.

(a) $\frac{10(s+1)}{s(s+2)(s+5)}$ (b) $\frac{60(s+2)(s^2+2s+5)}{(s+3)(s^2+4)(s^2+6s+25)}$

(c) $\frac{36(s+1)}{s(s+2)^2(s+3)}$ (d) $\frac{12(s+3)}{s^3(s+2)^2}$

Answers:

(a) $1 + \frac{5}{3}e^{-2t} - \frac{8}{3}e^{-5t}$

(b) $-\frac{30}{13}e^{-3t}+\frac{44}{13}\cos 2t+\frac{28}{13}\sin 2t-\frac{14}{13}e^{-3t}\cos 4t$

$+\frac{145}{13}e^{-3t}\sin 4t$

(c) $3-27e^{-2t}+24e^{-3t}+18te^{-2t}$

(d) $4.5t^2-6t+3.75-1.5te^{-2t}-3.75e^{-2t}$

A.7 PARSEVAL'S FORMULA

If $F(s)$ is the Laplace transform of $f(t)$, then

$$\int_0^\infty f^2(t)\,dt=\frac{1}{2\pi j}\int_{-j\infty}^{j\infty}F(s)F(-s)\,ds \tag{A.27}$$

This integral is very useful for calculating the integral square value of the error in a closed-loop system. The integral will be finite only if $F(s)$ has all its poles in the finite portion of left half of the s-plane, and in that case one may use the residue theorem to evaluate the integral on the right-hand side. Hence, one may write

$$\int_0^\infty f^2(t)\,dt=\text{sum of the residues of } F(s)\cdot F(-s) \text{ at all of its poles in the left half of the } s\text{-plane} \tag{A.28}$$

EXAMPLE A.5

Consider

$$f(t)=4e^{-2t} \tag{A.29}$$

Then

$$F(s)=\frac{4}{s+2} \tag{A.30}$$

and

$$F(s)F(-s)=\frac{4}{s+2}\cdot\frac{4}{-s+2}=\frac{-16}{(s+2)(s-2)} \tag{A.31}$$

From Eq. (A.28),

$$\int_0^\infty f^2(t)\,dt = \left[\frac{-16}{s-2}\right]_{s=-2} = 4 \tag{A.32}$$

This can be easily verified by actual integration of $16e^{-4t}$.

DRILL PROBLEM A.3

Use Parseval's theorem to evaluate the integral

$$\int_0^\infty f^2(t)\,dt$$

if

(a) $f(t) = 4e^{-2t}\cos 3t$
(b) $f(t) = 5e^{-3t} + 3e^{-t}\sin 2t$

Answers: (a) 2/13, (b) 31/240.

B Review of the Theory of Matrices

B.1 MATRICES VIEWED AS TRANSFORMATIONS

It is expected that the reader has come across the basic theory of matrices and is familiar with the rules for multiplication of matrices. It is common practice to define a matrix as a "collection of elements arranged in rectangular (or square) array." In general, a matrix will have n rows and m columns, where m and n are positive integers. A matrix with only one column (that is, $m = 1$) is called a column vector (or simply a vector). A matrix with one row is called a row vector.

It is often helpful to regard a matrix as a transformation. For example, consider the equation

$$y = Ax \tag{B.1}$$

where A is an $n \times m$ matrix, x is a column vector with m rows, and y is a column vector with n rows. We can say that the matrix A has changed the vector x into the vector y. In particular, if A is a square matrix with n rows and n columns, then both x and y are n-dimensional vectors.

A square matrix is said to be nonsingular if its inverse exists. This is possible if and only if one of the following conditions is satisfied:

1. The determinant of the matrix is not zero.
2. All the columns of the matrix are linearly independent.
3. All the rows of the matrix are linearly independent.

To calculate the inverse of a matrix, we first obtain its adjoint and then divide each element of the adjoint by the determinant of the matrix. The adjoint of a matrix is obtained by first transposing it, and then replacing each element of the transpose by the corresponding cofactor.

One should use a computer for inverting matrices of order higher than three. Such computer programs are part of most computer libraries.

B.2 EIGENVALUES AND EIGENVECTORS

For the case when A is square matrix it is of interest to find vectors that are not changed in direction by the transformation indicated by Eq. (B.1.). Such vectors are called the eigenvectors of A and represent the normal coordinates for the transformation A. For these, we have

$$y = Ax = \lambda x \tag{B.2}$$

where the scalar λ is called an eigenvalue of A.

We can rearrange Eq. (B.2) as

$$(\lambda I - A)x = 0 \tag{B.3}$$

where I is the identity matrix of dimension $n \times n$.

A trivial solution of Eq. (B.3) is the null vector, in which each component of x is zero. A nontrivial solution for x is obtained if the determinant, given by

$$\det(\lambda I - A) = 0 \tag{B.4}$$

Equation (B.4) is called the characteristic polynomial of A, and is a polynomial in λ of degree n. Hence it will have n roots. This implies that every $n \times n$ square matrix will have n eigenvalues, although some of these may be equal, corresponding to repeated roots of Eq. (B.4). This will be shown by the following example.

EXAMPLE B.1

Consider the matrix

$$A = \begin{bmatrix} 0 & 1 \\ -6 & -5 \end{bmatrix} \tag{B.5}$$

Here

$$\lambda I - A = \lambda \begin{bmatrix} 1 & 0 \\ 0 & 1 \end{bmatrix} - \begin{bmatrix} 0 & 1 \\ -6 & -5 \end{bmatrix}$$

$$= \begin{bmatrix} \lambda & -1 \\ 6 & \lambda + 5 \end{bmatrix} \tag{B.6}$$

and

$$\det(\lambda I - A) = \lambda^2 + 5\lambda + 6 = (\lambda + 2)(\lambda + 3) \tag{B.7}$$

with roots $\lambda_1 = -2$ and $\lambda_2 = -3$.

Hence the eigenvalues for this matrix are -2 and -3. We shall now determine the eigenvectors corresponding to the two eigenvalues. Let these be denoted by v_1 and v_2, respectively. Then

$$(\lambda_1 I - A)v_1 = \begin{bmatrix} -2 & -1 \\ 6 & 3 \end{bmatrix} \begin{bmatrix} v_{11} \\ v_{12} \end{bmatrix} = 0 \tag{B.8}$$

We note that the two equations for v_{11} and v_{12} are not independent. From these equations we have the relationship

$$2v_{11} + v_{12} = 0 \tag{B.9}$$

If we arbitrarily select $v_{11} = 1$, then we get $v_{12} = -2$; that is,

$$v_1 = \begin{bmatrix} 1 \\ -2 \end{bmatrix} \tag{B.10}$$

is an eigenvector of A corresponding to λ.

The reason for nonuniqueness of the values of v_{11} and v_{12} follows from the fact that replacing x by kx in Eq. (B.2), where k is a constant, simply scales the vector x without changing its direction. Hence any scalar multiple of v_1 is also an eigenvector of A, corresponding to λ_1.

Similarly, it can be verified that the other eigenvector is given by

$$v_2 = \begin{bmatrix} 1 \\ -3 \end{bmatrix} \tag{B.11}$$

A simple way to find the eigenvectors of the matrix A is first to obtain the adjoint of $(\lambda I - A)$. Then any column of the adjoint of $(\lambda_i I - A)$ will be an eigenvector corresponding to the eigenvalue λ_i.

EXAMPLE B.2

For the problem considered in Examplc B.1, the adjoint of $(\lambda I - A)$ is obtained as

$$\text{adj}\,(\lambda I - a) = \begin{bmatrix} \lambda + 5 & 1 \\ -6 & \lambda \end{bmatrix} \tag{B.12}$$

Hence an eigenvector corresponding to $\lambda_1 = -2$ is obtained either as

$$\begin{bmatrix} 3 \\ -6 \end{bmatrix} \quad \text{or} \quad \begin{bmatrix} 1 \\ -2 \end{bmatrix} \tag{B.13}$$

EXAMPLE B.3

Consider the third-order matrix

$$A = \begin{bmatrix} -2 & -5 & -5 \\ 1 & -1 & 0 \\ 0 & 1 & 0 \end{bmatrix} \tag{B.14}$$

In this case we have

$$\begin{aligned} \det\,(\lambda I - A) &= \begin{vmatrix} \lambda + 2 & 5 & 5 \\ -1 & \lambda + 1 & 0 \\ 0 & -1 & \lambda \end{vmatrix} \\ &= (\lambda + 2)(\lambda + 1)\lambda + (5\lambda + 5) \\ &= \lambda^3 + 3\lambda^2 + 7\lambda + 5 \\ &= (\lambda + 1)(\lambda^2 + 2\lambda + 5) \end{aligned} \tag{B.15}$$

Hence the eigenvalues are $\lambda_1 = -1$, $\lambda_2 = -1 + j2$, and $\lambda_2 = -1 - j2$.

The adjoint of $(\lambda I - A)$ is obtained as

$$\text{adj}\,(\lambda I - A) = \begin{bmatrix} \lambda^2 + \lambda & -(5\lambda + 5) & -(3\lambda + 5) \\ \lambda & \lambda^2 + 2\lambda & -5 \\ 1 & \lambda + 2 & \lambda^2 + 3\lambda + 7 \end{bmatrix} \tag{B.16}$$

The eigenvector corresponding to the three eigenvalues are easily calculated from any column of this matrix. Hence

$$v_1 = \begin{bmatrix} 0 \\ -1 \\ 1 \end{bmatrix} \qquad v_2 = \begin{bmatrix} -3 - j4 \\ -1 + j2 \\ 1 \end{bmatrix} \qquad v_3 = \begin{bmatrix} -3 + j4 \\ -1 - j2 \\ 1 \end{bmatrix} \tag{B.17}$$

From this example it is evident that the calculation of the eigenvalues and eigenvectors of a matrix is a tedious task for matrices of high order. The algorithm described in Chapter 11 (Leverrier's algorithm) can be utilized for obtaining the characteristic polynomial as well as the adjoint of $(\lambda I - A)$. In practice, the characteristic equation method is not a good general procedure for the computation of the eigenvalues of a matrix. For matrices of order up to 100, the Q-R algorithm is more suitable (for details see Wilkinson [1]). Standard program packages are available.

Some properties of the eigenvalues of a matrix are as follows:

1. A matrix of order n will have n eigenvalues, which may not all be distinct (i.e., some eigenvalues may be repeated).
2. A matrix has the same eigenvalues as its transpose.
3. The sum of the eigenvalues of a matrix is equal to the negative of its trace. (The trace of a matrix is defined as the sum of the elements on its main diagonal.)
4. If all the elements of a matrix are real, then each of its eigenvalues will either be real or one of a complex conjugate pair.
5. A symmetric matrix with real elements can have only real eigenvalues.
6. The product of the eigenvalues of an $n \times n$ matrix is equal to its determinant multiplied by $(-1)^n$.
7. The eigenvalues of a triangular (and hence also a diagonal) matrix are equal to the diagonal elements.
8. The matrix $B = P^{-1}AP$ has the same eigenvalues as A, where P is any nonsingular matrix.

B.3 DIAGONALIZATION OF A MATRIX

Let a square matrix A of order n have the eigenvalues $\lambda_1, \lambda_2, \ldots, \lambda_n$, and the corresponding eigenvectors be $v_1, v_2, \ldots, v_n$. Then we have

$$\left.\begin{aligned} Av_1 &= \lambda_1 v_1 \\ Av_2 &= \lambda_2 v_2 \\ &\vdots \\ Av_n &= \lambda_n v_n \end{aligned}\right\} \tag{B.18}$$

These equations can be combined together to obtain

$$A[v_1, v_2, \ldots, v_n] = [v_1, v_2, \ldots, v_n]\Lambda \tag{B.19}$$

where Λ is the diagonal matrix given by

$$\Lambda = \begin{bmatrix} \lambda_1 & 0 & 0 & \cdots & 0 \\ 0 & \lambda_2 & 0 & \cdots & 0 \\ 0 & 0 & \lambda_3 & \cdots & 0 \\ \vdots & \vdots & \vdots & & \vdots \\ 0 & 0 & 0 & \cdots & \lambda_n \end{bmatrix} \tag{B.20}$$

Let

$$M = [v_1 \quad v_2 \quad \cdots \quad v_n] \tag{B.21}$$

be defined as the modal matrix for A. Note that the columns of M are the n eigenvectors of A, and hence M is an $n \times n$ matrix. We may now write Eq. (B.19) as

$$AM = M\Lambda \tag{B.22}$$

Thus,

$$\Lambda = M^{-1}AM \tag{B.23}$$

is the diagonalized form of A.

It may be noted that this assumes that M is nonsingular; i.e., the various eigenvectors are linearly independent. This will, in general, be true only if either all the eigenvectors of A are diagonal or if A is a symmetric matrix. Matrices with repeated eigenvalues can be diagonalized if and only if for each eigenvalue λ_i of multiplicity m_i the rank of $(\lambda_i I - A)$ is equal to $n - m_i$. This will always be the case when A is a symmetric matrix.

EXAMPLE B.4

For the matrix in Example B.1 the eigenvalues were located at $\lambda_1 = -2$ and $\lambda_2 = 3$. The corresponding eigenvectors are found (using the procedure shown in Example B.2) as

$$v_1 = \begin{bmatrix} 1 \\ -2 \end{bmatrix} \quad \text{and} \quad v_2 = \begin{bmatrix} 1 \\ -3 \end{bmatrix} \tag{B.24}$$

Hence

$$M = \begin{bmatrix} 1 & 1 \\ -2 & -3 \end{bmatrix} \tag{B.25}$$

and

$$M^{-1} = \begin{bmatrix} 3 & 1 \\ -2 & -1 \end{bmatrix} \tag{B.26}$$

It is easily verified that

$$M^{-1}AM = \begin{bmatrix} -2 & 0 \\ 0 & -3 \end{bmatrix} \tag{B.27}$$

B.4 THE JORDAN FORM

If an eigenvalue λ_i of an $n \times n$ matrix A has multiplicity $m_i > 1$, then A cannot be diagonalized unless the rank of $(\lambda_i I - A)$ is equal to $(n - m_i)$. In such cases we can transform A to the Jordan form, which is very close to the diagonal form, by using generalized eigenvectors. Some typical Jordan matrices are shown below.

$$J = \left[\begin{array}{ccc|cc|c} \lambda_1 & 1 & 0 & 0 & 0 & 0 \\ 0 & \lambda_1 & 1 & 0 & 0 & 0 \\ 0 & 0 & \lambda_1 & 0 & 0 & 0 \\ \hline 0 & 0 & 0 & \lambda_2 & 1 & 0 \\ 0 & 0 & 0 & 0 & \lambda_2 & 0 \\ \hline 0 & 0 & 0 & 0 & 0 & \lambda_3 \end{array}\right] \tag{B.28}$$

$$J_2 = \left[\begin{array}{ccc|cc} \lambda_1 & 1 & 0 & 0 & 0 \\ 0 & \lambda_1 & 1 & 0 & 0 \\ 0 & 0 & \lambda_1 & 0 & 0 \\ \hline 0 & 0 & 0 & \lambda_1 & 1 \\ 0 & 0 & 0 & 0 & \lambda_1 \end{array}\right] \tag{B.29}$$

It may be noted that J_2 has two Jordan blocks for the same eigenvalues λ_1, which has multiplicity 5. This case will arise only if the rank of $(\lambda_i I - A)$ is $(n - 2)$ so that we get two independent eigenvectors. In general, the number of Jordan blocks for a given eigenvalue will be equal to r if the rank of $(\lambda_i I - A)$ is $(n - r)$.

A vector v is said to be a generalized eigenvector of rank k of A associated with λ_i if and only if

$$\left.\begin{aligned} &(\lambda_i I - A)^k v = 0 \\ &\text{and} \\ &(\lambda_i I - A)^{k-1} v \neq 0 \end{aligned}\right\} \tag{B.30}$$

It may be noted that for $k = 1$ we get v as an eigenvector, consistent with the definition given in Eq. (B.3). This justifies the use of the term "generalized eigenvectors."

In general, if the multiplicity of λ_i is m_i and the rank of $(\lambda_i I - A)$ is $(n - 1)$, then we should be able to get one eigenvector and $m_i - 1$ generalized eigenvectors that will be linearly independent. On the other hand, if the rank of $(\lambda_i I - A)$ is $(n - r)$, then we can get r eigenvectors, leading to r Jordan blocks. For each Jordan block we obtain sets of generalized eigenvectors following Eq. (B.30), and the total number of linearly independent generalized eigenvectors will be $m_i - r$.

The procedure will be illustrated by an example.

EXAMPLE B.5

Consider the matrix

$$A = \begin{bmatrix} 0 & 1 & 0 \\ 0 & 0 & 1 \\ -2 & -5 & -4 \end{bmatrix} \tag{B.31}$$

In this case, the characteristic polynomial is given by

$$\begin{aligned} \det(\lambda I - A) &= \begin{bmatrix} \lambda & -1 & 0 \\ 0 & \lambda & -1 \\ 2 & 5 & \lambda + 4 \end{bmatrix} \\ &= \lambda^3 + 4\lambda^2 + 5\lambda + 2 \\ &= (\lambda + 1)^2(\lambda + 2) \end{aligned} \tag{B.32}$$

Hence the eigenvalues are $\lambda_1 = -1$, with multiplicity two, and $\lambda_3 = -2$.

The adjoint of $(\lambda I - A)$ is given by

$$\operatorname{adj}(\lambda I - A) = \begin{bmatrix} \lambda^2 + 4\lambda + 5 & \lambda + 4 & 1 \\ -2 & \lambda^2 + 4\lambda & \lambda \\ -2\lambda & -5\lambda - 2 & \lambda^2 \end{bmatrix} \tag{B.33}$$

Hence the eigenvectors corresponding to λ_1 and λ_3 are obtained as

$$v_1 = \begin{bmatrix} 1 \\ -1 \\ 1 \end{bmatrix} \quad \text{and} \quad v_3 = \begin{bmatrix} 1 \\ -2 \\ 4 \end{bmatrix} \tag{B.34}$$

The vector v_2 must be a generalized eigenvector of order 2, corresponding to λ_1. First, we calculate

$$(\lambda_1 I - A)^2 = \begin{bmatrix} -1 & -1 & 0 \\ 0 & -1 & -1 \\ 2 & 5 & 3 \end{bmatrix} \begin{bmatrix} -1 & -1 & 0 \\ 0 & -1 & -1 \\ 2 & 5 & 3 \end{bmatrix}$$

$$= \begin{bmatrix} 1 & 2 & 1 \\ -2 & -4 & -2 \\ 4 & 8 & 4 \end{bmatrix} \tag{B.35}$$

It is seen by inspection that the rank of $(\lambda_1 I - A)^2$ is one, since any row (or column) of this matrix is a multiple of the first row (or column). Hence, we can find two linearly dependent nontrivial solutions of

$$(\lambda_1 I - A)^2 v = 0 \tag{B.36}$$

One obvious solution is v_1 given in Eq. (B.34), since Eq. (B.36) will be satisfied due to the fact that

$$(\lambda_1 I - A)v_1 = 0 \tag{B.37}$$

A second solution is obtained by noting that

$$(\lambda_1 I - A)v_2 = v_1 \tag{B.38}$$

satisfies Eq. (B.30) for $k = 2$. This gives one solution for v_2 as

$$v_2 = \begin{bmatrix} 0 \\ -1 \\ 2 \end{bmatrix} \tag{B.39}$$

It may be noted that this solution for v_2 is not unique, since any linear combination of v_1 and v_2 will also be a generalized eigenvector of A corresponding to the eigenvalue λ_1.

We can now construct the modal matrix as

$$M = [v_1 \quad v_2 \quad v_3] = \begin{bmatrix} 1 & 0 & 1 \\ -1 & -1 & -2 \\ 1 & 2 & 4 \end{bmatrix} \tag{B.40}$$

It is easily verified that

$$J = M^{-1}AM$$

$$= \begin{bmatrix} 0 & -2 & -1 \\ 2 & -3 & -1 \\ 1 & 2 & 1 \end{bmatrix} \begin{bmatrix} 0 & 1 & 0 \\ 0 & 0 & 1 \\ -2 & -5 & -4 \end{bmatrix} \begin{bmatrix} 1 & 0 & 1 \\ -1 & -1 & -2 \\ 1 & 2 & 4 \end{bmatrix}$$

$$= \left[\begin{array}{cc|c} -1 & 1 & 0 \\ 0 & -1 & 0 \\ \hline 0 & 0 & -2 \end{array}\right] \tag{B.41}$$

DRILL PROBLEM B.1

Determine the diagonal or Jordan form for the following matrices

(a) $\begin{bmatrix} 3 & 0 & 1 \\ 0 & 4 & 0 \\ 1 & 0 & 3 \end{bmatrix}$ (b) $\begin{bmatrix} 5 & 6 & 0 \\ 0 & 1 & 0 \\ -4 & 4 & 1 \end{bmatrix}$

Answers:

(a) $\begin{bmatrix} 2 & 0 & 0 \\ 0 & 4 & 0 \\ 0 & 0 & 4 \end{bmatrix}$ (b) $\begin{bmatrix} 5 & 0 & 0 \\ 0 & 1 & 1 \\ 0 & 0 & 1 \end{bmatrix}$

B.5 CAYLEY-HAMILTON THEOREM

This theorem states that if

$$\Delta(\lambda) \triangleq \det(\lambda I - A) = \lambda^n + a_{n-1}\lambda^{n-1} + \cdots + a_1\lambda + a_0 \tag{B.42}$$

then the matrix polynomial

$$\Delta(A) \triangleq A^n + a_1 A^{n-1} + \cdots + a_1 A + a_0 I = 0 \tag{B.43}$$

This theorem is very useful for calculating A^k when $k > n$, since we can express A^k as a linear combination of $\{A^{n-1}, A^{n-2}, \ldots, A, I\}$. This is done by simply dividing the polynomial λ^k by $\Delta(\lambda)$ and utilizing the remainder.

EXAMPLE B.6

Consider the matrix A given in Eq. (B.31). We shall use Cayley-Hamilton theorem to calculate A^{10}.

First, we note from Example B.5 that

$$\Delta(\lambda) = \lambda^3 + 4\lambda^2 + 5\lambda + 2 \tag{B.44}$$

Furthermore, we can write

$$\lambda^{10} = (\lambda^7 - 4\lambda^6 + 11\lambda^5 - 26\lambda^4 + 57\lambda^3 - 120\lambda^2 + 247\lambda - 502) \cdot (\lambda^3 + 4\lambda^2 + 5\lambda + 2) + 1013\lambda^2 + 2016\lambda + 1004 \qquad \text{(B.45)}$$

Note that the above is obtained simply by long division. Hence

$$A^{10} = 1013A^2 + 2016A + 1004I \qquad \text{(B.46)}$$

and is easily evaluated. The final answer is

$$A^{10} = \begin{bmatrix} 1004 & -2016 & 1013 \\ -2026 & -4061 & -2036 \\ 4072 & 8154 & 4083 \end{bmatrix} \qquad \text{(B.47)}$$

We can also use Cayley-Hamilton theorem to calculate the inverse of a matrix. This is done simply by multiplying both sides of Eq. (B.43) by A^{-1} and rearranging so that we can express A^{-1} as a linear combination of $\{A^{n-1}, A^{n-2}, \ldots, A, I\}$.

B.6 COMPUTATION OF exp (At)

The solution of the state equations

$$\dot{x} = Ax + Bu \qquad \text{(B.48)}$$

was obtained as

$$x(t) = e^{A(t-t_0)}x(t_0) + e^{At}\int_{t_0}^{t} e^{-A\tau}Bu(\tau)\,d\tau \qquad \text{(B.49)}$$

The integral on the right-hand side is the convolution integral and can be obtained using routine methods of numerical integration for any given $u(t)$ provided that the matrix e^{At} is known. Hence it is important to calculate e^{At} for any $n \times n$ matrix A with real elements. We shall consider several methods.

B.6.1 Use of Laplace transforms

It was seen in Chapter 3 that

$$e^{At} = \mathcal{L}^{-1}[(sI - A)^{-1}] \qquad \text{(B.50)}$$

Hence we must invert the matrix $(sI - A)$ and then obtain the Laplace transform of each element of the resulting matrix

$$\Phi(s) = (sI - A)^{-1} \qquad \text{(B.51)}$$

This method is quite suitable for hand computation of e^{At} when A is a 2×2 or 3×3 matrix. For matrices of higher order the inversion of $(sI - A)$ becomes tedious.

EXAMPLE B.7

Consider the second-order matrix

$$A = \begin{bmatrix} 0 & 1 \\ -2 & -3 \end{bmatrix} \tag{B.52}$$

Here,

$$(sI - A) = \begin{bmatrix} s & -1 \\ 2 & s+3 \end{bmatrix} \tag{B.53}$$

and

$$(sI - A)^{-1} = \frac{1}{s^2 + 3s + 2} \begin{bmatrix} s+3 & 1 \\ -2 & s \end{bmatrix}$$

$$= \begin{bmatrix} \dfrac{2}{s+1} - \dfrac{1}{s+2} & \dfrac{1}{s+1} - \dfrac{1}{s+2} \\[2ex] \dfrac{2}{s+2} - \dfrac{2}{s+1} & \dfrac{2}{s+2} - \dfrac{1}{s+1} \end{bmatrix} \tag{B.54}$$

Finally,

$$e^{At} = \begin{bmatrix} 2e^{-t} - e^{-2t} & e^{-t} - e^{-2t} \\ 2e^{-2t} - 2e^{-t} & 2e^{-2t} - e^{-t} \end{bmatrix} \tag{B.55}$$

For higher-order matrices the inverse of $(sI - A)$ can be obtained more conveniently by using Leverrier's algorithm, described in Chapter 3 (Sec. 2.2).

B.6.2 Use of diagonal or Jordan forms

First we note that if we have a diagonal matrix

$$\Lambda = \begin{bmatrix} \lambda_1 & 0 & \cdots & 0 \\ 0 & \lambda_2 & \cdots & 0 \\ \vdots & \vdots & & \vdots \\ 0 & 0 & \cdots & \lambda_n \end{bmatrix} \tag{B.56}$$

then

$$e^{\Lambda t} = \begin{bmatrix} e^{\lambda_1 t} & 0 & \cdots & 0 \\ 0 & e^{\lambda_2 t} & \cdots & 0 \\ & & & 0 \\ \vdots & \vdots & & \vdots \\ 0 & 0 & \cdots & e^{\lambda_n t} \end{bmatrix} \tag{B.57}$$

It is proved easily by recognizing that $e^{\Lambda t}$ is the inverse Laplace transform of $(sI - \Lambda)^{-1}$.

Hence, if

$$\Lambda = M^{-1}AM \tag{B.58}$$

where M is the modal matrix of A, then

$$e^{\Lambda t} = M^{-1}e^{At}M \tag{B.59}$$

or

$$e^{At} = Me^{\Lambda t}M^{-1} \tag{B.60}$$

The above is proved easily by noting that

$$\begin{aligned} sI - \Lambda &= M^{-1}sIM - M^{-1}AM \\ &= M^{-1}(sI - A)M \end{aligned} \tag{B.61}$$

Hence

$$(sI - \Lambda)^{-1} = M^{-1}(sI - A)^{-1}M \tag{B.62}$$

or

$$e^{\Lambda t} = M^{-1}e^{At}M \tag{B.63}$$

When A cannot be diagonalized, then we can use the corresponding Jordan form of A, so that

$$e^{At} = Me^{Jt}M^{-1} \tag{B.64}$$

Furthermore, it may be noted that e^{Jt} is easily determined from J, for example, by using Laplace transforms.

EXAMPLE B.8

Consider

$$J = \begin{bmatrix} \lambda_1 & 1 & 0 \\ 0 & \lambda_1 & 1 \\ 0 & 0 & \lambda_1 \end{bmatrix} \tag{B.65}$$

Then

$$(sI - J)^{-1} = \frac{1}{(s-\lambda_1)^3} \begin{bmatrix} (s-\lambda_1)^2 & (s-\lambda_1) & 1 \\ 0 & (s-\lambda_1)^2 & (s-\lambda_1) \\ 0 & 0 & (s-\lambda_1)^2 \end{bmatrix} \tag{B.66}$$

and

$$e^{Jt} = \begin{bmatrix} e^{\lambda_1 t} & te^{\lambda_1 t} & \frac{1}{2}t^2 e^{\lambda_1 t} \\ 0 & e^{\lambda_1 t} & te^{\lambda_1 t} \\ 0 & 0 & e^{\lambda_1 t} \end{bmatrix} \tag{B.67}$$

The main difficulty with this method is that one must determine the eigenvalues and eigenvectors (generalized if necessary) of the matrix A. Excellent numerical methods have been developed for this purpose. The most well-known is the Q-R algorithm. The corresponding computer program EISPACK is available in most computer libraries.

B.6.3 Use of Sylvester's interpolation formula

Following Cayley-Hamilton theorem, a convergent infinite series

$$f(A) = \sum_{k=0}^{\infty} c_k A^k \tag{B.68}$$

can be expressed as a unique polynomial in A of degree $n - 1$ or less, where n is the order of the characteristic polynomial of A.

Hence we may write

$$e^{At} = \alpha_0 I + \alpha_1 A + \alpha_2 A^2 + \cdots + \alpha_{n-1} A^{n-1} \tag{B.69}$$

and determine the α_k by solving the set of n equations

$$\begin{bmatrix} \lambda_1 & \lambda_1{}^2 & \lambda_1{}^3 & \cdots & \lambda_1^{n-1} \\ \lambda_2 & \lambda_2{}^2 & \lambda_2{}^3 & \cdots & \lambda_2^{n-1} \\ \vdots & & & & \\ \lambda_n & \lambda_n{}^2 & \lambda_n{}^3 & \cdots & \lambda_n^{n-1} \end{bmatrix} \begin{bmatrix} \alpha_0 \\ \alpha_1 \\ \\ \alpha_{n-1} \end{bmatrix} = \begin{bmatrix} e^{\lambda_1 t} \\ e^{\lambda_2 t} \\ \vdots \\ e^{\lambda_n t} \end{bmatrix} \tag{B.70}$$

where $\lambda_1, \lambda_2, \ldots, \lambda_n$ are the eigenvalues of A.

This will work if the eigenvalues are distinct. For the case of repeated eigenvalues the method needs a small modification, since the various rows of the matrix in Eq. (B.70) are no longer linearly independent. For example, if $\lambda_1 = \lambda_2$, then the first row will be replaced by its derivative with respect to λ_1, and the corresponding term on the right-hand side will be replaced by $te^{\lambda_1 t}$.

B.7 REFERENCES

1. J. H. Wilkinson, *The Algebraic Eigenvalue Problem*, Clarendon Press, Oxford, U.K., 1965.
2. K. Ogata, *State Space Analysis of Control Systems*, Prentice-Hall, Englewood Cliffs, N.J., 1967.

C

Review of the Theory of z Transforms

The role of the z transform in the analysis and design of sampled-data systems is similar to that of the Laplace transform in continuous-data systems. The main idea here is to replace a continuous-time signal by a sequence of equally spaced impulses, where the strength (or area) of each impulse is equal to the magnitude of the signal at the sampling instant. In this appendix we shall review the basic theory of the z transform.

C.1 DEFINITION

Consider a sequence of impulses of unit strength and period T, as shown in Fig. C.1 and defined by

$$\delta_T(t) = \sum_{n=0}^{\infty} \delta(t - nT) \tag{C.1}$$

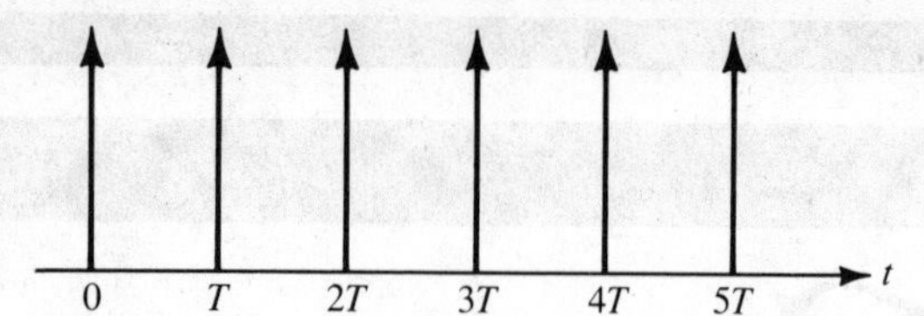

FIGURE C.1. Sequence of unit impulses T seconds apart.

The Laplace transform of the sequence of unit impulses is given by

$$I(s) = \mathscr{L}[\delta_T(t)] = \sum_{n=0}^{\infty} e^{-sTn} \tag{C.2}$$

Now consider a function $f(t)$ modulating these impulses as shown in Fig. C.2. This gives rise to the impulse train

$$f^*(t) = \sum_{n=0}^{\infty} f(nT)\delta(t - nT) \tag{C.3}$$

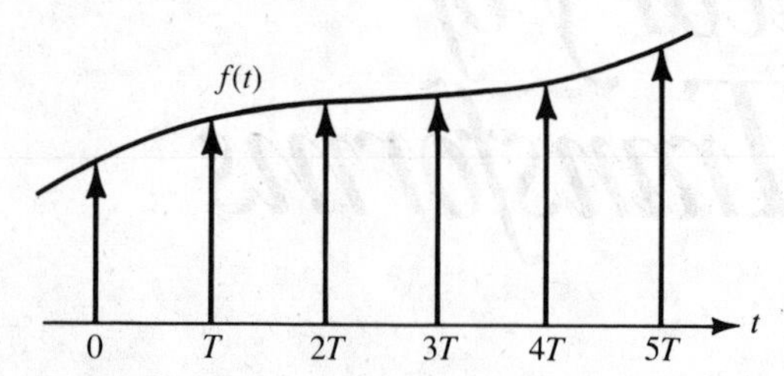

FIGURE C.2. Impulses modulated by $f(t)$.

Taking the Laplace transform, we get

$$F^*(s) = \mathscr{L}[f^*(t)] = \sum_{n=0}^{\infty} f(nT)e^{-sTn} \tag{C.4}$$

Note that both $I(s)$ and $F^*(s)$ involve e^{-sT}, making them irrational functions of s. A simplification is obtained if we use the substitution

$$z = e^{sT} \tag{C.5}$$

Then we may define the z transform of $f(t)$ as

$$F(z) \triangleq \mathscr{Z}[f(t)] = \mathscr{L}[f^*(t)]\big|_{z=e^{sT}} = \sum_{n=0}^{\infty} f(nT)z^{-n} \tag{C.6}$$

Evidently the z transform of $f(t)$ is the same as the Laplace transform of the sampled function $f^*(t)$, although it is disguised by the change in variable from s to z. It will be seen later that if $F(s)$ is a rational function of s, then $F(z)$

will be a rational function of z. As a matter of fact, the poles of $F(z)$ will be related to the poles of $F(s)$ through the transformation of Eq. (C.5). Furthermore, Eq. (C.5) may be utilized to interpret z as the unit advance operator (advancing by one sampling interval). We shall now obtain the z transforms of some standard functions.

C.2 SOME SIMPLE z-TRANSFORM PAIRS

These will be derived using the definition in Eq. (C.6).

(a) *Unit step*: This is defined as

$$f(t) = 1 \qquad \text{for } t \geq 0 \tag{C.7}$$

Its z transform is obtained as

$$F(z) = \sum_{n=0}^{\infty} f(nT)z^{-n} = 1 + z^{-1} + z^{-2} + \cdots$$

$$= \frac{1}{1 - z^{-1}} = \frac{z}{z - 1} \tag{C.8}$$

(b) *Exponential function*: Consider the exponential function

$$f(t) = e^{at} \tag{C.9}$$

Its z transform is obtained as

$$F(z) = \sum_{n=0}^{\infty} e^{naT}z^{-n} = 1 + e^{aT}z^{-1} + e^{2aT}z^{-2} + \cdots$$

$$= \frac{1}{1 - e^{aT}z^{-1}} = \frac{z}{z - e^{aT}} \tag{C.10}$$

(c) *Sine and cosine functions*: If in Eq. (C.9) we replace a by $j\omega$, we have

$$\mathscr{Z}[e^{j\omega t}] = \mathscr{Z}[\cos \omega t + j \sin \omega t] = \frac{z}{z - e^{j\omega T}}$$

$$= \frac{z}{(z - \cos \omega T) - j \sin \omega T}$$

$$= \frac{z(z - \cos \omega T) + jz \sin \omega T}{(z - \cos \omega T)^2 + (\sin \omega T)^2} \tag{C.11}$$

Hence, by separating the real and imaginary parts

$$\mathscr{Z}[\cos \omega t] = \frac{z(z - \cos \omega T)}{z^2 - 2z \cos \omega T + 1} \tag{C.12}$$

and

$$\mathscr{Z}[\sin \omega t] = \frac{z \sin \omega T}{z^2 - 2z \cos \omega T + 1} \tag{C.13}$$

(d) *Ramp function*: Consider the ramp function

$$f(t) = t \tag{C.14}$$

Its z transform is obtained as

$$\begin{aligned} F(z) &= \sum_{n=0}^{\infty} nTz^{-n} = Tz^{-1} + 2Tz^{-2} + 3Tz^{-3} + \cdots \\ &= Tz^{-1}(1 + 2z^{-1} + 3z^{-2} + \cdots) \\ &= \frac{Tz^{-1}}{(1 - z^{-1})^2} = \frac{Tz}{(z - 1)^2} \end{aligned} \tag{C.15}$$

(e) *Functions with exponential damping*: Consider the function

$$f_1(t) = e^{-at}f(t) \tag{C.16}$$

Its z transform is given by

$$\begin{aligned} F_1(z) &= \sum_{n=0}^{\infty} f(nT)e^{-naT}z^{-n} \\ &= \sum_{n=0}^{\infty} f(nT)(ze^{aT})^{-n} \\ &= F(ze^{aT}) \end{aligned} \tag{C.17}$$

This may be compared with the corresponding theorem in the Laplace-transform theory. That is,

$$L[e^{-aT}f(t)] = F(s + a) \tag{C.18}$$

which appears logical in view of the relationship given in Eq. (C.5).

DRILL PROBLEM C.1

Determine the z transform of the following if $T = 0.5$.
(a) $10 \cos (3t + 60°)$ (b) te^{-3t} (c) $10e^{-2t} \cos (3t + 60°)$

Answers:

(a) $\dfrac{5z^2 - 8.9922z}{z^2 - 0.1415z + 1}$ (b) $\dfrac{0.1116z}{(z - 0.2231)^2}$

(c) $\dfrac{5z^2 - 3.3081z}{z^2 - 0.052z + 0.1353}$

C.3 MAPPING BETWEEN THE s-PLANE AND THE z-PLANE

Equation (C3.5) may be considered as a mapping from the *s*-plane to the *z*-plane, as shown in Fig. C.3. In particular, it may be observed that the imaginary axis of the *s*-plane maps into the unit circle of the *z*-plane and the origin of *s*-plane maps into the point $1 + j0$ in the *z*-plane. We can also verify that the poles of $F(s)$ and $F(z)$ are related through Eq. (C.5) for each of the elementary functions considered so far. We can extend this to obtain the *z* transform of a function $F(z)$ if its Laplace transform $F(s)$ is known by simply using partial fractions. First, we note from Eq. (C.10) that

$$\mathscr{Z}\left[\frac{A}{s+a}\right] = \frac{Az}{z - e^{-aT}} \tag{C.19}$$

Hence, if

$$F(s) = \sum_{k=1}^{n} \frac{A_k}{s + a_k} \tag{C.20}$$

then

$$F(z) = \sum_{k=1}^{n} \frac{A_k z}{z - e^{-a_k T}} \tag{C.21}$$

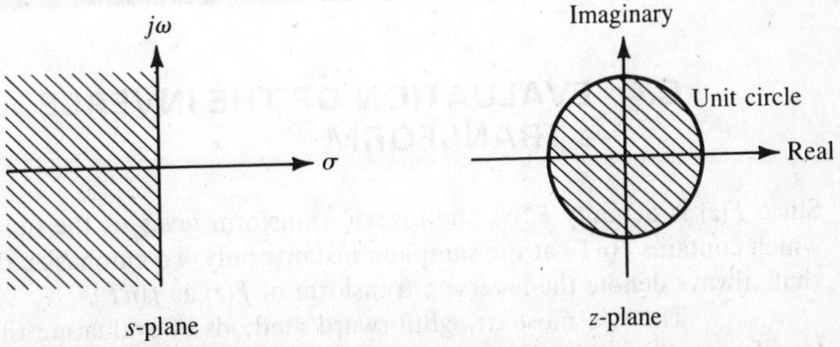

FIGURE C.3. Mapping from the s-plane to the z-plane.

Equation (C.21) provides us with a simple procedure for finding the z transform of a function if its Laplace transform is known. The only case where difficulty may arise is that of multiple poles. This is overcome by noting that

$$\frac{K}{(s+a)^{r+1}} = (-1)^r \frac{K}{r!} \frac{d^r}{da^r}\left(\frac{1}{s+a}\right) \tag{C.22}$$

Hence, if

$$F(s) = \frac{K}{(s+a)^{r+1}}$$

then

$$F(z) = (-1)^r \frac{K}{r!} \frac{d^r}{da^r}\left(\frac{z}{z-e^{-aT}}\right) \tag{C.23}$$

DRILL PROBLEM C.2

Determine the z transform of the following for $T = 0.2$.

(a) $\dfrac{10}{s(s+2)}$ (b) $\dfrac{4s}{(s+1)(s^2+4)}$

(c) $\dfrac{25(s+2)}{s(s+1)(s^2+6s+25)}$ (d) $\dfrac{1}{(s+3)^2}$

Answers:

(a) $\dfrac{1.6484z}{(z-1)(z-0.6703)}$ (b) $\dfrac{0.3934z^2+1.3452z}{(z-0.8187)(z^2-1.8421z+1)}$

(c) $\dfrac{0.3453z^3-0.0613z^2-0.1255z}{(z-1)(z-0.8187)(z^2-0.76472z+0.2019)}$

(d) $\dfrac{0.1098z}{(z-0.5488)^2}$

C.4 EVALUATION OF THE INVERSE z TRANSFORM

Since $F(z)$ is actually $F^*(s)$, the inverse transform gives us the function $f^*(t)$, which contains $f(nT)$ at the sampling instants only. To emphasize this fact, we shall always denote the inverse z transform of $F(z)$ as $f(nT)$.

The two most straightforward methods of evaluating the inverse z transform will now be discussed.

C.4.1. The method of partial fraction expansion

This is based on the use of Eq. (C.21). It may be noted that it is more desirable to expand $F(z)/z$ into partial fractions, in order that we may get z in the numerator after expansion. The following recognitions are useful.

If

$$F(z) = \frac{Kz}{z - a}$$

then

$$f(nT) = Ka^n \tag{C.24}$$

If

$$F(z) = \frac{(C + jD)z}{z - re^{j\phi}} + \frac{(C - jD)z}{z - re^{-j\phi}}$$

then

$$f(nT) = 2r^n(C \cos n\phi - D \sin n\phi) \tag{C.25}$$

If

$$F(z) = \frac{Kz}{(z - a)^r} \qquad r = 2, 3, \ldots$$

then

$$f(nT) = \frac{Kn(n-1)\cdots(n - r + 2)}{(r-1)!a^{r-1}} a^n \tag{C.26}$$

EXAMPLE C.1

Consider

$$F(z) = \frac{3z^2 - 4z}{(z - 2)(z^2 - 2z + 2)} \tag{C.27}$$

Then we have

$$\frac{1}{z} F(z) = \frac{3z - 4}{(z - 2)(z^2 - 2z + 2)}$$

$$= \frac{1}{z - 2} + \frac{-0.5 - j1}{z - \sqrt{2}e^{j\pi/4}} + \frac{-0.5 + j1}{z - \sqrt{2}e^{-j\pi/4}} \tag{C.28}$$

Hence

$$f(nT) = 2^n - 2^{n/2}(\cos n\pi/4 - 2 \sin n\pi/4) \tag{C.29}$$

C.4.2 The method of long division

Since $F(z)$ is a rational function of z, it can be expanded into a power series of the form

$$F(z) = \sum_{n=0}^{\infty} a_n z^{-n} \tag{C.30}$$

by long division, thus yielding $f(nT) = a_n$. This can be carried out very conveniently on either a computer or a programmable pocket calculator.

EXAMPLE C.2

We shall again consider the previous example.

$$F(z) = \frac{3z^2 - 4z}{(z-2)(z^2 - 2z + 2)} = \frac{3z^2 - 4z}{z^3 - 4z^2 + 6z - 4}$$

$$= 3z^{-1} + 8z^{-2} + 14z^{-3} + 20z^{-4} + 28z^{-5} + \cdots \tag{C.31}$$

It may be verified that the first five terms of Eq. (C.30) match exactly with the values of $f(nT)$ obtained from Eq. (C.29) for the corresponding values of n. The main difficulty with this method is that it is not convenient for obtaining the value of $f(nT)$ for large n.

DRILL PROBLEM C.3

Determine the inverse z transform of each of the following:

(a) $\dfrac{2z}{(z-1)(z-0.5)}$ (b) $\dfrac{4z^3 - 21z^2 + 29z}{(z-2)(z-3)^2}$

(c) $\dfrac{6z^2 - 5z}{(z-1)(z^2 - z + 1)}$ (d) $\dfrac{8z^2 - 3z}{(z-2)(z^2 - 0.4z + 0.16)}$

Answers:

(a) $4 - 2(0.5)^n$ (b) $3(2)^n + 3^n + \dfrac{2n}{3}(3)^n$

(c) $1 - \cos(\pi/3) + 6.3209 \sin(n\pi/3)$

(d) $3.8691(2)^n - 3.8691(0.4)^n \cos(1.0472n) - 2.9899(0.4)^n \sin(1.0472n)$

C.5 SOME USEFUL THEOREMS

We shall now discuss some theorems in z-transform theory. It is interesting to compare them with similar theorems from the theory of Laplace transforms.

C.5.1 The initial-value theorem

This enables us to determine the initial value $f(0)$ from the z transform $F(z)$ without actually finding the inverse z transform of $F(z)$. The theorem states that

$$f(0) = \lim_{z \to \infty} F(z) \tag{C.32}$$

Proof

$$F(z) = \sum_{n=0}^{\infty} f(nT)z^{-n} = f(0) + f(T)z^{-1} + f(2T)z^{-2} + \cdots$$

Hence

$$f(0) = \lim_{z \to \infty} F(z)$$

C.5.2 The final-value theorem

This enables us to determine the final value of the function $f(nT)$ as $n \to \infty$ from its z transform $F(z)$ without actually finding the inverse z transform. It may be stated as

$$\lim_{t \to \infty} f(t) = \lim_{z \to 1} (z - 1)F(z) \tag{C.33}$$

As in the case of Laplace-transform theory, this theorem must be applied with caution to only those cases where the final values exist. If $F(z)$ has any pole outside the unit circle, the final value will be infinite, and the theorem cannot be applied. Similarly, if there are poles on the unit circle, with the exception of a simple pole at $z = 1$, the final value is indeterminate, and the theorem is again inapplicable. If all the poles are strictly inside the unit circle (i.e., there is no pole at $z = 1$), then Eq. (C.33) gives zero final value. With a simple pole at $z = 1$, Eq. (C.33) is precisely the expression for the residue at that pole, and thus the final value.

C.5.3 Differentiation in the z domain

By definition,

$$F(z) = \sum_{n=0}^{\infty} f(nT)z^{-n} \tag{C.34}$$

Then

$$\frac{dF(z)}{dz} = -\sum_{n=0}^{\infty} nf(nT)z^{-n-1} \tag{C.35}$$

or

$$-z\frac{dF(z)}{dz} = \sum_{n=0}^{\infty} nf(nT)z^{-n} = \frac{1}{T}\sum_{n=0}^{\infty} nTf(nT)z^{-n}$$

$$= \frac{1}{T} Z[tf(t)]$$

Thus

$$Z[tf(t)] = -Tz\frac{dF(z)}{dz} \tag{C.36}$$

This theorem enables us to obtain the z transform of $tf(t)$ if the z transform of $f(t)$ is known. The reader should compare it with a similar theorem in the theory of Laplace transforms.

C.6 z TRANSFORMS AND DIFFERENCE EQUATIONS

Just as the Laplace transform is related to differential equations, we have a relationship between the z transform and difference equations. For example, the following z-transform relationship

$$Y(z) = (a_2z^2 + a_1z + a_0)X(z) - a_2(z^2x_0 + zx_1) - a_1zx_0 \tag{C.37}$$

is equivalent to the difference equation

$$y_k = a_2x_{k+2} + a_1x_{k+1} + a_0x_k \tag{C.38}$$

where

$$x_k \triangleq x(kT) \tag{C.39}$$

This follows immediately from the interpretation of z as the unit advance operator. Notice the similarity with the expression for the Laplace transform of the derivative dx/dt in terms of the Laplace transform of $x(t)$.

DRILL PROBLEM C.4

Determine the initial values of the time functions corresponding to each of the following z transforms.

(a) $\dfrac{4z^3 - 5z^2 + 8z}{(z-1)(z-0.5)^2}$ (b) $\dfrac{z(z-0.5)(z-0.1)}{(z^2-0.5z+1)(z^2-0.2z+0.1)}$

Answers: (a) 4, (b) 0.

DRILL PROBLEM C.5

Determine the final values of the time functions corresponding to the z transforms given in the previous problem.

Answers:
(a) 28, (b) indeterminate, due to pair of complex roots on the unit circle.

DRILL PROBLEM C.6

The difference equation for a discrete-time system is given by

$$y_{k+2} = 0.7y_{k+1} - 0.1y_k + 2x_{k+1} + 3x_k$$

Determine the z-transform relationship between $Y(z)$ and $X(z)$, assuming zero initial conditions.

Answer:
$(z^2 - 0.7z + 0.1)Y(z) = (2z + 3)X(z)$

C.7 INPUT/OUTPUT RELATIONSHIP

Consider the block diagram shown in Fig. C.4, where the input $x(t)$ is applied to a linear system with transfer function $G(s)$ after sampling; that is, the input to $G(s)$ is a sequence of impulses.

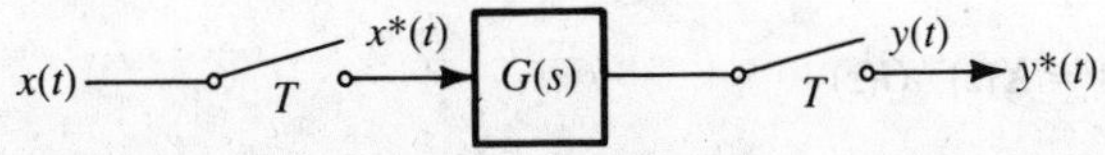

FIGURE C.4. A sampled-data system.

The output, $y(t)$, is a continuous function of time. A fictitious sampler has been introduced to obtain the sampled output $y^*(t)$. The two samplers have the same sampling rate and are assumed to be synchronized.

Since, by definition,

$$x^*(t) = \sum_{n=0}^{\infty} x(nT)\delta(t - nT) \tag{C.40}$$

the continuous output $y(t)$ may be regarded as the sum of the responses of the system to the sequence of impulses. This follows from the principle of superposition. Hence

$$\begin{aligned} y(t) &= x(0)g(t) + x(T)g(t - T)u_{t-T} + x(2T)g(t - 2T)u_{t-2T} + \cdots \\ &= \sum_{n=0}^{\infty} x(nT)g(t - nT)u_{t-nT} \end{aligned} \tag{C.41}$$

where $g(t)$ is the impulse response of the system—i.e., the inverse Laplace transform of $G(s)$—and u_{t-nT} is the unit step occurring at $t = nT$.

The output at the kth sampling instant is given by

$$y(kT) = \sum_{r=0}^{k} x(rT)g(kT - rT) \tag{C.42}$$

Note that this is a finite sum, since for any physical system $g(t)$ is 0 for $t < 0$. This follows from the fact that the system cannot respond to the impulse before it is applied. This is called the property of "causality"; that is, the cause must precede the effect. Define

$$G(z) = \sum_{n=0}^{\infty} g(nT)z^{-n} \tag{C.43}$$

as the z transform of the impulse response $g(t)$. Then

$$\begin{aligned} X(z)G(z) &= \{x(0) + x(T)z^{-1} + x(2T)z^{-2} + \cdots\}\{g(0) + g(T)z^{-2} + \cdots\} \\ &= x(0)g(0) + \{x(0)g(T) + x(T)g(0)\}z^{-1} + \cdots\} \\ &= \sum_{k=0}^{\infty}\left[\sum_{r=0}^{k} x(rT)g(kT - rT)\cdot z^{-k}\right] = \sum_{k=0}^{\infty} y(kT)z^{-k} \end{aligned} \tag{C.44}$$

Hence, we have the input/output relationship

$$Y(z) = X(z)\cdot G(z) \tag{C.45}$$

and $G(z)$ is called the z transfer function of $G(s)$.

EXAMPLE C.3

Determine $y(nT)$ for the system shown in the block diagram of Fig. C.5 if $x(t)$ is a unit step and $T = 0.1$ second. Assume that the samplers are synchronized.

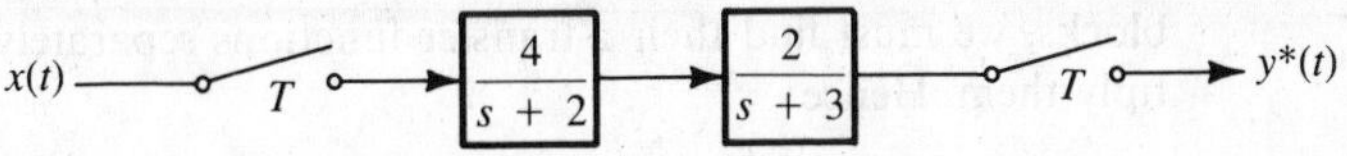

$$x(t) \xrightarrow{T} \boxed{\frac{4}{s+2}} \rightarrow \boxed{\frac{2}{s+3}} \xrightarrow{T} y^*(t)$$

FIGURE C.5. Block diagram for Example C.3.

Solution

A partial fraction expansion of the transfer function yields

$$G(s) = \frac{8}{(s+2)(s+3)} = \frac{8}{s+2} - \frac{8}{s+3} \tag{C.46}$$

Hence the z-transform function is given by

$$G(z) = \frac{8z}{z - e^{-2T}} - \frac{8z}{z - e^{-3T}} = \frac{8z(e^{-2T} - e^{-3T})}{(z - e^{-2T})(z - e^{-3T})} \tag{C.47}$$

The z transform of the input is obtained as

$$X(z) = \frac{z}{z-1} \tag{C.48}$$

Hence the z transform of the output is given by

$$Y(z) = G(z)X(z) = \frac{8z^2(e^{-0.2} - e^{-0.3})}{(z-1)(z - e^{-0.2})(z - e^{-0.3})} \tag{C.49}$$

The partial fraction expansion of $(1/z)Y(z)$ is obtained as

$$\frac{1}{z}Y(z) = \frac{13.2667}{z-1} - \frac{36.133}{z - e^{-0.2}} + \frac{22.866}{z - e^{-0.3}} \tag{C.50}$$

Hence, taking the inverse z transform,

$$y(nT) = 13.2667 - 36.133e^{-0.2n} + 22.866e^{-0.3n} \tag{C.51}$$

EXAMPLE C.4

Repeat Example C.3 for the block diagram shown in Fig. C.6.

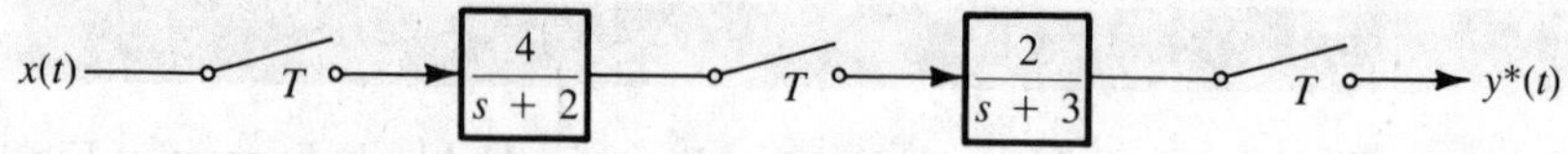

FIGURE C.6. Block diagram for Example C.4.

Solution

In this case, since we have a sampler between the two analog blocks, we must find their z-transfer functions separately and multiply them. Hence

$$G(z) = \frac{4z}{z - e^{-0.2}} \cdot \frac{2z}{z - e^{-0.3}} = \frac{8z^2}{(z - e^{-0.2})(z - e^{-0.3})} \qquad \text{(C.52)}$$

The z transform of the output is obtained as

$$Y(z) = G(z)X(z) = \frac{8z^3}{(z - 1)(z - e^{-0.2})(z - e^{-0.3})} \qquad \text{(C.53)}$$

The partial fraction expansion of $(1/z)Y(z)$ is given by

$$\frac{1}{z}\,Y(z) = \frac{170.279}{z - 1} - \frac{379.7}{z - e^{-0.2}} + \frac{217.421}{z - e^{-0.3}} \qquad \text{(C.54)}$$

and taking the inverse z transform

$$y(nT) = 170.279 - 379.7e^{-0.2n} + 217.421e^{-0.3n} \qquad \text{(C.55)}$$

These examples show that when a number of continuous-time systems are connected in cascade, without any sampler between them, the z transfer function of the combination is the z transform of the overall continuous-time transfer function and not the product of the z transfer functions of the individual elements.

DRILL PROBLEM C.7

Determine the transfer function of each of the sampled-data systems shown in Fig. C.7, assuming that the samplers are synchronized and $T = 0.2$ second.

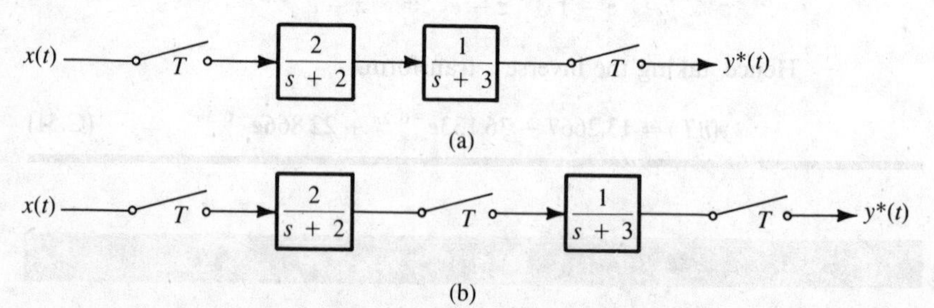

FIGURE C.7. Block diagrams for Drill Problem C.7.

Answers:

(a) $\dfrac{0.243z}{(z - 0.6703)(z - 0.5488)}$

(b) $\dfrac{4z^2}{(z - 0.6703)(z - 0.5488)}$

DRILL PROBLEM C.8

For each of the two systems in Drill Problem C.7 determine the response if the input is given by $x(t) = 10e^{-t}$.

Answers:

(a) $y(nT) = 49.6682e^{-0.2n} - 90.3331e^{-0.4n} + 40.6649e^{-0.6n}$

(b) $y(nT) = 16.7334e^{0.2n} - 24.9168e^{0.4n} + 9.1835e^{-0.6n}$

C.8 A SHORT TABLE OF LAPLACE AND z TRANSFORMS

$f(t)$	$F(s)$	$F(z)$
unit step	$\dfrac{1}{s}$	$\dfrac{z}{z-1}$
e^{-at}	$\dfrac{1}{s+a}$	$\dfrac{z}{z-e^{-aT}}$
t	$\dfrac{1}{s^2}$	$\dfrac{Tz}{(z-1)^2}$
$\cos \omega t$	$\dfrac{s}{s^2+\omega^2}$	$\dfrac{z(z-\cos \omega T)}{z^2 - 2z\cos \omega T + 1}$
$\sin \omega t$	$\dfrac{\omega}{s^2+\omega^2}$	$\dfrac{z \sin \omega T}{z^2 - 2z\cos \omega T + 1}$
$e^{-at}\cos \omega t$	$\dfrac{s+a}{(s+a)^2+\omega^2}$	$\dfrac{z(z-e^{-aT}\cos \omega T)}{z^2 - 2ze^{-aT}\cos \omega T + e^{-2aT}}$
$e^{-at}\sin \omega t$	$\dfrac{\omega}{(s+a)^2+\omega^2}$	$\dfrac{ze^{-at}\sin \omega T}{z^2 - 2ze^{-aT}\cos \omega T + e^{-2aT}}$
$tf(t)$	$-\dfrac{dF(s)}{ds}$	$-zT \cdot \dfrac{dF(z)}{dz}$
$e^{-at}f(t)$	$F(s+a)$	$F(ze^{aT})$

C.9 INVERSE z TRANSFORMS

$F(z)$	$f(nT)$
$\dfrac{Az}{z-a}$	$A(a)^n$
$\dfrac{(A+jB)z}{z-re^{j\phi}}+\dfrac{(A-jB)z}{z-re^{-j\phi}}$	$2(r)^n(A\cos n\phi - B\sin n\phi)$

C.10 REFERENCES

1. H. Freeman, *Discrete-Time System*, Wiley, New York, 1965.
2. E. I. Jury, *Sampled-Data Control Systems*, Wiley, New York, 1958.
3. B. C. Kuo, *Analysis and Synthesis of Sampled-Data Control Systems*, Prentice-Hall, Englewood Cliffs, N.J., 1963.
4. D. P. Lindroff, *Theory of Sampled-Data Control Systems*, Wiley, New York, 1965.
5. J. R. Ragazzini and G. F. Franklin, *Sampled-Data Control Systems*, McGraw-Hill, New York, 1958.
6. J. Tou, *Digital and Sampled-Data Control Systems*, McGraw-Hill, New York, 1959.

C.11 PROBLEMS

1. Find the z transform of each of the following.

a. $2e^{-3t}\cos 4t,$

b. $te^{-2t}\sin 3t$

c. $\dfrac{4s+3}{(s+1)(s^2+4s+8)}$

d. $\dfrac{3s+5}{(s+1)(s+2)^2}$

2. Find the inverse z transform of each of the following.

a. $\dfrac{10z(z+0.4)}{(z-1)(z-0.5)(z-0.2)}$

b. $\dfrac{4z(z-1)}{(z^2+0.5z+0.8)(z+0.3)}$

c. $\dfrac{2z(z+2z+1)}{(z^2-z+1)(z^2-2z+2)}$

d. $\dfrac{3z(z+0.3)}{(z-0.5)^2(z^2+0.8z+0.25)}$

3. Determine $c(nT)$ for the block diagrams of Fig. C.8 if $r(t)$ is a unit step.

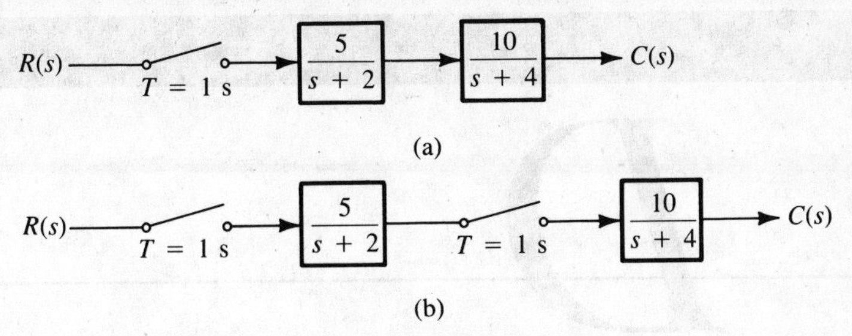

FIGURE C.8. Block diagrams for Problem 3.

4. A difference-equation model for estimating the population of the United States is as follows:

$$x_k = 2.00715x_{k-1} - 1.0078153x_{k-2}$$

where x_k = population in a given year (in millions)
x_{k-1} = population in the previous year (in millions)
Use the z-transform method to obtain the estimated populations for 1980 and 1984 if the population was 204.9 million in 1970 and 207 million in 1971.

D Computational Aspects

D.1 INTRODUCTION

In order to master thoroughly the various methods and algorithms for the analysis and design of control systems that have been presented in this book, it is necessary to work out many numerical problems related to each. In the past, this could not be possible without a great deal of drudgery because efficient computational hardware was not available to most students. The situation has changed completely during the past seven years. Today, practically all engineering undergraduates have programmable pocket calculators. Furthermore, a large number of them either own or have access to a personal computer. In most cases, however, good and reliable software for solving engineering problems is often unavailable.

In recognition of this fact, a number of programs will be listed in this appendix that can be used for solving all the problems given in the text, as well as the practical problems that a control engineer with a bachelor's degree is normally expected to solve. In view of the diversity in the nature of the available hardware, as well as programming languages, it was decided to select two of the most popular devices, the HP-41C (or CV) programmable calculator and the IBM Personal Computer (using Turbo-Pascal). Although Fortran is more popular with engineers, the choice of Pascal was based on its modular structure, which makes it more readable and easier to maintain or modify. Moreover, Turbo-Pascal is easily transported from one machine to another and is popular with the users of microcomputers. Translation of these programs to other dialects of Pascal is fairly straightforward. Furthermore, these programs can be used on an Apple *//* or Apple *///* computer with only a minor modification.

D.2 COMPUTATIONAL ALGORITHMS

An important aspect of the development of a computer program is the selection of an efficient algorithm that will give fairly accurate results in spite of round-off and truncation inside the computer. Although this is a very difficult task, every attempt has been made to ensure that all the programs listed here work under normal conditions. They have all been thoroughly tested with the examples given in the book; and have also been utilized for obtaining the answers for the drill problems. The programs have been made interactive so that the user does not have to memorize all the steps.

Programs have been listed for the following problems:

1. Frequency response calculations.
2. Transient response calculations.
3. Roots of polynomials.
4. Searching points on the root locus.
5. Design of lag, lead, and lag-lead compensators.
6. Stability analysis using the Routh criterion.
7. State transition matrices for discrete-time systems.
8. Determination of transfer functions from state equations and transformation to controllable canonical form.

Since the same algorithms have been used (with some minor modifications) for the programmable calculator as well as the IBM Personal Computer, we shall first briefly discuss these before presenting the actual programs.

D.2.1 Frequency response calculations

The main objective here is to calculate the gain and phase shift at various frequencies for a given transfer function, which is in the form of the ratio of two polynomials of s, with an additional delay term. It should also be possible for the user to specify that the gain be expressed in decibels, if desired. Another useful feature would be the ability to calculate the gain of the closed-loop system, assuming unity feedback.

The problem of evaluating $G(j\omega)$ for a given ω is solved by dividing the numerator and denominator polynomials by $s^2 + \omega^2$ and utilizing the remainder to calculate the value of each polynomial for the given ω. For example, if division of $P(s)$ by the factor $(s^2 + \omega^2)$ gives the remainder as $(cs + d)$, then $P(j\omega)$ is obtained as $d + jc\omega$. The main advantage of this approach is the reduction in arithmetic with complex numbers. This leads to higher accuracy in addition to faster computation.

D.2.2 Transient response calculations

The basic problem, in this case, is the calculation of the inverse Laplace transform of $C(s)$ in order to obtain $c(t)$, the transient response. Thus, it reduces to the evaluation of the residues at the various poles of $C(s)$. Hence, one is again faced with the problem of evaluating a rational function of s for a specified value of s, which may be either real or complex. The approach described in the previous section can be used again, with the difference that for $s = \sigma + j\omega$ the divisor should now be the quadratic $s^2 + as + b$, where $a = -2\sigma$, and $b = \sigma^2 + \omega^2$. The case of real s is simpler, since one may use Horner's rule for evaluating the polynomials.

This program is based on the assumption that the location of the poles of $C(s)$ is known. If this is not the case, then it will be necessary to evaluate the roots of the denominator polynomial. This is discussed in the next section.

Another assumption is that the poles of $C(s)$ are distinct. For the case of repeated poles, the procedure is more involved. One method of solving the problem is first to evaluate the residues at all other poles using the procedure described above. This may then be followed by calculating the residue at the multiple pole after removing all factors containing this term from the denominator, with the exception of one. This gives the numerator of the term with full multiplicity of the pole in the partial fraction expansion. The terms for lower multiplicity can then be obtained by reintroducing the various factors one by one, and taking advantage of the relationship

$$\frac{K}{(s+a)(a+b)} = \frac{K/(b-a)}{s+a} + \frac{K/(a-b)}{s+b} \tag{D.1}$$

In many cases one may utilize the initial-value theorem to reduce the amount of calculation required.

The following examples will illustrate the procedure.

EXAMPLE D.1

$$C(s) = \frac{4(s+1)}{(s+2)^2(s+3)(s^2+2s+5)}$$

First we obtain the residues of $C(s)$ at all poles except the multiple poles at $s = -2$. These are given below.

Pole	Residue
-3	-1
$-1 + j2$	$-0.14 - j0.02$
$-1 - j2$	$-0.14 + j0.02$

Next we evaluate the residue of

$$C_1(s) = \frac{4(s+1)}{(s+2)(s+3)(s^2+2s+5)}$$

at the pole $s = -2$. This is obtained as -0.8. Hence,

$$C(s) = \frac{-0.8}{(s+2)^2} + \frac{A}{s+2} + \frac{-1}{s+3} + \frac{-0.14 - j0.02}{s+1-j2} + \frac{-0.14 + j0.02}{s+1+j2}$$

Finally, the value of A is obtained to satisfy the initial condition for $c(t)$, which is determined by applying the initial-value theorem and is found to be zero. Hence $A = 1.28$.

Alternatively we could have first obtained the partial fraction expansion for $C_1(s)$, which is

$$C_1(s) = \frac{-0.8}{s+2} + \frac{1}{s+3} + \frac{-0.1 - j0.3}{s+1-j2} + \frac{-0.1 + j0.3}{s+1+j2}$$

Hence,

$$C(s) = \frac{1}{(s+2)} C_1(s)$$

$$= \frac{-0.8}{(s+2)^2} + \frac{1}{(s+2)(s+3)} - \frac{0.1 + j0.3}{(s+2)(s+1-j2)} - \frac{0.1 + j0.3}{(s+2)(s+1+j2)}$$

Further expansion is carried out easily by recalling Eq. (D.1). Hence

$$C(s) = \frac{-0.8}{(s+2)^2} + \left[\frac{1}{s+2} - \frac{1}{s+3}\right]$$
$$- \left[\frac{0.1 + j0.03}{(1 - j2 + 2)(s+2)} + \frac{0.1 + j0.3}{(2 - 1 + j2)(s + 1 - j2)}\right]$$
$$- \left[\frac{0.1 - j0.3}{(1 + j2 - 2)(s+2)} + \frac{0.1 - j0.3}{(2 - 1 - j2)(s + 1 + j2)}\right]$$
$$= \frac{-0.8}{(s+2)^2} + \frac{2.8}{s+2} - \frac{1}{s+3} - \frac{0.14 + j0.02}{s + 1 - j2}$$
$$- \frac{0.14 - j0.02}{s + 1 + j2}$$

EXAMPLE D.2

$$C(s) = \frac{4(s+1)}{(s+2)^3(s+3)(s^2+2s+5)}$$

As in the previous example, let

$$C_1(s) = \frac{4(s+1)}{(s+2)(s+3)(s^2+2s+5)}$$
$$= \frac{-0.8}{s+2} + \frac{1}{s+3} + \frac{-0.1 - j0.3}{s+1-j2} + \frac{-0.1 - j0.3}{s+1+j2}$$

Define

$$C_2(s) = \frac{1}{s+2} C_1(s)$$

Then, as in the previous example,

$$C_2(s) = \frac{-0.8}{(s+2)^2} + \frac{2.8}{s+2} - \frac{1}{s+3} - \frac{0.14 + j0.02}{s+1-j2}$$
$$- \frac{0.14 - j0.02}{s + 1 + j2}$$

Finally,

$$C(s) = \frac{1}{s+2} C_2(s)$$

$$= \frac{-0.8}{(s+2)^3} + \frac{2.8}{(s+2)^2} - \frac{1}{(s+2)(s+3)}$$

$$- \frac{0.14 + j0.02}{(s+2)(s+1-j2)} - \frac{0.14 - j0.02}{(s+2)(s+1+j2)}$$

$$= \frac{-0.8}{(s+2)^3} + \frac{2.8}{(s+2)^2} - \frac{0.928}{s+2} - \frac{1}{s+3} - \frac{0.036 - j0.052}{s+1-j2}$$

$$- \frac{0.036 + j0.052}{s+1+j2}$$

D.2.3 Roots of polynomials

Accurate determination of the roots of a polynomial is known to be a difficult problem. Direct methods for evaluating roots are known only for polynomials of degree not exceeding four. For degrees higher than four it is necessary to employ numerical methods based on search techniques like Newton-Raphson, secant, or similar methods. Let the polynomial be given by

$$P(s) = s^n + a_{n-1}s^{n-1} + \cdots + a_1 s + a_0 = 0 \tag{D.2}$$

If all the coefficients are real, then the n roots of $P(s)$ must be either real or occur in complex conjugate pairs.

The roots of the quadratic are determined using the well-known formula

$$s = \frac{1}{2} a_1 + \left(\frac{1}{4} a_1{}^2 - a_0\right)^{1/2} \tag{D.3}$$

Better accuracy is obtained if, in the case of real roots, the root with the largest absolute value is first evaluated. This is given by

$$s_1 = -\left[\left|\frac{1}{2} a_1\right| + \left(\frac{1}{4} a_1{}^2 - a_0\right)^{1/2}\right] \cdot \operatorname{sgn}(a_1) \tag{D.4}$$

The other real root is then obtained as

$$s_2 = \frac{a_0}{s_1} \tag{D.5}$$

The roots of a cubic can be directly evaluated by using Cardan's method. First we calculate one real root as follows: Let

$$Q = 3a_1 - a_2^{\,2} \tag{D.6}$$

$$R = \frac{1}{2}(9a_1a_2 - 27a_0) - a_2^{\,3} \tag{D.7}$$

If

$$Q^3 + R^2 < 0$$

then

$$s_1 = \frac{1}{3}\{[(R + (Q^3 + R^2)^{1/2}]^{1/3} + [(R - (Q^3 + R^2)^{1/2}]^{1/3} - a_2\} \tag{D.8}$$

If

$$Q^3 + R^2 < 0$$

then

$$s_1 = \frac{1}{3}\left[2(-Q)^{1/2}\cos\left[\frac{1}{3}\cos^{-1}\left(\frac{R}{-Q^{3/2}}\right)\right] - a_2\right] \tag{D.9}$$

The remaining roots are then determined by solving the quadratic obtained after $P(s)$ is divided by $(s - s_1)$.

For a quartic, the roots are obtained by first factoring it into two quadratics; that is,

$$\begin{aligned} P(s) &= s^4 + a_3s^3 + a_2s^2 + a_1s + a_0 \\ &= \{s^2 + (A + C)s + (B + D)\}\{s^2 + (A - C)s + (B - D)\} \end{aligned}$$

where

$$A = \frac{1}{2}a_3,\ B = \frac{1}{2}y_1,\ D = (B^2 - a_0)^{1/2}$$

$$C = \begin{cases} \dfrac{1}{D}\left(AB - \dfrac{1}{2}a_1\right) & \text{if } D \neq 0 \\ (A^2 - a_2 + y_1)^{1/2} & \text{if } D = 0 \end{cases} \tag{D.10}$$

and y_1 is the largest real root of the cubic

$$y^3 + b_2y^2 + b_1y + b_0 = 0 \tag{D.11}$$

with $b_2 = -a_2$, $b_1 = a_1a_3 - 4a_0$, $b_0 = a_0(4a_2 - a_3^2) - a_1^2$.

For the general case, the roots must be determined by search techniques. Real roots of a polynomial can be obtained rather simply by applying Newton-Raphson, secant, or reguli-falsi methods. The calculation of the value of $P(s)$ for a given s is carried out by using synthetic division. Another method, which generally converges faster is Laguerre's iteration method:

$$s_{k+1} = s_k - \frac{nP(s_k)}{P'(s) \pm \sqrt{H(s_k)}} \tag{D.12}$$

where

$$H(s) = (n-1)[(n-1)(P'(s))^2 - nP(s)P''(s)] \tag{D.13}$$

and n is the degree of the polynomial $P(s)$, $P'(s)$ is its first derivative with respect to s, and $P''(s)$ the second derivative. The sign in the denominator of Eq. (D.12) should be chosen such that the difference between s_{k+1} and s_k is as small as possible.

Laguerre's method has the property of global convergence for real roots and very good convergence for complex roots. For the general polynomial described by Eq. (D.2), one may start at $s_0 = 0 + j0$. This gives

$$s_1 = \frac{-na_0}{a_1 \pm \sqrt{H(s_0)}}$$

$$H(s_0) = (n-1)^2 a_1^2 - 2n(n-1)a_0 a_2$$

In particular, for $n = 2$, $H(s_0)$ is the discriminant of $P(s)$ and s_1 is the root with the least modulus, provided that the sign in the denominator of Eq. (D.12) is chosen as discussed in the previous paragraph.

After a root has been found, one may divide $P(s)$ by the factor $(s - s_k)$ if the root is real or by $(s^2 - 2as + a^2 + b^2)$ if the root is complex and equal to $(a + jb)$. The roots of the resulting lower-order polynomial may then be evaluated as before. This is called the process of "deflation." Clearly, deflation saves arithmetic operations. Another advantage of deflation is that the iterations cannot converge to the same root more than once unless it is a multiple root. However, there is some risk of error due to the fact that due to round-off errors, s_k may not be an exact root of $P(s)$. To be safe, one can first determine all the roots using deflation, and then for each root make a few more iterations using the original polynomial $P(s)$.

D.2.4 Searching points on the root locus

The root locus method gives us a set of rules for the approximate construction of the loci when a parameter varies from zero to infinity. To get a good plot we must calculate the location of other points on the loci. This can be done by searching in the s-plane, either along a line with constant real part or along a line with a given damping ratio, for a point where the angle condition is satisfied. The probability of finding the root is increased by starting with a reasonable guess for the location of the root. The program then calculates the value of the angle at that point. If the angle is not equal to 180°, then the original guess is changed by 5 percent and the angle is again evaluated. From these two points, the secant method is used for the next guess. This avoids the need for calculation of derivatives. The process is then continued until the difference between the angle and 180° is less than a specified tolerance. After the root is thus located, the value of the gain for this root is also displayed.

D.2.5 Design of lag, lead, and lag-lead compensators

The design procedure described in Chapter 10 is utilized to calculate the transfer function of a compensator from the specified values of the phase lead (or lag), gain, and the gain crossover frequency. If a compensator of a particular type does not exist, an error message is displayed. Since these designs require only the solution of quadratic equations, the computation is fairly straightforward.

D.2.6 Stability analysis using the Routh criterion

This program displays the Routh table (only the first column for the HP-41C) so that the information on stability can be obtained by examining the number of sign changes in the first column. There is also a provision for determining relative stability by using a transformation that shifts the axis in the s-plane by a desired amount.

It also includes a routine for bilinear transformation to the w-plane in order to determine the stability of discrete-time systems. Relative stability can, again, be determined by shifting the axis in the w-plane.

D.2.7 State transition matrices for discrete-time systems

Consider a linear system described by the state equation

$$\dot{x} = Ax + Bu \tag{D.14}$$

If we assume that the input to the system is allowed to vary only at the instants $t = kT$, where k is an integer and T is a constant, then we get the state transition equation

$$x[(k+1)T] = Fx(kT) + Gu(kT) \tag{D.15}$$

where $F = \exp(AT)$ and G is obtained from the integral of the matrix $\exp(At)$ over the interval 0 to T as discussed in Sec. 3.2.4.

This program allows the calculation of the exponential of a square matrix using the power series expansion. A matrix version of Horner's rule is used for reducing the number of arithmetic operations. The power series can be truncated after a specified number of terms. As long as the sampling interval has been selected to satisfy the criterion that all the eigenvalues of AT have magnitudes less than 0.5, the error in approximation is less than 0.000000001 if only 12 terms are retained. Furthermore, this power series can be integrated term by term to calculate the input matrix, G. The program gives both the matrices F and G for specified A, B, and T.

D.2.8 Transformation to the controllable canonical form

This program first calculates the characteristic polynomial of a single-input, single-output system from its controllability matrix. This then leads to the matrix for transformation to the controllable canonical form as well as the numerator of the transfer function.

D.3 PROGRAMS FOR HP-41C

The programs have divided into modules in order to avoid unnecessary duplication. Another objective was to minimize the number of magnetic cards on which the programs are stored. A brief description of the programs is given in the following subsections and is followed by the listing of the programs.

D.3.1 Program for polynomial inputs

This program, which can be stored on one magnetic card, consists of six subprograms. These can be used for storing two polynomials, viewing them, saving one of them in another set of registers and restoring it. These are required for many other programs, and it is suggested that this program be stored permanently in the calculator.

SUBPROGRAM I1. The coefficients of a polynomial of degree n are stored in registers 15 to $(15 + n)$, with the highest-degree coefficient in register 15. This should normally be used for storing the characteristic polynomial of a system, or the denominator of a transfer function. It will be called polynomial 1 in the future.

SUBPROGRAM V1. This is used for viewing the coefficients of polynomial 1.

SUBPROGRAM I2. The coefficients of a polynomial of degree m are stored in registers $(16 + n)$ to $(16 + n + m)$ by using this subprogram. This should be done only after the coefficients of polynomial 1 have already been stored. The numerator of a transfer function should be stored using this subprogram, and will be called polynomial 2.

SUBPROGRAM V2. This is used for viewing the coefficients of polynomial 2.

SUBPROGRAM SAVE. The coefficients of polynomial 1 are also stored in registers $(17 + m + n)$ to $(17 + m + 2n)$ by using this subprogram. It is convenient when polynomial 1 must be recalled after being changed in some way. This is utilized in the program for calculation of residues as well as for stability analysis using the Routh criterion.

SUBPROGRAM RESTORE. The save coefficients are restored in registers 15 to $(15 + n)$.

If a polynomial is given in the factored form then this program should be used in conjunction with the program for multiplication of polynomials in order to get the coefficients. The program listing is given below.

```
01♦LBL "V1"
02 SF 03
03 GTO 03
04♦LBL "I1"
05 15
06 ENTER↑
07 ENTER↑
08 "N=?"
09 PROMPT
10 STO 01
11 +
12 1 E-3
13 STO 00
14 *
15 +
16 STO 03
17♦LBL 03
18 RCL 03
19 RCL 01
20 GTO 01
21♦LBL "V2"
22 SF 03
23 GTO 04
24♦LBL "I2"
25 "M=?"
26 PROMPT
27 X=0?
28 SF 05
29 X=0?
30 1
31 STO 02
32 1
33 +
34 RCL 00
35 *
36 RCL 01
37 +
38 1
39 +
40 RCL 03
41 +
42 STO 04
43 0
44 FS? 05
45 STO IND Y
46♦LBL 04
47 SF 01
48 RCL 04
49 RCL 02
50 FS? 05
51 ISG Y
52 FS?C 05
53 DSE X
54♦LBL 01
55 CF 29
56 "a"
57 FS? 01
58 "b"
59 FIX 0
60 ARCL X
61 "├="
62 FC? 03
63 GTO 05
64 FIX 4
65 ARCL IND Y
66 AVIEW
67 PSE
68 GTO 06
69♦LBL 05
70 "├?"
71 PROMPT
72 FIX 4
73 STO IND Z
74 RDN
75♦LBL 06
76 1
77 -
78 ISG Y
79 GTO 01
80 CF 01
81 CF 03
82 SF 29
83 RTN
84♦LBL "RE"
85 SF 04
86♦LBL "SAVE"
87 RCL 03
88 RCL 02
89 RCL 01
90 +
91 2
92 +
93 ST+ Y
94 RCL 00
95 *
96 +
97 STO 05
98 RCL 03
99 FS?C 04
100 X<>Y
101♦LBL 02
102 RCL IND X
103 STO IND Z
104 RDN
105 1
106 +
107 ISG Y
108 GTO 02
109 END
```

D.3.2 Program for multiplying and dividing polynomials

This consists of four subprograms that can be used for multiplying or dividing the polynomials 1 or 2 by factors of the form $(s + c_0)$ or $(s^2 + c_1 s + c_0)$. The programs for multiplication are useful for determining the coefficients when a polynomial is given in the factored form. The division programs are useful for calculating residues as well as for deflating a polynomial after a root has been found. These have been called M1 (for multiplication by a linear factor), M2 (for multiplication by a quadratic factor), D1 (for division by a linear factor), and D2 (for division by a quadratic factor). The program is listed below.

```
01♦LBL "M1"
02 XEQ 07
03 RCL IND Y
04 1 E-3
05 ST+ IND 08
06 RDN
07 RCL IND X
08 ISG Y
09 ISG IND 06
10♦LBL 05
11 RCL IND Y
12 X<>Y
13 RCL 14
14 *
15 ST+ IND Z
16 RDN
17 ISG Y
18 GTO 05
19 RCL 14
20 *
21 STO IND Y
22 RTN
23♦LBL "M2"
24 XEQ 07
25 "c1=?"
26 PROMPT
27 STO 13
28 RCL IND Z
29 STO 12
30 2 E-3
31 ST+ IND 08
32 2
33 ST+ IND 06
34 RCL IND 12
35 STO 07
36 RCL 13
37 *
38 ISG 12
39♦LBL 06
40 RCL IND 12
41 STO 08
42 +
43 STO IND 12
44 RCL 07
45 RCL 14
46 *
47 RCL 08
48 STO 07
49 RCL 13
50 *
51 +
52 ISG 12
53 GTO 06
54 STO IND 12
55 RCL 08
56 RCL 14
57 *
58 ISG 12
59 CLA
60 STO IND 12
61 RTN
62♦LBL 07
63 "IN1 OR 2?"
64 AVIEW
65 STOP
66 STO 06
67 2
68 +
69 STO 08
70 "c0=?"
71 PROMPT
72 STO 14
73 RTN
74♦LBL "D1"
75 STO 14
76 RCL 03
77 1 E-3
78 -
79 STO 03
80 0
81 DSE 01
82♦LBL 13
83 ST- IND Y
84 RDN
85 RCL IND X
86 RCL 14
87 *
88 ISG Y
89 GTO 13
90 RTN
91♦LBL "D2"
92 2
93 ST- 01
94 1 E-3
95 *
96 ST- 03
97 RCL 03
98 1 E-3
99 +
100♦LBL 14
101 RCL IND X
102 RCL IND Y
103 RCL 13
104 *
105 ISG Z
106 ST- IND Z
107 RDN
108 RCL 14
109 *
110 ISG Y
111 GTO 02
112 RTN
113♦LBL 02
114 ST- IND Y
115 RDN
116 1
117 -
118 GTO 14
119 .END.
```

D.3.3 Roots of polynomials

This program, labeled "ROOTN," assumes that the polynomial has been stored using the program I1. It also requires that the programs D1 and D2 be stored in the calculator. Laguerre's iteration method is used for polynomials of degree greater than four. The program is listed below.

```
01♦LBL "ROOTN"
02 FS?C 05
03 GTO A
04 XEQ "I1"
05♦LBL A
06 ADV
07 4
08 RCL 01
09 X<=Y?
10 GTO D
11 "SF 4 IF D"
12 PROMPT
13 0
14 STO 09
15 STO 10
16 STOP
17♦LBL B
18 RCL 03
19 XEQ 09
20 STO 06
21 X<>Y
22 STO 05
23 R-P
24 RCL 01
25 ST* 05
26 ST* 06
27 RDN
28 FS? 04
29 TONE 9
30 FS? 04
31 PSE
32 1 E-7
33 X>Y?
34 GTO 01
35 SF 01
36 SF 02
37 RCL 03
38 2 E-3
39 -
40 XEQ 09
41 RCL 05
42 RCL 06
43 XEQ 11
44 STO 08
45 X<>Y
46 STO 07
47 RCL 03
48 1 E-3
```

```
 49 -
 50 SF 01
 51 XEQ 09
 52 STO 04
 53 X<>Y
 54 STO 00
 55 X<>Y
 56 RCL 00
 57 RCL 04
 58 XEQ 11
 59 RCL 01
 60 DSE X
 61 ST* Z
 62 *
 63 RCL 08
 64 -
 65 X<>Y
 66 RCL 07
 67 -
 68 X<>Y
 69 R-P
 70 RCL 01
 71 DSE X
 72 *
 73 SQRT
 74 STO 08
 75 X<>Y
 76 2
 77 /
 78 STO 07
 79 RCL 00
 80 RCL 04
 81 R-P
 82 RDN
 83 -
 84 ABS
 85 90
 86 X<=Y?
 87 SF 00
 88 RCL 07
 89 RCL 08
 90 FS?C 00
 91 CHS
 92 P-R
 93 ST+ 04
 94 X<>Y
 95 ST+ 00
 96 RCL 05
 97 RCL 06
 98 RCL 00
 99 CHS
100 RCL 04
101 XEQ 11
102 RCL 04
103 X↑2
104 RCL 00
105 X↑2
106 +
107 ST/ Z
108 /
109 ST- 09
110 RDN
111 ST- 10
112 GTO B
113+LBL 01
114 RCL 10
115 ABS
116 STO 10
117 1 E-9
118 X<=Y?
119 GTO 03
120 RCL 09
121 XEQ 02
122 CHS
123 XEQ "D1"
124 GTO A
125♦LBL 03
126 RCL 10
127 RCL 09
128 XEQ 00
129 CHS
130 ST+ X
131 STO 13
132 RCL 09
133 X↑2
134 RCL 10
135 X↑2
136 +
137 STO 14
138 XEQ "D2"
139 GTO A
140♦LBL 02
141 FC? 55
142 BEEP
143 FIX 4
144 "ROOT="
145 ARCL X
146 AVIEW
147 RTN
148♦LBL 00
149 FC? 55
150 BEEP
151 "COMPLEX ROOTS"
152 AVIEW
153 SF 13
154 CLA
155 FIX 3
156 ARCL X
157 "⊢ I"
158 ARCL Y
159 AVIEW
160 CF 13
161 RTN
162♦LBL 09
163 1 E-3
164 -
165 STO 02
166 RCL 01
167 STO 00
168 0
169 ENTER↑
170 ENTER↑
171♦LBL 10
172 RCL IND 02
173 FS? 01
174 XEQ 08
175 +
176 RCL 10
177 RCL 09
178 XEQ 11
179 DSE 00
180 CLA
181 ISG 02
182 GTO 10
183 CF 02
184 RCL IND 02
185 FS? 01
186 XEQ 08
187 +
188 CF 01
189 RTN
190♦LBL 08
191 RCL 00
192 *
193 FC? 02
194 RTN
195 DSE 00
196 RCL 00
197 *
198 ISG 00
199 CLA
200 RTN
201♦LBL 11
202 STO 13
203 RDN
204 STO 14
205 RDN
206 STO 11
207 RDN
208 STO 12
209 *
210 RDN
211 *
212 R↑
213 +
214 RCL 11
215 RCL 13
216 *
```

```
217 RCL 12
218 RCL 14
219 *
220 -
221 RTN
222♦LBL D
223 CF 04
224 RCL 03
225 RCL IND X
226 STO 08
227 RDN
228 ISG X
229 RCL 01
230 STO 05
231 DSE X
232♦LBL 06
233 RCL IND Y
234 RCL 08
235 /
236 STO IND Y
237 RDN
238 DSE X
239 CLA
240 ISG Y
241 GTO 06
242 2
243 RCL 05
244 CLD
245 X=Y?
246 GTO E
247 3
248 X=Y?
249 GTO e
250 RCL 01
251 STO 04
252 RCL 03
253 *
254 RCL 00
255 STO 05
256 4
257 *
258 -
259 STO 01
260 RCL 02
261 CHS
262 STO 02
263 CHS
264 4
265 *
266 RCL 03
267 X↑2
268 -
269 RCL 00
270 *
271 RCL 04
272 X↑2
273 -
274 STO 00
275 SF 03
276 XEQ e
277 RCL 14
278 2
279 /
280 STO 11
281 X↑2
282 RCL 05
283 -
284 1 E-7
285 X<>Y
286 X<=Y?
287 0
288 SQRT
289 STO 13
290 RCL 03
291 2
292 /
293 STO 10
294 RCL 11
295 *
296 RCL 04
297 2
298 /
299 -
300 RCL 13
301 X=0?
302 GTO 05
303 /
304 GTO 07
305♦LBL 05
306 RCL 10
307 X↑2
308 RCL 02
309 +
310 RCL 14
311 +
312 SQRT
313♦LBL 07
314 STO 12
315 RCL 10
316 +
317 STO 01
318 RCL 11
319 RCL 13
320 +
321 STO 00
322 XEQ E
323 RCL 11
324 RCL 13
325 -
326 STO 00
327 RCL 10
328 RCL 12
329 -
330 STO 01
331 GTO E
332 LBL e
333 RCL 01
334 3
335 *
336 RCL 02
337 X↑2
338 -
339 STO 06
340 3
341 *
342 RCL 02
343 X↑2
344 +
345 RCL 02
346 *
347 RCL 00
348 27
349 *
350 -
351 2
352 /
353 STO 07
354 X↑2
355 RCL 06
356 3
357 Y↑X
358 STO 14
359 +
360 X<0?
361 GTO 12
362 SQRT
363 STO 08
364 RCL 07
365 +
366 XEQ c
367 RCL 07
368 RCL 08
369 -
370 XEQ c
371 +
372 GTO 13
373♦LBL 12
374 RCL 07
375 RCL 14
376 CHS
377 SQRT
378 /
379 ACOS
380 3
381 /
382 COS
383 RCL 06
384 CHS
```

```
385 SQRT
386 *
387 2
388 *
389♦LBL 13
390 RCL 02
391 -
392 3
393 /
394 STO 14
395 FC? 03
396 XEQ 02
397 RCL 02
398 +
399 ENTER↑
400 ENTER↑
401 RCL 14
402 *
403 RCL 01
404 +
405 STO 00
406 RDN
407 STO 01
408♦LBL E
409 RCL 01
410 2
411 /
412 STO 07
413 X↑2
414 RCL 00
415 -
416 X<0?
417 GTO 14
418 SQRT
419 RCL 07
420 X>0?
421 SF 02
422 ABS
423 +
424 RCL 14
425 ABS
426 X<>Y
427 X>Y?
428 SF 01
429 FS?C 02
430 CHS
431 FS?C 01
432 STO 14
433 FS?C 03
434 RTN
435 XEQ 02
436 RCL 00
437 X<>Y
438 X=0?
439 /
440 GTO 02
441♦LBL 14
442 FS?C 03
443 RTN
444 ABS
445 SQRT
446 RCL 07
447 CHS
448 GTO 00
449♦LBL c
450 X<0?
451 SF 02
452 ABS
453 3
454 1/X
455 Y↑X
456 FS?C 02
457 CHS
458 .END.
```

D.3.4 Frequency response calculations

This program, labeled "FRSP," can be used for calculating the frequency response for a given transfer function. It is assumed that the polynomials have already been stored using I1 and I2 (and M1 and M2, if necessary). The program asks for the value of the delay. It also asks for setting flag 0 if the gain is to be in decibels and flag 4 if the gain of the closed-loop system is desired. The value of the angular frequency is then asked for, and the frequency response is calculated. To calculate the response for some other frequency it is not necessary to go through the entire process again. With the calculator in the user mode, it is only necessary to press the "F" key, and the value of the angular frequency is again requested. The program is listed below.

```
01♦LBL "FRSP"
02 "DELAY=?"
03 PROMPT
04 STO 11
05 "SF0 IF db"
06 PROMPT
07 "SF4 IF CL"
08 PROMPT
09♦LBL F
10 "W=?"
11 AVIEW
12 STOP
13 STO 10
14 VIEW X
15 RCL 04
16 XEQ 02
17 STO 07
18 RDN
19 STO 08
20 RCL 03
21 XEQ 02
22 ST/ 07
23 RDN
24 ST- 08
25 RCL 08
26 RCL 11
27 RCL 10
28 *
29 R-D
30 -
31 FS? 04
32 GTO 00
33 90
34 X<>Y
35 X>Y?
36 XEQ 04
37 FIX 3
38 "PHI="
39 XEQ D
40 RCL 07
41 FS? 00
42 GTO 08
43 "M="
44 GTO D
45♦LBL 08
46 LOG
47 20
48 *
49 SF 06
50 FIX 2
51 "GAIN="
52 GTO D
53♦LBL 04
54 360
55 -
56 RTN
```

```
57♦LBL 00
58 RCL 07
59 P-R
60 1
61 +
62 R-P
63 RCL 07
64 X<>Y
65 /
66 GTO 08
67♦LBL 02
68 1 E-3
69 -
70 0
71 ENTER↑
72 0
73♦LBL 03
74 RCL IND Z
75 +
76 X<>Y
77 RCL 10
78 ST* Z
79 ST* Y
80 RDN
81 CHS
82 ISG Z
83 GTO 03
84 RCL IND Z
85 +
86 R-P
87 RTN
88♦LBL "COMP"
89 "SF1 IF LL"
90 PROMPT
91 "PHASE=?"
92 PROMPT
93 STO 05
94 2
95 FS? 01
96 ST+ 05
97 "Wc=?"
98 PROMPT
99 STO 10
100 RCL 05
101 TAN
102 X↑2
103 STO 05
104 "GAIN=?"
105 PROMPT
106 10
107 /
108 10↑X
109 STO 06
110♦LBL 13
111 1
112 -
113 STO 07
114 RCL 05
115 RCL 06
116 *
117 STO 00
118 +
119 RCL 06
120 *
121 STO 12
122 RCL 05
123 RCL 07
124 -
125 FS? 01
126 X<> 12
127 ST/ 12
128 ST/ 00
129 RCL 00
130 X↑2
131 RCL 12
132 -
133 SQRT
134 RCL 00
135 -
136 STO 09
137 X<0?
138 SQRT
139 FS? 01
140 GTO 11
141 "a="
142 XEQ D
143 X↑2
144 RCL 06
145 -
146 RCL 07
147 /
148 SQRT
149 RCL 10
150 *
151 XEQ 04
152 RCL 09
153 /
154 CHS
155 SF 02
156 GTO 04
157♦LBL 11
158 X↑2
159 RCL 06
160 *
161 1
162 -
163 RCL 07
164 CHS
165 X<>Y
166 /
167 SQRT
168 RCL 10
169 *
170 STO 14
171 RCL 09
172 1
173 -
174 5
175 TAN
176 ST+ X
177 /
178 ENTER↑
179 X↑2
180 RCL 09
181 -
182 SQRT
183 -
184 RCL 10
185 *
186 STO 13
187 RCL 09
188 /
189 STO 00
190 RCL 14
191 RCL 09
192 *
193 STO 08
194 FS?C 04
195 GTO 12
196 RCL 14
197 RCL 13
198 +
199 RCL 10
200 *
201 RCL 14
202 RCL 13
203 *
204 RCL 10
205 X↑2
206 -
207 STO 12
208 R-P
209 RCL 00
210 RCL 08
211 +
212 RCL 10
213 *
214 RCL 12
215 R-P
216 ST/ Z
217 RCL 06
218 RCL T
219 /
220 X↑2
221 STO 06
222 SF 04
223 GTO 13
224♦LBL 12
225 CF 01
226 BEEP
227 XEQ 04
228 RCL 00
229 XEQ 04
230 RCL 14
231 SF 02
232 XEQ 04
233 RCL 13
234 SF 02
235♦LBL 04
236 CHS
237 "POLE="
238 FS?C 02
239 "ZERO="
240♦LBL D
241 ARCL X
242 FS?C 06
243 "├db"
244 AVIEW
245 PSE
246 PSE
247 END
```

D.3.5 Design of compensators

This program, labeled as "COMP," can be used for designing lead, lag, and lag-lead compensators using frequency response criteria. It has been listed with the program "FRSP."

D.3.6 Stability of dynamic systems

This program uses the Routh criterion for determining the stability of dynamic systems. It is expected that the coefficients of the characteristic polynomial have been stored using program I1. In addition to the Routh test, there are two other subprograms. The first, called "SA," can be used for shifting the axis in the *s*-plane by a specified amount and is required for determining relative stability from the Routh test. The shift in the axis is to the right for a positive input and to the left for a negative input. The coefficients of the new polynomial are displayed and also stored in place of polynomial 1. In case it is desired to return to the original polynomial, the subprogram "SAVE" should be used for saving it before either using the Routh test or shifting the axis. It can then be recovered through the subprogram "RESTORE." The other subprogram, labeled *Z-W*, performs the bilinear transformation necessary to apply the Routh test to a discrete-time system.

The program is listed below.

```
01♦LBL "ROUTH"
02 RCL 03
03 STO 06
04♦LBL 03
05 1 E-50
06 STO 07
07 RDN
08 RCL IND X
09 STO 09
10 RDN
11 ISG X
12 GTO 06
13 GTO 09
14♦LBL 06
15 RCL IND X
16 X=0?
17 X<> 07
18 STO 10
19 STO IND Y
20 RDN
21 ISG X
22♦LBL 04
23 RCL IND X
24 ISG Y
25 GTO 05
26 DSE Y
27 STO IND Y
28♦LBL 07
29 RCL 06
30 1
31 +
32 STO 06
33 GTO 03
34♦LBL 05
35 RCL IND Y
36 RCL 09
37 *
38 RCL 10
39 /
40 -
41 DSE Y
42 CLA
43 STO IND Y
44 RDN
45 1
46 +
47 ISG X
48 GTO 04
49 GTO 07
50♦LBL "Z-W"
51 RCL 01
52 RCL 00
53 *
54 1
55 +
56 STO 07
57♦LBL 10
58 RCL 03
59 STO 06
60 RCL 07
61 INT
62 ST+ 06
63 RCL IND 06
64 STO 10
65 RDN
66 ST- 06
67 RCL 00
68 *
69 STO 08
70 RDN
71 0
72♦LBL 11
73 RCL 07
74 INT
75 FACT
76 LASTX
77 RCL 08
78 INT
79 -
80 LASTX
81 FACT
82 X<>Y
83 X=0?
84 SF 01
85 FACT
86 *
87 /
88 -1
89 RCL 08
90 INT
91 Y↑X
92 *
93 RCL 10
94 *
95 +
96 RCL IND 06
97 X<>Y
98 FS? 01
99 STO IND 06
100 FC?C 01
101 ST+ IND 06
102 RDN
103 ISG 06
104 CLA
105 ISG 08
106 GTO 11
107 ISG 07
108 GTO 10
109♦LBL 09
110 BEEP
111 GTO "V1"
112♦LBL "SA"
113 "a=?"
114 PROMPT
115 STO 10
116 RCL 01
117 1
118 +
119 STO 07
120 RCL 03
121 STO 06
122♦LBL 08
123 RCL IND X
124 RCL 10
125 *
126 ISG Y
127 GTO 01
128 RCL 06
129 RCL 00
130 -
131 STO 06
132 DSE 07
133 GTO 08
134 GTO 09
135♦LBL 01
136 ST+ IND Y
137 RDN
138 GTO 08
139 END
```

D.3.7 Root locus calculations

This program assumes that the coefficients of the polynomials in the transfer function have been stored using the programs I1 and I2. The root can be searched in the *s*-plane either along a line with constant real part or along a line with constant damping ratio. The program is listed below.

```
01♦LBL "RLOC"
02 "SF 0 IF∠"
03 PROMPT
04 SF 04
05 FS? 00
06 XEQ 08
07 FC? 00
08 XEQ 10
09♦LBL 05
10 XEQ 06
11 ABS
12 VIEW X
13 180
14 -
15 STO 11
16 ABS
17 1 E-4
18 X>Y?
19 GTO 02
20 FS? 00
21 XEQ 03
22 FC?C 04
23 GTO 04
24 RCL 11
25 STO 06
26 D-R
27 1
28 +
29 RCL 10
30 STO 05
31 *
32 GTO 00
33♦LBL 04
34 RCL 05
35 RCL 11
36 *
37 RCL 10
38 STO 05
39 RCL 06
40 *
41 -
42 RCL 06
43 RCL 11
44 STO 06
45 -
46 CHS
47 /
48♦LBL 00
49 STO 10
50 FS? 00
51 XEQ 09
52 GTO 05
53♦LBL 02
54 CF 00
55 RCL 09
56 "SIGMA="
57 BEEP
58 XEQ 20
59 RCL 10
60 "W="
61 XEQ 20
62 RCL 07
63 1/X
64 "K="
65♦LBL 20
66 ARCL X
67 AVIEW
68 PSE
69 PSE
70 RTN
71♦LBL 03
72 RCL 10
73 RCL 09
74 CHS
75 R-P
76 STO 10
77 X<>Y
78 COS
79 STO 09
80 RTN
81♦LBL 08
82 "ZETA=?"
83 PROMPT
84 STO 09
85 "WN=?"
86 PROMPT
87 STO 10
88♦LBL 09
89 RCL 09
90 RCL 10
91 *
92 CHS
93 STO 09
94 X↑2
95 RCL 10
96 X↑2
97 -
98 CHS
99 SQRT
100 STO 10
101 RTN
102♦LBL 10
103 "SIGMA=?"
104 PROMPT
105 STO 09
106 X 2
107 "W=?"
108 PROMPT
109 STO 10
110 RTN
111♦LBL 06
112 RCL 04
113 XEQ 11
114 STO 07
115 RDN
116 STO 08
117 RCL 03
118 XEQ 11
119 ST/ 07
120 RDN
121 RCL 08
122 -
123 STO 08
124♦LBL 13
125 X>0?
126 XEQ 01
127 RTN
128♦LBL 01
129 360
130 -
131 GTO 13
132♦LBL 11
133 1 E-3
134 -
135 STO 11
136 0
137 ENTER↑
138 0
139♦LBL 12
140 RCL IND 11
141 +
142 STO 12
143 RCL 10
144 *
145 X<>Y
146 STO 13
147 RCL 09
148 *
149 +
150 RCL 12
151 RCL 09
152 *
153 RCL 10
154 RCL 13
155 *
156 -
157 ISG 11
158 GTO 12
159 RCL IND 11
160 +
161 FS? 03
162 RTN
163 R-P
164 RTN
165♦LBL "RES"
166 XEQ "SAVE"
167 "SF3 IF REAL"
168 PROMPT
169 FC?C 03
170 GTO C
171 "POLE=?"
172 PROMPT
173 STO 09
174 CHS
175 XEQ "D1"
176 0
177 STO 10
178 XEQ 06
179 1
180 ST+ 01
181 RCL 00
182 ST+ 03
183 XEQ "RE"
184 RCL 08
185 RCL 07
186 P-R
187 "RES="
188 GTO 20
189♦LBL C
190 XEQ 10
191 X 2
192 +
```

```
193 STO 14       207 90          221 XEQ 20       235 STO 09
194 RCL 09       208 RCL 10      222 CLA          236 RDN
195 CHS          209 SIGN        223 X<>Y         237 STO 10
196 ST+ X        210 *           224 "PHI="       238 XEQ 06
197 STO 13       211 +           225 GTO 20       239 CHS
198 XEQ "D2"     212 CHS         226♦LBL "DF"     240 XEQ 13
199 XEQ 06       213 RCL 07      227 "W=?"        241 RCL 07
200 2            214 RCL 10      228 PROMPT       242 LOG
201 ST+ 01       215 ST+ X       229 "T=?"        243 20
202 RCL 00       216 /           230 PROMPT       244 *
203 *            217 ABS         231 *            245 GTO 14
204 ST+ 03       218♦LBL 14      232 R-D          246 .END.
205 XEQ "RE"     219 "M="        233 1
206 RCL 08       220 BEEP        234 P-R
```

D.3.8 Residue calculations

This program, labeled "RES," can be utilized for calculating the residues at real or complex poles of rational functions of *s*. Only the case of distinct roots has been considered. For multiple poles one may use the procedure discussed in Sec. D.2.2. This program has many subroutines common with the program for root locus. Hence, the two have been listed together.

D.4 PROGRAMS IN TURBO-PASCAL

The listings for the various programs mentioned in Sec. D.2 are given below. Although these programs were written for use on the IBM Personal Computer, they will also work on an Apple *II*, *IIe*, or *III* computer that has a Turbo-Pascal compiler. Since Turbo-Pascal on an IBM Personal Computer (or compatible) works with the popular DOS operating system, it is particularly convenient. Moreover, it is rather inexpensive, and occupies less than 40 kilobytes of memory. All these programs have been thoroughly tested on an IBM Personal Computer as well as on a Texas Instruments Professional Computer.

```
PROGRAM FreqResponse;
(*$U+ *)
CONST  PI = 3.14159265358979323846264 3;
       space = ' ';
TYPE   COEFFICIENT = ARRAY [0..25] OF REAL;
       String14 = STRING[14];
VAR
  omega, im1, re1, im2, re2, gain, phase, step, delay,
                                 closegain, closephase : REAL;
  i, j, k, number, Field, Decimal  :  INTEGER;
  degreeN, degreeD  :  0..24;
  P,Q               : COEFFICIENT;
  Outfile  : TEXT;
  response, answer : CHAR;
  decibel, zero, flag  : BOOLEAN;
  FileName : String14;

(*$I Files.Pas *)
```

```
PROCEDURE Greeting;
BEGIN
  ClrScr;
  WRITELN;
  WRITELN;
  WRITELN;
  WRITELN;
  WRITELN ('This program calculates the frequency response of a linear');
  WRITELN ('system.  The transfer function can be either the ratio of ');
  WRITELN ('two polynomials or may be in the factored form, with a ');
  WRITELN ('constant term in the numerator.  Both may include a delay ');
  WRITELN ('term of the form exp (-sT).  The closed-loop gain and phase ');
  WRITELN ('shift for unity feedback can also be calculated. ');
  WRITELN;
  WRITELN ('Please press the <RETURN> key when ready. ');
  READLN;
END;

PROCEDURE Echo (VAR answer : CHAR);
BEGIN
  READ (KBD, answer);
  RESET (KBD);
  WRITELN (answer)
END;

PROCEDURE RequestToPrint;
VAR on : BOOLEAN;
BEGIN
  on := FileExists (OutFile, 'LST:');
  IF NOT on THEN
    WRITELN ('Please turn on the printer. ');
  WHILE NOT on DO
    on := FileExists (OutFile, 'LST:')
END;

PROCEDURE PrintOut;
VAR i : INTEGER;
BEGIN
  RequestToPrint;
  ASSIGN (Outfile, 'LST:');
  RESET (Outfile);
  WRITELN (OutFile, 'The numerator coefficients are ');
  FOR i := degreeN DOWNTO 0 DO
    WRITE (Outfile, 'A[', i, '] = ', P[i]:6:2, ',  ');
  WRITELN (Outfile);
  WRITELN (OutFile, 'The denominator coeficients are ');
  FOR i := degreeD DOWNTO 0 DO
    WRITE (Outfile, 'B[', i, '] = ', Q[i]:6:2, ',  ');
  WRITELN (Outfile);
  WRITE (Outfile, 'The transport delay is ');
  WRITELN (Outfile, delay:6:3, ' seconds.');
  WRITELN (Outfile);
  CLOSE (Outfile)
END;

PROCEDURE Verify;
VAR
  i, j   : INTEGER;
  answer : CHAR;
BEGIN
  ClrScr;
  REPEAT
    FOR i := degreeN DOWNTO 0 DO
           WRITE ('A[', i, '] = ', P[i]:6:2, ',  ');
```

```
    WRITELN;
    WRITE ('Are these entries correct?  (Y or N):  ');
    Echo (answer);
    IF answer IN ['N', 'n'] THEN
      BEGIN
        WRITE ('Which entry is wrong?   ');
        READLN (j);
        WRITE ('Enter correct value of A[', j,'] ');
        READLN (P[j]);
      END
  UNTIL answer IN ['Y', 'y'];
  REPEAT
    FOR i := degreeD DOWNTO 0 DO
            WRITE ('B[', i, '] = ', Q[i]:6:2, ',   ');
    WRITELN;
    WRITE ('Are these entries correct?  (Y or N):  ');
    Echo (answer);
    IF answer IN ['N', 'n'] THEN
      BEGIN
        WRITE ('Which entry is wrong?   ');
        READLN (j);
        WRITE ('Enter correct value of B[', j,'] ');
        READLN (Q[j]);
      END
  UNTIL answer IN ['Y', 'y'];
  WRITELN;
  REPEAT
    WRITELN ('The transport delay is ', delay:6:3, ' seconds.');
    WRITE ('Is this correct?  (Y or N):  ');
    Echo (answer);
    IF answer IN ['N', 'n'] THEN
    BEGIN
      WRITE ('What is the correct value of the transport delay? ');
      READLN (delay);
    END
  UNTIL answer IN ['Y', 'y'];
END;

PROCEDURE Polynomial;
VAR i : INTEGER;
BEGIN
  WRITELN ('Do you want to enter the coefficients from the keyboard');
  WRITE ('or write them from a diskfile? Enter  K or F ');
  Echo (response);
  IF response IN ['F', 'f'] THEN
    BEGIN
      WRITELN ('Entering the numerator  ');
      IF GetFile (OutFile) THEN
        BEGIN
          RESET (OutFile);
          READLN (OutFile, degreeN);
          FOR i := degreeN DOWNTO 0 DO
            READLN (OutFile, P[i]);
          CLOSE (OutFile)
        END
     ELSE response := 'K';
     WRITELN ('Entering the denominator  ');
       IF GetFile (OutFile) THEN
         BEGIN
           RESET (OutFile);
           READLN (OutFile, degreeD);
           FOR i := degreeD DOWNTO 0 DO
             READLN (OutFile, Q[i]);
           CLOSE (OutFile)
         END
```

```
      ELSE response := 'K';
    END;
  IF NOT (response IN ['F', 'f']) THEN
    BEGIN
      WRITE ('Enter degree of numerator polynomial:  ');
      READLN (degreeN);
      FOR i := degreeN DOWNTO 0 DO
        BEGIN
          WRITE ('A[', i, '] = ');
          READLN (P[i]);
        END;
      WRITE ('Enter degree of denominator polynomial:  ');
      READLN (degreeD);
      FOR i := degreeD DOWNTO 0 DO
        BEGIN
          WRITE ('B[', i, '] = ');
          READLN (Q[i]);
        END;
    END;
  WRITE ('Enter the value of transport delay:');
  READLN (delay);
  Verify
END;


PROCEDURE Factors;
VAR    M, N          : INTEGER;
       LIN           : ARRAY [0..25] OF REAL;
       QUAD1, QUAD2  : ARRAY [1..13] OF REAL;
PROCEDURE Linear (M : INTEGER; VAR R : COEFFICIENT);
BEGIN
  zero := FALSE;
  IF M <> 0 THEN  FOR i := 1 TO M DO
  BEGIN
    WRITE ('Enter the constant term in linear factor # ', i,':  ');
    READLN (LIN [i]);
    IF LIN [i] = 0 THEN zero := TRUE;
    R[i] := R[i-1];
    FOR j := i-1 DOWNTO 1 DO
      R[j] := R[j] * LIN [i] + R[j-1];
    R[0] := R[0] * LIN [i];
  END;
END;


PROCEDURE Quadratic (N, k : INTEGER; VAR R : COEFFICIENT);
BEGIN
  IF N <> 0 THEN FOR i := 1 TO N DO
  BEGIN
    WRITE ('Enter the constant term in quadratic factor # ', i,': ');
    READLN (QUAD1 [i]);
    WRITE ('Enter the coefficient of s in quadratic factor # ', i,': ');
    READLN (QUAD2 [i]);
    R[k+2] := R[k];
    IF k = 0 THEN R[k+1] := R[k] * QUAD2 [i]
      ELSE  BEGIN
          R[k+1] := R[k] * QUAD2 [i] + R[k-1];
          FOR j := k DOWNTO 2 DO
              R[j] := R[j] * QUAD1[i] + R[j-1] * QUAD2[i] + R[j-2];
          R[1] := R[1] * QUAD1[i] + R[0] * QUAD2[i];
        END;
    R[0] := R[0] * QUAD1[i];
    k := k + 2;
  END;
END;
```

```
BEGIN  (*Factors*)
  REPEAT
    WRITE ('Enter the constant term in the numerator:  ');
    READLN (LIN[0]);
    P[0] := LIN [0];
    WRITE ('How many linear factors are in the numerator?  ');
    READLN (M);
    degreeN := M;
    Linear (M, P);
    k := degreeN;
    WRITE ('How many quadratic factors are in the numerator?  ');
    READLN (N);
    Quadratic (N, k, P);
    degreeN := degreeN + N * 2;
    WRITELN ('The numerator has the following form : ');
    WRITE (LIN[0]:6:3);
    FOR i := 1 to M DO
      BEGIN
        IF NOT zero THEN WRITE (' (s + ', LIN[i]:6:3, ') ')
            ELSE WRITE  (' (s) ');
        zero := FALSE;
      END;
    FOR i := 1 to N DO
        WRITE (' (s^2 + ', QUAD2[i]:6:3, ' s + ', QUAD1[i]:6:3, ') ');
    WRITELN;
    WRITELN;
    WRITE ('Is this correct?  (Y or N) :  ');
    Echo (response)
  UNTIL response IN ['Y', 'y'];
  REPEAT
    Q[0] := 1;
    WRITE ('How many linear factors are in the denominator? ');
    READLN (M);
    degreeD := M;
    Linear (M, Q);
    k := degreeD;
    WRITE ('How many quadratic factors are in the denominator?  ');
    READLN (N);
    Quadratic (N, k, Q);
    degreeD := degreeD + N * 2;
    WRITELN ('The denominator has the following form : ');
    FOR i := 1 to M DO
      BEGIN
        IF NOT zero THEN WRITE (' (s + ', LIN[i]:6:3, ') ')
            ELSE WRITE  (' (s) ');
        zero := FALSE;
      END;
    FOR i := 1 to N DO
        WRITE (' (s^2 + ', QUAD2[i]:6:3, ' s + ', QUAD1[i]:6:3, ') ');
    WRITELN;
    WRITELN;
    WRITE ('Is this correct?  (Y or N) : ' );
    Echo (response)
  UNTIL response IN ['Y', 'y'];
  WRITE ('Enter the value of transport delay:  ');
  READLN (delay);
END;

PROCEDURE Input;
BEGIN
  ClrScr;
  WRITE ('Are the polynomials in the factored form?  (Y or N)  ');
  Echo (answer);
  IF answer IN ['N', 'n'] THEN Polynomial
    ELSE Factors;
  IF degreeN = 0 THEN
```

```
    BEGIN
      degreeN := 1;
      P[1] := 0;
    END;
  WRITE ('Do you want the coefficients printed?  (Y or N)  ');
  Echo (answer);
  IF answer IN ['Y', 'y'] THEN  PrintOut;
  WRITE ('Do you want to store the denominator coefficients on disk? ');
  WRITE ('  (Y or N)    ');
  Echo (answer);
  IF answer IN ['Y', 'y'] THEN
    BEGIN
      CreateFile (OutFile, FileName);
      REWRITE (OutFile);
      WRITELN (OutFile, degreeD);
      FOR i := degreeD DOWNTO 0 DO
        WRITELN (OutFile, Q[i]:Field:decimal);
      CLOSE (OutFile)
    END;
  WRITE ('Do you want to store the numerator coefficients on disk? ');
  WRITE ('  (Y or N)    ');
  Echo (answer);
  IF answer IN ['Y', 'y'] THEN
    BEGIN
      CreateFile (OutFile, FileName);
      REWRITE (OutFile);
      WRITELN (OutFile, degreeN);
      FOR i := degreeN DOWNTO 0 DO
        WRITELN (OutFile, P[i]:Field:decimal);
      CLOSE (OutFile)
    END;
  WRITE ('Do you want the gain in decibels?  (Y or N) ');
  Echo (response);
  IF response IN ['Y', 'y']  THEN decibel := TRUE
     ELSE  decibel := FALSE
END {Input};

PROCEDURE GetFrequencies;
  BEGIN
    WRITE ('Enter the lowest value of omega:  ');
    READLN (omega);
    WRITE ('Enter frequency-step:  ');
    READLN (step);
    WRITE ('Enter the number of steps desired:  ');
    READLN (number);
    WRITELN;
  END;

FUNCTION Sgn (X : REAL) : INTEGER;
BEGIN
  IF X >= 0 THEN Sgn := 1
     ELSE Sgn := -1
END;

FUNCTION arcTAN2 (X, Y : REAL) : REAL;
BEGIN
  IF (X = 0) AND (Y <> 0) THEN arcTAN2 := (PI/2) * Sgn (Y);
  IF X < 0 THEN  arcTAN2 := PI + arcTAN (Y/X);
  If X > 0 THEN  arcTAN2 := arcTAN (Y/X);
END;

PROCEDURE SynDiv (divid : COEFFICIENT;
                  degree : INTEGER;
                  omega : REAL;
                  VAR   re, im : REAL);
```

```
VAR  i : 1..24;
BEGIN
  FOR i := degree DOWNTO 2 DO
    divid[i-2] := divid[i-2] -SQR(omega) * divid[i];
  re := divid[0];
  im := divid[1] * omega;
END;  {SynDiv}

PROCEDURE Calculate;
VAR gain1, phase1, temp1, temp2 : REAL;
BEGIN
  SynDiv (P, degreeN, omega, re1, im1);
  SynDiv (Q, degreeD, omega, re2, im2);
  gain:= SQRT ((re1*re1 + im1*im1)/(re2*re2 + im2*im2));
  gain1 := gain;  (* Store gain as gain1 *)
  IF decibel THEN gain:= 20 * (Ln (gain) / Ln (10));
  phase1 := arcTAN2(re1,im1) - arcTAN2(re2,im2)- omega * delay;
  (* phase shift in radians *)
  phase := (180/PI) * phase1;   (* phase shift in degrees *)
  IF phase > 80 THEN
   phase := phase - 360;
  IF ABS (phase) > 360 THEN
   phase := phase - sgn(phase) * 360;
  temp1 := gain1 * cos (phase1);
  temp2 := gain1 * sin (phase1);
  closegain := gain1 / SQRT (1 + SQR(gain1) + 2 * temp1);
  IF decibel THEN
   closegain := 20 * Ln (closegain) / Ln (10);
  closephase := phase - (180/PI) * arcTan2 ((1 + temp1), temp2);
  IF closephase > 80 THEN
      closephase := closephase - 360;
  IF ABS(closephase) > 360 THEN
       closephase := closephase - sgn(closephase) * 360;
END;

PROCEDURE Output;
VAR  response : CHAR;
  BEGIN
    WRITE ('Do you want a printout of the ');
    WRITE ('frequency response?  (Y or N)  ');
    Echo (response);
    IF response IN ['Y','y'] THEN
      BEGIN
        RequestToPrint;
        ASSIGN (Outfile, 'LST:');
      END
    ELSE
      ASSIGN (Outfile, 'CON:');
    RESET (Outfile);
    flag := TRUE;
    ClrScr;
    WRITELN (Outfile);
    REPEAT
      GetFrequencies;
      IF flag THEN
        BEGIN
          IF decibel THEN
            WRITELN (Outfile, space:7, 'omega', space:5, 'gain(db)',
                space:3, 'phase (Deg)')
          ELSE WRITELN (Outfile, space:7, 'omega', space:7, 'gain',
                space:5, 'phase(Deg)');
          WRITELN (Outfile)
        END;
      flag := FALSE;
      FOR i := 1 TO number DO
```

```
        BEGIN
          Calculate;
          WRITELN (Outfile, space:5, omega:7:3, space:5,
                            gain:7:3, space:5, phase:7:2);
          omega := omega + step
        END;
      WRITE ('Do you want calculations for other frequencies?',
                                              ' (Y or N):  ');

      Echo (response)
    UNTIL response IN ['N','n'];
    flag := TRUE;
    WRITELN (Outfile);
    WRITE ('Do you want the closed-loop frequency response? (Y or N) ');
    Echo (response);
    IF response IN ['Y', 'y']  THEN
      BEGIN
        REPEAT
          GetFrequencies;
          IF flag THEN
            BEGIN
              IF decibel THEN
                  WRITELN (Outfile, space:7, 'omega', space:5, 'closed-',
                'loop gain(db)', space:3, 'closed-loop phase (Deg)')
                ELSE WRITELN (Outfile, space:7, 'omega', space:5,
                'closed-loop gain', space:3, 'closed-loop phase (Deg)');
                WRITELN (Outfile);
            END;
          flag := FALSE;
          FOR i := 1 TO number DO
           BEGIN
             Calculate;
             WRITELN (Outfile, space:5, omega:7:3, space:8,
                      closegain:7:3,  space:12,  closephase:7:2);
            omega := omega + step
          END;
        WRITE ('Do you want calcuations for other frequencies?',
                                         ' (Y or N): ');

        Echo (response)
      UNTIL response IN ['N', 'n'];
     END;
  WRITELN (Outfile);
  CLOSE (Outfile);
END;

BEGIN {MAIN PROGRAM}
  Greeting;
  Input;
  Output;
END.


PROGRAM Residue;
CONST  PI = 3.14159265358979323846264З; Field = 12; decimal = 6;
TYPE   COEFFICIENT = ARRAY [0..25] OF REAL;
       String14 = STRING[14];
VAR
  sigma, omega, ResReal, ResImag : REAL;
  i, j, k      :  INTEGER;
  degreeN, degreeD, degreeR  :   0..24;
  Numerator, Denominator, R :  COEFFICIENT;
  response, answer : CHAR;
  zero : BOOLEAN;
  OutFile : TEXT;
  FileName : String14;
```

```
  (*$I FIles.Pas *)
  (*$U+ *)
PROCEDURE Echo (VAR answer : CHAR);
BEGIN
  READ (KBD, answer);
  RESET (KBD);
  WRITELN (answer)
END;

PROCEDURE GeneralInstructions;
BEGIN
  ClrScr;
  WRITELN;
  WRITELN;
  WRITELN;
  WRITELN;
  WRITELN ('     This program calculates the residues at the poles of');
  WRITELN ('a given rational function of s for obtaining its inverse');
  WRITELN ('Laplace transform.  This  is  done  by  evaluating  the');
  WRITELN ('function at the pole after removing from the denominator');
  WRITELN ('a factor  containing  it.  Real as well as complex poles');
  WRITELN ('can be handled, but they  must  all be simple.  The case');
  WRITELN ('of multiple poles can be  solved by first obtaining  the');
  WRITELN ('partial  fraction  expansion with only one factor of the');
  WRITELN ('repeated root. The remaining  factors may then be added');
  WRITELN ('one by one and the partial fraction expansion  obtained.');
  WRITELN ('It is assumed that  either  the location of the poles is');
  WRITELN ('known or the denominator is given in the factored form.');
  WRITELN;
  WRITELN ('    Please press any key  when ready. ');
  Echo (answer)
END;

PROCEDURE RequestToPrint;
VAR on : BOOLEAN;
BEGIN
  on := FileExists (OutFile, 'LST:');
  IF NOT on THEN
    WRITELN ('Please turn on the printer.  ');
    WHILE NOT on DO
      on := FileExists (OutFile, 'LST:')
END;

PROCEDURE PrintOut;
VAR i : INTEGER;
BEGIN
  RequestToPRint;
  ASSIGN (OutFile, 'LST:');
  REWRITE (OutFIle);
  WRITELN (OutFile, 'The numerator coefficients are  ');
  FOR i := degreeN DOWNTO 0 DO
    WRITELN (OutFile, 'A[', i, ']', Numerator[i]:Field:decimal);
  WRITELN (OutFile);
  WRITELN (OutFile, 'The denominator coefficients are ');
  FOR i := degreeD DOWNTO 0 DO
    WRITELN (OutFile, 'B[', i, ']', Denominator[i]:Field:decimal);
  WRITELN (OutFile);
  CLOSE (OutFile)
END;

PROCEDURE Verify;
(* It verifies the input polynomials *)
VAR
  i, j   : INTEGER;
  answer : CHAR;
```

```
BEGIN
  ClrScr;
  REPEAT
    WRITELN ('The coefficients of the numerator are ... ');
    FOR i := degreeN DOWNTO 0 DO
        WRITELN ('A[', i, '] = ', Numerator[i]:Field:decimal, ',    ');
    WRITELN;
    WRITE ('Are these entries correct?  (Y or N): ');
    Echo (answer);
    IF answer IN ['N', 'n'] THEN
      BEGIN
        WRITE ('Which entry is wrong?   ');
        READLN (j);
        WRITE ('Enter correct value of A[', j,'] ');
        READLN (Numerator[j]);
      END
  UNTIL answer IN ['Y', 'y'];
  REPEAT
    WRITELN ('The coefficients of the denominator are ... ');
    FOR i := degreeD DOWNTO 0 DO
       WRITELN ('B[', i, '] = ', Denominator[i]:Field:decimal, ',    ');
    WRITELN;
    WRITE ('Are these entries correct?  (Y or N): ');
    Echo (answer);
    IF answer IN ['N', 'n'] THEN
      BEGIN
        WRITE ('Which entry is wrong?   ');
        READLN (j);
        WRITE ('Enter correct value of B[', j,'] ');
        READLN (Denominator[j]);
      END
  UNTIL answer IN ['Y', 'y'];
  WRITELN;
END;

PROCEDURE Polynomial;
(* It is used for the input of polynomials if they are not factored. *)
VAR i : INTEGER;
BEGIN
  WRITELN ('Do you want to enter the coefficients from the keyboard');
  WRITE ('or write them from a diskfile?  Enter  K or F  ');
  Echo (response);
  IF response IN ['F', 'f'] THEN
    BEGIN
      WRITELN ('Entering the numerator    ');
      IF GetFile (OutFile) THEN
        BEGIN
          RESET (OutFile);
          READLN (OutFile, degreeN);
          FOR i := degreeN DOWNTO 0 DO
            READLN (OutFile, Numerator[i]);
          CLOSE (OutFile)
        END
      ELSE response := 'K';
    WRITELN ('Entering the denominator   ');
      IF GetFile (OutFile) THEN
        BEGIN
          RESET (OutFile);
          READLN (OutFile, degreeD);
          FOR i := degreeD DOWNTO 0 DO
            READLN (OutFile, Denominator[i]);
          CLOSE (OutFile)
        END
      ELSE response := 'K';
    END;
```

```
    IF NOT (response IN ['F', 'f']) THEN
      BEGIN
        WRITE ('Enter degree of numerator polynomial:  ');
        READLN (degreeN);
        FOR i := degreeN DOWNTO 0 DO
          BEGIN
            WRITE ('A[', i, '] = ');
            READLN (Numerator[i]);
          END;
        WRITE ('Enter degree of denominator polynomial:  ');
        READLN (degreeD);
        FOR i := degreeD DOWNTO 0 DO
          BEGIN
            WRITE ('B[', i, '] = ');
            READLN (Denominator[i]);
          END;
      END;
    Verify
  END;

  PROCEDURE Factors;
  VAR    M, N            : INTEGER;
         LIN             : ARRAY [0..25] OF REAL;
         QUAD1, QUAD2    : ARRAY [1..13] OF REAL;

  PROCEDURE Linear (M : INTEGER; VAR R : COEFFICIENT);
  BEGIN
    zero := FALSE;
    IF M <> 0 THEN  FOR i := 1 TO M DO
    BEGIN
      WRITE ('Enter the constant term in linear factor # ', i,':   ');
      READLN (LIN [i]);
      IF LIN [i] = 0 THEN zero := TRUE;
      R[i] := R[i-1];
      FOR j := i-1 DOWNTO 1 DO
        R[j] := R[j] * LIN [i] + R[j-1];
      R[0] := R[0] * LIN [i];
    END;
  END;

  PROCEDURE Quadratic (N, k : INTEGER; VAR R : COEFFICIENT);
  BEGIN
    IF N <> 0 THEN FOR i := 1 TO N DO
    BEGIN
      WRITE ('Enter the constant term in quadratic factor # ', i,':   ');
      READLN (QUAD1 [i]);
      WRITE ('Enter the coefficient of s in quadratic factor #', i,': ');
      READLN (QUAD2 [i]);
      R[k+2] := R[k];
      IF k = 0 THEN R[k+1] := R[k] * QUAD2 [i]
          ELSE  BEGIN
              R[k+1] := R[k] * QUAD2 [i] + R[k-1];
              FOR j := k DOWNTO 2 DO
                  R[j] := R[j] * QUAD1[i] + R[j-1] * QUAD2[i] + R[j-2];
              R[1] := R[1] * QUAD1[i] + R[0] * QUAD2[i];
            END;
      R[0] := R[0] * QUAD1[i];
      k := k + 2;
    END;
  END;

  BEGIN  (*Factors*)
    REPEAT
      WRITE ('Enter the constant term in the numerator:  ');
      READLN (LIN[0]);
```

```
    Numerator[0] := LIN [0];
    WRITE ('How many linear factors are in the numerator?  ');
    READLN (M);
    degreeN := M;
    Linear (M, Numerator);
    k := degreeN;
    WRITE ('How many quadratic factors are in the numerator?  ');
    READLN (N);
    Quadratic (N, k, Numerator);
    degreeN := degreeN + N * 2;
    WRITELN ('The numerator has the following form : ');
    WRITE (LIN[0]:Field:decimal);
    FOR i:= 1 to M DO
      BEGIN
        IF NOT zero THEN WRITE (' (s + ', LIN[i]:Field:decimal, ') ')
           ELSE WRITE  (' (s) ');
        zero := FALSE;
      END;
     FOR i := 1 to N DO
        WRITE (' (s^2 + ', QUAD2[i]:Field:decimal, ' s +',
                                        QUAD1[i]:Field:decimal, ') ');
     WRITELN;
     WRITELN;
     WRITE ('Is this correct?  (Y or N) :  ');
     Echo (response)
   UNTIL response IN ['Y', 'y'];
   REPEAT
     Denominator[0] := 1;
     WRITE ('How many linear factors are in the denominator?  ');
     READLN (M);
     degreeD := M;
     Linear (M, Denominator);
     k := degreeD;
     WRITE ('How many quadratic factors are in the denominator?  ');
     READLN (N);
     Quadratic (N, k, Denominator);
     degreeD := degreeD + N * 2;
     WRITELN ('The denominator has the following form : ');
     FOR i := 1 to M DO
       BEGIN
         IF NOT zero THEN WRITE (' (s + ', LIN[i]:Field:decimal, ') ')
            ELSE WRITE  (' (s) ');
         zero := FALSE;
       END;
     FOR i := 1 to N DO
        WRITE (' (s^2 + ', QUAD2[i]:Field:decimal, ' s +',
                                        QUAD1[i]:Field:decimal, ') ');
      WRITELN;
      WRITELN;
      WRITE ('Is this correct?   (Y or N) : ' );
      Echo (response)
    UNTIL response IN ['Y', 'y'];
END;

PROCEDURE Input;
BEGIN
  ClrScr;
  WRITE ('Are the polynomials in the factored form?  (Y or N)  ');
  Echo (answer);
  IF answer IN ['N', 'n'] THEN Polynomial
     ELSE Factors;
  IF degreeN = 0 THEN
    BEGIN
      degreeN := 1;
      Numerator[1] := 0;
    END;
```

```
    WRITE ('Do you want to store the denominator coefficients on disk? ');
    WRITE ('   (Y or N)    ');
    Echo (answer);
    IF answer IN ['Y', 'y'] THEN
      BEGIN
        CreateFile (OutFile, FileName);
        REWRITE (OutFile);
        WRITELN (OutFile, degreeD);
        FOR i := degreeD DOWNTO 0 DO
          WRITELN (OutFile, Denominator[i]:Field:decimal);
        CLOSE (OutFile)
      END;
    WRITE ('Do you want to store the numerator coefficients on disk? ');
    WRITE ('   (Y or N)    ');
    Echo (answer);
    IF answer IN ['Y', 'y'] THEN
      BEGIN
        CreateFile (OutFile, FileName);
        REWRITE (OutFile);
        WRITELN (OutFile, degreeN);
        FOR i := degreeN DOWNTO 0 DO
          WRITELN (OutFile, Numerator[i]:Field:decimal);
        CLOSE (OutFile)
      END;
    WRITE ('Do you want the coefficients printed?  (Y or N)  ');
    Echo (answer);
    IF answer IN ['Y', 'y'] THEN
      PrintOut;
  END {Input};

  FUNCTION Sgn (X : REAL) : INTEGER;
  BEGIN
    IF X >= 0 THEN Sgn := 1
      ELSE Sgn := -1
  END;

  FUNCTION arcTAN2 (X, Y : REAL) : REAL;
  (* It gets the correct quadrant of the arc tangent *)
  BEGIN
    IF (X = 0) AND (Y <> 0) THEN arcTAN2 := (PI/2) * Sgn (Y);
    IF X < 0 THEN   arcTAN2 := PI + ArcTan (Y/X);
    If X > 0 THEN   arcTAN2 := ArcTan (Y/X);
  END;

  PROCEDURE Div1 (divid : COEFFICIENT; degree : INTEGER; sigma : REAL;
                  VAR  R : COEFFICIENT);
  (* This procedure divides the polynomial dividend by  (s + a)
     and stores the quotient as R  *)
  VAR  i : 0..24;
       a : REAL;
  BEGIN
    a := - sigma;
    FOR i := degree DOWNTO 1 DO
      BEGIN
        R [i-1] := divid [i];
        divid[i-1] := divid[i-1] -a * divid[i];
      END;
  END; (* Div1 *)

  PROCEDURE Div2 (divid : COEFFICIENT;  degree : INTEGER;
                  sigma, omega : REAL;  VAR   R : COEFFICIENT);
  (* This procedure divides the polynomial dividend by  (s^2 + as + b)
     and stores the quotient as R  *)
  VAR  i : 0..24;
       a, b : REAL;
```

```
BEGIN
  a := - 2 * sigma;
  b := SQR(sigma) + SQR(omega);
  FOR i := degree DOWNTO 2 DO
    BEGIN
      R [i-2] := divid [i];
      divid[i-1] := divid[i-1]-a * divid[i];
      divid[i-2] := divid[i-2] -b * divid[i];
    END;
END; (* Div2 *)

PROCEDURE Calculate (S : COEFFICIENT; degreeS : INTEGER; sigma : REAL;
                     VAR re : REAL);
   (* This procedure calculates S(s) for a real s  *)
VAR i : INTEGER;
BEGIN
  re := S[degreeS];
  IF degreeS > 0 THEN
    FOR i := degreeS DOWNTO 1 DO
       re := re * sigma + S[i-1];
END;

PROCEDURE SynDiv (divid : COEFFICIENT; degree : INTEGER;
                    sigma, omega : REAL; VAR   re, im : REAL);
(* This procedure divides the polynomial dividend by  (s^2 + as + b)
   and evaluates the remainder for (s = sigma + j omega) *)
VAR  i : 1..24;
     a, b : REAL;
BEGIN
  a := - 2 * sigma;
  b := SQR(sigma) + SQR(omega);
  FOR i := degree DOWNTO 2 DO
    BEGIN
      divid[i-1]:= divid[i-1] -a * divid[i];
      divid[i-2] := divid[i-2] -b * divid[i];
    END;
    re := divid[0] + divid[1] * sigma;
    im := divid[1] * omega;
 END;  {SynDiv}

 Procedure ComplexPole;
 (* Calculate the residue at a complex pole *)
 VAR  re1, re2, im1, im2, temp : REAL;
 BEGIN
   WRITE ('Enter the value of sigma:  ');
   READLN (sigma);
   WRITE ('Enter the value of omega:  ');
   READLN (omega);
   Div2 (Denominator, degreeD, sigma, omega, R);
   degreeR := degreeD - 2;
   SynDiv (Numerator, degreeN, sigma, omega, re1, im1);
   SynDiv (R, degreeR, sigma, omega, re2, im2);
   temp := (re2*re2 + im2*im2) * 2 * omega;
   ResReal := (im1*re2 - im2*re1)/temp;
   ResImag := -(re1*re2 + im1*im2)/temp;
   WRITELN ('The residue is ', ResReal:Field:decimal, ' + j ',
                                              ResImag:Field:decimal);

   WRITELN;
 END;

 PROCEDURE RealPole;
 (* Calculate the residue at a real pole *)
 VAR    re1, re2 : REAL;
 BEGIN
   WRITE ('Enter the value of real pole:  ');
```

```
  READLN (sigma);
  Div1 (Denominator, degreeD, sigma, R);
  degreeR := degreeD - 1;
  Calculate (Numerator, degreeN, sigma, re1);
  Calculate (R, degreeR, sigma, re2);
  ResReal := re1/re2;
  WRITELN ('The residue is ', ResReal:Field:decimal);
END;

BEGIN (* Main program *)
  GeneralInstructions;
  Input;
  REPEAT
    ClrScr;
    WRITE ('Is this a complex pole?  (Y or N) ');
    Echo (answer);
    IF answer IN ['Y', 'y'] THEN ComplexPole
       ELSE RealPole;
    WRITE ('Do you want any more residues? (Y or N)  ');
    Echo (response)
  UNTIL response IN ['N', 'n'];
END.


PROGRAM RootsOfRealPolynomials( Input , Output , OutFile );
(*$U+ *)
CONST DateOfThisVersion = '19 May 1985';
      PIby2 = 1.5707963267948966192313 22;
         PI = 3.14159265358979323846 2643;
          F = 20; (* Fieldwidth *)
          D = 15; (* Number of digits after the decimal point *)
      order = 50;
   (* Machine dependent constants: Turbo Pascal with 8087 support *)
           Base = 2.0;          (* machine number base *)
       Infinity = 1.0E+300; (* effective machine infinity *)

TYPE   subs = 0..order;
       STRING14 = STRING[14];
       COEFFICIENT = ARRAY [subs] OF REAL;

VAR
  sigma, omega, BestSigma, BestOmega, AbsP, error, Accuracy : REAL;
  bound, AssumedZero, Serror, alpha, beta : REAL;
  degreeP, degreeQ : INTEGER;
  P, Pprime, PdblPrime, Q, R, RE, IM : COEFFICIENT;
  OutFile  : TEXT;
  found, changed, complex, console : BOOLEAN;
  answer : CHAR;

(*$I Files.Pas *)        (* file handling module *)

FUNCTION MachineEpsilon : REAL ;
VAR a, b : REAL;
BEGIN
  a := 1.0 ;
  REPEAT
    a := a/Base ; b := 1.0 + a
  UNTIL  b = 1.0;
  MachineEpsilon := Base*a
END;

PROCEDURE Pause;
VAR
  answer : CHAR;
```

```
BEGIN
  WRITE ('Press <space> to continue... ');
  READ (KBD , answer );
  RESET (KBD);
  WRITELN
END;

PROCEDURE GeneralInstructions;
BEGIN
  ClrScr;
  WRITELN ('      This program calculates the roots of a polynomial');
  WRITELN;
  WRITELN ('                 n            n-1     ');
  WRITELN ('         P(s) = A[n]*s  + A[n-1]*s    + ... + A[1]*s + A[0]');
  WRITELN;
  WRITELN ('with  real  coefficients, of degree up to ',order:3,'.');
  WRITELN ('      The roots are evaluated  directly in case the degree');
  WRITELN ('is less than three.  For higher degrees, a search is used');
  WRITELN ('to  find either a real or a complex root.  For the latter');
  WRITELN ('case, the  conjugate  is also a root.  Hence, the two are');
  WRITELN ('combined to obtain a quadratic  with  real  coefficients.');
  WRITELN ('The polynomial is then deflated and the process continued');
  WRITELN ('until the degree is less than three.');
  WRITELN ('      At each stage, we try to find a root of a polynomial');
  WRITELN ('Q(s).  The number  z  is accepted as a  root of  Q(s)  if');
  WRITELN ('       ABS( delta(z)/z ) < Accuracy = MachineEpsilon.');
  WRITELN;
  WRITELN ('       Date of this version: ', DateOfThisVersion);
  WRITELN;
  WRITE (  '                    ');
  Pause
END;

PROCEDURE RequestToPrint;
VAR
  on : BOOLEAN;
BEGIN
  on := FileExists( OutFile , 'LST:' );
  IF  NOT on  THEN
    WRITELN('Please turn on the printer.  ');
  WHILE  NOT on  DO
    on := FileExists( OutFile , 'LST:' )
END;

PROCEDURE Echo( VAR answer : CHAR ) ;
BEGIN
  READ (KBD , answer);
  RESET (KBD);
  WRITELN (answer)
END;

PROCEDURE Verify; (* the polynomial's coefficients *)
VAR
  i, n, k, top, bottom : INTEGER;
  answer : CHAR;
BEGIN
  changed := FALSE;
  IF ( ( degreeP + 1 )  MOD  15 ) = 0  THEN
    n := ( degreeP + 1 ) DIV 15
  ELSE
    n := ( ( degreeP + 1 ) DIV 15 ) + 1;
  FOR  k := 1  TO  n  DO
    BEGIN
      bottom := degreeP - ( k - 1 )*15 - 14;
```

```
      IF  bottom < 0  THEN
        bottom := 0;
      top := degreeP - ( k - 1 )*15;
      REPEAT
        ClrScr;
        WRITELN ('The coefficients are:');
        WRITELN;
        FOR  i := top  DOWNTO  bottom  DO
          WRITELN ('  A[', i, '] = ', P[i]:F:D);
        WRITELN;
        IF  bottom > 0  THEN
          WRITELN ('The remaining coefficients will also be verified.');
        WRITE('Are these entries correct?  (Y or N):  ');
        Echo (answer);
        IF  NOT (answer  IN  ['Y', 'y'])  THEN
          BEGIN
            changed := TRUE;
            REPEAT
              WRITE ('Which entry is wrong? ');
              WRITE ('Enter the subscript number: ');
              READLN (i);
              IF  NOT (i  IN  [ bottom..top ] )  THEN
                WRITELN (' *** Error: invalid subscript entered')
            UNTIL  i  IN  [ bottom..top ];
            WRITE ('Enter the correct value of A[', i ,']:  ');
            READLN (P[i] )
          END
      UNTIL answer IN ['Y', 'y'];
    END;
  WRITELN
END;

PROCEDURE PrintCoefficients;
VAR i : INTEGER;
BEGIN
  WRITELN (OutFile, 'The coefficients of the original polynomial are:');
  WRITELN (OutFile);
  FOR  i := degreeP  DOWNTO  0  Do
    WRITELN (OutFile, '    A[', i, '] = ', P[i]:F:D);
  WRITELN (OutFile);
  WRITELN (OutFile);
  CLOSE (OutFile )
END;

PROCEDURE PrintOneRoot;
BEGIN
  WRITE (sigma);
  IF  omega >= 0.0  THEN
    WRITE (' + j ')
  ELSE
    WRITE (' - j ');
  WRITE (ABS(omega) )
END;

PROCEDURE PutCoefficients;
VAR  answer : CHAR;
BEGIN
  WRITE ('Do you want the coefficients printed?  (Y or N) :  ');
  Echo (answer );
  IF  answer  IN  ['Y', 'y']  THEN
    BEGIN
      RequestToPrint;
      PrintCoefficients
    END
END;
```

```
PROCEDURE GetCoefficients;
VAR
  i : INTEGER;
BEGIN
  REPEAT
    WRITE ('Enter the degree of the polynomial:  ');
    READLN (degreeP);
    IF  NOT ( degreeP  IN  [ 1..order ] )  THEN
      WRITELN (' *** Error: invalid degree entered' );
  UNTIL  degreeP  IN  [ 1..order ] ;
  FOR i := degreeP DOWNTO 0 DO
    REPEAT
      WRITE ('Enter   A[', i, '] :  ');
      READLN ( P[i] );
      IF  i = degreeP  THEN
        IF  P[i] = 0.0  THEN
          WRITELN (' *** Error: degree of the polynomial is ', degreeP,
                  ' but  A[', degreeP , '] = 0.0' )
    UNTIL  P[degreeP] <> 0.0;
  Verify;
  changed := TRUE;
  PutCoefficients
END;

FUNCTION Max( a , b : REAL ) : REAL;
BEGIN
  IF  a < b  THEN
    Max := b
  ELSE
    Max := a
END;

FUNCTION Cabs( a , b : REAL ) : REAL; (* absolute value of  a + jb *)
BEGIN
  IF  ( a <> 0.0 ) AND ( b <> 0.0 )  THEN
    Cabs := SQRT( a*a  +  b*b )
  ELSE
    IF  a = 0.0  THEN
      Cabs := ABS(b)
    ELSE
      Cabs := ABS(a)
END;

PROCEDURE DataIn; (* the degree and coefficients of the polynomial *)
VAR answer : CHAR;
    i : INTEGER;
    FileName : STRING14;
BEGIN
  ClrScr;
  WRITELN ('Do you want to enter the coefficients from the keyboard');
  WRITE ('or read them from a file?  Enter  K or F   ');
  Echo (answer );
  IF  answer  IN  ['F' , 'f']  THEN
    IF  GetFile (OutFile)  THEN
      BEGIN
        RESET (OutFIle);
        READLN (OutFile, DegreeP);
        FOR i := DegreeP DOWNTO 0 DO
            READLN (OutFile , P[i]);
        CLOSE (OutFile);
        Verify;
        PutCoefficients
      END
    ELSE
      answer := 'K';
```

```
  IF  NOT (answer  IN  [ 'F' , 'f' ])  THEN
    GetCoefficients;
  IF  changed  THEN
    BEGIN
      WRITE ('Do you want to write the coefficients to a file?');
      WRITE ('  (Y or N)  ');
      Echo (answer);
      IF  answer  IN  ['Y' , 'y']  THEN
        BEGIN
          CreateFile (OutFile , FileName);
          REWRITE (OutFile);
          WRITELN (OutFIle, DegreeP);
          FOR  i := degreeP  DOWNTO  0  DO
             WRITELN (OutFile , P[i]);
          CLOSE (OutFile )
        END
    END;
  (* Save the original polynomial coefficients for future checking *)
           Q := P;
      degreeQ := degreeP;
  (* Find a bound for the magnitude of the roots *)
      bound := 0.0;
      FOR  i := 0  TO  degreeP - 1  DO
        bound := Max (bound , ABS( P[i]/P[degreeP]));
      bound := 2.0 * (bound + 1.0)
END;

FUNCTION ArcTan2( x, y : REAL ) : REAL;
(* It gives the arctan of y/x in the correct quadrant on the range
              -PI  to  PI *)
BEGIN
  ArcTan2 := 0.0;
  IF ( x = 0.0 ) AND ( y <> 0.0 )  THEN
    IF  y >= 0.0  THEN
      ArcTan2 := PIby2
    ELSE
      ArcTan2 := - PIby2;
  IF  x < 0.0  THEN
    IF  y >= 0.0  THEN
      ArcTan2 := ArcTan(y/x) + PI
    ELSE
      ArcTan2 := ArcTan(y/x) - PI;
  IF  x > 0.0  THEN
    ArcTan2 := ArcTan(y/x)
END;

PROCEDURE Div1 (VAR P : COEFFICIENT ; degree : INTEGER ;
                   sigma : REAL;   VAR R : COEFFICIENT);
(* This procedure divides a polynomial by  (s - sigma)
    and stores the quotient as R *)
VAR  i : INTEGER;
BEGIN
  R [degree] := 0.0;
  FOR  i := degree  DOWNTO  1  DO
    R [i-1] := P [i]  +  sigma * R [i]
END;

PROCEDURE Div2 (VAR P : COEFFICIENT ; degree : INTEGER ;
                   sigma, omega : REAL ; VAR  R : COEFFICIENT);
(* This procedure divides a polynomial by  (s*s -as + b)
    and stores the quotient as R *)
VAR  i : INTEGER;
     a, b : REAL;
```

```
BEGIN
  a := sigma + sigma;
  b := sigma * sigma  +  omega * omega;
  R [degree] := 0.0 ;
  R [degree-1] := 0.0 ;
  FOR  i := degree  DOWNTO  2  DO
    R [i-2] := P [i]  +  (a * R [i-1]  -  b * R [i] );
  alpha := P [1]  + (a * R[0] - b * R [1]);
  beta := P [0] - b * R [0]
END;

PROCEDURE Quadratic (a2,a1,a0 : REAL ; VAR re1, re2, im1, im2 : REAL);
(* Finds the roots of a quadratic *)
VAR
  b, d, e : REAL;
BEGIN
  b := a1/2.0;
  IF  ABS(b) >= ABS(a0)  THEN
    BEGIN
      e := 1.0 - ( a2/b )*( a0/b );
      d := SQRT( ABS(e) ) * ABS(b)
    END
  ELSE
    BEGIN
      IF  a0 >= 0.0  THEN
        e := a2
      ELSE
        e := -a2;
      e := b * ( b/ABS(a0) ) - e;
      d := SQRT (ABS(e)) * SQRT (ABS(a0))
    END;
  IF  e >= 0.0  THEN
    BEGIN (* Real roots *)
      IF  b >= 0.0  THEN
        d := -d;
      re1 := ( d - b )/a2;
      IF  re1 <> 0.0  THEN
        re2 := ( a0/re1 )/a2
      ELSE
        re2 := 0.0;
      im1 := 0.0 ; im2 := 0.0
    END
  ELSE
    BEGIN (* Complex roots *)
      re2 := -b/a2 ; re1 := re2 ;
      im2 := ABS( d/a2 ) ; im1 := -im2
    END
END;

PROCEDURE ComplexDiv (x1, y1, x2 , y2 : REAL ; VAR x3 , y3 : REAL);
(* This procedure divides  x1 + j(y1)  by  x2 + j(y2)
                   to get  x3 + j(y3)                   *)
VAR
  temp : REAL;
BEGIN
  IF  ( x2 <> 0.0 ) AND ( y2 <> 0.0 )  THEN
    BEGIN
      temp := x2 * x2  +  y2 * y2;
      x3 := ( x1*x2 + y1*y2 )/temp;
      y3 := ( x2*y1 - x1*y2 )/temp
    END
  ELSE
    IF  x2 = 0.0  THEN
```

```
      BEGIN
        x3 :=  y1/y2;
        y3 := -x1/y2
      END
    ELSE
      BEGIN
        x3 := x1/x2;
        y3 := y1/x2
      END
END;

PROCEDURE ComplexMult(x1, y1, x2, y2 : REAL ; VAR x3, y3 : REAL);
(* This procedure multiplies  x1 + j(y1)  and  x2 + j(y2)
                          to get  x3 + j(y3)                    *)
BEGIN
  x3 := x1*x2 - y1*y2;
  y3 := x1*y2 + y1*x2
END;

PROCEDURE NewGuess;
VAR  answer : CHAR;
BEGIN
  WRITELN;
  WRITE ('Do you want to use a different value for the root?');
  WRITE ('    (Y or N)   ');
  Echo( answer );
  IF  answer  IN  [ 'Y' , 'y' ]  THEN
    BEGIN
      WRITE ('Enter the guess for the real part:   ');
      READLN ( sigma ) ;
      WRITE ('Enter the guess for the imaginary part:   ');
      READLN ( omega )
    END
END;

PROCEDURE Notify( VAR answer : CHAR );
BEGIN
  WRITELN ('The current best value of the "root" is ');
  PrintOneRoot;
  WRITELN (' with  P(s) = ' , AbsP:10 );
  WRITELN;
  WRITE ('Do you want to continue the search process?   (Y or N)   ');
  Echo( answer );
  IF  NOT ( answer  IN  [ 'N' , 'n' ] )  THEN
    BEGIN
      answer := 'y';
      NewGuess
    END
END;

PROCEDURE Derivative( VAR P, Pprime : COEFFICIENT ; degree : INTEGER );
VAR
  i : INTEGER;
BEGIN
  FOR  i := degree  DOWNTO  1  DO
    Pprime[i-1] := i * P[i]
END;

PROCEDURE Initialize;
(* This procedure asks for a guess for the root.  The default starting
   point is at 0 + j0 .  The coefficients of P'(s) and P"(s) are also
     obtained *)
BEGIN
  WRITELN;
  sigma := 0.0;
  omega := 0.0 ;
```

```
  WRITE ('The initial guess for the root will be taken as ');
  WRITELN ('  0.0 + j0.0');
  NewGuess;
  Derivative (P,Pprime,degreeP);
  Derivative (Pprime,PdblPrime,degreeP-1)
END;

PROCEDURE NewSigmaNewOmega (dSigma , dOmega : REAL );
VAR
  RootMag : REAL;
BEGIN
  RootMag := Cabs (sigma , omega );
  Serror  := Cabs (dSigma , dOmega );
  IF  RootMag > 0.0  THEN
    Serror := Serror/RootMag;
  (* The new root location is now obtained *)
        sigma := sigma - dSigma;
        omega := omega - dOmega;
  IF  ABS (omega) <= AssumedZero  THEN
    omega := 0.0
END;

PROCEDURE Deflate;
(* If a root is found then the polynomial is deflated by dividing
   it by a linear factor in the case of a real root, or a quadratic
   factor in the case of a complex root. *)
BEGIN
  IF  ABS (omega) <= AssumedZero  THEN
    BEGIN
      omega := 0.0;
      Div1 (P, degreeP, sigma, R);
      RE [degreeP] := sigma;
      IM [degreeP] := 0.0;
      degreeP := degreeP - 1;
      P := R
    END
  ELSE
    BEGIN
      Div2 (P, degreeP, sigma, omega, R);
      RE [degreeP] := sigma;
      RE [degreeP - 1] := sigma;
      IM [degreeP] := ABS (omega);
      IM [degreeP - 1] := - IM [degreeP];
      degreeP := degreeP - 2;
      P := R
    END;
  WRITELN;
  WRITELN ('The new deflated polynomial has degree ', degreeP)
END;

PROCEDURE RemoveRootsAtZero;
VAR  i, k : INTEGER;
     terminate : BOOLEAN;
BEGIN
  k := -1 ;
  terminate := FALSE ;
  REPEAT
    IF  P [k+1] = 0.0  THEN
      BEGIN
        k := k + 1;
        RE [degreeP - k] := 0.0;
        IM [degreeP - k] := 0.0
      END
    ELSE
      terminate := TRUE
  UNTIL  terminate  OR  ( k = ( degreeP - 1 ) );
```

```
   IF  k > -1 THEN
     BEGIN
       degreeP := degreeP - k - 1;
       FOR  i := 0  TO  degreeP  DO
         R [i] := P [i+k+1];
       P := R
     END
 END;

 PROCEDURE Evaluate (VAR Poly : COEFFICIENT ; degree : INTEGER ;
                          sigma, omega : REAL ; VAR  re, im : REAL);
 (*  This procedure evaluates a polynomial at  s = sigma + j(omega)  *)
 VAR  k : INTEGER;
 BEGIN
   IF  omega = 0.0  THEN
     BEGIN
       re := Poly [degree];
       im := 0.0;
       FOR  k := degree - 1  DOWNTO  0  DO
         re := re * sigma  +  Poly [k];
       error := ABS (re)
     END
   ELSE
     BEGIN
       Div2 (Poly, degree, sigma, omega, R);
       re := beta + alpha*sigma;
       im := alpha*omega;
       error := Cabs (re , im)
     END
 END;

 PROCEDURE DumpIt; (* The roots are displayed or printed *)
 VAR
   i, k, count : INTEGER;
   reP, imP : REAL;
 BEGIN
   WRITELN (OutFile, 'The roots are:');
   WRITELN (OutFile);
   i := degreeP;
   count := 0 ;
   WHILE  i >= 1  DO
     BEGIN
       Evaluate (P, degreeP, RE[i], IM[i], reP, imP);
       k := degreeP + 1 - i;
       IF  IM[i] = 0.0  THEN
         WRITELN (OutFile,'[',k:2,']',RE[i]:F:D,' ':27,'P(root) = ',
                                                              error:10)
       ELSE
         BEGIN
           WRITELN (OutFile,'[',k:2,']',RE[i]:F:D, ' + j ', IM[i]:F:D,
                                      '  ',     'P(root) = ',error:10);
           WRITELN (OutFile,'[',k+1:2,']',RE[i]:F:D, ' - j ',
                         IM[i]:F:D, '  ',  'P(root) =',error:10);
           i := i - 1;
           count := count + 1 ;
         END;
       i := i - 1;
       count := count + 1 ;
       IF  ( degreeP > 20 ) AND console AND ( count >= 20 )  THEN
         BEGIN
           count := 0;
           Pause
         END
     END;
   CLOSE (OutFile)
 END;
```

```
PROCEDURE Output; (* The roots are displayed at the console *)
BEGIN
  ClrScr;
  ASSIGN (OutFile , 'CON:');
  RESET (OutFile);
  console := TRUE;
  DumpIt
END;

PROCEDURE Accept (i : INTEGER);
BEGIN
  WRITE ('The "root" is ');
  PrintOneRoot;
  WRITELN;
  RE[i] := sigma;
  IM[i] := ABS(omega);
  IF  complex  THEN
    BEGIN
      RE[i-1] := sigma;
      IM[i-1] := -IM[i]
    END
END;

PROCEDURE Locate ( i : INTEGER );
BEGIN
  sigma := RE[i];
  omega := IM[i];
  WRITE ('Root[',degreeP + 1 - i,'] = ');
  PrintOneRoot;
  WRITELN;
  IF  omega = 0.0  THEN
    complex := FALSE
  ELSE
    complex := TRUE
END;

PROCEDURE QueryPrinting;
VAR
  answer : CHAR;
BEGIN
  WRITELN;
  WRITE ('Do you want the roots printed?  (Y or N) :  ');
  Echo (answer );
  IF  answer  IN  ['Y', 'y']  THEN
    BEGIN
      RequestToPrint;
      console := FALSE;
      DumpIt
    END
END;

PROCEDURE Laguerre (RealP , ImP : REAL);
(* Using Laguerre's method this procedure calculates the
   new guess for the root *)
VAR  RealPp, ImPp, RealPdblP, ImPdblP, RealH, ImH : REAL;
     RealPr, ImPr, RealDenom, ImDenom, RealPpS, ImPpS : REAL;
     MagRootH, theta, dSigma, dOmega : REAL;

BEGIN
  (* First we calculate P'(s) and P"(s) at the current guess *)
      Evaluate (Pprime, degreeP-1, sigma, omega, RealPp, ImPp);
      Evaluate (PdblPrime, degreeP-2, sigma, omega, RealPdblP, ImPdblP);

  (* Now calculate the square of P'(s) and
      store as  RealPpS + j(ImPpS) *)
      ComplexMult (RealPp, ImPp, RealPp, ImPp, RealPpS, ImPpS);
```

```
(* Now calculate P(s)*P"(s) and store as  RealPr + j(ImPr) *)
     ComplexMult (RealP, ImP, RealPdblP, ImPdblP, RealPr, ImPr);

  (* Calculate the square root of H(s) *)
     RealH := (degreeP-1)*RealPpS - degreeP*RealPr;
     ImH := (degreeP-1)*ImPpS - degreeP*ImPr;
     theta := ArcTan2(RealH, ImH)/2.0;
     MagRootH := SQRT ((degreeP - 1 ) * Cabs (RealH , ImH));
     RealH := COS (theta);
     ImH := SIN (theta);

     IF  (RealPp * RealH + ImPp * ImH) >= 0.0  THEN  (* use  theta *)
       BEGIN
         RealDenom := RealPp   +  MagRootH * RealH;
           ImDenom  :=  ImPp    +  MagRootH * ImH
       END
     ELSE  (* use  theta + PI  *)
       BEGIN
         RealDenom := RealPp   -  MagRootH * RealH;
           ImDenom  :=  ImPp    -  MagRootH * ImH
       END;

     IF  ( RealDenom = 0.0 ) AND ( ImDenom = 0.0 )  THEN
       BEGIN
         WRITELN ('The denominator is 0.0');
         WRITE ('Please enter a new guess for the real part of the root: ');
         READLN (sigma);
         WRITE ('Please enter a new guess for the imaginary part of the ',
                'root: ');
         READLN (omega)
       END
     ELSE
       BEGIN  (* Divide nP(s) by P'(s) + RootH(s) *)
         ComplexDiv (degreeP*RealP, degreeP*ImP, RealDenom,
                                          ImDenom, dSigma,dOmega);

         IF  NOT complex  THEN
           dOmega := 0.0;
         NewSigmaNewOmega (dSigma , dOmega)
       END
END;

PROCEDURE Iteration; (* main iteration procedure *)
CONST
  MaxIterations = 40;
VAR
  RealP, ImP : REAL;
  k : INTEGER;
BEGIN
  found := FALSE;
  k := 1;
  AbsP := Infinity;
  Serror := 1.0;
  WHILE  (k <= MaxIterations) AND (NOT found)  DO
    BEGIN
      Evaluate (P, degreeP, sigma, omega, RealP, ImP);
      IF  error <= AbsP THEN
        BEGIN
          BestSigma := sigma;
          BestOmega := omega;
          AbsP := error
        END;
      WRITELN ('[',k:2,']   s = ',sigma:F:D,' + j ',omega:F:D,
                                              '    P(s) = ',error:10);
      IF  (Serror <= Accuracy) OR (error = 0.0)  THEN
        found := TRUE
      ELSE
```

```
          BEGIN
            IF  Max (ABS(sigma) , ABS(omega)) < bound  THEN
              BEGIN
                Laguerre (RealP , ImP);
                k := k + 1
              END
             ELSE
              BEGIN
                WRITELN;
                WRITELN ('The method appears to be diverging');
                WRITELN ('A new starting value should be entered');
                k := MaxIterations + 1
              END
          END
      END
END;

PROCEDURE Search; (* to try to find a root *)
VAR
  answer : CHAR;
BEGIN
  ClrScr;
  REPEAT
    Iteration;
    IF  found  THEN
      BEGIN
        WRITE ('A root of the');
        IF  degreeP < degreeQ  THEN
          WRITE (' deflated')
        ELSE
          WRITE (' original');
        WRITELN (' polynomial was found')
      END
     ELSE
      BEGIN
        sigma := BestSigma;
        omega := BestOmega;
        Notify (answer);
        IF  NOT (answer  IN  [ 'Y' , 'y' ])  THEN
          BEGIN
            found := TRUE;
            WRITE ('The best approximation was accepted as');
            WRITE (' a "root" of the ');
            IF  degreeP < degreeQ  THEN
              WRITE ('deflated')
            ELSE
              WRITE ('original');
            WRITELN (' polynomial')
          END
      END
  UNTIL  found;
  WRITE ('The "root" is ');
  PrintOneRoot;
  WRITELN
END;

PROCEDURE HighDegree;
(* Utilizes the procedures Initialize, Search, and Deflate, until
   the degree is less than three.  *)
VAR  i : INTEGER;
BEGIN
  WHILE  degreeP > 2  DO
    BEGIN
      IF  degreeP < degreeQ  THEN
        BEGIN
          WRITELN;
```

```
          WRITELN ('Beginning the search for the next root')
        END;
      Initialize;
      Search;
      Deflate
    END
END;

PROCEDURE Refine (i : INTEGER);
(* to try to improve the roots using the original polynomial *)
VAR  answer : CHAR;
BEGIN
  REPEAT
    Iteration;
    IF  found  THEN
      Accept(i)
    ELSE
      BEGIN
        sigma := BestSigma;
        omega := BestOmega;
        Notify (answer);
        IF  NOT  (answer  IN  [ 'Y' , 'y' ])  THEN
          Accept(i)
      END
  UNTIL  found  OR  (NOT  (answer  IN  [ 'Y' , 'y' ]))
END;

PROCEDURE CleanUp;
VAR  i : INTEGER;
BEGIN
  i := degreeP;
  WHILE  i >= 1  DO
    BEGIN
      IF  i < degreeP  THEN
        Pause;
      ClrScr;
      Locate (i);
      Refine (i);
      IF  complex  THEN
        i := i - 2
      ELSE
        i := i - 1
    END;
  Pause
END;

BEGIN (* Main Program *)
  AssumedZero := MachineEpsilon;
  Accuracy := AssumedZero;
  GeneralInstructions;
  REPEAT
    DataIn;
    complex := TRUE;
    RemoveRootsAtZero;
    IF  degreeP > 2  THEN
      HighDegree;
    IF  degreeP = 2  THEN
      Quadratic (P[2], P[1], P[0], RE[1], RE[2], IM[1], IM[2])
    ELSE
      BEGIN
        RE[1] := -P[0]/P[1];
        IM[1] := 0.0
      END;
    IF  degreeQ > 2  THEN
      BEGIN      (* Restore P(s) to the original polynomial *)
        P := Q;
```

```
          degreeP := degreeQ;
          Derivative (P, Pprime, degreeP);
          Derivative( Pprime, PdblPrime, degreeP-1);
          WRITELN;
          WRITELN('All roots have now been found; however, some may ',
                  ' be in error due to');
          WRITELN('the deflation process.');
          WRITELN;
          Pause;
          Output;
          QueryPrinting;
          WRITELN;
          WRITE ('Do you want to "refine" the roots using the original',
                    ' polynomial?  (Y or N)  ');
          Echo (answer);
          IF  answer  IN  [ 'Y' , 'y' ]  THEN
            BEGIN
              CleanUp;
              Output
            END
        END
      ELSE
        BEGIN
          answer := 'y' ;
          Output
        END;
      IF  answer  IN  [ 'Y' , 'y' ]  THEN
        QueryPrinting;
      WRITELN;
      WRITE ('Do another polynomial?  (Y or N) ');
      Echo (answer)
    UNTIL  NOT (answer  IN  [ 'Y' , 'y' ])
END.

PROGRAM RootLocus;
CONST  PI = 3.1415926535897932384626433;
       tolerance = 0.0001;
       Field = 9;
       decimal = 4;
       space = ' ';
TYPE   COEFFICIENT = ARRAY [0..25] OF REAL;
       String14 = STRING[14];
VAR
  sigma, omega, omega1, omega2, modulus, angle, zeta, omegaN : REAL;
  i, j, k      : INTEGER;
  degreeN, degreeD  :  0..24;
  P,Q,R  : COEFFICIENT;
  OutFile  : TEXT;
  response, answer : CHAR;
  FileName : String14;
  radial, zero, found  : BOOLEAN;

(*$U+ *)
(*$I Files.Pas *)

PROCEDURE GeneralInstructions;
BEGIN
  ClrScr;
  WRITELN;
  WRITELN;
  WRITELN;
  WRITELN;
  WRITELN ('     This program determines points on the locus of the');
  WRITELN ('poles of the transfer function of a  closed loop system');
  WRITELN ('as the gain is varied from  zero to  infinity.  This is');
```

```
    WRITELN ('done by searching for a point where the angle condition');
    WRITELN ('is satisfied.  The  search can be carried out on either');
    WRITELN ('a line with  constant  real  part or  on a line  with a');
    WRITELN ('constant damping ratio.  The  transfer  function can be');
    WRITELN ('either in the factored or expanded form. ');
    WRITELN;
    WRITELN ('   Please press the <RETURN> key when ready. ');
    READLN;
  END;

  PROCEDURE Echo (VAR answer : CHAR);
  BEGIN
    READ (KBD, answer);
    RESET (KBD);
    WRITELN (answer)
  END;

  PROCEDURE RequestToPrint;
  VAR on : BOOLEAN;
  BEGIN
    on := FileExists (OutFile, 'LST:');
    IF NOT on THEN
      WRITELN ('Please turn on the printer. ');
    WHILE NOt on DO
      on := FileExists (OutFile, 'LST:')
  END;

  PROCEDURE PrintOut;
  VAR i : INTEGER;
  BEGIN
    RequestToPrint;
    ASSIGN (OutFile, 'LST:');
    RESET (OutFIle);
    WRITELN (OutFile, 'The numerator coefficients are ');
    FOR i := degreeN DOWNTO 0 DO
      WRITELN (OutFIle, 'A[', i, '] = ', P[i]:Field:decimal, ',  ');
    WRITELN (OutFile);
    WRITELN (OutFile, 'The denominator coefficients are ');
    FOR i := degreeD DOWNTO 0 DO
      WRITELN (OutFIle, 'B[', i, '] = ', Q[i]:Field:decimal, ',  ');
    WRITELN (OutFile);
    CLOSE (OutFile)
  END;

  PROCEDURE Verify;
  (* It verifies the input polynomials *)
  VAR
    i, j : INTEGER;
    answer : CHAR;
  BEGIN
    ClrScr;
    REPEAT
      FOR i := degreeN DOWNTO 0 DO
        WRITELN ('A[', i, '] = ', P[i]:Field:decimal, ',  ');
      WRITELN;
      WRITE ('Are these entries correct? (Y or N):  ');
      Echo (answer);
      IF answer IN ['N', 'n'] THEN
        BEGIN
          WRITE ('Which entry is wrong?  ');
          READLN (j);
          WRITE ('Enter correct value of A[', j,'] ');
          READLN (P[j]);
        END
    UNTIL answer IN ['Y', 'y'];
    REPEAT
```

```
      FOR i := degreeD DOWNTO 0 DO
        WRITELN ('B[', i, '] = ', Q[i]:Field:decimal, ',  ');
      WRITELN;
      WRITE ('Are these entries correct?  (Y or N): ');
      Echo (answer);
      IF answer IN ['N', 'n'] THEN
        BEGIN
          WRITE ('Which entry is wrong? ');
          READLN (j);
          WRITE ('Enter correct value of B[', j,'] ');
          READLN (Q[j]);
        END
    UNTIL answer IN ['Y', 'y'];
    WRITELN;
  END;

  PROCEDURE Polynomial;
        (* Polynomials in the expanded form are entered. *)
  VAR i : INTEGER;
      response : CHAR;
  BEGIN
    WRITELN ('Do you want to enter the coefficients from the keyboard');
    WRITE (' or enter them from a diskfile? Enter K or F :   ');
    Echo (response);
    IF response IN ['F', 'f'] THEN
      BEGIN
        WRITELN ('Entering the numerator ');
        IF GetFile (OutFile) THEN
          BEGIN
            RESET (OutFile);
            READLN (OutFile, degreeN);
            FOR i := degreeN DOWNTO 0 DO
               READLN (OutFile, P[i]);
            CLOSE (OutFile)
          END
        ELSE response := 'K';
        WRITELN ('Entering the denominator ');
          IF GetFile (OutFile) THEN
            BEGIN
              RESET (OutFile);
              READLN (OutFile, degreeD);
              FOR i := degreeD DOWNTO 0 DO
                READLN (OutFile, Q[i]);
              CLOSE (OutFile)
            END
          ELSE  response := 'K';
      END;
    IF NOT (response IN ['F', 'f']) THEN
      BEGIN
        WRITE ('Enter degree of numerator polynomial:  ');
        READLN (degreeN);
        FOR i := degreeN DOWNTO 0 DO
          BEGIN
            WRITE ('A[', i, '] = ');
            READLN (P[i]);
          END;
        WRITE ('Enter degree of denominator polynomial:  ');
        READLN (degreeD);
        FOR i := degreeD DOWNTO 0 DO
          BEGIN
            WRITE ('B[', i, '] = ');
            READLN (Q[i]);
          END;
      END;
    Verify
  END;
```

```
PROCEDURE Factors;
(* This procedure allows input of polynomials in factored form *)
VAR    M, N      : INTEGER;
       LIN       : ARRAY [0..25] OF REAL;
       QUAD1, QUAD2 : ARRAY [1..13] OF REAL;

PROCEDURE Linear (M : INTEGER; VAR R : COEFFICIENT);
(* It allows input of linear factors of the form s + a *)
BEGIN
  zero := FALSE;
  IF M <> 0 THEN  FOR i := 1 TO M DO
  BEGIN
    WRITE ('Enter the constant term in linear factor # ', i,':  ');
    READLN (LIN [i]);
    IF LIN [i] = 0 THEN zero := TRUE;
    R[i] := R[i-1];
    FOR j := i-1 DOWNTO 1 DO
      R[j] := R[j] * LIN [i] + R[j-1];
    R[0] := R[0] * LIN [i];
  END;
END;

PROCEDURE Quadratic (N, k : INTEGER; VAR R : COEFFICIENT);
(* This procedure allows input of factors like s^2 + as + b *)
BEGIN
  IF N <> 0 THEN FOR i := 1 TO N DO
  BEGIN
    WRITE ('Enter the constant term in quadratic factor # ', i,':  ');
    READLN (QUAD1 [i]);
    WRITE ('Enter the coefficient of s in quadratic factor #
    ',
                                                          i,':  ');
    READLN (QUAD2 [i]);
    R[k+2] := R[k];
    IF k = 0 THEN R[k+1] := R[k] * QUAD2 [i]
 ELSE  BEGIN
         R[k+1] := R[k] * QUAD2 [i] + R[k-1];
         FOR j := k DOWNTO 2 DO
           R[j] := R[j] * QUAD1[i] + R[j-1] * QUAD2[i] + R[j-2];
         R[1] := R[1] * QUAD1[i] + R[0] * QUAD2[i];
       END;
    R[0] := R[0] * QUAD1[i];
    k := k + 2;
  END;
END;

BEGIN  (*Factors*)
  REPEAT
    WRITE ('Enter the constant term in the numerator:  ');
    READLN (LIN[0]);
    P[0] := LIN [0];
    WRITE ('How many linear factors are in the numerator?  ');
    READLN (M);
    degreeN := M;
    Linear (M, P);
    k := degreeN;
    WRITE ('How many quadratic factors are in the numerator?  ');
    READLN (N);
    Quadratic (N, k, P);
    degreeN := degreeN + N * 2;
    WRITELN ('The numerator has the following form : ');
    WRITE (LIN[0]:Field:decimal);
    FOR i := 1 to M DO
      BEGIN
        IF NOT zero THEN
          WRITE (' (s + ', LIN[i]:Field:decimal, ') ')
```

```
      ELSE WRITE (' (s) ');
      zero := FALSE;
    END;
   FOR i := 1 to N DO
     WRITE (' (s^2 + ', QUAD2[i]:Field:decimal, ' s + ',
                                      QUAD1[i]:Field:decimal, ') ');

   WRITELN;
   WRITELN;
   WRITE ('Is this correct?  (Y or N) : ');
   Echo (response)
 UNTIL response IN ['Y', 'y'];
 REPEAT
   Q[0] := 1;
   WRITE ('How many linear factors are in the denominator? ');
   READLN (M);
   degreeD := M;
   Linear (M, Q);
   k := degreeD;
   WRITE ('How many quadratic factors are in the denominator?  ');
   READLN (N);
   Quadratic (N, k, Q);
   degreeD := degreeD + N * 2;
   WRITELN ('The denominator has the following form : ');
   FOR i := 1 to M DO
     BEGIN
       IF NOT zero THEN
         WRITE (' (s + ', LIN[i]:Field:decimal, ') ')
       ELSE WRITE (' (s) ');
       zero := FALSE;
     END;
   FOR i := 1 to N DO
      WRITE (' (s^2 + ', QUAD2[i]:Field:decimal, ' s + ',
                                      QUAD1[i]:Field:decimal, ') ');

   WRITELN;
   WRITELN;
   WRITE ('Is this correct?  (Y or N) : ' );
   Echo (response)
 UNTIL response IN ['Y', 'y'];
END;

PROCEDURE Input;
BEGIN
  ClrScr;
  WRITE ('Are the polynomials in the factored form?  (Y or N)  ');
  Echo (answer);
  IF answer IN ['N', 'n'] THEN Polynomial
     ELSE Factors;
  IF degreeN = 0 THEN
    BEGIN
      degreeN := 1;
      P[1] := 0;
    END;
  WRITE ('Do you want the coefficients printed?  (Y or N)  ');
  Echo (answer);
  IF answer IN ['Y', 'y'] THEN  PrintOut;
  WRITE ('Do you want to store the denominator coefficients ');
  WRITE ('on disk?  (Y or N)  ');
  Echo (answer);
  IF answer IN ['Y', 'y'] THEN
    BEGIN
      CreateFile (OutFile, FileName);
      WRITELN (OutFile, degreeD);
      FOR i := degreeD DOWNTO 0 DO
      WRITELN (OutFile, Q[i]:Field:decimal);
      CLOSE (OutFile)
    END;
```

```
   WRITE ('Do you want to store the numerator coefficients ');
   WRITE (' on disk?  (Y or N)    ');
   Echo (answer);
   IF answer IN ['Y', 'y'] THEN
     BEGIN
       CreateFile (OutFile, FileName);
       WRITELN (OutFile, degreeN);
       FOR i := degreeN DOWNTO 0 DO
         WRITELN (OutFile, P[i]:Field:decimal);
       CLOSE (OutFile)
     END;
 END {Input};

 FUNCTION Sgn (X : REAL) : INTEGER;
 BEGIN
   IF X >= 0 THEN Sgn := 1
      ELSE Sgn := -1
 END;

 FUNCTION arcTAN2 (X, Y : REAL) : REAL;
 (* It gives the arctangent in the correct quadrant *)
 BEGIN
   IF (X = 0) AND (Y <> 0) THEN arcTAN2 := (PI/2) * Sgn (Y);
   IF X < 0 THEN  arcTAN2 := PI + arcTAN (Y/X);
   If X > 0 THEN  arcTAN2 := arcTAN (Y/X);
 END;


 PROCEDURE Evaluate (Q, P : COEFFICIENT; sigma, omega : REAL;
                          VAR   modulus, angle : REAL);
 VAR       re1, im1, re2, im2   : REAL;
 (* This procedure evaluates G(s) where s = sigma + j omega *)


 PROCEDURE SynDiv (divid : COEFFICIENT; degree : INTEGER;
                        sigma, omega : REAL; VAR   re, im : REAL);
 (* This procedure divides the polynomial dividend by  (s^2 + as + b)
    and ealuates the remainder for   (s = sigma + j omega) *)
 VAR  i : 1..24;
      a, b : REAL;
 BEGIN
   a := - 2 * sigma;
   b := SQR(sigma) + SQR(omega);
   FOR i := degree DOWNTO 2 DO
     BEGIN
       divid[i-1] := divid[i-1] -a * divid[i];
       divid[i-2] := divid[i-2] -b * divid[i];
     END;
     re := divid[0] + divid[1] * sigma;
     im := divid[1] * omega;
 END;  {SynDiv}


 BEGIN  (* Main Evaluate *)
   SynDiv (Q, degreeD, sigma, omega, re1, im1);
   SynDiv (P, degreeN, sigma, omega, re2, im2);
   modulus:= SQRT ((re1*re1 + im1*im1)/(re2*re2 + im2*im2));
   angle := (180/PI) * (arcTAN2(re1,im1) - arcTAN2(re2,im2));
   IF angle > 80 THEN angle := angle - 360;
 END; (* Evaluate *)

 PROCEDURE Initialize;
 (* This procedure gets the line of search and the initial guess *)
 BEGIN
   ClrScr;
   WRITE ('Is the damping ratio to be constant?   (Y or N)    ');
```

```
    Echo (answer);
    found := FALSE;
    IF answer IN ['Y', 'y'] THEN
      BEGIN
        radial := TRUE;
        WRITE ('Enter the value of the damping ratio :  ');
        READLN (zeta);
        WRITE ('Enter a good guess for the undamped natural ');
        WRITE ('frequency : ');
        READLN (omegaN);
        sigma := - zeta * omegaN;
        omega := omegaN * SQRT (1 - SQR(zeta));
      END
    ELSE
      BEGIN
        radial := FALSE;
        WRITE ('Enter the desired value of sigma:  ');
        READLN (sigma);
        WRITE ('Enter a good guess for omega: ');
        READLN (omega);
      END;
    omega1 := omega
  END;

  PROCEDURE Search;
  (* It searches for the root satisfying the angle condition. *)
  VAR  error, error1 : REAL;
       i : INTEGER;

  PROCEDURE Output;  (* Output of the final result *)
  BEGIN
    WRITELN ('The root has been located at ', sigma:Field:decimal,
                                    ' + j ', omega1:Field:decimal);
    WRITELN ('The gain for this root is ', modulus:7:4);
  END;

  BEGIN (* Main Search *)
    Evaluate (Q, P, sigma, omega, modulus, angle);
    WRITE ('sigma = ',sigma:Field:decimal, ',  omega = ',
                                          omega:Field:decimal);
    WRITELN (',  angle = ', angle:9:4);
    error := ABS(angle) - 180;
    IF ABS(error) <= tolerance  THEN  found := TRUE;
    IF NOT found THEN
      BEGIN
        omega1 := omega * (1 + sgn (error)/20);
        i := 1;
        IF radial THEN
          sigma := - omega1 * zeta/SQRT(1 - SQR(zeta));
        WHILE (i <= 15) AND (NOT found) DO
          BEGIN
            Evaluate (Q, P, sigma, omega1, modulus, angle);
            WRITE ('sigma = ',sigma:Field:decimal, ',  omega = ',
            omega1:Field:decimal);
            WRITELN (',  angle = ', angle:Field:decimal);
            error1 := ABS(angle) - 180;
            IF ABS(error1) <= tolerance  THEN
                 found := TRUE;
            i := i + 1;
            IF NOT found THEN
             BEGIN
               omega2 := omega1 - (omega1 - omega)
                               * error1/(error1 - error);
               omega := omega1;
               omega1 := omega2;
               error := error1;
```

```
              IF radial THEN
              sigma := - omega1 * zeta/SQRT(1 - SQR(zeta));
            END;
        END;
     END;
     IF found THEN Output;
     IF (NOT found) AND (k > 15) THEN
       BEGIN
         WRITE ('No root found after 15 iterations. The final value');
         WRITELN (' of the angle is ', angle:Field:decimal);
         WRITELN (' for the point ', sigma:Field:decimal, ' + j ',
                                      omega1:Field:decimal);
       END;
END;  (* Search *)


BEGIN {MAIN PROGRAM}
  GeneralInstructions;
  Input;
  REPEAT
    Initialize;
    Search;
    WRITE ('Do you want more points ?  (Y or N) :  ');
    Echo (response);
    ClrScr
  UNTIL response in ['N', 'n'];
END.


PROGRAM Compensator;
(*$U+ *)

(* This program can be used for designing lead, lag and laglead
   compensators which will provide specified phase shift and gain
   at a given frequency. *)

CONST  Pi = 3.14159265359; phi_2= 2;
VAR  pole, zero, alpha, phase, gain, omega, pole_2, zero_2,
           tan_phi, ratio, a, b, c, p, q, r :  REAL;
    answer                          :   STRING [80];
    response                        :   CHAR;
    lead, lag, laglead              :   BOOLEAN;

PROCEDURE Initialize;
BEGIN
  lag := FALSE;
  lead := FALSE;
  laglead := FALSE;
END;

PROCEDURE Echo (VAR response : CHAR);
BEGIN
  READ (KBD, response);
  WRITELN (response)
END;

PROCEDURE GeneralInstructions;
BEGIN
  ClrScr;
  GOTOXY (0, 4);
  WRITELN ('  This program allows ths design of lead, lag and');
  WRITELN ('laglead compensators which will provide specified');
  WRITELN ('phase shift and gain at a given frequency. ');
  WRITELN;
```

```
  WRITELN ('   Please press the <RETURN> key when ready. ');
  ECHO (response);
END;

PROCEDURE GetData;
BEGIN
  ClrScr;
  WRITE ('What type of compensator do you want ?');
WRITE(' (lead, lag or laglead) ');
  READLN (answer);
  IF (answer = 'lead') OR (answer = 'LEAD') THEN
     lead := TRUE;
  IF (answer = 'lag') OR (answer = 'LAG') THEN
     lag := TRUE;
  IF (answer ='LAGLEAD') OR (answer =  'laglead') THEN
     laglead := true;
  IF lag  THEN
    BEGIN
      WRITE ('What is the desired phase-lag?  (in degrees) ');
      READLN (phase);
      phase := - phase;
      REPEAT
        WRITE ('What is the gain (in dB) for this phase lag?  ');
        READLN (gain);
        IF gain > 0 THEN
          WRITELN (CHR(7),  'Gain must be negative',
                                   ' for a lag compensator!')
      UNTIL gain < 0
    END;
  IF lead THEN
    BEGIN
      WRITE ('What is the desired phase-lead?  (in degrees)  ');
      READLN (phase);
      REPEAT
        WRITE ('What is the gain (in dB) for this phase-lead?  ');
        READLN (gain);
        IF  gain < 0 THEN
           WRITELN (CHR(7), 'Gain must be positive,'
                                    ' for a lead compensator!')
      UNTIL gain > 0
    END;
  IF laglead THEN
    BEGIN
      WRITE ('What is the desired phase lead? (in degrees)  ');
      READLN (phase);
      REPEAT
        WRITE ('What is the gain (in dB) for this phase lead?  ');
        READLN (gain);
        IF gain > 0 THEN
          WRITELN (CHR(7) 'Gain must be negative',
                                  ' for a laglead compensator!')
      UNTIL gain < 0
    END;
  WRITE ('At what frequency?  (rad/sec)    ');
  READLN (omega);
END;

PROCEDURE LeadOrLag;
BEGIN
  phase := phase * pi / 180;
  tan_phi := SIN (phase) / COS (phase);
  ratio := EXP (LN (10) * (gain / 10));
  a := SQR (tan_phi) - ratio + 1;
  b := 2 * SQR (tan_phi) * ratio;
  c := (SQR (tan_phi) * ratio + ratio - 1) * ratio;
```

```
    IF ((lead) AND (a > 0) ) OR ((lag) and (c > 0)) THEN
       WRITELN ('A solution is not possible!  ')
    ELSE
      BEGIN
        IF lead THEN
          alpha := (- b - SQRT (SQR (b) - 4 * a * c)) /(2 * a);
        IF lag THEN
          alpha := (- b + SQRT (SQR (b) - 4 * a * c)) /(2 * a);
        zero := (omega / alpha) * SQRT((SQR (alpha) - ratio)/ (ratio - 1));
        pole := zero * alpha;
        WRITELN ('The transfer function of the compensator is  ');
        WRITELN (alpha:8:5, '(s + ', zero:8:5, ') / (s + ', pole:8:5,')')
      END
  END;

  PROCEDURE LeadAndLag;
  BEGIN
    phase := (phase + phi_2) * pi / 180;
    tan_phi := SIN (phase) / COS (phase);
    ratio := EXP (LN (10) * (gain / 10));
    c := SQR (tan_phi) - ratio + 1;
    b := 2 * SQR (tan_phi) * ratio;
    a := (SQR (tan_phi) * ratio + ratio - 1) * ratio;
    IF a > 0 THEN
       WRITELN ('A solution is not possible!  ')
    ELSE
      BEGIN
        alpha := (- b - SQRT (SQR (b) - 4 * a * c)) /(2 * a);
        zero := omega * SQRT ((1 - ratio)/(SQR(alpha) * ratio - 1));
        pole := zero * alpha;
        p := (alpha - 1) * COS (-phi_2 * pi/180) / sin (-phi_2 * pi/180);
       q := - p/2 + SQRT (SQR (p/2) - alpha);
       r := alpha / q;
       zero_2 := omega * r;
       pole_2 := zero_2 / alpha;
       WRITELN ('The transfer function of the compensator is    ');
       WRITE ('(s + ', zero:8:5, ')(s + ', zero_2:8:5);
      WRITELN (') / (s + ', pole:8:5,')(s + ', pole_2:8:4, ')')
    END;
  END;

  BEGIN  (* Main program *)
    GeneralInstructions;
    REPEAT
      Initialize;
      GetData;
      IF lead OR lag THEN LeadOrLag ELSE LeadAndLag;
      WRITE ('Do you want any more compensators calculated? ');
      WRITE (' (Y or N)  ');
      Echo (response)
    UNTIL  response IN ['N', 'n'];
  END.



  PROGRAM Stability;
  CONST  epsilon  = 1.0E-20;
         F = 8; (*Fieldwidth*)
         D = 4; (*Decimel*)
         blank = ' ';
  TYPE   COEFFICIENT = ARRAY [-2..20] OF REAL;
         STRING14 = STRING[14];
  VAR
    i, j, k, m, n  :  INTEGER;
    degree  :  0..20;
    P, Q, R  :  COEFFICIENT;
```

```
  response, answer : CHAR;
  OutFile : Text;
  FileName : STRING[14];

(*$I  Files.pas *)

PROCEDURE Echo (VAR answer : CHAR);
BEGIN
  READ (KBD, answer);
  RESET (KBD);
  WRITELN (answer)
END;

PROCEDURE GeneralInstructions;
BEGIN
  ClrScr;
  GoToXY (0, 5);
  WRITELN ('               This program  determines the stability of a');
  WRITELN ('dynamic system by applying the  Routh criterion to find');
  WRITELN ('the number of roots of the characteristic polynomial in');
  WRITELN ('the right half of the s-plane.  The  relative stability');
  WRITELN ('can be determined by  shifting  the axis to the left in');
  WRiTELN ('the s-plane by a specified amount.  It can also be used');
  WRITELN ('for determining the stability of  sampled-data  systems');
  WRITELN ('by transforming from the z-plane to the w-plane.');
  WRITELN;
  WRITELN ('   Please press any key when ready. ');
  Echo (answer);
END;

PROCEDURE Verify;
VAR
  i, j   : INTEGER;
  answer : CHAR;
BEGIN
  REPEAT
    ClrScr;
    FOR i := degree DOWNTO 0 DO
            WRITELN ('A[', i, '] = ', P[i]:F:D, ',');
    WRITELN;
    WRITE ('Are these entries correct?  (Y or N):  ');
    Echo (answer);
    IF answer IN ['N', 'n'] THEN
      BEGIN
        WRITELN ('Which entry is wrong?   ');
        READLN (j);
        WRITELN ('Enter correct value of A[', j,'] ');
        READLN (P[j]);
      END
  UNTIL answer IN ['Y', 'y'];
  WRITELN;
END;

PROCEDURE Routh;
BEGIN
  P[-1] := 0;
  P[-2] := 0;
  FOR i := degree DOWNTO (degree-1) DO
    BEGIN
      j := i;
      k := i;
      WRITE ('s^', j:2, blank:5);
      WHILE j >= 0 DO
        BEGIN
          WRITE (P[j]:F:D, blank:4);
          j := j - 2;
        END;
```

```
      WRITELN;
    END; (* The first two rows have been obtained. *)
  FOR i := (degree-2) DOWNTO 0 DO
     (* Now we obtain the remaining rows *)
    BEGIN
      j := i;
      k := i + 1;
      WRITE ('s^', j:2, blank:5);
      IF P[k] = 0 THEN P[k] := epsilon;
      (* If the pivot = 0 then replace it by epsilon *)
      WHILE j >= 0 DO
        BEGIN
          P[j] :=  P[j] - P[k+1] * P[j-1]/P[k];
          WRITE (P[j]:F:D, blank:4);
          j := j - 2;
        END;
      WRITELN;
    END;
  END;

PROCEDURE ShiftAxis;
VAR sigma : REAL;
BEGIN
  P := Q;  (* Restore the original polynomial *)
  WRITE ('Enter the desired shift (negative for leftwards):  ');
  READLN (sigma);
  k := degree;
  j := 0;
  WHILE j <= k DO
    BEGIN
      FOR i := (k-1) DOWNTO j DO
      P[i] := P[i] + sigma * P[i+1];
      j := j +1;
      END;
END;

FUNCTION factorial (n : INTEGER) : INTEGER;
(* It calculates the factorial of an integer *)
VAR  temp, i, j : INTEGER;
BEGIN
  IF n > 0 THEN
    BEGIN
      temp := n;
      j := n;
      FOR i := n DOWNTO 2 DO
        BEGIN
          temp := temp * (j - 1);
          j := j - 1
        END;
      factorial := temp;
    END
  ELSE  factorial := 1;
END;

FUNCTION combination (i,k : INTEGER) : REAL;
(* It calculates the combinations of i objects taken k at a time *)
BEGIN
  combination := factorial(i)/(factorial(k) * factorial(i-k));
END;

FUNCTION Signum (i : INTEGER) : INTEGER;
(* It is +1 for even i and -1 for odd i *)
BEGIN
  IF Odd (i) THEN
    Signum := -1
  ELSE Signum := 1
END;
```

```
PROCEDURE zToW;
(* It performs the bilinear transformation from the z-plane to
   the w-plane for applying Routh test to sampled-data systems *)
VAR i, j, k : INTEGER;
BEGIN
  n := degree;
  j := 1;
  R := P;
  FOR k := 0 TO 1 DO
    R[n-k] := P[n] + Signum(k) * P[n-1];
  P := R;
  FOR i := 1 TO (n - 1)  DO
    BEGIN
      j := j + 1;
      R[n] := P[n] + P[n-j];
      FOR k := 1 TO (j-1) DO
          R[n-k] := P[n-k+1] + P[n-k]
                       + Signum(k) * P[n-j] * combination (j,k);
      R[n-j] := P[n-j+1] + Signum(j) * P[n-j];
      P := R;
    END;
 END;

PROCEDURE Input;
(* For the input of the characteristic polynomial. It also includes
   z-to-w transformation for sampled-data systems *)
VAR      answer : CHAR;
         i, degreeP : INTEGER;
         FileName : STRING14;
BEGIN
  ClrScr;
  WRITELN ('Do you want to get the coefficeints from the keyboard ');
  WRITE ('or to read them from a file?  Enter K or F   ');
  Echo (answer);
  IF answer IN ['F', 'f'] THEN
    BEGIN
      IF GetFile (OutFile) THEN
        BEGIN
          degreeP := - 1;
          WHILE NOT EOF (OutFile) DO
            BEGIN
              degreeP := degreeP + 1;
              READLN (OutFIle, P[degreeP])
            END;
          degree := degreeP;
          CLOSE (OutFile);
          Verify
        END
    END
  ELSE
    BEGIN
      WRITE ('Enter degree of the polynomial:  ');
      READLN (degree);
      FOR i := degree DOWNTO 0 DO
        BEGIN
          WRITE ('Enter   A[', i, '] :  ');
          READLN (P[i]);
        END;
      Verify;
     WRITE ('Do you want to store these in a file?  Enter Y or N ');
     Echo (answer);
     IF answer IN ['Y', 'y'] THEN
       BEGIN
         CreateFile (OutFile, FileName);
         REWRITE (OutFile);
         FOR i := 0 TO degree DO
           WRITELN (OutFile, P[i]);
```

```
        CLOSE (OutFile)
      END;
    END;
  WRITE ('Is this a polynomial in z?   (Y or N) :  ');
  Echo (response);
  IF response in ['Y','y'] THEN zToW;
  Q := P; (* Save P(s) as Q(s)   *)
END; (* Input *)


BEGIN (* MAIN PROGRAM *)
  GeneralInstructions;
  Input;
  ClrScr;
  Routh;
  REPEAT
    WRITE ('Do you want to shift the axis?   (Y or N) ');
    Echo (response);
    IF response in ['Y', 'y'] THEN
      BEGIN
        ShiftAxis;
        Routh;
      END
  UNTIL response in ['N', 'n']
END.


PROGRAM Transition;
(*$U+ *)
{This program calculates the transition matrices F and G
   given A,B and T}

CONST     MaxN = 25;
          MaxM = 25;
          Field   = 10 {Fieldwidth};
          Decimal = 6 {Number of digits after decimal point};

TYPE      Matrix = ARRAY [1..MaxN, 1..MaxM] OF REAL;
          Vector = ARRAY [1..MaxN] OF REAL;
          String14 = STRING[14];
          String80 = STRING[80];

VAR    A, B, P, C, F, G  :  Matrix;
       NumberOfTerms, rowA, columnA, rowB, columnB, i, j, k : INTEGER;
       T          : REAL;
       OutFile    : TEXT;
       answer     : CHAR;
       Title      : String80;
       square     : BOOLEAN;
       FileName   : STRING14;

(*$I Files.Pas *)
(*$I Matrices.Pas *)

PROCEDURE Input;
BEGIN
  ClrScr;
  WRITELN ('This program calculates the state transition matrices');
  WRITELN ('for a linear system from the matrices A and B and the');
  WRITELN ('sampling interval T by seris approximation.');
  WRITELN;
  WRITELN ('Press any key to continue... ');
  Echo (answer);
  WRITELN ('Get matrix A ');
  WRITELN ('Press any key to continue...');
```

```
  Echo (answer);
  square := TRUE;
  GetMatrix (A, rowA, columnA, MaxN, MaxM);
  WRITELN ('Get matrix B ');
  WRITELN ('Press any key to continue...');
  Echo (answer);
  square := FALSE;
  GetMatrix (B, rowB, columnB, MaxN, MaxM);
  WRITE ('Do you want to print the matrices A and B? (Y or N)  ');
  Echo (answer);
  IF answer IN ['Y', 'y'] THEN
    BEGIN
      Title := 'The matrix  A:';
      PutMatrix (A, rowA, columnA, Title, 'P');
      WRITELN (OutFile);
      WRITELN (OutFile);
      Title := 'The matrix  B:';
      PutMatrix (B, rowB, columnB, Title, 'P');
      WRITELN (OutFile);
      WRITELN (OutFile);
    END;
  WRITE ('Enter the value of the sampling interval T:      ');
  READLN (T);
  FOR i := 1 TO rowA DO
    BEGIN
      FOR j := 1 TO columnA DO
        A [i,j] := A [i,j] * T
    END;
  FOR i:= 1 TO rowB DO
    BEGIN
      FOR j := 1 TO columnB DO
        B [i,j] := B [i,j] * T
    END;
  WRITE ('Enter the number of terms to be used in the series:    ');
  READLN (NumberOfTerms);
END;  (* Input *)

PROCEDURE AddI(M1: Matrix; VAR M2 : Matrix);
{This procedure adds the identity matrix to M1}
 VAR      i, j : INTEGER;
 BEGIN
   FOR i := 1 TO rowA DO
     BEGIN
       FOR j:= 1 TO rowA DO
         M2[i,j] := M1[i,j];
       M2[i,i] := M1[i,i] + 1
     END;
 END;

PROCEDURE Scale(M1 : Matrix;  k : INTEGER; VAR  M2: Matrix);
{This procedure divides the elements of M1 BY k to obtain M2}
VAR       i,j  : INTEGER;
BEGIN
  FOR i := 1 TO rowA DO
    BEGIN
      FOR j := 1 TO columnA DO
        M2[i,j] := M1[i,j]/k
    END;
END;

PROCEDURE Multiply (M1,M2 : Matrix;  row_1, column_1, row_2,
                     column_2 :INTEGER;  VAR PRODUCT : Matrix);
 (* This procedure multiplies two conformable matrices M1 and M2 to
   get the matrix PRODUCT = M1 . M2 *)
VAR    i, j, k : INTEGER;
         temp  : REAL;
```

```
BEGIN
   FOR i := 1 TO row_1 DO
     BEGIN
       FOR j := 1 TO column_2 DO
         BEGIN
           temp := 0;
           FOR k := 1 TO column_1 DO
             IF (M1 [i,k] <> 0) AND (M2 [k,j] <> 0) THEN
               temp := temp + M1 [i,k] * M2 [k,j];
           PRODUCT[i,j] := temp
         END;
     END;
END;

PROCEDURE Output;
{This procedure outputs the matrix [M]}
VAR       ANSWER : CHAR;
BEGIN
  Title := 'The matrix F:';
  WRITE ('Do you want the output sent to the Printer or the ');
  WRITE ('Console?  (P or C)  ');
  Echo (answer);
  PutMatrix (F, rowA, columnA, Title, answer);
  WRITELN ('Press any key to continue... ');
  Echo (answer);
  WRITE ('Do you want to print the matrix F?  (Y or N)  ');
  Echo (answer);
  IF answer IN ['Y', 'y'] THEN
    PutMatrix (F, rowA, columnA, Title, 'P');
  WRITELN ('Press any key to continue... ');
  Echo (answer);
  Title := 'The matrix G: ';
  PutMatrix (G, rowB, columnB, Title, answer);
  WRITELN ('Press any key to continue... ');
  Echo (answer);
  WRITE ('Do you want to print the matrix G?  (Y or N)  ');
  Echo (answer);
  IF answer IN ['Y', 'y'] THEN
    PutMatrix (G, rowB, columnB, Title, 'P');
  WRITELN ('Press any key to continue... ');
  Echo (answer);
  WRITE ('Do you want to store the matrices F and G on the disk?');
  WRITE ('  (Y or N) ');
  Echo (answer);
  IF answer IN ['Y', 'y'] THEN
    BEGIN
      Title := 'The matrix  F:';
      WRITE ('Storing matrix F.  Press any key to continue... ');
      Echo (answer);
      PutMatrix (F, rowA, columnA, Title, 'F');
      Title := 'The matrix G:';
      WRITE ('Storing marix G.  Press any key to continue... ');
      Echo (answer);
      PutMatrix (G, rowB, columnB, Title, 'F');
    END;
  END; {Output}


BEGIN {MAIN PROGRAM}
  Input;
  Scale (A, NumberOfTerms, C);
  AddI (C,F);
  FOR k:= (NumberOfTerms - 1)  DOWNTO 2 DO
    BEGIN
      Scale (A,k,P);
      Multiply (F, P, rowA, columnA, rowA, columnA, C);
```

```
      AddI(C,F)
    END;
  Multiply (F, B, rowA, columnA, rowB, columnB, G);
  Multiply (F, A, rowA, columnA, rowA, columnA, C);
  AddI(C,F);
  Output;
END.


PROGRAM TransferFunction;
(*$U+ *)
(* This program calculates the transfer function of a linear
   system from the state equations, given A,B and C *)

CONST     MaxN = 10;
          MaxM = 10;
          Field      = 14 {Fieldwidth};
          decimal    = 7 {Number of digits after decimal point};
          assumedZero     = 1.0E-13;
          blank  =     ' ';

TYPE      Matrix = ARRAY[1..MaxN, 1..MaxM] OF REAL;
          Vector = ARRAY [1..MaxN] OF REAL;
          String14 = STRING [14];
          String80 = STRING [80];

VAR       A, U, M, N , V , W, T, P, Q         : Matrix;
          B, C, y , Z                         : Vector;
          rowA, columnA, columnU,
             i, j, k, s, r, rank              : INTEGER;
          big, ab, pivot, tol                 : REAL;
          found                               : BOOLEAN;
          OutFile                             : TEXT;
          answer                              : CHAR;
          Title                               : String80;
          FileName                            : String14;
          square                              : BOOLEAN;

(*$I Files.Pas *)
(*$I Matrices.Pas *)

PROCEDURE GeneralInstructions;
BEGIN
  ClrScr;
  GOTOXY (1, 4);
  WRITELN ('           This program calculates the transfer function');
  WRITELN ('of a single-input  single-output system from its state');
  WRITELN ('equations.  Given the matrix A and vectors B and C, it');
  WRITELN ('gives the  controllabilty  matrix, the  characteristic');
  WRITELN ('polynomial  and  the  numerator  polynomial.  It  also');
  WRITELN ('calculates the  matrix which will  transform the given ');
  WRITELN ('state equations to the phase-variable canonical form. ');
  WRITELN;
  WRITELN ('        Please press the <ENTER> key when ready. ');
  READLN;
END;

PROCEDURE Input;
(* This procedure inputs all the data needed *)
VAR
     i, j : INTEGER;
BEGIN
  ClrScr;
  GoToXY(5,5);
```

```
    WRITELN ('Get matrix A.  Press any key to continue ...');
    Echo (answer);
    square := TRUE;
    ClrScr;
    GoToXY(5,5);
    GetMatrix (A, rowA, columnA, MaxN, MaxM);
    WRITELN ('Get vector B.  Press any key to continue ...');
    Echo (answer);
    square := FALSE;
    Getvector (B, rowA, MaxN);
    ClrScr;
    GoToXY(5,5);
    WRITELN ('Get vector C.  Press any key to continue ...');
    Echo (answer);
    Getvector (C, columnA, MaxN);
    ClrScr;
    GoToXY(5,5);
    WRITE ('Do you want to print A, B and  C?  (Y or N)  ');
    Echo (answer);
    IF answer IN ['Y', 'y'] THEN
      BEGIN
        Assign (OutFile, 'LST:');
        RESET (OutFIle);
        Title := 'The matrix A:';
        PutMatrix (A, rowA, columnA, Title, 'P');
        WRITELN (OutFile);
        WRITELN (OutFile);
        Title := 'The vector  B:';
        Putvector (B, rowA,  Title, 'P');
        WRITELN (OutFile);
        WRITELN (OutFile);
        Title := 'The vector C:';
        Putvector (C, columnA, Title, 'P');
        WRITELN (OutFile);
        WRITELN (OutFile)
      END;
   ClrScr;
 END; (* Input *)

 PROCEDURE MatrixMultiply (M1,M2 : Matrix;  row_1,column_1,row_2,
                    column_2 :INTEGER;  VAR PRODUCT : Matrix);
  (* This procedure multiplies two conformable matrices M1 and M2
     to get the matrix PRODUCT = M1 . M2 *)
  VAR     i, j, k : INTEGER;
          cross   : REAL;
  BEGIN
    FOR i := 1 TO row_1 DO
      FOR j := 1 TO column_2 DO
        BEGIN
          cross := 0;
          FOR k := 1 TO column_1 DO
            IF (M1 [i,k] <> 0) AND (M2 [k,j] <> 0) THEN
              cross := cross + M1[i,k] * M2[k,j];
          product[i,j] := cross;
        END;
 END;
 PROCEDURE ColumnMultiply (M : Matrix;  V : Vector;
                                  row, column : INTEGER;
                           VAR Product : Vector);
  (* This procedure multiplies the matrix M with the column vector V
     to get the vector Product = M . V *)
  VAR     i, j : INTEGER;
          cross   : REAL;
  BEGIN
    FOR i := 1 TO row DO
```

```
        BEGIN
          cross := 0;
          FOR j := 1 TO column DO
            IF (M [i,j] <> 0) AND (V [j] <> 0) THEN
              cross := cross + M[i,j] * V[j];
          Product[i] := cross;
        END;
END;

PROCEDURE RowMultiply (V : Vector;  M : Matrix;
                       row, column : INTEGER;  VAR Product : Vector);
 (* This procedure multiplies the row vector V with the matrix M to
    get the vector Product = V M  *)
 VAR    i, j : INTEGER;
        cross   : REAL;
 BEGIN
   FOR i := 1 TO column DO
       BEGIN
         cross := 0;
         FOR j := 1 TO row DO
           IF (M [j,i] <> 0) AND (V [j] <> 0) THEN
             cross := cross + M[j,i] * V[j];
         Product[i] := cross;
       END;
END;

PROCEDURE Controllability (A : Matrix; B : Vector; VAR U : Matrix);
(* This procedure calculates the n+1 columns of the controllability
   matrix, which are later used for determining the characteristic
   polynomial. *)
VAR    i, j   : INTEGER;
       M, N   : Vector;
BEGIN
  columnU := columnA + 1;
  M := B;
  FOR j := 1 TO rowA DO
        U [j, 1] := M [j];
  FOR i := 2 TO columnU DO
    BEGIN
      ColumnMultiply (A, M, rowA, columnA, N);
      M := N;
      FOR j := 1 TO rowA DO
          U [j, i] := M [j];
    END;
  V := U; (* Store it as the matrix V as well. *)
  Title := 'The controllability matrix : ';
  PutMatrix (U, rowA, columnA, Title, 'C');
  WRITE ('Do you want it printed?  (Y or N)    ');
  Echo (answer);
  IF answer in ['Y', 'y'] THEN
  BEGIN
    PutMatrix (U, rowA, columnA, Title, 'P');
    WRITELN (OutFile);
    WRITELN (OutFile)
  END;
  ClrScr;
  WRITE ('Do you want to store the controllability matrix on disk?');
  WRITE ('  (Y or N)  ');
  Echo (answer);
  IF answer in ['Y', 'y'] THEN
      PutMatrix (U, rowA, columnA, Title, 'F');
  WRITE ('Press any key to continue ...');
  Echo (answer);
  ClrScr;
END;
```

```
(* We now give a series of procedures which are used for trans-
   forming the controllability matrix with (n+1) columns to the
   Hermite normal form. For a controllable system the last column
   gives the coefficients of the characteristic polynomial.    *)
FUNCTION max (a,b: REAL): REAL;
(* It determines the larger of a and b *)
BEGIN
  IF a < b THEN   max := b
  ELSE  max := a
END;
PROCEDURE Norm;  (* Test for zero number, tolerance *)
VAR
  i, j:INTEGER;
BEGIN
  tol := 0;
  FOR i := 1 TO rowA DO
    BEGIN
      FOR j := 1 TO columnU DO
        tol := Max (tol, ABS(V[i,j]))
    END;
  tol := tol * assumedZero
END;

PROCEDURE Search;
(* Procedure for searching for the pivot
     in the pth row and lth column *)
VAR
  j: INTEGER;
BEGIN
  big := tol;
  WHILE (s <= columnU) AND (big <= tol) DO
    BEGIN
      FOR j:= k TO rowA DO
        BEGIN
          ab := ABS (V[j,s]);
          IF big < ab THEN
            BEGIN
              big := ab;
              r := j
            END;
        END;
      IF big <= tol THEN
        BEGIN
          FOR j := k TO rowA DO
            V[j,s] := 0;
          s := s + 1
        END
    END;
  IF big > toL THEN
    BEGIN
      rank := rank + 1;
      found := TRUE;
    END
    ELSE found := FALSE;
END;


(* Copy row with pivot and interchange rows k and p *)
PROCEDURE InterChange;
VAR
  j : INTEGER;
BEGIN
  FOR j := 1 TO columnU DO
    BEGIN
      P[1,j] := V[r,j];  (* Row with pivot *)
```

```
      V[r,j] := V [k,j];
      V[k,j] := P [1,j];
    END
END;

(* Obtain column for outer product *)
PROCEDURE Column;
VAR
  i : INTEGER;
BEGIN
  FOR i := 1 TO rowA DO
    Q[i,1] := V[i,s];
  Q[k,1] := Q[k,1] - 1
END;
(*  Add outer product to transform the sth column *)
PROCEDURE AddOuterPr;
VAR
  i, j : INTEGER;
BEGIN
  pivot := V[k,s];
  FOR i := 1 TO rowA DO
    BEGIN
      FOR j := s TO columnU DO
        IF (Q[i,1] <> 0) AND (P[1,j] <> 0) THEN
          BEGIN
            V[i,j] := (V[i,j] * pivot - (Q[i,1]*P[1,j]))/pivot;
            IF ABS (V[i,j]) <= tol THEN
              V[i,j] := 0
          END
    END
END;

(* Transform the V-matrix to the Hermite normal form by
   adding the outer product for each pivot *)

PROCEDURE Transform (VAR  V : Matrix; rowA, columnU : integer);
BEGIN
  Norm;
  s := 1;
  k := 1;
  rank := 0;
  WHILE (k <= rowA) AND (s <= columnU) DO
    BEGIN
      Search;
      IF found THEN
        BEGIN
          InterChange;
          Column;
          AddOuterPr;
          s := s + 1;
          k := k + 1;
        END;
    END;
END;


(* The characteristic polynomial is obtained from the last column
   of the transformed matrix *)

PROCEDURE Characteristic;
  VAR    answer, ch  : CHAR;
         i           : INTEGER;
  BEGIN
    Transform (V, rowA, columnU);
    WRITELN;
```

```
WRITELN ('The rank of U is  ', rank,'! ');
IF rank <> rowA THEN
(* The method does not work if the system is uncontrollable. If
  it is observable, the transfer function can be obtained by
  transposing A and interchanging B and C *)
  BEGIN
    WRITE ('The system is uncontrollable! Transpose A and  ');
    WRITELN ('interchange B and C. ');
    READLN
 END;
FOR i := 1 TO rowA DO
  y[i] := V[i, columnU];
WRITELN ('The characteristic polynomial is of degree ', rowA);
WRITE ('The coefficients, starting from the next ');
WRITELN ('highest degree term, are: ');
FOR i := 0 to (rowA - 1)  DO
   WRITELN (-y[rowA -i] :Field:decimal);
WRITELN;
WRITELN ('Do yo want to store these on the disk?  (Y or N) ');
Echo (answer);
IF answer IN ['Y', 'y'] THEN
  BEGIN
    CreateFile (OutFile, FileName);
    WRITELN (OutFile, rowA);
    FOR i := 0 TO (rowA - 1)  DO
      WRITELN (OutFile, -y[rowA - i]:FIELD:DECIMAL);
    CLOSE (OutFile)
  END;
END;

(* We now obtain the transformation matrix T and then the numerator
   polynomial from CT *)

PROCEDURE Numerator;
  VAR  i, j : INTEGER;
  BEGIN
  (* First we form the matrix W from the characteristic polynomial as
     the inverse of the  controllability matrix  for the controllable
     canonical form. *)
    FOR j := 1 TO (columnA - 1) DO
      W [1, j] := - y [j + 1];
    W [1, columnA] := 1;
    FOR i := 2 TO rowA DO
      BEGIN
        FOR j := 1 TO (columnA - i + 1) DO
             W [i,j] := W [i - 1, j + 1];
        FOR j := (columnA - i + 2) to columnA  DO
            W [i,j] := 0;
      END;
   (* Premultiply W by U to get the transformation matrix T *)
    MatrixMultiply (U, W, rowA, columnA, rowA,columnA, T);
    Title := 'The matrix for transformation to phase-variable form : ';
    PutMatrix (T, rowA, columnA, Title, 'C');
    WRITE ('Do you want it printed?  (Y or N)    ');
    Echo (answer);
    IF answer in ['Y', 'y'] THEN
      BEGIN
        PutMatrix (T, rowA, columnA, Title, 'P');
        WRITELN (OutFile);
        WRITELN (OutFile)
      END;
    WRITE ('Do you want it stored on the disk?  (Y or N)  ');
    Echo (answer);
    IF answer in ['Y', 'y'] THEN
        PutMatrix (T, rowA, columnA, Title, 'F');
    WRITE ('Press any key to continue ...');
```

```
    Echo (answer);
    RowMultiply (C, T, columnA, rowA, Z);
    (* Premultiply T by C to get Z which has the
      numerator coefficients *)
    WRITELN ('The degree of the numerator is ', rowA - 1, '.');
    WRITE ('The coefficients, starting with the  ');
    WRITELN ('highest degree term, are :  ');
    FOR i := columnA DOWNTO 1 DO
      WRITELN (Z [i] :Field: decimal);
    WRITE ('Do yo want to store these on the disk?  (Y or N) ');
    Echo (answer);
    IF answer IN ['Y', 'y'] THEN
      BEGIN
        CreateFile (OutFile, FileName);
        WRITELN (OutFile, rowA - 1);
        FOR i := rowA DOWNTO 1  DO
          WRITELN (OutFile, z[r]:FIELD:DECIMAL);
        CLOSE (OutFile)
      END;
    END;

BEGIN (* Main program *)
  GeneralInstructions;
  Input;
  Controllability (A, B, U);
  Characteristic;
  Numerator;
END.


{This program, named Files.Pas, is called as an include file in most of
  the programs that have been listed.  It enables creation of new files,
  as well as, writing to or reading from a disk file}

FUNCTION FileExists (VAR SourceFile : TEXT; FileName : STRING14) : BOOLEAN;
(* It determines whether a called file already exist *)
BEGIN
  ASSIGN (SourceFile , FileName);
  {$I-}  RESET (SourceFile);  {$I+}
  IF  IOresult = 0  THEN
    BEGIN
      FileExists := TRUE;
      CLOSE (OutFile)
    END
  ELSE
    FileExists := FALSE
END; { FileExists }

FUNCTION GetFile (VAR SourceFile : TEXT) : BOOLEAN;
(* If a file already exists then this procedure allows reading of data from
  that file.  If a called file does not exist then it allows the entry of
  the correct name of the file without causing the program to crash.  *)
VAR
  answer : CHAR;
  found, abort : BOOLEAN;
  FileName : STRING14;
BEGIN
  REPEAT
    WRITE ('Enter the name of the input file:  ');
    READLN (FileName );
    IF FileExists (SourceFile , FileName)  THEN
      BEGIN
        found := TRUE;
        abort := FALSE
      END
```

```
    ELSE
      BEGIN
        found := FALSE;
        WRITELN(' *** The file ', FileName , ' cannot be found. ***');
        WRITE ('Do you want to try a different file name? (Y or N)  ');
        READ (KBD , answer) ;
        WRITELN (answer) ;
        IF  answer  IN ['Y' , 'y']  THEN
          abort := FALSE
        ELSE
          abort := TRUE
      END
  UNTIL found OR  abort;
  GetFile := found
END; { GetFile }

PROCEDURE CreateFile (VAR OutFile : TEXT ; VAR FileName : STRING14) ;
(*  This procedure allows the creation of a new file of desired name.  If a
    file with that name already exists then the user is asked if the old file
    is to be destroyed.  *)
VAR
  answer : CHAR;
  Create : BOOLEAN;
BEGIN
  REPEAT
    WRITE('Enter the name of the output file:  ');
    READLN( FileName );
    IF FileExists( OutFile , FileName )  THEN
      BEGIN
        WRITELN(' *** The file ', FileName , ' already exists. *** ');
        WRITE('Do you want to destroy it?  (Y or N)  ');
        READ( KBD , answer );
        WRITELN( answer ) ;
        IF  answer  IN ['Y' , 'y']  THEN
          Create := TRUE
        ELSE
          Create := FALSE
      END
    ELSE
      Create := TRUE;
    IF Create THEN
      REWRITE( OutFile )
  UNTIL Create
END; { CreateFile }


{ This program, named Matrices.Pas, is called as an include file in
  most programs that utilize matrices.  It enables writing and
  reading of matrices and vectors from disk-files.

  The following must be declared in the calling program

     $I Files.Pas;

     TYPE
       Matrix = ARRAY [1..MaxN , 1..MaxM] OF REAL;
       Vector = ARRAY [1..MaxN] OF REAL;
       STRING14 = STRING[14];
       STRING80 = STRING[80];

     VAR
       Field, Decimal : REAL  (the output field width and number of
                               decimal places of the entries.)

       square : BOOLEAN;}
```

```
PROCEDURE Echo (VAR answer : CHAR);
BEGIN
  READ (KBD, answer);
  RESET (KBD);
  WRITELN (answer)
END;

PROCEDURE PutMatrix (A : Matrix; n, m : INTEGER;
                          Title : STRING80; Place : CHAR);
VAR   i, j, NumberPerLine : INTEGER;
      SourceFile : TEXT;
      FileName : STRING14;
BEGIN
  IF Place IN ['F', 'f'] THEN
    BEGIN
      CreateFile (SourceFile, FileName);
      WRITELN (SourceFile, Title);
      WRITELN (SourceFIle, n, '  ', m)
    END
  ELSE
    IF Place IN ['P', 'p'] THEN
      BEGIN
        ASSIGN (SourceFile, 'LST:' );
        RESET (SourceFile);
        ClrScr;
        WRITELN (SourceFile, Title)
      END
    ELSE
      BEGIN
        ASSIGN (SourceFile, 'CON:' );
        RESET (SourceFile);
        ClrScr;
        WRITELN (SourceFile, Title)
     END;
 NumberPerLine := 80 DIV (Field + 1);
 FOR i := 1 TO n DO
   BEGIN
     FOR j := 1 TO m DO
       BEGIN
         WRITE (SourceFile, A[i,j] : Field : Decimal, ' ');
         IF ((j MOD NumberPerLine) = 0) OR (j = m) THEN
           WRITELN (SourceFile)
       END  {j-loop}
  END;  {i-loop}
  CLOSE (SourceFile)
END;

PROCEDURE GetMatrix( VAR A : Matrix ; VAR n, m : INTEGER ;
                                    MaxN, MaxM : INTEGER );
      { Enter a matrix and store it in a file, if desired }

VAR
  KeyBoard : BOOLEAN;
  answer : CHAR;

PROCEDURE SelectInputMode;
VAR
  answer : CHAR;
BEGIN
  ClrScr;
  WRITE ('Do you want to input the matrix from the keyboard or from');
  WRITE (' a file?  (K or F)  ');
  Echo (answer);
  IF  answer  IN  ['F' , 'f']  THEN
    KeyBoard := FALSE
```

```
    ELSE
      KeyBoard := TRUE
  END;

  PROCEDURE RowAndColumnDimensions;
  BEGIN
    ClrScr;
    REPEAT
      WRITELN ('The dimensions of the matrix:');
      WRITE (' Number of rows =     ');
      READLN (n);
      IF square THEN
        m := n
      ELSE
        BEGIN
          WRITE (' Number of columns =     ');
          READLN (m)
        END
    UNTIL  ( m  IN  [ 1..MaxM ] ) AND ( n  IN  [ 1..MaxN ] )
  END;

  PROCEDURE MatrixIn; { Type in a matrix }
  VAR
    i, j : INTEGER;
  BEGIN
    FOR  i := 1  TO  n  DO
      BEGIN
        WRITELN ('Row ', i);
        FOR  j := 1  TO  m  DO
          BEGIN
            WRITE (' ', j ,':   ');
            READLN (A[i,j])
          END
      END
  END;

  PROCEDURE Verify; { Verify the matrix that was typed in }
  VAR
    i, j : INTEGER;
    answer : CHAR;
    Title : String80;
  BEGIN
    REPEAT
      ClrScr; {Clear screen}
      Title := 'The following matrix was entered: ';
      PutMatrix (A, n, m, Title, 'C');
      WRITE ('Are the entries correct?  (Y or N)  ');
      Echo (answer);
      WRITELN;
      IF  NOT (answer  IN  [ 'Y' , 'y' ])  THEN
        BEGIN
          WRITE ('Which entry is wrong?  Row(space)Column   ');
          READLN (i, j);
          WRITE ('entry (', i , ',' , j , ') =  ' );
          READLN (A[i,j])
        END;
    UNTIL  answer  IN  ['Y' , 'y']
  END;

  PROCEDURE ReadMatrix;  { from a file }
  VAR
    i, j : INTEGER;
    SourceFile : TEXT;
    Title : String80;
```

```
BEGIN
  WRITELN;
  IF  GetFile (SourceFile)  THEN
    BEGIN
      RESET (SourceFile);
      READLN (SourceFile, Title);
      READLN (SourceFile , n , m);
      FOR  i := 1  TO  n  DO
        BEGIN
          FOR  j :=  1  TO  m  DO
            BEGIN
              READ (SourceFile , A[i,j]);
              IF  EOLN (SourceFile) OR (j = m)  THEN
                READLN (SourceFile)
            END {j-loop}
        END; {i-loop}
      CLOSE (SourceFile)
    END
  ELSE
    KeyBoard := TRUE
END;

BEGIN { GetMatrix }
  SelectInputMode;
  IF  NOT KeyBoard  THEN
    ReadMatrix
  ELSE
    BEGIN
      RowAndColumnDimensions;
      MatrixIn
    END;
  Verify;
  WRITELN;
  Title := 'The matrix: ';
  WRITE ('Do you want to write the matrix into a file?  (Y or N)  ');
  Echo (answer);
  IF  answer  IN  ['Y' , 'y']  THEN
   PutMatrix (A, n, m, Title, 'F');
END;

PROCEDURE PutVector (A:Vector; n:INTEGER; Title:String80; Place:CHAR);
VAR    i : INTEGER;
       Sourcefile : TEXT;
       FileName : String14;
BEGIN
  IF Place IN ['F', 'f'] THEN
    BEGIN
      CreateFile (SourceFile, FileName);
      WRITELN (SourceFile, Title);
      WRITELN (SourceFIle, n)
    END
  ELSE
    IF Place IN ['P', 'p'] THEN
      BEGIN
        ASSIGN (SourceFile, 'LST:' );
        RESET (SourceFile);
        ClrScr;
        WRITELN (SourceFile, Title)
      END
    ELSE
      BEGIN
        ASSIGN (SourceFile, 'CON:' );
        RESET (SourceFile);
        ClrScr;
        WRITELN (SourceFile, Title)
      END;
```

```
    FOR i := 1 TO n DO
      WRITELN (SourceFile, A[i] : Field : Decimal, ' ');
    CLOSE (SourceFile)
END;

PROCEDURE GetVector  (VAR  A : Vector; VAR n : INTEGER; MaxN : INTEGER);
   {Enter a vector and store it in a file, if desired }

VAR  KeyBoard : BOOLEAN;
     answer : CHAR;

PROCEDURE SelectInputMode;
VAR
  answer : CHAR;
BEGIN
  ClrScr;
  WRITE ('Do you want to input the vector from the keyboard or from');
  WRITE (' a file?  (K or F)  ');
  Echo (answer);
  IF  answer  IN  ['F' , 'f']  THEN
    KeyBoard := FALSE
  ELSE
    KeyBoard := TRUE
END;

PROCEDURE RowDimension;
BEGIN
  ClrScr;
  REPEAT
    WRITE ('The dimension of the vector =  ');
    READLN (n)
  UNTIL  n IN [1..MaxN]
END;

PROCEDURE VectorIn;    {Type in a vector}
VAR  i : INTEGER;
BEGIN
  FOR i := 1 TO n DO
    BEGIN
      WRITE ('Row ', i, ':  ');
      READLN (A[i])
    END
END;

PROCEDURE VerifyVector; { Verify the vector that was typed in }
VAR   i : INTEGER;
      answer : CHAR;
BEGIN
  REPEAT
    ClrScr; {Clear screen}
    Title := 'The following vector was entered: ';
    PutVector (A, n, Title, 'C');
    WRITE ('Are the entries correct?  (Y or N)  ');
    Echo (answer);
    IF  NOT (answer  IN  [ 'Y' , 'y' ])  THEN
      BEGIN
        WRITE ('Which entry is wrong?  Row   ');
        READLN (i);
        WRITE ('entry (', i , ') =  ' );
        READLN (A[i])
      END;
  UNTIL  answer  IN  ['Y' , 'y']
END;
```

```
PROCEDURE ReadVector; { from a file }
VAR
  i : INTEGER;
  SourceFile : TEXT;
  Title : String80;
BEGIN
  WRITELN;
  IF  GetFile (SourceFile)  THEN
    BEGIN
      RESET (SourceFile);
      READLN (SourceFile, Title);
      READLN (SourceFile , n);
      FOR  i := 1  TO  n  DO
        READLN (SourceFile , A [i]);
      CLOSE (SourceFile)
    END
  ELSE
    KeyBoard := TRUE
END;

BEGIN { GetVector }
  SelectInputMode;
  IF  NOT KeyBoard  THEN
    ReadVector
  ELSE
    BEGIN
      RowDimension;
      VectorIn
    END;
  VerifyVector;
  WRITELN;
  Title := 'The vector: ';
  WRITE ('Do you want to write the vector into a file?  (Y or N)  ');
  Echo (answer);
  IF  answer  IN  ['Y' , 'y']  THEN
      PutVector (A, n, 'The vector is ', 'F')
END;
```

Answers to Selected Problems

Chapter 2

1. **a.** $\dfrac{s^2 + 3s + 2}{s^2 + 3.2s + 2}$, **b.** $\dfrac{1}{s^3 + 5s^2 + 6s + 1}$

2. $\dfrac{1}{(s+1)^3} = \dfrac{1}{s^3 + 3s^2 + 3s + 1}$

3. **a.** $2 + 2.83 \cos(2t + 0.542)$ **b.** $2 + 0.155 \cos(2t - 2.41)$

4. $G(s) = \dfrac{7.259}{s(s + 0.578)}$

5. $G(s) = \dfrac{10/3}{s^2 + 2s + 4/3}$, $x(t) = 5 - 5e^{-t} \cos \dfrac{t}{\sqrt{3}} - 3\sqrt{3}\, e^{-t} \sin \dfrac{t}{\sqrt{3}}$

6. $\dfrac{K_m}{s(sL_f + R_f)(Js + D)}$

7. $\dfrac{1}{(s+5)(0.12s^2 + 2.803s + 0.08)}$

8. $\dfrac{K_1}{Js^2 + Ds + K_2}$

9. $K_2 = \omega_0^2 M_2$

10. $\dfrac{100}{s^3 + 4s^2 + 5s + 15}$

11. a. $\dfrac{G_1G_2G_3 - G_4(1 + G_2H_2 + G_1G_2H_1)}{1 + G_2H_2 + G_1G_2H_1 + G_2G_3H_3}$

b. $\dfrac{G_4 + G_1G_2G_3}{1 + G_1G_2H_3 + G_2G_3H_2 + G_1G_2G_3H_1 + G_4H_1 - G_4H_2G_2H_3}$

12. $\dfrac{1}{s^2 + (2D/M)s + 2K/M}$

Chapter 3

1. Let $y_1 = x_1$, $y_2 = \dot{x}_1$, $y_3 = x_2$, and $y_4 = \dot{x}_2$. Then

$$\dot{y}_1 = y_2$$

$$\dot{y}_2 = \frac{1}{M_1} f - \frac{K_1}{M_1} y_1 - \frac{D_1}{M_1} y_2 - \frac{K_2}{M_1}(y_1 - y_3)$$

$$\dot{y}_3 = y_4$$

$$\dot{y}_4 = -\frac{K_2}{M_2}(y_3 - y_1)$$

2. $\dot{x}_1 = -10x_2$
$\dot{x}_2 = -10x_3$
$\dot{x}_3 = -0.15x_1 + 0.5x_2 - 4x_3 - r$

$c = -x_1$

$G(s) = \dfrac{100}{s^3 + 4s^2 + 5s + 15}$

3. a. $\dfrac{s^2 + 5s + 4}{s^3 + 6s^2 + 10s + 2}$

b. $F = \begin{bmatrix} 0.431758 & 0.048130 & 0.249768 \\ 0.153509 & 0.230119 & 0.048130 \\ 0.297898 & 0.201639 & 0.681526 \end{bmatrix}$ $\quad G = \begin{bmatrix} 0.407755 \\ 0.068705 \\ 0.497036 \end{bmatrix}$

5. $\dot{x} = \begin{bmatrix} -1 & -2 & 0 & 0 \\ 2 & -1 & 0 & 0 \\ 0 & 0 & 0 & 0 \\ 0 & 0 & 0 & -2 \end{bmatrix} x + \begin{bmatrix} 1 \\ 0 \\ 1 \\ 1 \end{bmatrix} u \qquad y = [1 \quad 0.5 \quad 1 \quad 2]\, x$

6. $\dot{x} = \begin{bmatrix} -2 & 1 & 0 & 0 \\ 0 & -2 & 1 & 0 \\ 0 & 0 & -2 & 0 \\ 0 & 0 & 0 & -1 \end{bmatrix} x + \begin{bmatrix} 0 \\ 0 \\ 1 \\ 1 \end{bmatrix} u \qquad y = [2 \quad 4 \quad 5 \quad 3]\, x$

7. a. $\dot{x} = \begin{bmatrix} 0 & 1 & 0 & 0 \\ 0 & 0 & 1 & 0 \\ 0 & 0 & 0 & 1 \\ -8 & -20 & -18 & -7 \end{bmatrix} x + \begin{bmatrix} 0 \\ 0 \\ 0 \\ 1 \end{bmatrix} u \qquad y = [54 \quad 90 \quad 47 \quad 8] x$

b. $\dot{x} = \begin{bmatrix} 0 & 0 & 0 & -8 \\ 1 & 0 & 0 & -20 \\ 0 & 1 & 0 & -18 \\ 0 & 0 & 1 & -7 \end{bmatrix} x + \begin{bmatrix} 54 \\ 90 \\ 47 \\ 8 \end{bmatrix} u \qquad y = [0 \quad 0 \quad 0 \quad 1] x$

9. $\frac{d}{dt}\begin{bmatrix} i_f \\ c_g \\ \omega \end{bmatrix} = \begin{bmatrix} -5 & 0 & 0 \\ \frac{200}{3} & -\frac{70}{3} & -\frac{10}{3} \\ 0 & 1 & -40 \end{bmatrix} \begin{bmatrix} i_f \\ c_g \\ \omega \end{bmatrix} + \begin{bmatrix} 5 \\ 0 \\ 0 \end{bmatrix} v(t)$

10. $G(s) = \dfrac{\frac{1000}{3}}{s^3 + \frac{205}{3} s^2 + \frac{3760}{3} s + \frac{14{,}050}{3}}$

11. $\frac{d}{dt}\begin{bmatrix} \theta \\ \omega \end{bmatrix} = \begin{bmatrix} 0 & 1 \\ 0 & -\frac{K_2 + D}{J} \end{bmatrix} \begin{bmatrix} \theta \\ \omega \end{bmatrix} + \begin{bmatrix} 0 \\ \frac{K_1}{J} \end{bmatrix} v(t)$

12. $\dfrac{X_1(s)}{U(s)} = \dfrac{s^2 + 3}{s^4 + 4s^2 - 9}$

13. $F = \begin{bmatrix} 1.059803 & 0.198701 & -0.007936 & 0.039471 \\ 0.596104 & 0.980861 & -0.118413 & 0.389466 \\ -0.007936 & -0.039471 & 0.941390 & 0.190765 \\ -0.118413 & -0.389466 & -0.572294 & 0.862448 \end{bmatrix} \qquad G = \begin{bmatrix} 0.019934 \\ 0.198701 \\ -0.002645 \\ -0.039471 \end{bmatrix}$

14. $G(s) = \begin{bmatrix} \dfrac{2s^2 - 16s + 8}{s^3 + 8s^2 + 16s - 8} & \dfrac{5s^2 - 34s + 16}{s^3 - 8s^2 + 16s - 8} \\ \dfrac{4s}{s^3 - 8s^2 + 16s - 8} & \dfrac{-s^2 + 10s}{s^3 - 8s^2 + 16s - 8} \end{bmatrix}$

15. a.

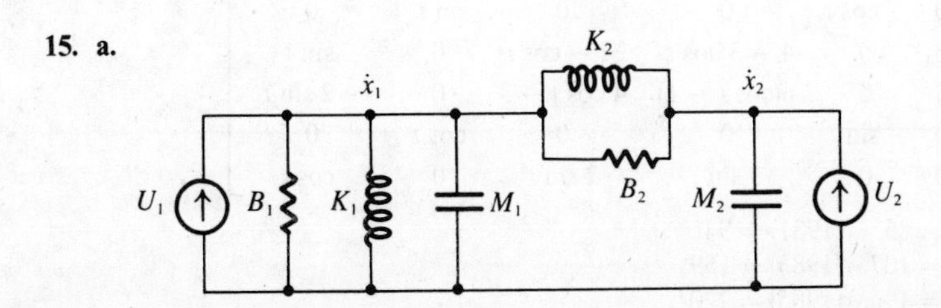

b. $$\frac{d}{dt}\begin{bmatrix} x_1 \\ \dot{x}_1 \\ x_2 \\ \dot{x}_2 \end{bmatrix} = \begin{bmatrix} 0 & 1 & 0 & 0 \\ -\frac{(K_1+K_2)}{M_1} & -\frac{(B_1+B_2)}{M_1} & \frac{K_2}{M_1} & \frac{B_2}{M_2} \\ 0 & 0 & 0 & 1 \\ \frac{K_2}{M_2} & \frac{B_2}{M_2} & -\frac{K_2}{M_2} & -\frac{B_2}{M_2} \end{bmatrix} \begin{bmatrix} x_1 \\ \dot{x}_1 \\ x_2 \\ \dot{x}_2 \end{bmatrix} + \begin{bmatrix} 0 & 0 \\ \frac{1}{M_1} & 0 \\ 0 & 0 \\ 0 & \frac{1}{M_2} \end{bmatrix} \begin{bmatrix} u_1 \\ u_2 \end{bmatrix}$$

16. **a.** Let $x_1 = \theta_1$, $x_2 = \theta_2$, $x_3 = \theta_3$, $x_4 = \dot{\theta}_1$, $x_5 = \dot{\theta}_2$, and $x_6 = \dot{\theta}_3$. Then

$$\dot{x} = \begin{bmatrix} 0 & 0 & 0 & 1 & 0 & 0 \\ 0 & 0 & 0 & 0 & 1 & 0 \\ 0 & 0 & 0 & 0 & 0 & 1 \\ 0 & 0 & 0 & 0 & 0 & \omega_0 \\ 0 & 0 & 0 & 0 & 0 & 0 \\ 0 & 0 & 0 & \omega_0 & 0 & 0 \end{bmatrix} x + \begin{bmatrix} 0 & 0 & 0 \\ 0 & 0 & 0 \\ 0 & 0 & 0 \\ 1/I & 0 & 0 \\ 0 & 1/I & 0 \\ 0 & 0 & 1/I \end{bmatrix} \begin{bmatrix} L_1 \\ L_2 \\ L_3 \end{bmatrix}$$

$$y = \begin{bmatrix} 1 & 0 & 0 & 0 & 0 & 0 \\ 0 & 1 & 0 & 0 & 0 & 0 \\ 0 & 0 & 1 & 0 & 0 & 0 \end{bmatrix} x$$

b. $$G(s) = \begin{bmatrix} \frac{1}{I(s^2+\omega_0^2)} & 0 & \frac{\omega_0}{Is(s^2+\omega_0^2)} \\ 0 & \frac{1}{Is^2} & 0 \\ \frac{-\omega_0}{Is(s^2+\omega_0^2)} & 0 & \frac{1}{I(s^2+\omega_0^2)} \end{bmatrix}$$

17. $$e^{At} = \begin{bmatrix} 1 & 0 & 6(\sin t - t) & 4\sin t - 3t & 0 & 2(\cos t - 1) \\ 0 & \cos t & 0 & 0 & \sin t & 0 \\ 0 & 0 & 4 - 3\cos t & 2(t - \cos t) & 0 & \sin t \\ 0 & 0 & 6(\cos t - 1) & 4\cos t - 3 & 0 & -2\sin t \\ 0 & -\sin t & 0 & 0 & \cos t & 0 \\ 0 & 0 & 3\sin t & 2\sin t & 0 & \cos t \end{bmatrix}$$

18. $y(1980) = 86$, $y(1981) = 94$
$y(1982) = 107$, $y(1983) = 159$
$y(1984) = 196$, $y(1985) = 220$
$y(1986) = 249$, $y(1987) = 281$
$y(1988) = 290$

19. $$G(z) = \frac{0.473263}{z^4 - 0.3z^3 + 0.0316z^2 - 0.00132z + 0.000016}$$

20. $$G(s) = \frac{30s^2 + 382.5}{s^5 + 18s^4 + 176.75s^3 + 481.5s^2 + 2006.5s + 1692}$$

21. $$F = \begin{bmatrix} 1 & 0 & -0.000098 & -0.009769 & 0.097642 & -0.058395 \\ 0.099999 & 0.999993 & 0.000506 & -0.098449 & 0.000064 & 0.000353 \\ -0.004748 & 0 & 0.999061 & -0.003898 & -0.000428 & 0.000153 \\ 0.000031 & 0.000138 & -0.010172 & 0.959430 & 0.095862 & -0.068429 \\ 0.000005 & 0.000004 & -0.001604 & -0.192698 & 0.949979 & -1.693861 \\ 0 & 0 & 0 & 0 & 0 & 7.389056 \end{bmatrix}$$

$$G = [-0.031867 \quad 0.000195 \quad 0.000069 \quad -0.038919 \quad -1.167901 \quad 6.389056]^T$$

22.

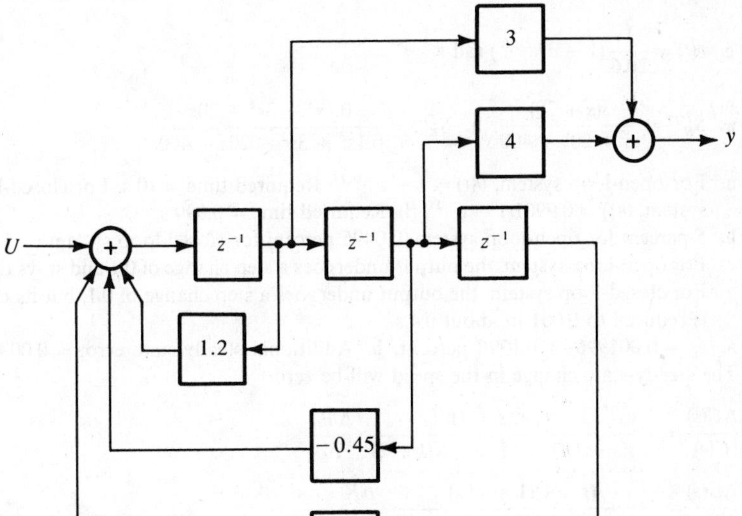

23. $$x(k+1) = \begin{bmatrix} 0 & 1 & 0 \\ 0 & 0 & 1 \\ 0.06 & -0.5 & 1.2 \end{bmatrix} x(k) + \begin{bmatrix} 0 \\ 0 \\ 1 \end{bmatrix} u(k) \qquad y(k) = [0 \quad 0.65 \quad 0.2]\, x(k)$$

(This is one of many possible solutions.)

24. $$G(s) = \frac{1}{s + 0.4} \qquad e^{At} = \begin{bmatrix} e^{-0.4t} & 1.25(e^{-0.4t} - e^{-0.8t}) & 0 \\ 0 & e^{-0.8t} & 0 \\ 0 & \frac{3}{13}(e^{0.5t} - e^{-0.8t}) & e^{0.5t} \end{bmatrix}$$

Chapter 4

1. (i) $S_{K_e}^{T(s)} = \dfrac{60s^2 + 23s + 1}{60s^2 + 23s + 12K_e + 1}$

(ii) $\dfrac{C(s)}{D(s)} = G_d(s) = \dfrac{K_e(1 + 3s)}{60s^2 + 23s + 12K_e + 1}$

$C(s) = T(s)R(s) + G_d(s)D(s)$

(iii) $c_{ss} = 59.95$ km/h

2. **a.** $\Delta c_{ss} = 0.0025$ km/h, **b.** $\Delta c_{ss} = 2.995$ km/h

3. **a.** $S_{K_a}^{G(s)} = 1 \quad S_{K_a}^{T(s)} = \dfrac{s + 0.6}{s + 20.6}$

b. $e_{ss} = 0.029$ rad/s,

c. $e(t) = \dfrac{1}{20.6}(1 - e^{-20.6t})$ rad/s

4. **a.** $S_K^{T(s)} = \dfrac{s(s + 20)}{s^2 + 20s + 400}$, **b.** $S_K^{T(s)} = \dfrac{0.1s^3 + 3s^2 + 20s}{0.1s^3 + 3s^2 + 20s + 400}$

5. **a.** For open-loop system, $\theta(t) = 1 - e^{-0.1t}$. Required time = 10 s. For closed-loop system, $\theta(t) = 0.9901(1 - e^{-10.1t})$. Required time = 0.099 s.
   b. 5 percent for open-loop system, 0.0495 percent for closed-loop system.
   c. For open-loop system, the output undergoes a step change of 0.1 and stays there. For closed-loop system, the output undergoes a step change of 0.1, but its effect is reduced to 0.001 in about 0.4 s.
6. **a.** $e_{ss} = 0.001996$ or 0.1996 percent, **b.** Additional steady-state error = 0.003992.
7. The steady-state change in the speed will be zero.

8. $\dfrac{\Delta C(s)}{C(s)} = -\dfrac{T_0 s^2(s + 1)}{s(s + 1)(1 + sT_0) + K(s + 2)} \cdot \dfrac{\Delta T_0}{T_0}$

9. $\dfrac{\Delta C(s)}{C(s)} = \dfrac{s(s + 1)(1 + sT_0)}{s(s + 1)(1 + sT_0) + K(s + 2)} \cdot \dfrac{\Delta K}{K}$

10. **a.** $e_{ss} = -1$, **b.** $\Delta e_{ss} = 0.05$ (or 5.0 percent), **c.** $\Delta e_{ss} = 0.05$

11. $S_K^{T(s)} = \dfrac{s(s + 2)}{s^2 + 3.4s + 10} \quad S_p^{T(s)} = \dfrac{-2s}{s^2 + 3.4s + 10} \quad S_\alpha^{T(s)} = \dfrac{-1.4s}{s^2 + 3.4s + 10}$

12. **a.** −0.01, **b.** 0.01, **c.** 0.007

Chapter 5

1. **a.** The second specification (on settling time) cannot be satisfied.
   b. $K = 2.8615$ will satisfy the specification on maximum overshoot. Settling time = 4 s, time to reach the first peak = 2.3026 s.
2. **a.** (i) 0, (ii) 0.2, (iii) ∞
   b. (i) 0, (ii) 0, (iii) $\frac{4}{9}$.
3. **a.** (i) 0.6, (ii) ∞; **b.** (i) 0, (ii) $\frac{16}{9}$
4. $\alpha = 0.1739$, $e_{ss} = 0.3739$

5. **a.** Settling time = 8 s, maximum overshoot = 63.2 percent, $e_{ss} = \frac{1}{12}$.
   b. Settling time = 2.667 s, maximum overshoot = 22.11 percent, $e_{ss} = 0.25$.
   c. $\alpha = 3.1569$, settling time = 1.9245 s, maximum overshoot = 9.478 percent, $e_{ss} =$ 0.3464.
6. $K_1 = 10.4$, $K_2 = 1$, settling time = 0.8 s, maximum overshoot = 16.3 percent.
7. **a.** $K = 4$, $z = 2$; **b.** 0.25; **c.** 0.125.
8. $K = 100$, $\alpha = 0.08$
11. $K = \dfrac{106}{17}$, $p = \dfrac{4}{17}$, $z = \dfrac{25}{106}$, $e_{ss} = 1.02$.
12. **a.** $1 - 0.083969e^{-0.340618t} - 0.916032e^{-0.00469908t} \cos 0.38311472t$
   $- 0.085886e^{-0.00469908t} \sin 0.38311472t$
   b. 45.365; **c.** 0.5.
13. **a.** $1 - 0.209504e^{-0.312209t} - 0.790498e^{-0.143895t} \cos 0.373421t$
   $+ 0.479772e^{-0.143895t} \sin 0.373421t$
   b. 2.7375; **c.** 2.5.
14. **a.** $1 - e^{-10t}(\cos 17.3205t + 0.57735 \sin 17.3205t)$
   b. 0.05; **c.** 0.05; **d.** 1000
15. **a.** $1 - 1.118284e^{-6.5045458t} + 0.118284e^{-61.4954542t}$
   b. 0.082353; **c.** 0.17; **d.** a solution does not exist for positive K.
16. The responses are very close. For $t > 3$ s, the difference is less than 1.3 percent.

Chapter 6

1. **a.** Stable, **b.** unstable, **c.** unstable, **d.** unstable
2. **a.** $0 < K < 66$; **b.** $0 < K < 20$; **c.** unstable for all K.
3. **a.** $4.125 < K < 25.875$, **b.** not possible, **c.** not possible
4. **a.** $0 < K < 12$, **b.** not possible
5. **a.** $0 < \alpha < 1$, **b.** $0.1772 < \alpha < 0.605$
6. **a.** $K > 0$, **b.** $K > 3.75$
7. Unstable.
8. **a.** None, **b.** none.
9. $K > -4.4635$
10. $-3.8297 < K < 3.636$
11. $K > 3.4884$
12. **a.** $K > 0$, **b.** not possible for any K.
13. $y_{ss} = 3$
14. $y_{ss} = 3t - 17$

Chapter 7

2. $\alpha = 0.1162$, $S_{\alpha}^{q_1} = 0.5774e^{j\pi/6}$
3. **b.** There are two possible solutions: (i) $z = 14.454$, (ii) $z = 0.22304$
   c. (i) $1 - 0.4036e^{-238.314t} - 0.5964e^{-3.115t} \cos 7.1378t$
   $+ 0.2747e^{-3.115t} \sin 7.1378t$
   Maximum overshoot = 22.1 percent
   (ii) $1 - 0.9894e^{-101.04t} - 0.0106e^{-0.5943t} \cos 1.3617t$
   $+ 0.0151e^{-0.5943t} \sin 1.3617t$
   Maximum overshoot = 0.78 percent

4. **a.** 0.6
   b. (i) 0.7438, (ii) 0.5115
   c. (i) $1 - 1.2768e^{-1.4476t} + 0.4149e^{-2.2762t} \cos(3.9426t + 0.8401)$
   (ii) $1 - 1.2157e^{-2.6091t} + 0.9796e^{-1.6955t} \cos(2.9366t + 1.3488)$
   d. (i) $S_{\alpha}^{q_1} = 3.1979e^{j1.8872}$
   (ii) $S_{\alpha}^{q_1} = 2.8807e^{j2.396}$
5. $\alpha = 0.4537$, $e_{ss} = 0.3102$
6. $\alpha = 1.2731$ $\quad S_{\alpha}^{q_1} = 1.2193e^{-j0.0691}$
7. $K = 9.6192$ $\quad c(t) = 0.5869 - 0.4006e^{-37.0184t} + 0.792e^{-4.61t} \cos(4.61t - 1.8082)$
8. There are two solutions.
   (i) $p = 9.3259$
   $c(t) = 1 - 0.096e^{-0.8497t} + 1.0484e^{-5.2381t} \cos(9.0726t + 2.6106)$
   (ii) $p = 78.9228$
   $c(t) = 1 + 0.0155e^{-77.7365t} + 1.0265e^{-1.5932t} \cos(2.7594t - 3.2885)$
9. The damping ratio is maximum for $K = 3.6$, with $\zeta = 0.0558$.

$$c(t) = 1 + 0.172967e^{-1.6377t} - 0.056373e^{-4.2823t} - 1.116612e^{-0.04t} \cos 0.71535t - 0.003886e^{-0.04t} \sin 0.71535t$$

10. The best damping (considering all four poles) is obtained for $\alpha = 0.318$.

$$c(t) = 1 + 0.275813e^{-0.392425t} - 0.117687e^{-4.823171t} - 1.158126\, e^{-0.392202t} \cos 1.3646648t - 0.669476e^{-0.392202t} \sin 1.3646648t$$

11. There are two solutions.
    (i) $k_v = 0.37769$, $e_{ss} = 1.17769$, maximum overshoot = 20 percent
    (ii) $k_v = 30.73034$, $e_{ss} = 31.53$, no overshoot.
12. $k_a = 7.20397$
    $c(t) = 1 - 0.000021e^{-77.6809t} - 0.999979e^{-0.1794t} \cos 0.310723t - 0.582694e^{-0.1794t} \sin 0.310723t$
    Maximum overshoot ≈ 18 percent
13. $k_v = 11.05$, $e_{ss} = 3.5125$.
14. $K = 17.1$ causes the complex poles to have damping ratio 0.707. The dominant pole, then, is real, at $s = -1.166$. The complex poles are at $-5.417 \pm j5.417$.
15. Sensitivity of the real dominant pole to K is -1.0225. Sensitivity of the upper complex pole to K is $0.5119 + j1.3196$.

Chapter 8

2. $G(s) = \dfrac{100}{s(s^2 + 5s + 100)}$

5. $\dfrac{31.62s(s + 300)^2}{(s + 20)(s + 50)(s + 5692.1)}$

6. $\dfrac{5(s + 2)(s + 10)}{s}$

8. 13 dB

Chapter 9

1. **a.** 2.14 dB, 6.85°; **b.** 3.93 dB, 16.88°
2. **a.** 28.18, **b.** 17.95

3. **a.** Stable, **b.** unstable
4. **b.** Gain must be increased by 8.98 dB. Then, $M_m = 3.24$ dB, $\omega_m = 9.5$ rad/s.
   c. Gain must be increased by 8.87 dB to make $M_m = 3$ dB. Then, $\omega_m = 9.48$ rad/s.
5. 11.63
6. **a.** 39°, **b.** 1.85, **c.** 1.55, **d.** 8 rad/s.
7. Reduce the gain by a factor of approximately 0.9.
8. **a.** Stable with gain margin of 7 dB, **b.** 2.088.
9. $K_p = 3.9$ for M_m to be 3 dB. Phase margin = 41.5°.
10. **a.** Gain margin is ∞ for all $K > 0$, **b.** 28.28, **c.** 34.
11. **a.** Stable with gain margin 12 dB. **b.** K_p should be increased to 8.8 to obtain phase margin of 45°.
12. **a.** 0.889, **b.** 2.7 dB.
13. 1.35, 41.6°
14. 0.122, 42°
15. 1.2, 52.5°
16. 142, 43°
17. 132.83, 2.48 dB
18. Stable

Chapter 10

1. **a.** $K = 28.45$, $\omega_m = 1.5$ rad/s
   b. $K = 3327$, $\omega_m = 17.8$ rad/s
2. **a.** Lag compensator, $G_c(s) = \dfrac{s + 0.06}{s + 0.01}$
   b. Lag compensator, $G_c(s) = \dfrac{s + 0.5}{s + 0.1}$
3. **a.** $G_c(s) = \dfrac{6.6819(s + 3.9669)}{s + 26.506}$ with $K = 250$
   b. With $K = 250$,
   $$G_c(s) = \frac{0.4329s + 0.1674}{s + 0.1674}$$
4. $G_c(s) = \dfrac{7.6573(s + 0.58)}{s + 4.4416}$
5. **a.** $K = \dfrac{224}{3}$, $\alpha = 24$
   b. $1 - 0.1026e^{28t} + 1.2943e^{-4t}\cos(4\sqrt{3}t - 3.9462)$
6. **a.** Gain margin = 15.25 dB, phase margin = 52°, $M_m = 1.1$ dB, $\omega_m = 5.5$ rad/s (with compensation)
   b. Gain margin = 7 dB, phase margin = 12°, $M_m = 12$ dB, $\omega_m = 2.4$ rad/s (without compensation)
7. $K = 250$
   a. $\dfrac{5.9648(s + 4.8926)}{s + 29.1832}$, **b.** $\dfrac{0.3749(s + 0.6593)}{s + 0.2472}$

8. One solution is $G_c(s) = \dfrac{(s + 3)(s + 5)}{(s + 40)^2}$ with the resulting closed-loop transfer function

$$T(s) = \frac{24{,}000}{(s + 60)(s^2 + 20s + 400)}$$

Sensitivity of the upper complex pole to K

$$S_K^{q_1} = 13.093e^{j1.2373}$$

9. **a.** 4.2 to 7 rad/s, **b.** 0.25 to 2.3 rad/s, **c.** 2.4 to 3.4 rad/s

10. $G_u(s) = \dfrac{3.35s + 487.09}{s^2 + 9s + 20}$ $\quad G_c(s) = \dfrac{-33.803s^2 + 3.4s + 11.76}{s^2 + 9s + 20}$ $\quad K = \dfrac{40}{3}$

11. $K = 128{,}000$, $G_c(s) = \dfrac{(s + 4)(s + 8)}{s^2 + 120s + 6400}$

12. $K = 36.19$, $G_c(s) = \dfrac{0.05632s + 0.012656}{s + 0.012656}$ $\quad (\omega_c = 1.0 \text{ rad/s})$

13. $K = 200$

$$G_c(s) = \frac{0.05271(s + 0.3541)}{s + 0.01866} \quad (\omega_c = 1 \text{ rad/s})$$

14. $K = 200$, $G_c = \dfrac{3(s + 2.5)}{s + 7.5}$ makes $M_m = 2.21$ dB, $\omega_m = 5.8$ rad/s

$$c(t) = 1 + 0.26046e^{-2.5589t} - 1.026046e^{-3.4706t} \cos 10.2564t$$
$$- 0.3407e^{-3.4706t} \sin 10.2564t$$

15. $G_p = \dfrac{20}{s(s + 2)}$, $G_c = \dfrac{20(s + 2)}{s + 20}$ $\quad G_pG_c(s) = \dfrac{400}{s(s + 20)}$, $T(s) = \dfrac{400}{s^2 + 20s + 400}$

$$c(t) = 1 - e^{-10t} \cos 17.321t - 0.57735e^{-10t} \sin 17.321t$$

16. $K_p = 9.5$, $G_c(s) = \dfrac{5.5977s + 14.2803}{s + 14.2803}$ $\quad (\omega_c = 5 \text{ rad/s})$

17. $K_A = 10.823$, $G_c = \dfrac{4.90882(s + 8.0339)}{s + 39 \cdot 42{,}269}$ $\quad (\omega_c = 25 \text{ rad/s})$

18. $K = 100$, $G_c = \dfrac{7200(s + 25)(s^2 + 2s + 26)}{s(s + 4)(s + 12)(s^2 + 244s + 28{,}800)}$ (cascade compensator)

19. $T(s) = \dfrac{7.8125(s + 4)(s + 12)}{(s + 7.5)(s^2 + 10s + 50)}$ $\quad K = 3.90625$, $G_u(s) = \dfrac{-1.693764s - 39.136718}{(s + 10)^2}$

$$G_c(s) = \frac{-1.086841s^2 + 11.099683s - 5.496355}{(s + 10)^2}$$

20. $K = 2.748244$ $\quad c(t) = 1 - 0.1187e^{-0.2561t} + 0.0733e^{-0.574t}$

$$- 0.9547e^{-1.1841t} \cos 1.1841t$$
$$+ 1.3773e^{-1.1841t} \sin 1.1841t$$

$$e_{ss} = 0.158$$

21. $K = 2.4496644$, $\alpha = 0.6794521$

$c(t) = 1 - 0.13581e^{-0.2496644t} - 0.86419e^{-t}\cos t + 1.55156e^{-t}\sin t$

$e_{ss} = 0.2$

22. $K = 256$, $\alpha = 0.0546875$, lag compensator is

$$G_c(s) = \frac{s + 0.0625}{s + 0.01}$$

Chapter 11

1. a. 10.3445 **b.** $1 - (0.9536)^n\{\cos(0.4445n) - 0.10289\sin(0.4445n)\}$

2. $$\frac{10.322(z^2 - 1.8187z + 0.8187)}{(z-1)(z+1)(z+0.99442)}$$

3. a. $$\frac{0.3075(z - 0.6988)}{z - 0.9074}$$

b. $c(nT) = 0.9085 - 0.0664(0.7188)^n - 0.8421(-0.4148)^n$

4. a. $K < 2.449$, **b.** $G_c(z) = \dfrac{3.41782(z - 0.739915)}{z - 0.1110772}$

5. $$G_c(z) = \frac{(z - 0.9048374)(z - 0.9801987)}{(z^2 - 1)(0.04805z + 0.046166)}$$

7. $G_u(z) = \dfrac{-0.912398z - 0.009719}{(z - 0.1)^2}$ $\quad G_y(z) = \dfrac{0.2552z^2 - 0.308158z + 0.131258}{(z-0.1)^2}$

8. $$G_c(z) = \frac{(1067.76/K)(z - 0.81873)}{(z+1)(z+0.9355254)}$$

9. $K = 100$, $G_c(z) = \dfrac{3.865(z - 0.740775)}{z + 0.0019126}$

$c(nT) = 1 + 0.269159(0.697283)^n - (0.597724)^n\, 1.27096\cos(0.884738n)$
$\quad -(0.597724)^n\, 0.745658\sin(0.884738n)$

10. Adjust K_A to make open-loop dc gain equal to 20.

$$G_c(z) = \frac{0.213903\,(z - 0.316758)}{z - 0.853868}$$

$c(nT) = 0.9524 - 0.019885(0.775205)^n - (0.57901)^n\, 0.972268\cos(0.856964n)$
$\quad - (0.57901)^n\, 0.50331\sin(0.856964n)$

11. $G_u(z) = \dfrac{-252.034571}{z - 0.15}$ $\quad G_c(z) = \dfrac{13103.015z - 12191.77}{z - 0.15}$

12. $G_u(z) = \dfrac{0.685196}{z - 0.15}$ $\quad G_c(z) = \dfrac{2.378z - 1.505981}{z - 0.15}$

$c(nT) = 0.96987 - (0.28284)^n\, 0.96987\cos(n\pi/4) - (0.28284)^n\, 2.19772\sin(n\pi/4)$

13. $K = 0.000464$ for phase margin of 45°.

14. $K = 0.0125, \quad G_c(z) = \dfrac{0.00211965(z - 0.994557372)}{z - 0.999988506}$

$$c(nT) = 1 - 0.0693(0.8901)^n + 0.5691(0.9923)^n - (0.983835)^n\,1.52868\cos(3.12093n) - (0.983835)^n\,1.36802\sin(3.12092n)$$

Chapter 12

1. Uncontrollable, observable, stable.

2. $D(z) = \dfrac{1.58199 - 0.58199z^{-1}}{1 + 0.41801z^{-1}}$

3. $k^T = [-4.64 \quad -0.96 \quad 5.36]$

4. $G_c = \dfrac{9s^2 + 100s + 298}{12.5(3s^2 + 16s + 24)}$

5. $G(s) = \dfrac{s^2 - 3s - 4}{s^3 - 2s^2 + 4s - 4} \qquad Kk^T = [12/11 \quad 54/11 \quad -8]$

6. $Kk^T = [0.169991 \quad 1.100326 \quad 5.546071]$

7. $L = [-7996.0392 \quad 47.3508 \quad 13.598]^T$

8. $L = [149.8421 \quad 57.3158 \quad -34.2632]^T$

9. $G_u(s) = \dfrac{9s^2 + 35338.788s + 83596.95}{s^3 + 30s^2 + 325s + 1250}$

$$G_c(s) = \frac{-933.941s^2 - 2738.893s + 684.3537}{s^3 + 30s^2 + 325s + 1250}$$

10. $G_u(s) = \dfrac{4927.737s + 12294.68}{s^2 + 20s + 125}$

$$G_c(s) = \frac{-131.1663s^2 - 409.2758s + 42.2021}{s^2 + 20s + 125}$$

11. $u(0) = 12.924718, \; u(T) = -28.185228, \; u(2T) = 15.145333$

$$D(z) = \frac{12.924718 - 28.185228z^{-1} + 15.145333z^{-2}}{1 + 0.5701885z^{-1} + 0.5468854z^{-2}}$$

12. $Kk^T = [6.6230232 \quad -7.7059972 \quad 30.5092806]$

$$c(nT) = 1.022359(0.25)^n - 0.000182 - 2(0.282843)^n[0.511089\cos(n\pi/4) + 0.044294\sin(n\pi/4)]$$

13. $L = [-6235.6 \quad 17.245 \quad 2.8121]^T$

14. $G_u(z) = \dfrac{-9.478287z - 11.047222}{z^2 - 0.22z + 0.012}$

$$G_c(z) = \frac{-0.06565z^2 + 0.128342z - 0.067038}{z^2 - 0.22z + 0.12}$$

15. $Kk^T = [-4/9 \quad -10/9 \quad -184/9 \quad -55/9]$
16. $L = [14 \quad 83 \quad -308 \quad -974]^T$
17. $Kk^T = [13/3 \quad 5 \quad 5/2 \quad 5/6]$
18. $Kk^T = [1867.9004 \quad 19.26785 \quad -844.4561 \quad 815.3443]$

19. $$G_u(z) = \frac{-74.896422z^2 - 307.428192z - 77.20793}{z^3 - 0.3z^2 + 0.04z + 0.002}$$

$$G_c(z) = \frac{-235155.27z^3 + 690222.197z^2 - 688017.59z + 231931.53}{z^3 - 0.3z^2 + 0.04z + 0.002}$$

20. The system is uncontrollable.
21. $Kk^T = [-450 \quad 46.8 \quad 843.75]$
22. $Kk^T = [-551.25 \quad 64.8 \quad 843.75]$
23. The state x_3 is uncontrollable. All states are observable.

24. **a.** $$x(k+1) = \begin{bmatrix} 0 & 1 & 0 \\ 0 & 0 & 1 \\ -0.01 & 0.021 & 0.8 \end{bmatrix} x(k) + \begin{bmatrix} 0 \\ 0 \\ 1 \end{bmatrix} u(k)$$

$$y(k) = [0 \quad \alpha \quad 1] x(k)$$

b. Unobservable if α is -1, -0.041421 or 0.241421. (The transfer function has pole-zero cancellation for these values of α.)

c. $$x(k+1) = \begin{bmatrix} 0 & 0 & -0.01 \\ 1 & 0 & 0.21 \\ 0 & 1 & -0.8 \end{bmatrix} x(k) + \begin{bmatrix} 0 \\ \alpha \\ 1 \end{bmatrix} u(k)$$

$$y(k) = [0 \quad 0 \quad 1] x(k)$$

d. Uncontrollable if α is -1, -0.041421 or 0.241421. (The transfer function has pole-zero cancellation for these values of α.)

26. $Kk^T = [-9.380385 \quad -1.7]$.

Chapter 13

1. $$1 + \frac{4}{\pi} \sin^{-1}\left(\frac{10}{A}\right) - \frac{76}{\pi A^2}(A^2 - 100)^{0.5}$$

2. **a.** Unstable, **b.** $K < 5.682$ for stability.

3. $$\left(\frac{3}{4} A^2\omega^3 + \frac{1}{4} A^2\omega\right) e^{j\pi/2}$$

4. $$\frac{2}{\pi}\left[\sin^{-1}\frac{15}{A} + \frac{15}{A^2}(A^2 - 225)^{0.5}\right]$$

5. **a.** Unstable, **b.** $K < 8.52$ for stability.
9. Stable.

11. $$\frac{40}{\pi}\left[\sin^{-1}\frac{1}{A} - \sin^{-1}\frac{1}{2A} + \frac{1}{A^2}(A^2 - 1)^{0.5}\right]$$

12. $\pi/6$, $2\pi/3$, 0.49.

14. $P = \begin{bmatrix} 0.110577 & 0.052885 \\ 0.052885 & 0.201923 \end{bmatrix}$, stable since P is positive definite.

15. $P = \begin{bmatrix} -0.113070 & -0.374164 & -0.230095 \\ -0.374164 & -0.245593 & -0.062918 \\ -0.230395 & -0.062918 & -0.255927 \end{bmatrix}$, unstable.

16. Stable.

19. Stable.

Appendix C

1. a. $\dfrac{2z(z - e^{-3T}\cos 4T)}{z^2 - 2ze^{-3T}\cos 4T + e^{-6T}}$ b. $\dfrac{Tz(z^2 - e^{-4T})e^{-2T}\sin 3T}{(z^2 - 2ze^{-2T}\cos 3T + e^{-4T})^2}$

   c. $\dfrac{0.2z^2 - ze^{-2T}(0.2\cos 2T - 2.1\sin 2T)}{z^2 - 2ze^{-2T}\cos 2T + e^{-4T}} - \dfrac{0.2z}{z - e^{-T}}$

   d. $\dfrac{2z}{z - e^{-T}} - \dfrac{2z}{z - e^{-2T}} + \dfrac{Tze^{-2T}}{(z - e^{-2T})^2}$

2. a. $35 - 60(0.5)^n + 25(0.2)^n$

   b. $-\dfrac{260}{37}(-0.3)^n + 2(0.8944)^n[3.515\cos(1.8541n) + 2.1243\sin(1.8541n)]$

   c. $6\cos(n\pi/3) - 2\sqrt{3}\sin(n\pi/3) - (\sqrt{2})^n[6\cos(n\pi/4) - 8\sin(n\pi/4)]$

   d. $\dfrac{8}{3}n(0.5)^{n-1} - 2(0.5)^n + (0.5)^n[2\cos(2.4981n) + \dfrac{26}{9}\sin(2.4981n)]$

3. a. $c(nT) = 3.4465 - 3.9129e^{-2n} + 0.4664e^{-4n}$

   b. $c(nT) = 58.9048 - 9.0508e^{-2n} + 0.146e^{-4n}$

4. 220.2 millions in 1980 and 222.5 millions in 1984.

Index